Illustrated Plant Pathology
Basic Concepts

Illustrated
Plant Pathology
BASIC CONCEPTS

By :
H.LEWIN DEVASAHAYAM
Associate Professor of Plant Pathology (Retd.)
Tamil Nadu Agricultural University
Coimbatore, Tamil Nadu
&
L.DARWIN CHRISTDHAS HENRY
Senior Lecturer in Plant Pathology
Annamalai University
Chidambaram, Tamil Nadu

New India Publishing Agency
Pitam Pura, New Delhi-110 088

Published by
Sumit Pal Jain *for*

New India Publishing Agency

101, Vikas Surya Plaza, CU Block, L.S.C. Mkt.,
Pitam Pura, New Delhi- 110 088, (India)
Phone: 011-27341717, Fax: 011-27341616
Mobile : 09717133558
E-mail: newindiapublishingagency@gmail.com
Web: www.bookfactoryindia.com

ISBN : 978-93-80235-08-0

Composed and Designed by NIPA

FOREWORD

Among the various constraints faced in crop husbandry, plant diseases play a very important role. Under favorable conditions some diseases may cause even cent per cent loss in commercial crops. Instances of huge economic losses and even famines have been reported due to occurrence of some serious diseases in major crops. In this respect disease management plays a crucial role in enhancing crop production. With limitations in many input parameters, such as cultivable land, water, fertilizer etc., it is imperative that measures have to be taken to increase crop production with the available resources. For this, knowledge of diseases and measures to control them are of vital importance.

Though various strategies to increase crop production are being adopted, scientific plant protection measures are considered to be very important. Such measures, besides controlling the diseases lead to increase in crop production and income to the farmers, besides making our country self sufficient in food and other commodities.

The book 'Illustrated Plant Pathology' deals mainly with the fundamental aspects of Plant Pathology and the various methods of control of plant diseases. The authors have given detailed account of the life cycle of many major fungal pathogens. The symptoms caused by different pathogenic microorganisms, their survival, mode of spread etc. have also been dealt with in detail. Methods of conducting plant disease surveillance, assessment of disease intensity and yield loss have also been incorporated. In principles of plant disease management, various measures of disease control under evasion, exclusion, eradication, protection and immunization have been discussed elaborately. A glossary of scientific terms has also been included, which may serve as a ready reckoner. Further, the text is substantiated with beautiful, hand-drawn illustrations, all drawn by the authors.

The book, which has been compiled as per the undergraduate syllabus of the Agricultural Universities, will be useful for the postgraduate students and also for the farmers who are interested in scientific agriculture. I am quite sure that this book will be of immense help for students in furthering their knowledge in the field of Plant Pathology and Plant Protection. I congratulate the authors Thiru. H. Lewin Devasahayam and Dr. L. Darwin Christdhas Henry for their highly commendable work.

Chidambaram
04-03-2008

(Dr. G. KUPPUSWAMY)
Dean
Faculty of Agriculture
Annamalai University

PREFACE

The ultimate aim of crop husbandry is to cultivate crops and obtain maximum returns according to their genetic potential. Pathogenic microorganisms stand in the way of crops attaining their full growth and maturity. Some plant pathogens, as a result of infection of the host plant may cause death of a part or the entire plant and cause total loss, while some others affect some parts of the plant resulting in reduction of yield. Some pathogens may affect the stored produce and cause spoilage. So, it becomes imminent that plant pathogens have to be controlled by adopting appropriate control measures so as to minimize the losses and maximize the yield.

The causative agents of diseases in plants include pathogenic microorganisms, such as fungi, bacteria, viruses, mycoplasma, protozoa and nematodes; phanerogamic parasites, algae and lichens of the Plant Kingdom; abiotic agents or unfavorable environmental factors, such as deficiency due to lack of certain macro or micro nutrients or toxicity due to excess of certain macro and micro nutrients; moisture, light or heat stress, as well as the presence of some toxic chemicals in the air, soil or water.

Plant pathology deals with the study of plant diseases caused by various pathogenic agents and environmental factors; the mechanisms by which these agents or factors induce diseases in plants; occurrence and mode of spread of the diseases and the methods of preventing or controlling them, thereby reduce the damage and consequent loss caused by them. As a separate field of Biological Science, Plant Pathology helps to enhance our knowledge about plant diseases, at the same time it tends to develop methods, equipments and materials through which plant diseases can be avoided or controlled.

In an ecological community, the majority of living organisms are plants. Plants have the ability to convert solar energy into chemical energy through the process of photosynthesis and this chemical energy is stored in the form of carbohydrates, proteins and fats in the plant cells. All vertebrates including humans, animals etc. and invertebrates, such as insects, nematodes etc. depend entirely on these plant materials for their survival and existence.

It has been established that microorganisms were present on this earth several thousands of years before humans and plants came into existence and had intruded into the world of microorganisms. These microorganisms, which had been living entirely on the organic matter for their sustenance, had gradually changed their mode of life and had become parasites on plants. These parasites, which cause various types of plant diseases, are known as plant pathogens.

Infection by plant pathogens invariably exhibits certain specific abnormalities in the host plants called disease symptoms. Each and every pathogenic microorganism produces quite different and distinct symptoms. These symptoms may be visible to the naked eye or in some cases may not be visible externally. From such characteristic symptoms, most of the diseases, such as leaf spots, blights, rusts, smuts, wilts, rots etc. can be identified. Further, in the case of several diseases, the spores, spore bodies, sporophores or such other structures produced by the pathogens can also be seen and the diseases can be identified. So, a basic knowledge about the disease symptoms, the causal organisms responsible for causing the diseases, their life cycle, reproductive capacity, mode of survival and spread of the pathogens etc. goes a long way in adopting appropriate control measures to combat the diseases. Diseases may spread through air, water, soil, seed or vegetative propagating materials. Many virus and mycoplsma diseases are spread through insect and non-insect vectors actively or passively. Control measures and plant protection chemicals to be used for the control of different diseases differ depending upon the causal organisms and other characteristics of the pathogens.

Various measures under evasion, exclusion, eradication and protection help to keep in check many plant diseases to a large extent. However, under certain circumstances, especially when the disease intensity crosses the economic threshold level, direct control measures by the use of plant protection chemicals have to be resorted to so as to control the disease and avoid economic losses. In this respect, knowledge about plant protection chemicals is also of vital importance.

To attain self-sufficiency in food production, plant protection is an essential and powerful tool. To use this tool effectively, it is quite important that we should acquaint ourselves with the various aspects concerning pathogenic microorganisms. This knowledge will serve as a basis for adopting suitable control measures at the appropriate time, thereby pave the way to maximize production.

In this context, we hope that this book, which has been brought out as per the syllabus of B.Sc.(Ag.) Degree course of the Agricultural Universities will be of immense help and guidance to the students of Agriculture, people working in the Department of Agriculture, people involved in Scientific Agriculture and the General Public.

Numerous illustrations have been given to enable the reader to understand the text easily and to make the study more interesting'

The authors are highly thankful to Dr.G.Kuppuswamy, Dean, Faculty of Agriculture, Annamalai University, Chidambaram, Tamil Nadu for having given the foreword to the book.

Acknowledgements are also due to friends and colleagues who have inspired, helped and encouraged the authors in their endeavor to complete the book successfully. The authors wish to express their sincere thanks to M/S. New India Publishing Agency, 101, Vikas Surya Plaza, Pitam Pura, New Delhi - 110 088 for the excellent manner in which this book has been brought out.

H.Lewin Devasahayam
L.Darwin Christdhas Henry

LIST OF FIGURES

Figure 1. Fungal mycelium 22
Figure 2. Rhizomorph 22
Figure 3. Intercellular mycelium 22
Figure 4. Intracellular mycelium 22
Figure 5. Types of haustoria 22
Figure 6. Stroma 24
Figure 7. Sclerotium 24
Figure 8. Fragmentation of somatic hyphae (Arthrospores) 24
Figure 9. Chlamydospores 24
Figure 10. Fission (Budding) of somatic cells 24
Figure 11. Simple sporophore (Conidiophore) 28
Figure 12. Sporodochium 28
Figure 13. Coremium (Synnema) 28
Figure 14. Various types of conidia 28
Figure 15. Life cycle of *Olpidium viciae* 41
Figure 16. Life cycle of *Synchytrium endobioticum* 44
Figure 17. Life cycle of *Physoderma zeae maydis* 46
Figure 18. Life cycle of *Urophlyctis alfalfae* 48
Figure 19. Life cycle of *Plasmodiophora brassicae* 51
Figure 20. Life cycle of *Spongospora subterranea* 53
Figure 21. Life cycle of *Aphanomyces euteichus* 55
Figure 22. Life cycle of *Pyhtium debaryanum* 57
Figure 23. Life cycle of *Phytophthora infestans* 60
Figure 24. Life cycle of *Plasmopara viticola* 62
Figure 25. Sporangiophore characteristics of the Genera of Family Peronosporaceae and method of germination of stporangia 64

Figure 26. Life cycle of *Albugo candida* 66
Figure 27. Genus - *Mucor* 69
Figure 28. Genus - *Choanephora* 69
Figure 29. Life cycle of *Rhizopus stolonifer* 71
Figure 30. Genus - *Entomophthora* 74
Figure 31. Life cycle of *Saccharomyces cereviciae* 74
Figure 32. Sexual reproduction and ascus formation in the 76
Ascomycetes
Figure 33. Different types of ascocarps 78
Figure 34. Various types of asci 82
Figure 35. Various types of ascal openings 82
Figure 36. Life cycle of *Taphrina deformans* 84
Figure 37. Form genus - *Aspergillus* 88
Figure 38. Form genus - *Penicillium* 88
Figure 39. Conidiophore and conidia development in ....
Erysiphaceae 91
Figure 40. Types of cleistothecia 93
Figure 41. Sexual reproduction in Erysiphaceae 95
Figure 42. Life cycle of *Claviceps purpurea* 100
Figure 43. Genus - *Leucostoma (Valsa)* 102
Figure 44. Genus - *Nectria* 102
Figure 45. Life cycle of *Sclerotinia sclerotiorum* 106
Figure 46. Life cycle of *Mycosphaerella musicola* 112
Figure 47. Mechanism of dikaryotic cell division by clamp
connection 117
Figure 48. Typical hymenium of a Basidiomycete 117
Figure 49. Stages in the development of a biasidium 117
Figure 50. Edible mushrooms *(Morchella, Agaricus,*
Volvariella and *Pleurotus)* 124
Figure 51. Life cycle of *Puccinia graminis tritici* 134
Figure 52. Morphological variations of telial stages of Uredinales 139
Figure 53. Life cycle of *Ustilago maydis* 141
Figure 54. Life cycle of *Tilletia caries* 143
Figure 55. Pycnidia and conidia of important form genera of
Sphaeropsidales 147
Figure 56. Acervuli and conidia of important genera of
Melanconiales 149
Figure 57. Conidiophores and conidia of important Genera of
Moniliaceae 155

Figure 58. Conidiophores and conidia of important form genera of Dematiaceae and Tuberculariaceae 160
Figure 59 Mode of entry of fungal pathogens into the host 169
Figure 60 Manipulation of the disease triangle 187
Figure 61 Grade chart for disease assessment (Rice blast) 201
Figure 62 Grade chart for disease assessment (Pearl millet) 202
Figure 63 Structure of a typical bacterium 208
Figure 64 Reproduction of bacteria by binary fission............. 208
Figure 65 Endospore formation in bacteria 208
Figure 66 Types of bacteria 209
Figure 67 Types of flagella in bacteria 209
Figure 68 Morphology of some species of bacteria 218
Figure 69 Symptoms of bacterial diseases 220/22
Figure 70 Mode of entry of bacterial pathogens into the host .. 224
Figure 71 Shapes of some viruses as seen through Electron microscope 230
Figure 72 Diagramatic representation of viral RNA replication .. 234
Figure 73 Systemic distribution of viruses in plants after .... mechanical inoculation 236
Figure 74 Structure of *Tobacco mosaic virus* 248
Figure 75 Infection by plus and minus strands of ssRNA viruses . 248
Figure 76 Schematic diagram of genera of plant viruses 250
Figure 77 Symptoms of virus diseases 261
Figure 78 Symptoms of virus diseases 263
Figure 79 Phanerogamic parasites - Stem parasites.............. (Genus - *Cuscuta*) *273*
Figure 80 Stem parasite (Genus - *Dendrophthoe*)................. 273
Figure 81 Phanerogamic parasites - Root parasite (Genus - *Orobanche*) 275
Figure 82 Root parasite (Genus - *Striga*) 275
Figure 83 Bucket sprayer 351
Figure 84 Knapsack sprayer 351
Figure 85 Rocker sprayer 351
Figure 86 Foot sprayer 351
Figure 87 'Marut' sprayer 355
Figure 88 Hand atomizer or Flit pump............................. 355
Figure 89 Power sprayer 355
Figure 90 Hand rotary duster 355

LIST OF TABLES

Table 1. Scale for evaluating the disease intensity of potato late blight 197

Table 2. Scale for evaluating the disease intensity of rice blast 198

Table 3. Diseases introduced into India from other countries 284

Table 4. Examples of some disease resistant crop varieties 307

Table 5. List of antibiotics produced by microorganisms for disease control 340

Table 6. Quantity of spray fluid required for spraying on different crops 342

Table 7. Quantity of fungicidal formulationto be mixed with water to get spray fluid of desired concentration 343

Table 8. Compatibility of fungicides with other agrochemicals 358

Table 9. List of few pathogenic strains resistant to specific plant protection chemicals 361

CONTENTS

Foreword v
Preface vii
List of Figures xi
List of Tables xv

Chapter - 1
FUNDAMENTALS OF PLANT PATHOLOGY **1-276**

Definition of Plant Pathology 1
History of Plant Pathology 2
Pathogens 10
  Fungi 10
  Bacteria 11
  Viruses 11
  Viroids 12
  Mollicutes 13
  Phytoplasmas 13
  Spiroplasmas 15
  Rickettsia-like organisms 16
  Algae 17
  Phanerogamic parasites 18
Koch's postulates or Koch's rules 19
General characters of Fungi 19
Types of parasitism 26
Major symptoms of fungal diseases 27
Taxonomy of Fungi 33
Kingdom - Myceteae (Fungi) 38
  Division 1 - Gymnomycota 38

Subdivision 1 - Acrasiogymnomycotina 38
Class - Acrasiomycetes 38
Subdivision 2 - Plasmodiogymnomycotina 38
Class 1 - Protosteliomycetes 39
Class 2 - Myxomycetes 39
Division II - Mastigomycota 39
Subdivision 1 - Haplomastigomycotina................... 39
Class 1 - Chytridiomycetes 39
Order - Chytridiales 39
Class 2 - Hyphochytridiomycetes 47
Class 3 - Plasmodiophoromycetes 47
Subdivision 2 - Diplomastigomycotina 50
Class - Oomycetes 52
Order - Saprolegniales 52
Order - Peronosporales 54
Division III - Amastigomycota 65
Subdivision 1 - Zygomycotina 67
Class - Zygomycetes 67
Order - Mucorales....................................... 67
Order - Entomophthorales 70
Subdivision 2 - Ascomycotina............................. 72
Class - Ascomycetes 72
Order - Protomycetales 79
Order - Endomycetales 79
Order - Taphrinales 81
Order - Eurotiales.. 83
Order - Microascales 87
Order - Erysiphales 89
Order - Xylariales 96
Order - Clavicepitales 97
Order - Helotiales.. 104
Order - Pezizales 108
Order - Dothidiales 109
Order - Pleosporales.................................... 113
Subdivision 3 - Basidiomycotina.......................... 114
Class - Basidiomycetes 114
Order - Aphyllophorales 118
Order - Agaricales 120
Order - Exobasidiales.................................... 125
Order - Tulasnellales 125
Order - Uredinales 126

Order - Ustilaginales 135
Subdivision 4 - Deuteromycotina 142
Survival of plant pathogens 163
Mode of entry of fungal pathogens.......................... 168
Dissemination or spread of plant pathogens 172
Active or autonomous dispersal 173
Dispersal through soil 173
Dispersal through seed 175
Dispersal through plants and vegetative plant propagants 175
Passive or indirect dispersal.................................. 176
Dispersal through humans 176
Dispersal through insects................................. 176
Dispersal through nematodes 178
Dispersal through birds and animals 178
Dispersal through water.................................. 178
Dispersal through wind 179
Physiological specialization of pathogenic fungi 181
Epidemiology of crop diseases............................... 183
Factors responsible for the development of epidemics ... 184
Plant disease surveillance....................................... 190
Methods of conducting disease surveillance 190
Scrutiny of the survey reports............................... 192
Assessment of disease intensity............................. 193
Assessment of disease intensity in the case of other diseases 196
Sampling for assessment of disease intensity 198
Calculation of disease intensity 199
Disease forecasting.. 203
Forewarning of crop disease occurrence 205
Bacteria 206
General characters of bacteria 206
Morphology of bacteria.................................... 207
Reproduction in bacteria.................................. 210
Classification of bacteria 211
Symptoms of bacterial diseases.............................. 217
Mode of entry of bacterial pathogens 223
Dissemination of bacterial pathogens 225
Plant viruses 226
Definition of viruses... 227
Nature of viruses 228

Properties of viruses 229
Multiplication of viruses 233
Translocation of viruses in plants 235
Transmission of plant virus diseases 235
Transmission by vegetative propagation 235
Mechanical or SAP transmission........................... 237
Seed transmission 237
Pollen transmission 238
Insect transmission 239
Mite transmission 242
Nematode transmission 242
Fungus transmission 243
Dodder transmission 243
Classification of plant viruses................................. 243
General characteristics of important genera of plant viruses 247
Common symptoms of virus diseases 257
Common symptoms of viroid diseases 265
Symptoms of Phytoplasma and Spiroplasma diseases...... 266
Symptoms of diseases caused by fastidious vascular bacteria 267
Symptoms of diseases caused by algal parasites 268
Phanerogamic parasites 269

Chapter - 2
PRINCIPLES OF PLANT DISEASE MANAGEMENT.......... 277-393

Evasion 278
Exclusion 282
Eradication 285
Protection.. 290
Immunization.. 293
Mechanism of disease resistance in plants 298
Methods of selection of resistant genotypes............... 305
Other general methods of protection 308
Direct control of plant pathogens by chemical formulations 310
Formulation of fungicides...................................... 311
Mode of action of fungicides 313
Classification of fungicides.................................... 314
Methods of application of plant protection chemicals..... 315
Qualities of an ideal fungicide 319

Chemicals used as fungicides 319
Quantity of fungicidal fluid required for spraying.......... 341
Preparation of fungicidal spray fluid 342
Plant protection appliances 344
Sprayers 344
Dusters 354
Compatibility of fungicides with other agrochemicals 356
Phytotoxicity of fungicides 357
Resistance of plant pathogens to plant protection chemicals 359
Precautions and safety measures in handling fungicides.. 360
Management of seed-borne diseases 363
Management of soil-borne diseases 366
Management of foliar diseases 373
Simple diagnostic techniques for identification of diseases 376
Seed health-testing methods 379
Biological control of crop diseases 382
Biotechnological approaches in plant disease management 386
Use of plant products in plant disease management 388
Use of antiviral principles in plant disease management . 392

References *395*

Glossary of Scientific Terms *399*

Subject Index *433*

1

Fundamentals of Plant Pathology

Definition of Plant Pathology

'Plant pathology' may be defined as 'the study of microorganisms and of the environmental factors that cause disease in plants; of the mechanisms by which these factors induce disease in plants; and of the methods of preventing or controlling disease and reducing the damage it causes'. Plant Pathology or **'Phytopathology'** is a branch of Biological Science, which exclusively deals with plant diseases and various aspects connected with plant diseases. The term Phytopathology had been derived from the Latin language. The Latin word **'PHYTON'** means plants; **'PATHOS'** means suffering and **'LOGOS'** means Science. So, Phytopathology means the Science dealing with suffering plants. Trees, shrubs, grasses etc., whether cultivated or wild are termed as plants and they form the majority of the earth's living environment. These plants, which are so vital for humans, animals and insects are vulnerable to attack by numerous microorganisms, called **'plant pathogens'**. Some environmental factors are also detrimental to the normal growth of plants, resulting in diseases. Diseased plants exhibit some type of abnormalities and these abnormalities are termed as **'disease symptoms'**.

Depending upon the pathogens attacking the plants or adverse environmental factors the disease symptoms vary markedly.

The Science of Plant Pathology has four main objectives:

1. To acquire knowledge about biotic plant pathogens and abiotic environmental factors, which are responsible for causing diseases.
2. To know more about the occurrence and mode of spread of the pathogens on a large scale.
3. To study the interaction between the host and the pathogen with respect to environmental factors.
4. To adopt suitable methods of preventing or controlling the diseases, thereby reducing the damage caused by the pathogens.

Systematic knowledge about Plant Pathology will be of much help in protecting the plants from attack by the pathogens, thereby allowing the plants to grow and attain maturity according to their genetic potential.

History of Plant Pathology

Ever since human beings started growing plants to meet their requirements of food and various other related commodities from plants, plant pathogens have also started competing with humans to obtain their share of food from plants. In this conflicting struggle for survival, the pathogens have been constantly attacking crops raised by man and causing diseases, resulting in crop damage and yield loss. Inspite of man's superiority, vast potential and knowledge, it has not been possible to control the diseases and overcome the menace caused by these microorganisms.

The quest to know the causes of plant diseases and measures to control them had begun with the advent of civilization several decades before the birth of Christ. Mention about natural havoc in crop plants is found in the ancient Greek civilization and in our ancient Epics and Vedas. Existence of microbial organisms, although unseen at that time was indirectly recognized by the ancient Greek civilization (3400 BC), ancient Hebrews (1500 BC) and the ancient Hindu culture (1500 BC). This is evident from the elaborate arrangements for ventilation, drainage, latrines etc. found in the excavated ruins of that time and the use of fire for purifying things in Hindu rituals. Though there was lack of information about the cause of diseases, a few ancient Roman writers (**Homer**, 1000 BC) had mentioned the therapeutic value of sulfur on plant diseases.

Symptoms of plant diseases are mentioned in **Hindu Mythology**, **Jataka of Bhuddism**, **Raghuvamsha of Kalidas** and other ancient Indian literature. Rusts, smuts, downy mildews, powdery mildews, blights etc. are quoted very often in the **Bible**. In **Rigveda**, besides the classification of plant diseases as blight, wilt, root rot etc., the **'Germ Theory'** of diseases was also advocated. The learned men in the Vedic period were aware of the fact that diseases were caused by microorganisms. **Vriksha Ayurveda**, a book written by Surapal in ancient India, is the first book in which good light has been thrown on plant diseases. In this, plant diseases were divided into two groups, internal and external. Internal diseases were supposed to be due to disorders in the plant system, while external diseases were supposed to be due to attack by microorganisms and insects.

Rigveda, **Atharvaveda** (1500 - 500 BC), the **Artha shashtra of Kautilya** (321 - 186 BC), **Sushruta samhita** (200 - 500 AD), **Vishnu Purana** (500 AD), **Agnipurana** (500 - 700 AD), **Vishnudharmottara** (500 - 700 AD) etc. are ancient books from India, where plant diseases and other enemies of plants and methods to control them have been mentioned. In the Rigveda, '**krimi**' (microbes) have been mentioned as part of the living world. In Atharvaveda, these organisms have been divided into '**shista**' (beneficial) and '**ashishta**' (harmful) krimi. The ashishta krimi was identified as the ones, which spoiled milk, grains etc. and caused diseases in cattle and humans. In ancient India, different treatments, such as hygiene, tree surgery, protective covering with pastes, certain cultural practices etc. were adopted for treating different diseases caused by different agencies based on both superstition and scientific observation. Honey, ghee, milk, barley flour, herbal pastes, plant extracts, medicinal plants etc. were used for chemical treatment. Oil cakes of mahua, mustard, sesamum, castor etc. were used for the control of root diseases.

It has been suggested by many that the streams of knowledge on the science of Botany and Plant Pathology or at least some ideas on these sciences went from the East to the West. The Greek philosopher **Theophrastus** (286 BC) in his book **'Enquiry into plants'** had recorded some observations on diseases of trees, cereals and legumes. According to him, plant diseases were more severe in lowlands than on hilly tracts and some diseases, such as rusts were more common on cereals than on legumes and vine crops. However, in the ancient times, diseases were attributed to many causes, such as divine power, religious beliefs, occultation, superstition, effect of stars and moon, bad wind, wrath of God etc. The Romans believed that God controlled the weather and brought about diseases as a manifestation of His wrath on people, who continued to sin. In order to appease the Gods and save the crops from rust disease of grain crops, they went to the extent of creating a Rust God, **'Robigo'** and offered special sacrifices. These beliefs continued for several decades.

During the eleventh and twelfth century (AD), diseases affecting human beings, such as leprosy, bubonic plague, syphilis etc., now known to be caused by bacteria were recognized.

During the thirteenth century, simple lens was first invented by **Roger Bacon** (1267 AD). In 1590, **Zacharias Janssan** of Holland invented the first compound lens system, which had no arrangements for focussing. **Anton van Leeuwenhock** (1632 - 1723) also of Holland invented the first simple microscope with focussing device. **Leeuwenhock** recorded the first authentic observations on bacteria in 1675. He discovered several forms of bacteria in water, various infusions, faeces and teeth scrapings and presented his observations in a series of more than 200 letters to 'The Royal Society of London'. He called these microorganisms as **'animalcules'** or **'little animals'**. He is also credited with having seen for the first time the spermatozoa, red blood cells, protozoa etc. Although these microorganisms were seen associated with diseased plants, they were not considered to be the causal agents of diseases. It was believed that diseases and the associated microorganisms were of spontaneous origin (abiogenesis), which was contradicted by many other scientists at that time.

The Italian botanist **Micheli** (1679 - 1737), who is considered as the **'Founder of the Science of Mycology'**, was the first scientist to study fungi and to view their spores in 1729. He also proved that, if these spores were placed on a piece of fruit, they grew into new thallus of the fungus. This was also not universally accepted at that time.

In 1755, the French botanist **Tillet** published a paper on bunt or stinking smut of wheat, based on well-planned field experiments and proved that wheat seeds containing a black powder on their surface produced more diseased plants than clean seeds or seeds treated with copper sulfate. With his findings, **Tillet** actually showed that wheat smut is a contagious plant disease. However, he believed that it was a poisonous or toxic substance contained in the smut dust rather than living microorganisms that caused the disease. In 1807, **Prevost** another French scientist showed that the black powder contained on the wheat seeds was the fungus spores and the spores were the cause of smut. But 'The French Academy of Science' did not accept his contention, as scientists throughout the world still believed that microorganisms and their spores were the result and not the cause of diseases.

It was the invention of the microscope and the discovery of microorganisms that enabled scientists to think seriously about the origin of life. **L. Spallanzani** (1765) was the first to provide evidence that life (microorganisms) do not arise spontaneously in organic infusions. He boiled beef broth for one hour and then sealed the flasks. No microbes appeared in the broth. He concluded that microorganisms entered the broth only through air, thus contradicting the doctrine of 'Spontaneous generation'.

The devastating epidemics of late blight of potato in Northern Europe, particularly Ireland in the 1840s, completely destroyed potato crop in Ireland in 1845 and 1846 and caused widespread famine that resulted in the death of hundreds of thousands of people. This evoked tremendous interest in finding out the cause and methods to control the disease. The German scientist **Anton de Bary** (1831 - 1888), who laid the foundation of modern 'Experimental Plant Pathology', conducted detailed study of potato late blight fungus. In 1861, he established that a fungus *(Phytophthora infestans)* was the cause of late blight of potato, a disease closely resembling the mildews. **De Bary** also showed conclusively that the smut and rust fungi are the cause and not the result of plant diseases and that some rust diseases require two alternate host plants. He also suggested the role of enzymes in the interaction between pathogen and the host.

It was during this period that **Louis Pasteur** (1860 - 1863) provided irrefutable evidence that microorganisms arise only from pre-existing microorganisms and that fermentation is a biological phenomenon and not a chemical one. Although Pasteur's conclusions were not generally accepted at that time, the proof of involvement of microorganisms (germs) in fermentation and disease development put an end to the theory of spontaneous generation and provided the basis for the germ theory of diseases. **Pasteur** also discovered life in the absence of oxygen (anaerobic growth), formulated the methods of pasteurization, discovered antirabies vaccine, suggested methods of control of pebrine disease of silk worm and isolated the organism responsible for chicken cholera. Because of his contributions to the science of Microbiology, he is called the **'Father of Microbiology'**. However, **John Tyndall** (1877) gave the final blow to spontaneous generation doctrine. He conducted several experiments and proved that air, free of dust contained no microorganisms and that, if no dust is present, sterile broth will remain free from microbial growth for an indefinite period.

Davaine (1863) made the first satisfactory demonstration of the probable relationship of microorganisms and disease (**The germ theory of disease**).

He found that the blood of animals affected by anthrax disease contained rod-shaped microorganisms and that transfer of such blood to a healthy animal could transmit the disease.

Later on **Robert Koch** (1875), a German physician confirmed the germ theory of disease and based on several experiments conducted by him, he put forward a hypothesis known as **'Koch's postulates'**, which is advocated even now to confirm the pathogenecity of a particular microorganism. His contributions in the field of Microbiology include the introduction of aniline dyes for staining bacteria, use of agar agar and gelatin to prepare solid culture media, stressing the need for pure cultures in the study of pathogens and establishing Koch's postulates, discovering the causal agents of anthrax *(Bacillus anthracis)* and tuberculosis *(Mycobacterium tuberculosis)*. Because of his outstanding contributions in the field of methodology, he is called the **'Father of Microbial Techniques'**.

Brefeld (1875 - 1912), a colleague of **Anton de Bary**, discovered the methods of artificial culture of microorganisms. With these methods, the study of infectious microorganisms became easier.

In 1878, the downy mildew of grapevine was introduced into Europe from America. The disease almost wiped out grapevines in France, which in turn almost finished the wine industry. Professor **Millardet** discovered Bordeaux mixture for the control of this disease. Though the discovery of Bordeaux mixture was accidental, it proved to be a very effective fungicide in controlling downy mildew of grapevine and late blight of potato.

The establishment of fungal etiology in plant disease (potato late blight) by **Anton de Bary** (1861), discovery of fungal disease in animals (silk worm) by **A. Bassi** of Italy (1836), discovery of bacterial etiology of higher animal diseases (anthrax) by **Robert Koch** (1876), discovery of leprosy bacterium *(Mycobacterium leprae)* by **Hensen** (1874), abscesses bacterium (*Staphylococcus* sp.) by **Ogston** (1881), diphtheria bacterium *(Corynebacterium diptheriae)* by **Klebs** (1883), tetanus bacterium *(Clostridium titani)* and plague bacterium *(Pasteurella pestis)* by **Kitasato** (1880 - 1894), discovery of immunization technique against tetanus and diphtheria by **Von Bahring** (1890), discovery of defense mechanism of white blood cells by **Mitchnikoff** (1890), discovery of some arsenical substances like **'salvarsan'** for destroying syphilis bacteria in the body by **Paul Ehrlich** (1854 - 1915), which made him the **'Founder of Chemotherapy'**, and many other findings during this period made the latter half of the nineteenth century an exciting period in the history of Microbiology and of infectious diseases.

T.J.Burrill of USA (1878 - 1882) for the first time demonstrated that fire blight of pear and apple was caused by a bacterium *(Erwinia amylovora)*. By 1900, over a score of plant diseases had been conclusively shown to be caused by bacteria. **E.F.Smith** of USA (1899 - 1901) was the main contributor to the discovery of most of these bacterial plant diseases, besides introduction of methodologies for the study of such diseases. Because of his vast contribution in the field of bacterial plant diseases, he is rightly called the **'Father of Phytobacteriology'**.

Berkeley and **Schacht** (1857) discovered root knot nematode and cyst nematode of beet. **Cobb** (1913 - 1932) studied the structure of nematodes and classified them. The economic loss due to nematode infestation of plants was realized and soil fumigants were developed to control nematodes. Later on '**Nematology**' became a separate branch of Biological Science. **Cobb** also introduced the **'Cobb scale'**, a system of grading diseases to assess the intensity of diseases quantitatively.

The discovery of *Tobacco mosaic virus* by **Adolf Meyer** (1886), was the beginning of studies on viruses as causal agents of diseases. **Meyer** proved that this disease could be transmitted from diseased to healthy plants by the sap from leaves showing mosaic symptoms. **E.F.Smith** of USA (1891), working with 'peach yellows' demonstrated for the first time that budding or grafting could be another method of transmission of plant viruses. **Ivanowski** and **Beijerinck** are considered to be the pioneers for starting studies on plant viruses. In 1892, **Ivanowski** demonstrated that *Tobacco mosaic virus* could pass through filters that retained even the smallest bacterial cells. He suggested that viruses were smaller than the smallest known bacterium and still he believed them to be some sort of very small bacterium. In 1898, **Beijerinck** proved that the virus producing tobacco mosaic is not a microorganism and he termed it as **'contagium vivum fluidum'** (infectious living fluid) and coined the name **'virus'**. He also observed that new growing tissues were more prone to infection than old tissues and that the virus moves in the xylem and phloem. The contributions made by him in the field of Virology earned him the name **'Father of Plant Virology'**. In 1901, **Beijerinck** found the free-living, nitrogen-fixing bacterium *Azotobacter* and described its usefulness in promoting soil fertility. **Herelle** (1917) discovered **'bacterial viruses'** (bacteriophages), which are capable of attacking and destroying bacteria.

In the last decade of the eighteenth century and the first decade of the nineteenth century, much importance was given to genetics of plant disease resistance and efforts to develop resistant varieties in all crops were made.

However, only a few crops had varieties that possess permanent resistance to one or more diseases. One of the causes for short-lived nature of resistance was found to be due to the variability among pathogens. The Swedish scientist **Erikson** (1894) first discovered this phenomenon of variability. He found the existence of '**physiologic races**' in the rust fungus. In the second decade of the nineteenth century, the famous American Plant Pathologist, **E.C.Stakman** continued this study and came to the conclusion that due to continuous evolution of races and biotypes in species of the rust fungus, its pathogenic capability undergoes changes and as a result the resistance capability of the host also shows changes. **Hansen** and **Smith** (1932) gave the first demonstration of the origin of physiologic races through heterokaryosis.

Stanley published an important discovery about viruses in 1935. He treated the sap from mosaic affected leaves of tobacco with ammonium sulfate and obtained a crystalline protein, which when placed on healthy tobacco leaves, could produce the disease. This discovery finally proved that viruses are not living microorganisms, because no living form after it is chemically treated and crystallized can still remain viable. In 1936, **Bawden** and his co-workers in England found that the crystalline powder of the virus contains protein and ribonucleic acid.

In 1936, the '**Electron microscope**' was invented and with this, it was possible to see virus particles in the infected sap and study them in detail.

Gierrer and **Schramm** (1956) proved that the nucleic acid fraction of the virus is actually the infectious agent.

Before 1967, only fungi, bacteria, nematodes and viruses were considered to be the main causal agents of plant diseases. **Doi** et.al. (1967) of Japan were the first to report that sieve elements of many plants showing 'witches' broom' and 'yellows' symptoms, on examination with electron microscope, revealed the presence of 'mycoplasmal' bodies, which were not present in healthy plants. These phloem inhabiting organisms causing plant diseases are termed as '**mycoplasma-like organisms**' (MLO) and because of their similarity to the pleuropneumonia agent, they are also known as '**pleuropneumonia-like organisms**' (PPLO). These diseases were earlier thought to be caused by viruses. The mycoplasmal bodies were found to be transmitted mostly by leafhoppers. There was remission of symptoms when Tetracycline was applied to the plant. Subsequently, in more than 200 plant species showing symptoms of 'witches' broom' and 'yellows', mycoplasma-like bodies were found in the vascular tissues. These organisms have not been successfully cultured and hence it has not been possible to apply Koch's postulates. These microorganisms have been placed in the Family - Mycoplasmataceae.

In 1969, certain MLOs were found to possess helical (spiral) structure and were motile in nature. They were called as '**spiroplasmas**'. The spiroplasmas causing '*Citrus* stubborn disease' and 'corn stunt disease' have been cultured and Koch's postulates proved. These organisms are placed under the Family - Spiroplasmataceae and the phloem inhabiting causal agent of '*Citrus* stubborn disease' is named *Spiroplasma citri.*

Diener (1971) showed that a small, naked, single stranded, circular molecule of infectious RNA, which he called a '**viroid**', caused the 'potato spindle tuber disease'. Viroids seemed to be the smallest infectious nucleic acid. So far, more than 20 viroids have been found to infect plants. '*Chrysanthemum* stunt disease', '*Chrysanthemum* chlorotic mottle disease', 'coconut cadang cadang disease', 'tomato bunchy top disease' etc. are some of the important diseases caused by viroids.

In 1973, another group of fastidious prokaryotes was discovered in *Citrus* greening diseased plants. This organism was found to have definite cell wall, gram-negative in character and susceptible to both Penicillin and Tetracycline. Hence, these organisms were thought to be bacteria. Similar organisms were found in 'Pierces' disease' of grapevine and in many other diseases. These gram-negative bacteria are usually referred to as '**rickettsia-like bacteria**' (RLB) or '**rickettsia-like organisms**' (RLO). 'Ratoon stunting disease' of sugarcane, which was thought to be a virus disease, was later found to be due to a RLB (**Maramorosch**, 1973). In 1974, a coryneform bacterium was reported as the causal agent of this disease by **Teakle** et.al. **Davis** et.al. (1980) cultured this gram-positive, xylem inhabiting coryneform bacterium on synthetic medium and proved Koch's postulates. This bacterium has now been named as *Clavibacter xyli* sub sp. *xyli.*

Randles et.al. (1981), discovered '**virusoids**'. The virusoids are similar to viroids, but smaller than viroids. They have low molecular weight and the RNA is circular in structure. They are always associated with larger RNA molecule of another virus. According to **Robertson** et.al. (1983), some virusoids are necessary for the replication of RNA of the virus with which they are associated and may form part of the viral genome. Most virusoids are found to be more like a satellite i.e. extra RNA of a virus and are capable of replication only in the presence of the associated virus.

Mc Coy and **Martinez-Lopez** (1982) discovered a flagellated '**protozoan**' associated with 'heart rot' of coconut palm. This phloem inhabiting protozoa, *Phytomonas staheli* is associated with coconut and oil palm diseases in South America.

Although the existence of microorganisms and the role played by them in causing plant diseases were known to the ancient Indians and measures to control some of the plant diseases were practiced by them as mentioned earlier, organized researches on plant pathogens and plant diseases were started in this country only in the first decade of the 20th century. The British Government that ruled India at that time established **'The Imperial Agricultural Research Institute'** at Pusa, Bihar. It was in this Institute that **E.J.Butler**, by 1910 initiated elaborate study of Indian fungi and diseases caused by them and continued the work for the next 20 years. As such, he is considered to be the **'Father of Modern Plant Pathology'** in India. He was instrumental in moulding several world renowned Plant Pathologists in India. **J.F.Dastur** (1886 - 1971), a colleague of **Butler** was the first Indian Plant Pathologist to conduct detailed study of fungi and plant diseases. **B.B.Mundkur** did significant work on smut, cotton wilt etc. He is responsible for starting **'The Indian Phytopathological Society'** in 1948 and for bringing out the journal '**Indian Phytopathology**'. He also wrote the textbook 'Fungi and Plant Diseases'. **K.C.Mehta**, **J.C.Luthra**, **S.N.Dasgupta**, **T.S.Sadasivam** and many others had contributed much in the field of Plant Pathology. The names of **M.V.Patel**, **V.P.Bhida** and **G.Rangaswami** can be cited for pioneering the work on bacterial plant diseases. Following the initial establishment of Indian Universities in 1857 at Calcutta, Bombay and Madras, many other Universities had also been established subsequently and these Universities have now taken up both Plant Pathological Research and Teaching.

Pathogens

Fungi

The '**fungi**' (sing. - fungus) are **Thallophytes**, which do not possess chlorophyll (achlorophyllous), even in those members that are generally green-colored. The chlorophyllous plants use solar energy and carbon dioxide present in the air and synthesize their own requirements of food through the process of photosynthesis, whereas the fungi are incapable of photosynthesis. They live on organic matter or food already prepared. Many fungi are parasitic on plants, while some others parasitize animals, birds etc. and derive their requirements of food from these hosts. Many other fungi are saprophytic in nature and live on organic matter alone.

Most of the more than 1,00,000 species of fungi recorded are strictly saprophytic (obligate saprophytes) and they live only on dead organic matter, which they help to decompose. More than 10,000 species of fungi can cause diseases in plants. All plants are attacked by some species of fungi and some fungal species can attack more than one plant species. Some fungi known as

'obligate parasites' or **'biotrophs'** continue their entire life in association with their host plants. Others known as **'non-obligate'** parasites require a host plant for part of their life cycles but can complete their life cycles on dead organic matter, as well as on living plants.

Bacteria

'Bacteria' (sing. - bacterium) are microscopic organisms and like fungi they are, either parasitic or saprophytic in nature. But, a few bacteria are photosynthetic and are capable of manufacturing their own requirements of food. The scientist, **Anton van Leewenhock** (1676) discovered bacteria and because of their motility, they were considered to be of animal origin. Characteristics, such as possession of unicellular body, absence of well-defined nucleus, capacity to reproduce asexually etc. are more akin to blue-green algae of the Plant Kingdom and hence they were treated as fungi. The German scientist, **Anton de Bary** (1831 - '88) separated bacteria from fungi and classified them under the Class - Schizomycetes. **Ehrenberg** (1829) formed a separate Division - Bacterium. **Louis Pasteur** (1822 - '55), conducted detailed research on bacteria and reported several facts, such as the ability of bacteria to ferment many products, and ability to cause several diseases in plants, animals and humans. Now bacteria and mollicutes are classified under a different group called **'Prokaryotes'**. Bacteria are ubiquitous. They are found in the air, soil and water, as well as in animals, plants, humans etc. They are highly resistant to any type of stress. Even under conditions of very low temperature (-19°C) or high temperature (75°C), severe drought and other adverse conditions, they can survive and continue their life.

Viruses

From rabies-affected humans or dogs, when no pathogenic fungi or bacteria could be detected, **Pasteur** (1876) suggested that some pathogen, even smaller than the already known microscopic organisms might have been associated with the disease. But, only after several years, his suggestion was proved to be correct. **Mayer** (1886) described a kind of disease in tobacco and named it as 'tobacco mosaic disease'. He found out that this disease could be easily transmitted to healthy tobacco plants. Elaborate research work conducted by the Russian scientist, **Iwanowski** (1864 - 1920) led to a clear understanding of viruses. He took the sap from mosaic affected plants, passed the sap through a bacterial filter and then examined the filtered sap under microscope. The sap was free from any fungi or bacteria. Even when the sap was kept for several days, there was no fungal or bacterial growth. However, when the sap was rubbed on the surface of healthy leaves of tobacco plants, mosaic symptoms appeared. During this period, results of several such research work were also published. The fact that this type of diseases was caused by a different type pathogen was then established. **Beijerinck** (1898) named the pathogen as '**virus**'. In Latin, the word virus means '**poison potion**' or '**poison**'.

Herelle (1917) found out that some pathogens are capable of attacking bacteria and named them as '**bacterial viruses**', '**bacterial eaters**' or '**bacteriophages**'. **Stanley** (1935) was the first to suggest that viruses were a different type of pathogen and could produce specific symptoms.

After the invention of 'electron microscope' in 1936, several other sophisticated scientific equipments were introduced, which made it possible to view and photograph viruses and also to study their various characteristics. The plant pathogenic spherical virus particles have a diameter of 20 mμ and the cylindrical virus particles are 15 mμ in width and of varying lengths. A total of about one million virus particles can be accommodated inside a single bacterial cell. Now, viruses are known to be '**mesobiotic factors**', which possess the characters of both living organisms and non-living entities.

Viroids

In 1971, **Diener** ascertained that a small, naked, single - stranded, circular molecule of infectious ribonucleic acid (RNA), which he called a '**viroid**', caused the 'potato spindle tuber disease'. These viroids are the smallest and simplest agents of infectious diseases. Viroids are low molecular weight RNA that can infect plant cells. They can grow and replicate themselves or reproduce in nature only in living hosts. So far, 20 plant diseases have been shown to be caused by viroids. The most important viroid plant diseases are 'cadang-cadang disease of coconut', 'potato spindle tuber disease', '*Chrysanthemum* stunt' and a few others. No animal or human disease has been shown to be caused by a viroid yet. Viroids differ from viruses in at least two main characteristics : **(i)** the size of RNA is much smaller and consists of 250 - 370 nucleotides (bases) in viroids, as against 4,000 - 20,000 nucleotides (4 - 20 kb) in viruses **(ii)** Viroids lack a protein coat and apparently exist as free RNA, while the virus RNA is enclosed in a protein coat. Because of their small size, viroids lack sufficient information to code for even a coat protein or for a replicase enzyme required for its multiplication.

Viroids are circular, single-stranded RNA molecules. However, there is some type of base pairing mechanism, which makes the RNA stable. The viroids possess many of the properties of ssRNAs and are about 40 nm. in length, and have the thickness of double-stranded DNA. The viroids are known to replicate by direct RNA copying, in which all components required for viroid replication are provided by the host. Viroid diseases produce a variety of symptoms that resemble those caused by virus infection. They may interfere with the host metabolism in ways resembling those of viruses. Viroids may activate the enzyme 'protein kinase', which in turn may trigger viroid pathogenesis and disease development in the host.

Viroids are spread from diseased to healthy plants, primarily through sap carried on hands or tools during propagation, mainly through vegetative propagation or cultural operations. Some, such as '*Potato spindle viroid*', '*Chrysanthemum stunt viroid*' and '*Chrysanthemum chlorotic mottle viroid*' are transmitted through sap, while others, such as '*Coconut cadang-cadang viroid*', '*Tomato bunchy top viroid*' and '*Apple scar skin viroid*' may be transmitted through pollen and seed. No specific insect or other vectors of viroids are known.

Viroids survive in nature, outside the host or in dead plant refuse for a few minutes to a few months. They may overwinter or oversummer in their perennial hosts. Viroids are usually highly resistant to high temperatures and cannot be inactivated in infected plants or plant materials by heat treatment.

Mollicutes

Phytoplasmas

The Japanese scientists, **Doi** (1967) and **Ishue** (1967) were the first to suggest that '**mycoplasma-like microorganisms**' caused some type of yellowing diseases in plants. In that year, some wall-less microorganisms were detected under the electron microscope in the phloem tissues of plants infected with some type of yellowing diseases and in insect vectors of these diseases. The causal organism of these diseases, which were till then thought to be viruses, differed from viruses in their characteristics and hence they were subsequently called as **mycoplsma-like organisms** (MLO), because of their superficial resemblance to mycoplasmas. It was later shown that they were not true mycoplasmas. Though all these microorganisms belong to mollicutes, most of them are round to elongate and they are presently known as '**phytoplasmas**', while a few of them have a helical structure and are called '**spiroplasmas**'.

More than 200 distinct plant diseases affecting several types of plants have been found to be caused by phytoplasmas. Some very destructive diseases of trees and vines, such as 'pear decline', 'grape yellows', 'coconut lethal yellowing', 'apple proliferation' etc., as well as some such diseases of herbaceous annuals and perennials, such as 'aster yellows' of vegetables and ornamentals are caused by phytoplasmas. Some of the diseases, such as 'gingelly phyllody', 'brinjal little leaf', 'rice yellow dwarf' and 'sugarcane grassy shoot', which are known to have been caused by mycoplasma-like organisms may be due to phytoplasmas. So far, only a few diseases, such as '*Citrus* stubborn' and 'corn stunt' are known to be caused by spiroplasmas. The main characteristics of such yellow type diseases are gradual and uniform yellowing or reddening of the leaves, smaller leaves, shortening of the internodes, stunting of the plant, excessive proliferation of shoots, formation of 'witches' brooms', greening and sterility of flowers, reduced yields, die-back, decline, and ultimately death of the plant. Often root abnormalities and necrosis precede the aboveground symptoms.

The true nature of phytoplasmas and their taxonomic position among lower organisms is still uncertain. Morphologically, the phytoplasmas resemble typical mycoplasmas found in animals and humans and those living saprophytically. However, their genomes are only distantly related to true mycoplasmas. It had not been possible to grow phytoplasmas on artificial nutrient media and so far, it had not been possible to reproduce the disease on healthy plants inoculated with the phytoplasmas obtained from diseased plants. However, it had been possible to grow spiroplasmas on artificial nutrient media and to reproduce the disease through inoculation by vectors injected with the organism from culture.

The Class - Mollicutes has a single Order - Mycoplasmatales. Under this Order, there are three Families - Mycoplasmataceae, Acholeplasmataceae and Spiroplasmataceae. Each Family has a single Genus - *Mycoplasma*, *Acholeplasma* and *Spiroplasma* respectively. Though phytoplasmas are classified under the Order - Mycoplasmatales, no separate Family or Genus has been assigned to them.

The phytoplasmas resemble mycoplasmas in all morphological characteristics. However, genetically they are more related to *Acholeplasma* than to Mycoplasma. The phytoplasmas lack cell wall, are bounded by a unit membrane and have cytoplasm, ribosome and strands of nuclear material. They have no flagella and reproduce by budding and by binary, transverse fission of cells and they produce no spores. They are usually spheroid to ovoid or irregularly tubular to filamentous. The size of spherical phytoplasmas may vary from one to a few microns, while the filamentous forms may range in length from a few microns to 150 μm.

Phytoplasmas, as well as spiroplasmas are generally present in the sap of a small number of phloem sieve tubes. They are mostly transmitted by plant hoppers from diseased to healthy plants, but some are transmitted by psyllids and plant hoppers. Plant mollicutes also grow in the alimentary canal, haemolymph and salivary glands and intercellularly in different body organs of their vectors. Insect vectors can acquire the pathogen after feeding on infected plants, especially from young leaves and stems of infected plants for several hours or days or if they are injected with extracts from infected plants or vectors, which had already acquired the pathogen. Only after an incubation period of 10 to 45 days, depending upon the prevailing temperature, the vector can transmit the mollicutes and not immediately after feeding on the infected plant. The shortest incubation period occurs at 30°C and the longest at about 10°C. Incubation period is required for the multiplication and distribution of the mollicutes within the insect vector. After acquisition, the mollicute multiplies first in the intestinal cells of the vector, then it passes into the haemolymph and infects the internal organs and finally the brain and the salivary glands. Only after the concentration of the mollicutes in the salivary glands reaches a certain level, the insect can transmit the pathogen to healthy plants and continue to be a vector for the rest of its life. In some cases, the vectors are also affected adversely by the mollicutes and exhibit severe pathological symptoms. Nymphs can acquire the mollicutes more readily than adult hoppers and this ability to transmit the mollicutes continues even after subsequent moults. But, the pathogen is not passed on to the next generation through the eggs of the vectors.

Numerous attempts to culture phytoplasmas on artificial culture media have not been successful. However, phytoplasmas have been extracted from their hosts and from the vectors in almost pure form and for most of them antiserum including antibodies have been obtained. Specific antibodies, DNA probes etc. are being successfully used for detection and identification of the pathogen and for controlling the diseases through production of pathogen-free propagating materials. Nowadays, serological and nucleic acid techniques are widely used for the detection of mollicute infections.

Mollicutes are sensitive to antibiotics, particularly those of the Tetracycline group. When infected plants are treated with Tetracycline, either by immersion or through injection, the symptoms disappear, but reappear, if the treatment is stopped. Dormant propagative plant organs can be completely freed from mollicutes by hot water treatment at 30° - 50°C. The time required for hot water treatment is 72 hours at the lower temperature and 10 minutes at the higher temperature.

Spiroplasmas

'Spiroplasmas' are helical mollicutes. The spiroplasma cells vary in shape from spherical to slightly ovoid, the diameter ranging from 100 - 240 nm, to helical and branched, non-helical filaments, measuring about 120 nm in diameter and 2.0 - 4.0 μm in length. In later stages of growth, they may reach lengths up to 15 μm. Unlike the phytoplasmas, spiroplasmas can be isolated from their host plants or their insect vectors and cultured on nutrient media. In liquid media, they produce mostly helical forms. They multiply by fission. They do not have a true cell wall, but are bounded by a unit membrane. They are non-flagellate. But, the helical filaments are capable of movement by a quick, rotary or screw-type motion of the helix. Spiroplasmas are resistant to Penicillin, but inhibited by Tetracycline.

Culture of spiroplasmas obtained from the host plants can be fed or injected into their insect vectors. After acquisition, the vectors on feeding on the host plants, transmit the pathogen to the plants. Such infected host plants develop typical symptoms of the disease in due course.

So far, spiroplasmas are known to cause the 'stubborn disease' in *Citrus* plants *(Spiroplasma citri)*. This pathogen has also been found in many other dicots, such as crucifers, lettuce and peach. They are also found to infect the leafhopper vectors. Many kinds of spiroplasmas are known to infect honeybees and several other insects. Many others live saprophytically on flowers and other plant surfaces.

Rickettsia - Like Organisms

'The fastidious vascular bacteria' known earlier as **'rickettsia-like organisms'** (RLO), which cause plant diseases cannot be cultured on nutrient media. Some such bacteria are yet to be identified, named and classified. In 1972, such phloem-limited bacteria were observed in the phloem tissues of clover, affected with the 'clover club leaf disease' and later in *Citrus* plants affected with the 'greening disease'. In 1973, similar xylem-limited bacteria were observed in the xylem vessels of grape plants affected with 'Pierces' disease'. Subsequently, in more than 20 such diseases, one or more xylem-limited organisms were found **(e-g)** sugarcane ratoon stunting, plum leaf scald, almond leaf scorch etc.

The fastidious vascular bacteria are generally rod-shaped cells and 0.2 - 0.5 μm x 1.0 - 4.0 μm in size. A cell membrane and a cell wall bound them, but in the case of the phloem-inhabiting bacteria, the cell wall appears more like a second membrane than as a cell wall. They are non-flagellate and the cells are usually undulated or rippled. Almost all of them are gram-negative. Most of the xylem-limited bacteria are placed under the new Genus - *Xylella*. However, the xylem-inhabiting bacteria causing 'sugarcane ratoon stunting' and 'Bermuda grass stunting' are gram-positive and are classified under the Genus - *Corynebacterium (Clavibacter)*. The fastidious vascular bacteria cannot be grown on ordinary bacteriological media, but the xylem-inhabiting fastidious bacteria can be grown on complex nutrient media.

Xylem-feeding insects, such as sharpshooter-leafhoppers and spittlebugs transmit all gram-negative, xylem-inhabiting fastidious bacteria. The vectors are capable of acquiring and transmitting the bacteria in less than 2 hours. The adult vectors can transmit the bacteria for life, but it is not passed on to the next generation. For the gram-positive, xylem-inhabiting fastidious bacteria, no insect vector is found so far. But this type of bacteria, which cause 'sugarcane ratoon stunting', is transmitted mechanically by cutting implements during harvest. Leafhoppers and psyllids are the vectors of 'clover club leaf' and '*Citrus* greening bacteria' respectively. The clover club leaf bacterium multiplies in its leafhopper vector and the bacterium is passed on to the next generation by transovarial transmission.

The symptoms of diseases caused by fastidious xylem-inhabiting bacteria are marginal necrosis of leaves, stunting, general decline and low yields. The symptoms caused may be due to plugging of the xylem vessels by bacterial cells and by a matrix material partly of bacterial and partly of plant origin. In the case of 'sugarcane ratoon stunting', the visible symptoms are stunting and internal discoloration of the stem. This disease, which is a very important one, is caused by the xylem-limited, gram-positive bacteria *Clavibacter xyli* sub sp. *xyli*, while. *Xylella fastidiosa*, which is a fastidious xylem-limited, gram-negative bacterium is the causal organism of diseases, such as '*Citrus* variegation chlorosis', 'Pierces' disease of grape', 'plum leaf scald', 'leaf scorch of oak, almond, mulberry' etc. On the other hand, the symptoms of diseases caused by fastidious phloem-inhabiting bacteria are often associated with leaf stunting and clubbing and sometimes shoot proliferation. The phloem-limited bacteria cause '*Citrus* greening disease' and some other minor diseases.

Fastidious vascular bacteria are sensitive to many antibiotics, such as Tetracycline and Penicillin, as well as to high temperatures. Heat treatment of entire plants or the propagative organs by immersing them in hot water at 50°C for 2 hours or hot air at 54°C for 8 hours eliminates the bacteria. Thermotherapy has been found to cure grapevines from 'Pierces' disease' and sugarcane from 'ratoon stunting disease'.

Algae

Some '**algae**' belonging to the Plant Kingdom are also responsible for causing some type of diseases in plants. Most of the algae possess chlorophyll.

So, they can manufacture their own requirements of food by the process of photosynthesis, while some algae lack chlorophyll, as such they depend on higher plants for their requirements of food and live as parasites, thereby causing diseases in plants.

The most important algal pathogen, which causes diseases in plants, belongs to the Genus - *Cephaleuros*. It is classified under the Family - Chlorophyceae. The mycelium of this pathogen is found intercellularly in the host. Organisms belonging to this Genus infect hill crops, such as coffee and tea and horticultural crops, such as mango, jack, guava, *Citrus* etc. The disease caused by this pathogen in tea crop is known as 'red rust'. However, this rust is completely different from the rust disease caused by fungi on wheat, pearl millet, groundnut, beans etc. The symptoms of red rust are the formation of round, orange or red colored, raised or sunken, thick spots, which look like encrustation. The algal spots are seen sticking on to the leaf or stem surfaces. Sometimes, the bark of the affected branches breaks and splits, leading to tip drying of the branches and consequently the leaves turn yellow, wither and are shed prematurely, leading to severe loss in yield. **(e-g)** *Cephaleuros parasiticus* - causes 'red rust of tea'; *C. virescens* - causes 'red rust of mango, guava and citrus'; *C. mycoidea* - causes 'red rust of tea and black berry of pepper'.

Phanerogamic Parasites

More than 2,500 species of '**phanerogamic parasites**' belonging to the higher plants are known to live parasitically on other plants. These parasitic plants produce flowers and seeds and belong to several widely separated botanical families. Their dependence on the host plants varies greatly. Some phanerogamic parasites live exclusively on the stem region of the host plants, while some others live on the root region of the host plants. The type of parasitism also varies.

Types of Phanerogamic Parasites

Stem parasites. Among the stem parasites there are '**holo stem parasites**' and '**semi stem parasites**'. In the case of holo stem parasites, the leaves are diminutive, yellow or brownish in color and lack chlorophyll. As such, they are unable to manufacture their own requirements of sugar and starch through the process of photosynthesis. So, they attach themselves to the stem region of their host and derive all their requirements of nutrients directly from the host by means of haustoria **(e-g)** *Cuscuta gronovii*

In the case of semi stem parasites, they have functional leaves with chlorophyll and hence, they are capable of manufacturing their own requirements of sugar and starch through photosynthesis, but take only mineral nutrients and water from the host plants by means of haustoria. **(e-g)** *Dendrophthoe falcata.*

Root parasites. Among the root parasites also there are both '**holo root parasites**' and '**semi root parasites**'. They parasitize the root region only. The holo root parasites depend entirely on the host plants for all their requirements of nutrients and water **(e-g)** *Orobanche cernua, O. ramosa.*

The semi root parasites depend on the host plants only for their requirements of mineral nutrients and water **(e-g)** *Striga densiflora.*

Koch's Postulates or Koch's Rules

If a pathogen found on a host seems to be the cause of a particular disease, but no previous reports are available to support this, then the following procedures are followed to verify the hypothesis that the isolated pathogen is the cause of the disease:

1. The pathogen associated with the disease must be found in all the diseased plants examined
2. In the case of non-obligate parasites, the pathogen must be isolated, grown as pure culture on nutrient media and its characteristics studied. In the case of obligate parasites, it must be inoculated on a susceptible host plant and its characteristics studied.
3. The pathogen, when inoculated on healthy plants of the same species or variety must produce the same disease on the inoculated plants.
4. From the inoculated plants, the pathogen must again be reisolated in pure culture and its characters must be identical to the original culture obtained in step 2.

If all the above steps, which are known as '**Koch's rules**' or '**Koch's postulates**' are followed and proved true, then the isolated pathogen is identified as the causal agent of the disease. However, in the case of some viruses, phytoplasmas, some phloem-inhabiting bacteria and protozoa that cannot be cultured, purified and reintroduced to other healthy plants, Koch's rules are not applicable.

General Characters of Fungi

The fungi belong to a group of **Thallophytes**, which are devoid of chlorophyll. With few exceptions, they resemble simple plants in that, they have definite cell walls, they are non-motile, although they may have motile reproductive cells and they reproduce by means of spores. A '**spore**', which is synonym to a seed of higher plants, is a simple, minute, propagating unit without an embryo that serves in the production of a new individual of the same species. However, fungi do not possess stem, roots or leaves and they do not have a vascular system as in plants. Fungi are usually filamentous and multicellular. The filaments, which are the body or '**soma**' of a fungus, elongate by apical growth, but most parts of a fungus are capable of growth and a minute fragment from almost any part of the fungus is able to produce a new growing point and produce a new individual. The fungal thallus or **mycelium** typically consists of microscopic threads of filaments that branch in all directions, spreading over or within the substratum utilized for food. The individual branches or filaments of the mycelium are known as '**hyphae**' (sing. - hypha). The hyphae are generally uniform in thickness, usually about 1.0 - 2.0 µm. in diameter, but in some fungi may be more than 100 µm. thick. The length may vary from a few micrometers in some fungi, to several meters in some others. A hypha is made of a thin, transparent, tubular wall, filled with protoplasm. In some fungi the hypha is divided into separate cells by means of cross walls or '**septa**' (sing. - septum) and each cell may contain one or two nuclei. In others the hyphae contain many nuclei which may or may not have septa. The hyphae without septa are called '**coenocytic**', '**non-septate**' or '**aseptate**' and those with septa are called '**septate**' **(Fig. 1)**. Reproductive structures are differentiated from somatic structures and exhibit a wide variety of forms, which serve as a basis for classification of the fungi. The nucleus is bounded by two nuclear membranes.

Some lower fungi lack true mycelium, but produce strands, which are grossly dissimilar and of varying diameter called '**rhizomycelium**'. They are non-nucleate and resemble mycelium superficially. Some microorganisms previously thought to be primitive fungi, but now classified under the Kingdom - Protozoa (Myxomycota, Plasmodiophoromycetes), produce a naked, amoeboid, multinucleate body called 'plasmodium' instead of mycelium. The mycelium of some fungi form thick strands called '**rhizomorphs**' **(Fig. 2)**. Here the unit hyphae lose their individuality and form complex tissues. This string-like mass has a thick, hard cortex and a growing tip, like that of a root. Rhizomorphs are resistant to adverse environmental conditions and remain dormant. When favorable conditions return, their growth is resumed. The advanced forms of fungi, especially the Basidiomycetes, usually produce these types of structures.

The mycelium of parasitic fungi grows on the surface of the host or within the host tissues. In the case of mycelium, which grows in between the host cells **(intercellular) (Fig. 3)**, food is absorbed through the host cell wall. If the mycelium penetrates into the cells **(intracellular) (Fig. 4)**, the hyphae come into direct contact with the host protoplasm and obtain their nourishment directly. Intercellular hyphae of many fungi, especially the obligate parasites obtain their nourishment through specialized absorbing organs called '**haustoria**' (sing. - haustorium) **(Fig. 5)**, which are outgrowths of the somatic hyphae. The haustoria enter into the host cells through minute pores punctured in the cell wall. Haustoria may be knob-shaped, elongated or branched like a miniature root system. The hyphae of saprophytic fungi come in intimate contact with the substratum and obtain food by direct diffusion through the hyphal walls, causing disintegration of the organic matter that they utilize.

Many fungi produce various types of somatic and reproductive structures. Two such important somatic structures are the '**stroma**' (pl.- stromata) and the '**sclerotium**' (pl.- sclerotia). A stroma is a compact, somatic structure, much like a mattress or cushion made up of loosely woven hyphae more or less parallel to one another known as 'prosenchyma' tissue. Usually, fructifications are formed on or in the stroma **(Fig. 6)**. On the other hand, a sclerotium consists of closely packed, more or less isodiametric or oval cells known as 'pseudoparenchyma' tissue. It is a hard resting body, resistant to unfavorable conditions and may remain dormant for long periods of time and germinate when favorable conditions return. Species of *Rhizoctonia, Sclerotium* and *Claviceps* produce sclerotia **(Fig. 7)**.

Most fungi are capable of growth between 0° - 35°C, but the optimum temperature range is 20° - 30°C. However, there are many thermophilic fungal species, which can grow at higher temperatures of 50°C or more and at lower temperatures of 20°C or less. Fungi prefer an acid medium for growth, with a pH of 6.0 being the optimum for most species. Although light is not essential for the growth of fungi, some light is needed for sporulation in many species. However, light plays an important part in spore dispersal, since the spore-bearing organs of many fungi are positively phototrophic and discharge their spores towards light. Fungal hyphae are capable of indefinite growth under favorable conditions. The hypha grows only at its tip leading to hyphal elongation. The mycelium of a fungus generally begins as a short germ tube emerging from a germinating spore and grows more or less equally in all directions from a central point and develops a spherical colony.

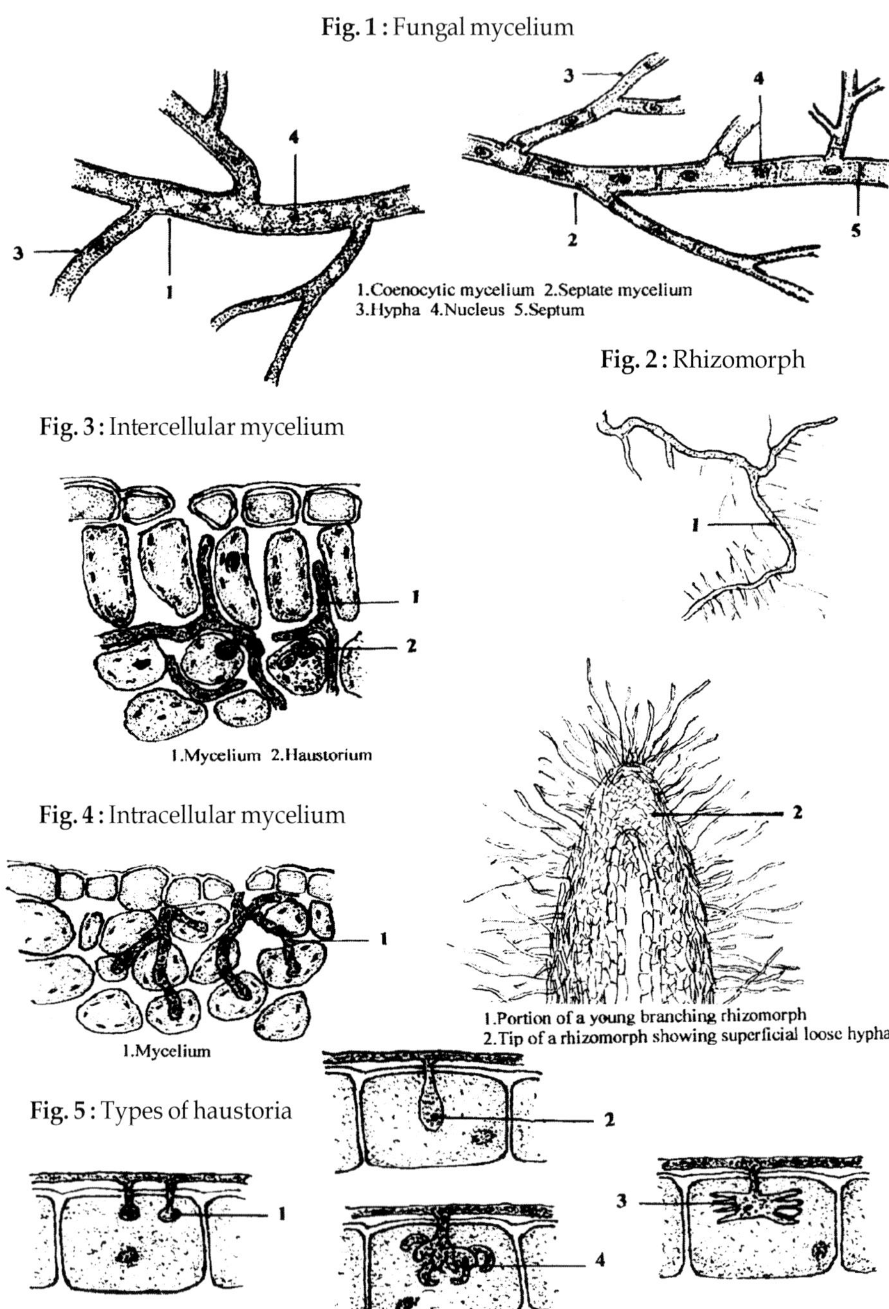

Fig. 1 : Fungal mycelium

Fig. 2 : Rhizomorph

Fig. 3 : Intercellular mycelium

Fig. 4 : Intracellular mycelium

Fig. 5 : Types of haustoria

In fungi, two types of reproduction namely, '**sexual reproduction**' and '**asexual reproduction**' are recognized. In asexual reproduction (somatic or vegetative reproduction), union of nuclei, sex cells or sex organs is not involved. In sexual reproduction union of two nuclei occurs. In some fungi, in the formation of either sexual or asexual reproductive organs, the entire thallus is converted into reproductive structures. Fungi that follow this pattern are called '**holocarpic**'. In the case of majority of other fungi, the reproductive organs arise from only a portion of the thallus, while the remainder continues its normal somatic life. Fungi belonging to this category are called '**eucarpic**'. Asexual and sexual reproduction is typical in fungi. In general, asexual reproduction is more important for the propagation of the species, as it produces numerous individuals simultaneously. Further, asexual cycle is usually repeated several times during the season, while sexual reproduction is carried out once in a season. Sexually produced spores are mostly resting spores and they serve to carry over the disease from one season to the following seasons.

Four types of asexual reproduction are found in fungi: **(1) Fragmentation of somatic hyphae:** In some fungi, the hypha may break up into their component cells that behave as spores known as '**arthrospores**' **(Fig. 8)**. If the cells get enveloped in a thick wall before they separate from each other or from the parent hyphae, they are called '**chlamydospores**'. Chlamydospores may be terminal or intercalary. Sometimes one or more cells of a conidium may be transformed into chlamydospores, as in the case of *Fusarium* **(Fig.9)**. **(2) Fission of somatic cells:** In this type of reproduction, a single hyphal cell splits into two daughter cells by constriction and formation of a cell wall. This is characteristic of many simple organisms, including some yeasts. **(3) Budding of somatic cells or spores: 'Budding'** is the production of small outgrowth (bud) from a parent cell. As the bud is formed, the nucleus of the parent cell divides and one daughter nucleus migrates into the bud. The bud develops in size while still attached to the parent cell and finally detaches and forms a new individual, as in the case of yeasts **(Fig. 10)**. **(4) Production of spores:** The most common method of asexual reproduction in fungi is by means of spores called '**conidia**'. Spores vary in color from hyaline (colorless) to green, yellow, orange, red, brown and black. They also vary greatly in size, shape and in number of cells **(Fig. 14)**.

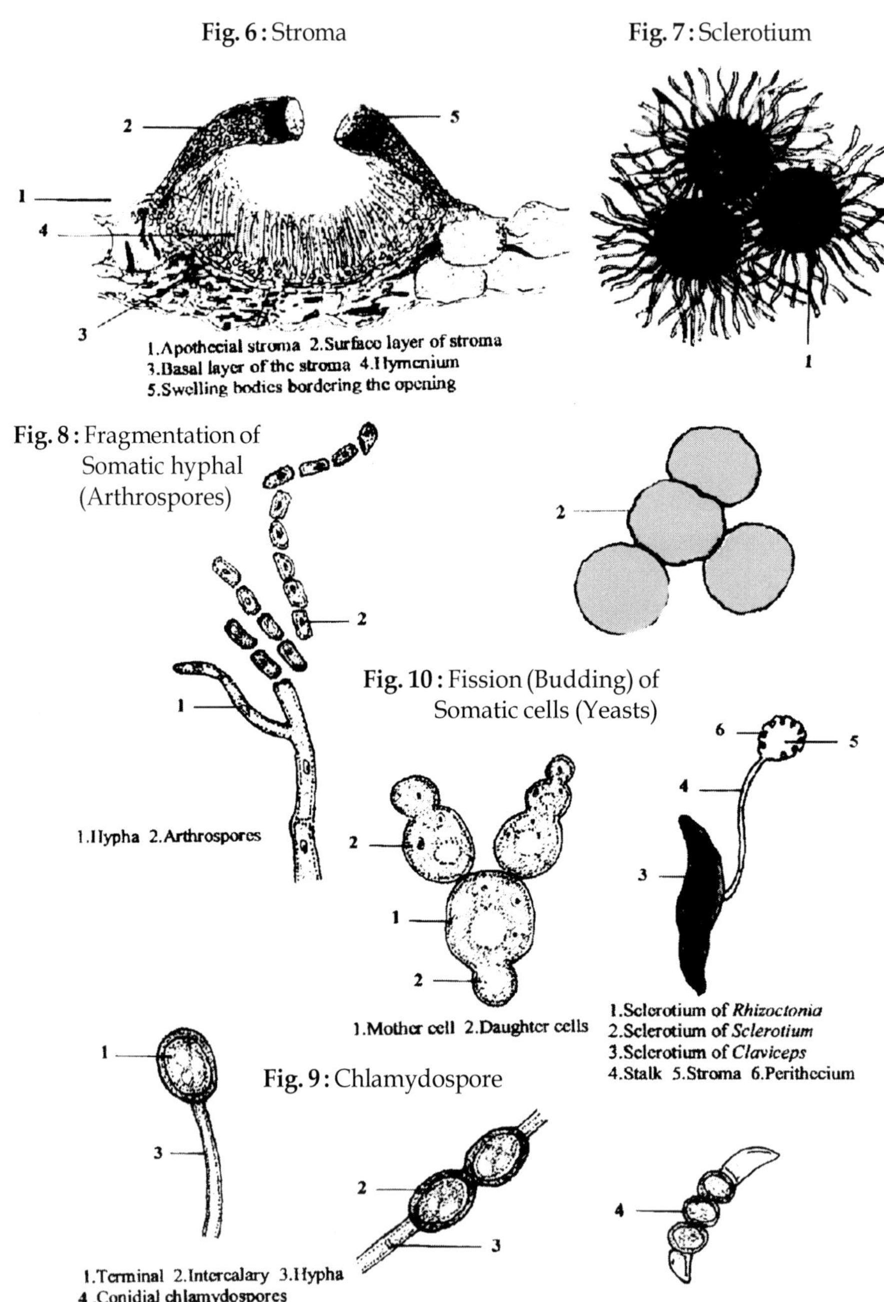
Fig. 6 : Stroma
1.Apothecial stroma 2.Surface layer of stroma
3.Basal layer of the stroma 4.Hymenium
5.Swelling bodies bordering the opening
Fig. 7 : Sclerotium
Fig. 8 : Fragmentation of Somatic hyphal (Arthrospores)
1.Hypha 2.Arthrospores
Fig. 10 : Fission (Budding) of Somatic cells (Yeasts)
1.Mother cell 2.Daughter cells
1.Sclerotium of *Rhizoctonia*
2.Sclerotium of *Sclerotium*
3.Sclerotium of *Claviceps*
4.Stalk 5.Stroma 6.Perithecium
Fig. 9 : Chlamydospore
1.Terminal 2.Intercalary 3.Hypha
4.Conidial chlamydospores

Some fungi produce only one type of spore, while others produce as many as four types. Asexually produced spores are borne in **'sporangia'** (sing. - sporangium) and are called **'sporangiospores'**. A sporangium is a sac-like structure, whose entire contents are converted through cleavage to form many sporangiospores, which are either motile or non-motile. The motile ones are called **'zoospores'** and the non-motile ones **'aplanospores'**. The zoospores may have one or two **'flagella'** (sing. - flagellum). Sporangia germinating directly by producing germ tubes instead of producing zoospores are also referred to as conidia. Spores are also produced at the tip or sides of special hyphae in various ways and are called **'conidia'** (sing. - conidium). The spore-bearing hyphae are called **'conidiophores' (Fig. 11)**. Some species produce spores on specialized, cushion-shaped stroma consisting of bundles or complex masses of hyphae (compound sporophores) that merely remain near together without lateral union and produce spores (conidia) from the hyphal tips. These are called **'sporodochia'** (sing. - sporodochium) **(Fig.12)**. In some other cases, a group of conidiophores unite, more or less closely and cemented together into columns and these are called **'coremia'** (sing. - coremium) or **'synnemata'** (sing. - synnema) **(Fig.13)**. Conidia are also produced inside thick-walled, flask-shaped structures called **'pycnidia'** (sing. - pycnidium) or inside inverted saucer-shaped structures called **'acervuli'** (sing. - acervulus).

Sexual reproduction in fungi involves the union of two compatible nuclei. The process of sexual reproduction consists of three distinct phases. In the first phase called **'plasmogamy'**, union of protoplasts of two cells brings the nuclei close together within a single cell. In the case of more advanced fungi fusion of the two nuclei does not take place immediately, resulting in a binucleate cell containing one nucleus from each parent. Such a pair of nuclei is called **'dikaryon'**. During cell division of the binucleate cell, the dikaryotic condition may continue from cell to cell by the simultaneous mitotic division of the two nuclei and so, in each cell there will be one nucleus from each of the parent. Eventually nuclear fusion takes place. The fusion of the two nuclei brought together by plasmogamy into one diploid or zygote nucleus is the second phase, which is called **'karyogamy'**. This is followed by meiosis of the diploid nucleus, resulting in the changeover to the haploid condition, which is the third phase of sexual reproduction. In sexual reproduction, some species produce distinguishable male and female sex organs on the same thallus. These species are called **'hermaphroditic'** or **'monoecious'**. In other species, there are separate male and female thalli. Here the male thalli produce only male sex organs, while the female thalli produce only female sex organs. These species are called **'dioecious'**. Single dioecious fungi cannot reproduce sexually by itself. The sex organs of fungi are called **'gametangia'** and these may form different sex cells called **'gametes'**. The gametangia and gametes, which are morphologically indistinguishable are called **'isogametangia'** and **'isogametes'** respectively, whereas the male and female gametangia and gametes, which are morphologically different, are called **'heterogametangia'** and **'heterogametes'** respectively. In heterogamous fungi, the male gametangium is called the **'antheridium'** and the female the **'oogonium'**. In fungi there are different methods by which compatible nuclei are brought together in the process of plasmogamy. In some fungi, two gametangia, which are morphologically identical, unite and produce a zygote called a **'zygospore'**. In fungi belonging to Ascomycetes the sexual spores, usually eight in number are produced within a sac-like cell, the **'ascus'** and the spores are called **'ascospores'**. In some others, no definite gametes are produced, but instead one mycelium may unite with other compatible mycelium. This happens in the fungi belonging to Basidiomycetes. In fungi belonging to Basidiomycetes, sexual spores are produced on club-like zygote cell called **'basidium'** and the spores are called **'basidiospores'**. However, in a large group of fungi belonging to Deuteromycetes (Fungi Imperfecti), no sexual reproduction occurs and they reproduce only asexually. Most of the fungi depend for their spread on several agents, such as wind, water, birds, insects, other mammals and humans, primarily by means of asexual or sexual spores produced by them. **'Zoospores'** are the only fungal spores that are motile, but can move only for short distances of a few millimeters or centimeters. However, zoospores require free water for movement and germination.

Types of Parasitism

In nature fungi live, either as a **'parasite'** or as a **'saprophyte'** (saprobe). The parasites obtain their food by infecting living organisms and the saprophytes obtain their food from dead organic matter. Many others form symbiotic relationships with plants as in lichens and in mycorrhizae. Different types of parasitism are noticed in fungi. Fungi that live solely on dead organic matter and incapable of infecting living organisms are known as **'obligate saprophytes'** (obligate saprobes), while those fungi, which cannot live except on living organisms are known as **'obligate parasites'**. Some fungi though naturally are parasitic on living hosts, in the absence of suitable hosts start leading a saprophytic life. These fungi are called **'facultative saprophytes'** (facultative saprobes), while some other fungi though they are saprophytes normally, become parasites, when environmental conditions are favorable to them and when suitable hosts are available. These are known as **'facultative parasites'**. A living organism infected by a parasite is called the **'host'**. The saprophytes, as well as the facultative saprophytes and facultative parasites can be successfully grown in artificial nutrient media. The enzymes, a fungus is capable of producing govern to a large extent its ability to break down complex and bigger molecules present in the food substances into smaller molecules, so as to assimilate them ultimately.

Major Symptoms of Fungal Diseases

Plants infected by fungal pathogens, most often cause certain abnormal changes, which are manifested in one or more parts of the plant or in the entire plant. Such abnormal changes are called **'symptoms'**. The symptoms caused by the fungal pathogens are specific and in many cases, the diseases can be identified by the very symptoms. Mostly, the symptoms develop on the outside and are quite visible; while in some cases the symptoms can be seen only on the inside and are not quite visible outside. In general, fungi cause local or general **'necrosis'** or killing of plant tissues, resulting in stunted growth of plant parts or entire plants. Some pathogens cause malformations in one or more parts, while some others cause excessive growth of infected plants or plant parts. In a number of cases, the fructifications of the pathogens are seen on the outside, which also aid in identifying the pathogens. The most common symptoms produced by fungi are detailed below:

Leaf spots. A large number of fungal pathogens, especially those belonging to Class - Deuteromycetes produce leaf spots. The leaf spots vary considerably in size, shape, color etc. depending upon the species, which cause them. These localized lesions, which are produced mostly on the leaves, consist of dead and collapsed cells. The spots may remain isolated till the end or they may coalesce to form bigger spots, sometimes covering the entire leaf surface. In some cases, a yellow or orange halo is formed around the spots. Pycnidia, acervuli or conidia may be produced on these spots at a later stage. Formation of such spots results in loss of chlorophyll, which in turn impairs the normal functioning of the leaves and ultimately the leaves dry and fall off prematurely leading to loss in yield.

(e-g) *Phyllosticta* leaf spot of banana; gray leaf spot of coconut; brown leaf spot of rice.

Shot holes. Some fungal pathogens after infection cause leaf spots on the leaves initially. In due course, the tissues in the central portion of the spots die, turn black, disintegrate and are blown off by wind, leaving behind holes. The holes thus formed appear to be made by bullet shots. Depending upon the severity of infection, the size of the holes may vary

(e-g) Tapioca shot holes.

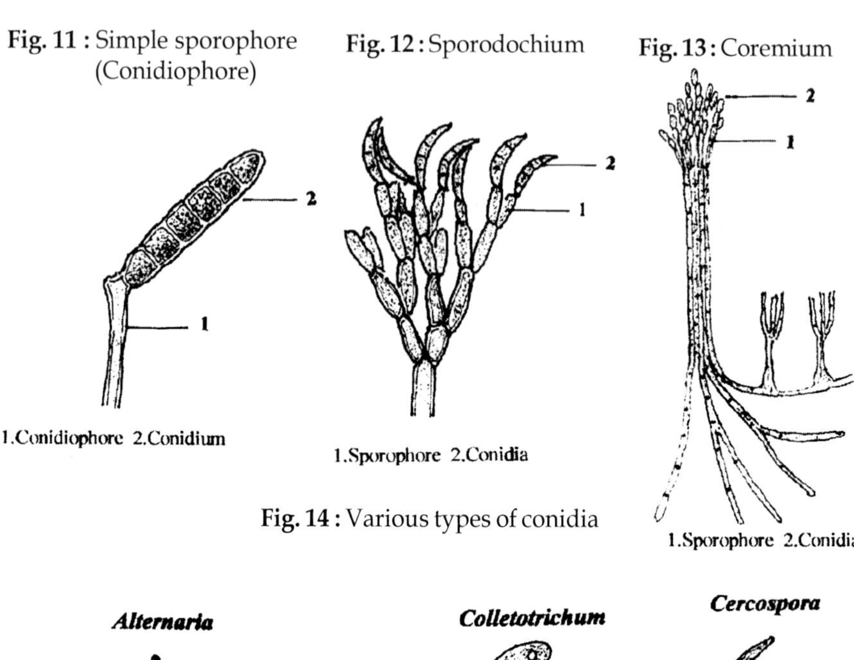

Fig. 11 : Simple sporophore (Conidiophore)

2
1

1.Conidiophore 2.Conidium

Fig. 12 : Sporodochium

2
1

1.Sporophore 2.Conidia

Fig. 13 : Coremium

2
1

1.Sporophore 2.Conidia

Fig. 14 : Various types of conidia

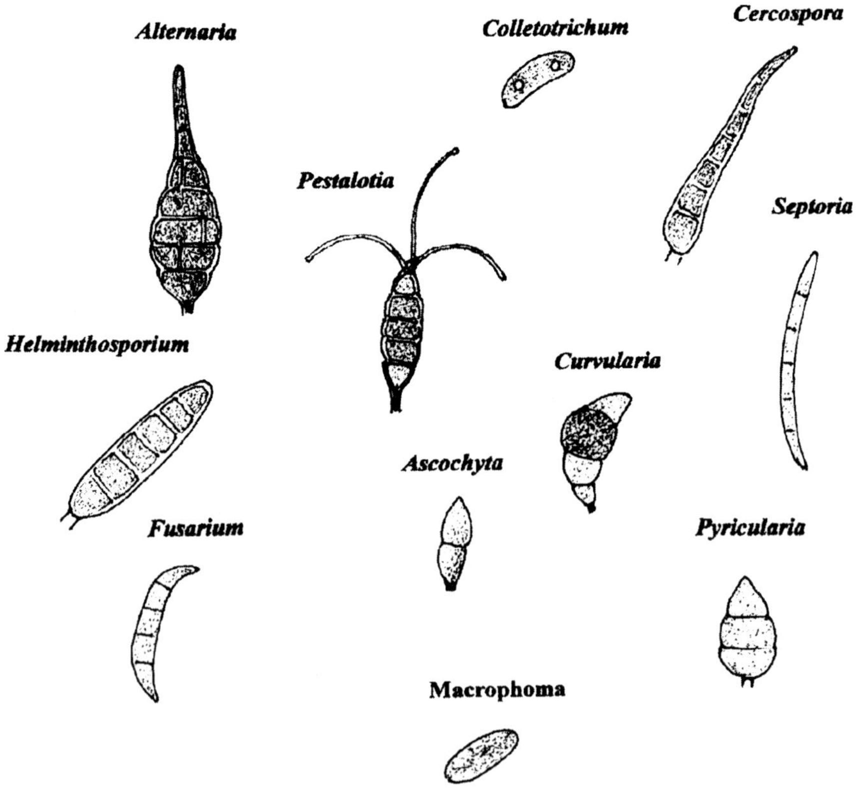

Alternaria
Colletotrichum
Cercospora
Pestalotia
Septoria
Helminthosporium
Curvularia
Ascochyta
Fusarium
Pyricularia
Macrophoma

Blight, leaf blight and sheath blight. Some pathogens cause general and extremely rapid browning, drying and death of leaves, branches, twigs and floral parts. Such premature death of affected plant parts results in total loss

(e-g) Late blight of potato

In some cases, large portions of the infected leaves turn dark-brown, the tissues in the affected areas dry and die rapidly

(e-g) Sorghum leaf blight.

In some other cases, the blighted symptoms appear on the outer leaf sheath regions and rapidly spread to all the leaf sheaths and leaves resulting in the death of the entire plant

(e-g) Sheath blight of rice.

Powdery mildew. Spots or patches of white to grayish, powdery coating or powdery mildew growth appear on a portion or on the entire surface of leaves, young shoots and other organs. Powdery mildew is most commonly seen on the upper surface of leaves, but the infection may also be seen on the under surface of leaves, young shoots and stems, buds, flowers and young fruits

(e-g) Powdery mildew of grapes; Powdery mildew of cucurbits; Powdery mildew of lady's finger.

Downy mildew. As a result of infection, white to grayish, visible mat of fungal downy growth appears on the under surface of leaves, while the corresponding areas on the upper surface of the leaves show pale yellow or yellow, irregular patches. Later the affected areas are killed and turn brown. The patches often enlarge and coalesce to form large, dead areas, sometimes covering the entire leaf surface resulting in defoliation

(e-g) Downy mildew of grapes; Downy mildew of millets; Downy mildew of cucurbits; Downy mildew of onion; Downy mildew of crucifers.

Rust. The rust fungi mostly attack the leaves and stems of plants. On the leaves rust infection appears, either on the upper surface or on the lower surface as numerous, rusty, yellow, orange, red, brown, black or even white, slightly elevated spots. The spots are elliptical, oval or round in shape and are strictly localized, but rarely the spots extend. Later on, these spots develop into rust sori beneath the epidermis. On maturity of the spores, the sori rupture and black, dark-brown or orange-colored spores come out in large numbers and are deposited on the leaf surface giving the appearance of rust

(e-g) Black, orange and yellow rusts of wheat; Groundnut rust; Coffee rust; Bean rust.

Smut. Most smut fungi attack the ovaries of grains of several cereal crops and the grains are transformed into smut sori containing large number of black-colored, powdery, smut spores. Mostly, the spores are held together temporarily by a thin, filmy membrane or in some cases by a more hardy membrane. When the membrane ruptures, the dark, powdery spores are blown off by wind. However, many smut fungi attack and produce smut sori on the leaves, stems or floral parts

(e-g) Smuts of cereals; Covered smut or bunt of wheat; Kernel smut of rice; Leaf smut of rice.

Wilt. The symptoms of wilt are characterized by generalized loss of turgidity, drooping of leaves and shoots and rapid death of the infected plants. The stem of affected plant, when split open shows longitudinal brown or black, intermittent or continuous streaks along the vascular region

(e-g) Red gram wilt; Cotton wilt; Banana wilt.

Stem rot. The initial symptoms appear as small, dark-brown or black spots on the basal region of the stem. These spots enlarge and girdle the stem in course of time. The tissues in the basal stem portion are destroyed and disintegrated, as a result the leaves start drying and ultimately the entire plant dies. In the disintegrated stem region, white, cottony growth of mycelium and minute, black sclerotia are also seen in large numbers

(e-g) Stem rot of rice; Stem rot of tomato.

Sooty mould. Sooty moulds appear on the leaves, shoots and fruits of several types of plants as a superficial, black, mycelial growth in the form of a thin, filmy layer or crust. These fungi are not actually parasitic, but live on the honeydew secreted by sucking insect pests. The fungal growth gives the affected parts a black, sooty appearance, which interferes with the photosynthetic function of the leaves. The mycelium sometimes peels off as a black, papery layer

(e-g) Sooty mould of mango, cashew and guava.

Root rot. The infected plants show disintegration or decay of parts or of the entire root system. In the disintegrated root regions, white or gray, cottony mycelial growth and minute, black sclerotia, which are embedded in the mycelium in large numbers can be seen.

(e-g) Root rot of groundnut, cotton, pulses and gingelly.

Die-back. The symptoms appear mostly at the time of flowering. Extensive necrosis of one or more shoots, starting from the tips and advancing toward the base is the characteristic symptom. On the dried twigs, large number of minute, pinhead-like dots, which represent the fruiting bodies (acervuli) of the pathogen, can be seen

(e-g) Die-back of chillies.

Damping off of seedlings. Pre-emergence and post-emergence damping off occurs in susceptible hosts. In **'pre-emergence damping off'**, the pathogens inhabiting the soil attack the seeds, as a result they fail to germinate, become soft and mushy, turn brown, shrink and finally disintegrate. Young seedlings, before emergence are attacked and from the point of infection the disease spreads rapidly and the invaded cells collapse, leading to the death of seedlings in large numbers. In **'post-emergence damping off'**, the seedlings that have already emerged are attacked at the collar region. The invaded regions become water-soaked, discolored and the basal part of the stem of seedlings become soft and much thinner, as a result the seedlings collapse and fall over on the soil. The pathogen continues to invade the fallen seedlings, which quickly wither and die in large numbers in patches. The symptoms are much pronounced in the nurseries

(e-g) Damping off of seedlings of chillies, eggplant, tobacco and tomato.

Anthracnose. Symptoms appear on the leaves, stems, floral parts or fruits of the host plants as irregularly circular, necrotic, sunken and ulcer-like lesions. The lesions have a dark-brown margin around them and in the central region of the lesions, fructifications (acervuli) are seen as black, minute pinhead- like dots.

(e-g) Anthracnose of grapes, bean, peas etc.

Ergot or sugary disease. The symptoms appear only on the inflorescence. The first symptoms appear as creamy droplets of a sticky fluid from young florets of infected earheads. The droplets are soon replaced by a hard, horn-like, purplish-black mass of fungal mycelium. These are the sclerotia called **'ergot'** that develop in the place of kernels. They are about a few millimeters in diameter and up to about 0.5 cm. in length

(e-g) Ergot of pearl millet; Ergot of rye.

Green ear. The symptoms appear commonly in the earheads of bajra, rye etc., which are infected by the downy mildew pathogen. All the floral parts of the invaded earheads are transformed into long, twisted, green, leafy structures. Sometimes, the symptoms are manifested in the entire earhead. In other cases, only the basal portion of the earhead alone exhibits such abnormal growth, while in the top portion seeds may be present. The leafy structures contain a large number of sexual spores (oospores), which carry over the disease during the following seasons

(e-g) Green ear of pearl millet (bajra); Green ear of rye.

Soft rot and **dry rot**. The symptoms appear on succulent parts of host plants, such as fruits, bulbs, tubers, fleshy leaves etc. In the case of **'soft rot'**, infected areas of fleshy organs at first become water-soaked and very soft. More often, the mycelium grows outward through wounds and covers the affected portions. Gradually the fungal growth extends to the surface of healthy portions also. Finally the entire organ breaks down and disintegrates in a watery rot

(e-g) Soft rot of fruits and vegetables.

In the case of **'dry rot'**, the rotting starts from a wound and spreads rapidly. In the infected area, the skin is slightly sunken and dark in color. As the infection spreads and the disease enters deeper into the fleshy tissues, the overlying skin develops a number of concentric wrinkles around the point of infection. With progressive rotting and loss of water, the affected organ shrinks gradually, dries and becomes rather hard **(e-g)** Dry rot of potato.

Stem bleeding. The symptoms are seen on the stem portion and are more common in older plants. In the infected stem, several small cracks develop and through these cracks a reddish-brown fluid oozes out, which turns black after drying.

(e-g) Coconut stem bleeding.

Canker. Cankers are formed mostly on the stem region of the plants and sometimes, on the fruits also. Cankers on the main stem are formed mostly at the base of a lateral branch, which had died, or near a dormant bud. At first, they seem to be simple swellings, but later become open wounds with dead wood at the bottom, which are surrounded by raised tiers of swollen tissues. Thick resin comes out from these wounds and flows down the stem

(e-g) Canker of apple; Stem canker of rose; Stem canker of potato; Canker of peach.

Tar spots. Numerous, small, black, elliptical or oval or circular, thick, raised spots are seen on the upper surface of leaves. These localized spots are shiny and look like coal tar droplets.

(e-g) Tar spot of cholam; Tar spot of grasses.

Blister blight. Young leaves of infected plants show fairly big blisters on the under surface. The blisters that are white in color initially turn brown later on. The upper surface of the leaf corresponding to the blisters on the lower surface is sunken and yellowish in color

(e-g) Blister blight of tea.

Scab. The general symptoms are characterized by localized lesions on host fruits, tubers, leaves etc., which are usually raised or sunken and cracked, giving a scabby or corky appearance. The common scab shows concentric, wrinkled layers of the skin around a central depression. In some forms of scabs, deep depressions, bordered by torn skin are seen, while some others produce small, pimple-like growths

(e-g) Common scab of potato; Apple scab; *Citrus* scab.

Club root. The roots of affected plants enlarge abnormally and become spindle-like, spherical, knobby or club-shaped swellings. The swellings may be few and isolated or they may coalesce and cover the entire root system. The pathogen lives in the host cells, but the host is not killed for a long time. The invaded and adjacent cells are stimulated by the pathogen to enlarge and multiply, which results in the formation of club roots

(e-g) Club root of crucifers, such as cabbage and cauliflower.

Wart. Affected plants develop whitish warts or galls around the base of the stem at or below soil level. This dry, hard, irregular growth or protuberances are produced in clumps on tubers and stems.

(e-g) Crown wart of lucerne, Wart disease of potato.

All the symptoms mentioned above, whether they occur in a particular part or certain parts of the plant cause pronounced stunting or death of the infected parts or plants in some cases and ultimately result in loss of yield.

Taxonomy of Fungi

The fungi do not embrace a single phylogenetically related group. The fungal-like organisms called **'pseudo fungi'** and the **'true fungi'** differ fundamentally in the structure of their vegetative thalli. The evolutionary development of the fungi has resulted from disturbances of the nutritional physiology of their algal ancestors. Following the original step in their evolution, the fungi continued to evolve along many independent lines. The most important evolutionary trend of the fungi has been towards the development of specialized sexual organs and towards the ultimate separation of plasmogamy from karyogamy. These sexual processes occur at different times in the life history of the higher types of fungi; plasmogamy occurring at the beginning of the life of the individual and karyogamy at the end. The morphological specialization of the fungi has led to increasing complexity of the thallus and to the production of highly specialized fructifications. The fructifications reach their highest evolutionary development in the Basidiomycetes.

The organisms referred to as the 'pseudo fungi' or 'lower fungi' or 'fungal-like organisms' are now placed under the Kingdom - Protozoa or Kingdom - Chromista. These so called pseudo fungi, which probably represent the degenerate flagellates that have become parasitic and are responsible for causing many important plant diseases are also taken into consideration along with the true fungi in this discussion. So, all the fungi including the slime moulds are placed in the Kingdom - Myceteae of the Superkingdom - Eukaryonta. The grouping used in the classification of the fungi are shown below :

Super Kingdom
- Kingdom
  - Division
    - Class
      - Order
        - Family
          - Genus
            - Species

Each of these groups may be further divided into Subgroups, Subdivision, Subclass and Suborder, whenever necessary. The classification of fungi followed in this book is based on the classification given by **'Alexopoulos and Mims'**.

Super Kingdom - Eukaryonta
- Kingdom - Myceteae (Fungi)
  - Division I - Gymnomycota
    - Subdivision 1 - Acrasiogymnomycotina
      - Class - Acrasiomycetes
    - Subdivision 2 - Plasmodiogymnomycotina
      - Class 1 - Protosteliomycetes
      - Class 2 - Myxomycetes
        - Subclass 1 - Ceratiomyxomycetidae
        - Subclass 2 - Myxogastromycetidae
        - Subclass 3 - Stemonitomycetidae
  - Division II - Mastigomycota
    - Subdivision 1 - Haplomastigomycotina
      - Class 1 - Chytridiomycetes

Order 1 - Chytridiales
Family 1 - Olpidiaceae
Family 2 - Synchytriaceae
Family 3 - Cladochytriaceae
Family 4 - Chytridiaceae
Order 2 - Harpochytridiales
Order 3 - Blastocladiales
Order 4 - Monoblepharidales
Class 2 - Hypochytridiomycetes
Order - Hypochytridiales
Class 3 - Plasmodiophoromycetes
Order - Plasmodiophorales
Family - Plasmodiophoraceae
Subdivision 2 - Diplomastigomycotina
Class - Oomycetes
Order 1 - Saprolegniales
Family - Saprolegniaceae
Order 2 - Leptomitales
Order 3 - Lagenidiales
Order 4 - Peronosporales
Family 1 - Pythiaceae
Family 2 - Peronosporaceae
Family 3 - Peronophytophthoraceae
Family 4 - Albuginaceae
Division III - Amastigomycota
Subdivision 1 - Zygomycotina
Class - Zygomycetes
Order 1 - Mucorales
Family 1 - Mucoraceae
Family 2 - Choanephoraceae
Order 2 - Entomophthorales
Family - Entomophthoraceae

Subdivision 2 - Ascomycotina

Class - Ascomycetes

Subclass 1 - Hemiascomycetidae

Order 1 - Protomycetales

Family - Protomycetaceae

Order 2 - Endomycetales

Family 1 - Endomycetaceae
Family 2 - Saccharomycetaceae
Family 3 - Spermophthoraceae

Order 3 - Taphrinales

Family - Taphrinaceae

Subclass 2 - Plectomycetidae

Order 1 - Eurotiales

Family - Eurotiaceae

Order 2 - Microascales

Family - Ophiostomataceae

Subclass 3 - Hymenoascomycetidae

Order 1 - Erysiphales

Family - Erysiphaceae

Order 2 - Xylariales

Family - Polystigmataceae

Order 3 - Clavicepitales

Family - Clavicepitaceae

Order 4 - Diaporthales

Family - Diaporthaceae

Order 5 - Hypocreales

Family 1 - Nectriaceae
Family 2 - Hypocreaceae

Order 6 - Helotiales

Family 1 - Sclerotiniaceae
Family 2 - Dermataceae

Order 7 - Pezizales

Family - Morchellaceae

Subclass 4 - Loculoascomycetidae

Order 1 - Dothideales

Family 1 - Dothideaceae
Family 2 - Capnodiaceae

Order 2 - Pleosporales

Family 1 - Pleosporaceae
Family 2 - Venturiaceae

Subdivision 3 - Basidiomycotina

Class - Basidiomycetes

Subclass 1 - Holobasidiomycetidae

Order 1 - Agaricales

Family 1 - Tricholomataceae
Family 2 - Volvariaceae
Family 3 - Agaricaceae

Order 2 - Exobasidiales

Family - Exobasidiaceae

Order 3 - Tulasnellales

Family - Ceratobasidiaceae

Subclass 2 - Teliomycetidae

Order 1 - Uredinales

Family 1 - Pucciniaceae
Family 2 - Melampsoraceae

Order 2 - Ustilaginales

Family 1 - Ustilaginaceae
Family 2 - Tilletiaceae

Subdivision 4 - Deuteromycotina

Form class - Deuteromycetes

Form subclass 1 - Coelomycetidae

Form order 1 - Sphaeropsidales

Form family 1 - Sphaeropsidaceae
Form family 2 - Nectrioidaceae

Form order 2 - Melanconiales

Form family - Melanconiaceae

Form subclass 2 - Hyphomycetidae

Form order 1 - Moniliales

Form family 1 - Moniliaceae
Form family 2 - Dematiaceae
Form family 3 - Tuberculariaceae

Form order 2 - Agonomycetales

In naming an organism, binomial nomenclature is followed, the first name denoting the generic name and the second one which follows denoting the species name. Binomials are usually descriptive of the organisms and are derived from Greek or Latin languages.

Kingdom-Myceteae (Fungi)

Achlorophyllous; saprophytes or parasites; unicellular or mostly with filamentous thallus (soma); usually possess cell walls containing chitin and other carbohydrates; nutrition by absorption, except in the slime moulds where it is **'phagotrophic'** (feeding by engulfing the food); propagation by means of spores produced by various types of sporophores; asexual and sexual reproduction usually present. The Kingdom is subdivided into three major Divisions: Gymnomycota, Mastigomycota and Amastigomycota.

Division-I. Gymnomycota

Mostly single-celled organisms without cell wall; phagotrophic.

Subdivision - 1. Acrasiogymnomycotina. Cellular slime moulds whose thallus is a myxamoeba, which is transformed into a pseudoplasmodium that culminates into a sporocarp, bearing walled spores in a mucoid droplet at the tip of each branch. There is a single Class - Acrasiomycetes.

Class - Acrasiomycetes. Except in one species, cells non-flagellate; cells uninucleate, naked, haploid myxamoeba that feed by engulfing bacteria; myxamoeba aggregate to form pseudoplasmodia; sporocarps called as **'sorocarps'** usually have cellular stalks; sexual reproduction, if present through macrocysts. The organisms belonging to this Class are not of agricultural importance.

Subdivision - 2. Plasmodiogymnomycotina. Thallus a simple myxamoeba mostly with pseudopodia or a true plasmodium terminating into one or more sporophores. There are two Classes.

Class - 1. Protosteliomycetes (The protostelids). Myxamoeba forming sporocarps directly or after developing into plasmodia; no sexual reproduction; biflagellate cells may be produced. The organisms belonging to this Class are not of agricultural importance.

Class - 2. Myxomycetes (The true slime moulds). Their body is a naked, amorphous (without a definite shape) plasmodium; the zoospores (swarm cells) are biflagellate and heterokont (flagella unequal in length); most slime moulds live in cool, shady, moist places in the woods, on decaying logs, dead leaves or other moist organic matter; they do not infect plants; the swarm cell eventually withdraws its flagella and changes into a myxamoeba. They feed on bacteria, protozoa and other minute organisms. Rarely slime moulds creep over ornamentals and cover parts of low-lying plants, but they do not infect plants. So, in general they are of little economic importance.

Division-II. Mastigomycota

Fungi with centrioles functioning during nuclear division; typical flagellate cells produced during the life cycle; nutrition absorptive; thallus varied, from unicellular to an extensive, filamentous, coenocytic mycelium, sometimes forming pseudoseptae; unicellular organism is converted into a spore case or sporangium; asexual reproduction typically by zoospores; sexual reproduction by various means. There are two Subdivisions:

Subdivision - 1. Haplomastigomycotina. Flagellate fungi, producing either uniflagellate or biflagellate zoospores; life cycle either haplobiontic (haploid) or diplobiontic (diploid); some are aquatic, while others are endoparasitic slime moulds.

Class - 1. Chytridiomycetes. Produce zoospores, each with single, posterior, whiplash flagellum. The coenocytic thallus, multinucleate, globose, oval, elongated simple hypha or well developed mycelium; zygote converted into a resting spore or resting sporangium; chitin is the chief constituent of cell walls, but glucans are also present; more prevalent in aquatic habitats, but many of them inhabit the soil. There are four Orders under this Class : Chytridiales, Harpochytridiales, Blastocladiales and Monoblephariales.

Order-Chytridiales (Chytrids). Water or soil-inhabiting organisms; many soil-inhabiting ones parasitize algae and water moulds, while many of the soil-inhabiting ones are parasitic on vascular plants; mostly endobiotic (living entirely within host cells), whereas some are epibiotic (producing only their reproductive organs on the outside of the host); asexual reproduction by zoospores borne in sporangia having one or more papillae; sexual reproduction, when occurs results in a thick-walled resting spore or resting sporangium. The Order - Chytridiales is subdivided into several Families based on sporangial form and development

Inoperculate Chytridiales: The inoperculate series of the Chytridiales do not form a well-defined, circular cap (operculum) at the tip of the discharge papilla, but discharge their zoospores through a pore in the wall of the sporangium or discharge tube formed, when the discharge papilla dissolves.

Family - Olpidiaceae. Holocarpic organisms parasitic on algae, fungi, mosses, pollen grains and vegetative tissues of flowering and crop plants; thallus develops into a single inoperculate sporangium or resting sporangium; zoospores with one terminal flagellum.

Genus - *Olpidium*. (e-g) *Olpidium viciae* - parasitic on the leaves and stems of *Vicia unijuga*; *O. brassicae* - parasitic on the roots of cabbage and some other plants. It transmits plant viruses also.

Life cycle of *Olpidium viciae*. During wet weather, the zoospores escape from the zoosporangium through an exit tube. After a period of swarming, they encyst on the surface of the host. Infection takes place through a minute pore digested in the host cell wall and the protoplast enters the host cell, leaving the empty cyst outside. It secretes a membrane around itself and grows into a sporangium, while its nuclei divide repeatedly. Zoospores are soon formed, escape and repeat the asexual life cycle. In the sexual phase, two zoospores behaving as planogametes fuse. Copulation of the two gametes results in a motile zygote. The zygote infects a host cell just as a zoospore and develops into a thick-walled resting sporangium. It does not germinate immediately and is capable of overwintering. The resting sporangium is binucleate, but before germination karyogamy takes place followed by meiosis. Several nuclear divisions result in the formation of a multinucleate structure. The protoplast also undergoes cleavage and surrounds each nucleus forming uninucleate zoospores, which escape and reinfect host plants **(Fig. 15)**

Family - Synchytriaceae. Parasitic, holocarpic organisms; produce inoperculate sporangia; thallus divides into several sporangia or gametangia, which are enveloped in a common membrane to form a sorus; motile zoospores are released from the sporangia and gametes are released from the gametangia; zoospores with one terminal flagellum; whether the motile cells will be asexual zoospores or sex cells (gametes) depend on the presence or absence of sufficient water at a critical point in their development.

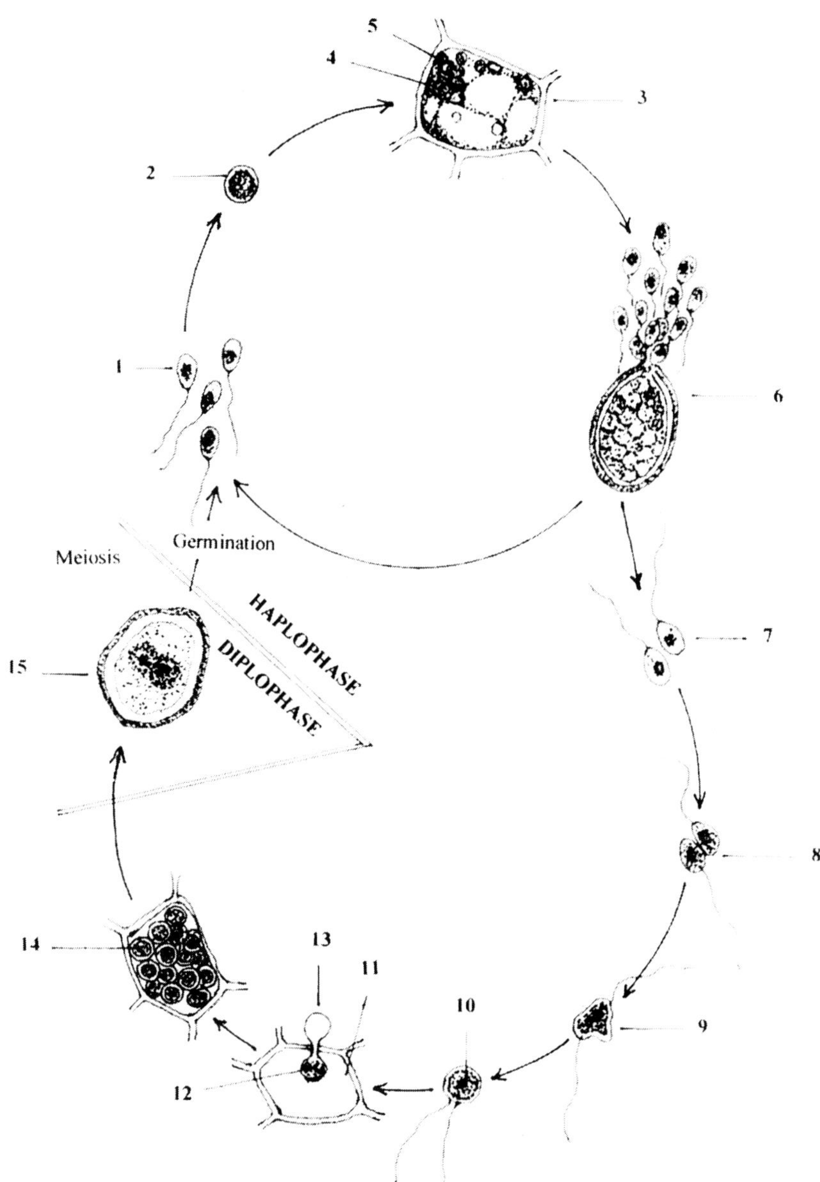

1. Zoospores. 2. Encysted zoospore. 3. Host infection. 4. Host nucleus. 5. Formation of sporangium within the host cell. 6. Germinating sporangium releasing zoospores or gametes. 7. Planogametes. 8. Copulation of gametes. 9. Plasmogamy. 10. Zygote. 11. Infection of host cell. 12. Naked vegetative cell of the parasite within the host cell. 13. Empty zygote shell. 14. Resting sporangia within host cell. 15. Karyogamy in resting sporangium.

Fig. 15 : Life cycle of *Olpidium viciae*

Genus - *Synchytrium* (e-g) *Synchytrium endobioticum* - a serious parasite of potato tubers causing black wart disease.

Life cycle of *Synchytrium endobioticum*. The pathogen is an obligate parasite. In potato tubers, the fungus produces dark warts due to hyperplasia and hypertrophy of the host cells. The cells within these warty tissues contain resting sporangia. Infection of the tuber and other underground parts of the plant takes place by posteriorly uniflagellate, naked zoospores, which are released by sporangia or resting spores present in the wart tissues of infected tubers, plant debris or in the soil. The thick-walled resting sporangia remain viable in the soil for many years and serve as prosorus. The outer wall (epispore) of the resting sporangium bursts open by an irregular aperture and the inner wall (endospore) expands into a balloon-like vesicle, within which a single sporangium is formed. Zoospores are released from this sporangium. The zoospores can swim for about 2 hours in the soil water. When a zoospore comes in contact with the host, such as eyes of potato tubers, it comes to rest, withdraws its flagellum and bodily enters the host cell. It settles down at the bottom of the cell and enlarges in size. Hypertrophy of the host cell occurs. The surrounding cells also expand and form the tumor or wart. The enlarging thallus fills the lower part of the infested host cell and becomes spherical. A double layered, brown, chitinous wall is formed around the thallus and it becomes the prosorus of the summer spore. The prosorus, which is uninucleate up to this stage, on maturity germinates within the host cell by protrusion of the inner wall through a pore in the outer wall. The inner wall then expands as a vesicle and fills the upper half of the host cell. The contents of the prosorus pass into the vesicle. During its passage into the vesicle the nucleus divides repeatedly to form several nuclei. The cytoplasmic contents of the vesicle become cleaved and each nucleus is surrounded by a portion of the cytoplasm to form a sporangium, thus forming a sorus containing about 4 to 9 sporangia. Each nucleus undergoes further divisions, so that 200 to 300 nuclei are formed in each sporangium or gametangium. The cytoplasm around each nucleus rounds up to form a zoospore or gamete. As the sporangium ripens, it absorbs water and swells, as a result the host cell bursts open by means of a small slit through which the zoospores escape. As many as 500 to 600 zoospores may be present in a sporangium. These zoospores can swim in soil-water for about 20 hours. When suitable host tissues are available, they encyst and cause infection. Sometimes a single host cell may be penetrated by several zoospores. Within the host cell the fungal thallus enlarges, forms prosorus and produces a fresh crop of sporangia during the growing season of the host crop.

In dry conditions or at the end of the crop season, the zoospores function as planogametes. They fuse in pairs to form motile zygotes. The flagella are retained and the zygote swims in the soil-water for sometime. The gametes from the same sporangium or gametangium do not fuse, because they are physiologically similar. Fusion occurs between gametes from different gamatangia, which may be from the same prosorus. The biflagellate zygote settles on the host surface and causes infection in the same manner as the zoospores. Nuclear fusion occurs in the zygote before penetration. While zoospore infection is mainly responsible for hypertrophy, the zygote infection mainly causes hyperplasia. The endobiotic zygote now settles down at the bottom of the host cell. Repeated cell divisions of the host cover the zygote deep in the warty tissue. The zygote thallus enlarges and is surrounded by a two-layered wall, the outer thick, dark- brown wall with folds and ridges and the inner thin, hyaline layer surrounding the granular cytoplasm. This is the resting spore or resting sporangium **(Fig. 16)**

Family - Cladochytriaceae. Vegetative thallus consists of a mycelium of thin-walled, delicate hyphae; produces rhizoids; zoospores released from the zoosporangium through an apical papilla.

Genus - *Physoderma* (e-g) *Physoderma zeae-maydis* - causes gummy tumors and brown spots on stems and leaves of maize.

Life cycle of *Physoderma zeae-maydis*. It is a parasite of maize and teosinte. It produces gummy tumors on the stems and leaves of the hosts. There are two distinct types of vegetative thalli produced during its life cycle, an ephemeral, monocentric, epibiotic thallus, which is followed by the development of a persistent, polycentric, endobiotic thallus. The resting spores, which appear as brown or reddish-brown powder, escape from the disintegrating tissues of the host during the spring season. During germination, the endospore escapes from the enclosing wall through a pore, which is covered with a lid. The entire protoplast then forms the zoosporangium. After swimming about for several hours, a zoospore comes to rest on the epidermis of the host plant, forms a cell wall and then puts out a knob-like protuberance called the **'apophysis'**. Rhizoids develop from the apophysis and penetrate the outer wall of the host. The zoospore remains on the outer surface of the host, absorbing nourishment from it. After the zoospore has increased considerably in size, it develops into a thin-walled, extramatrical zoosporangium. This zoosporangium releases several zoospores when mature through an apical papilla and the zoospores soon reinfect other epidermal cells of the host. Under favorable conditions, many successive generations may follow in a single season.

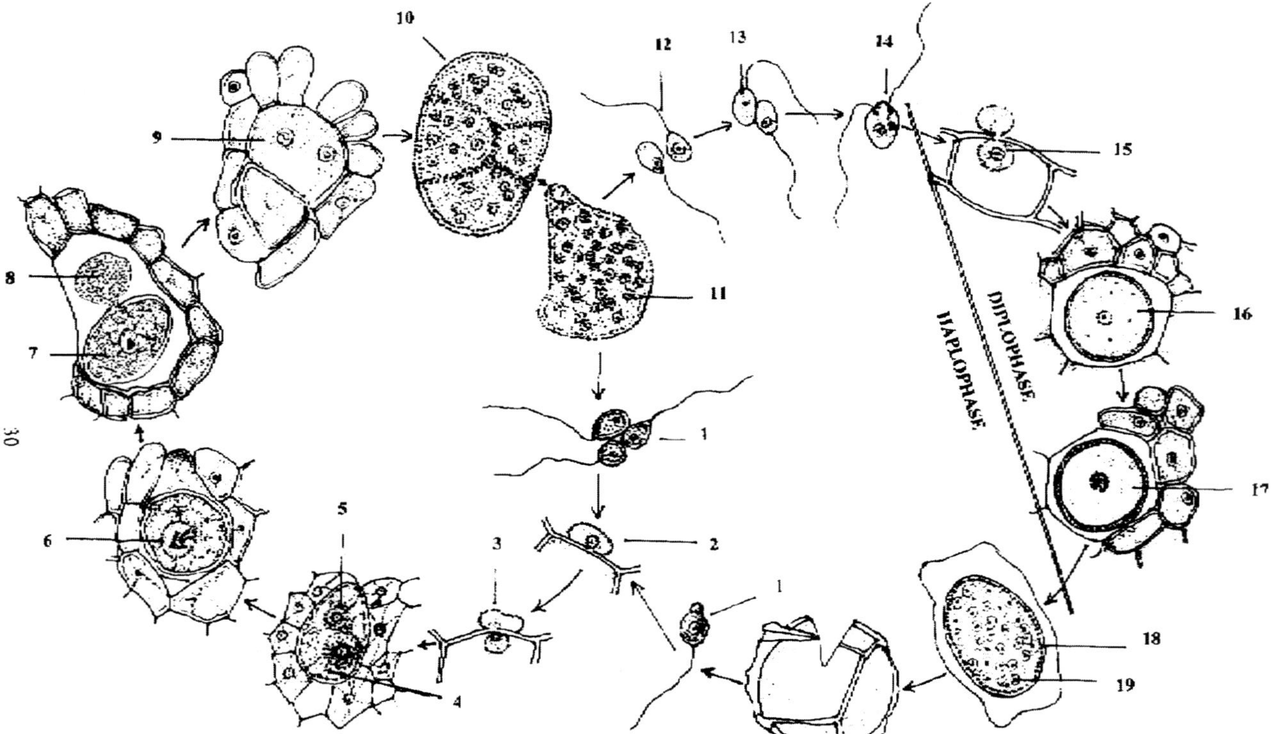

1. Zoospores. 2., 3. Penetration of the host epidermal cell. 4. Young protoplast within a hypertrophied epidermal cell. 5. Host nucleus. 6. Mature spore. 7. Germination of spore 8. Immature sorus. 9. Nuclear division in sorus. 10. Sorus extruded from host cell. 11. Gametangium or Sporangium. 12. Gametes. 13. Copulation of two planogametes and plasmogamy. 14. Karyogamy. 15. Penetration of zygote into the host cell. 16. Young resting sporangium. 17. Mature resting sporangium. 18. Resting spore within host cell during maturation of zoospore primordia. 19. Zoospore primordia. 20. Resting spore emptied of zoospores.

Fig. 16 : Life cycle of *Synchytrium endobioticum*

The parasite enters an endobiotic stage and develops an intramatrical mycelium. This mycelium produces resting spores and pustules on the surface of the host. A change in the nuclear condition may initiate the change from extramatrical to intramatrical growth. Two zoospores may copulate and the zygote penetrates the wall of the host and develops the persistent, diploid, intramatrical mycelium. After a period of growth the hyphae develop in the enlarged areas, which later become the turbinate cells. The nucleus remains in the enlarged portion of the hypha and is soon cut off from the mother hypha by a septum. While the turbinate cell is developing, the tip of the hypha resumes its growth and the distal daughter cell produces another turbinate cell. This process may be repeated, forming a group of turbinate cells linked together by delicate hyphae called **'rhizomycelium'**. The turbinate cells of the mycelium may become resting spores. Meiosis may occur during the germination of the resting spore **(Fig. 17)**

Genus - *Urophlyctis* (e-g) *Urophlyctis alfalfae* - causes crown wart of alfalfa.

Life cycle of *Urophlyctis alfalfae*. The parasite causes crown wart or gall-like tumors on young seedlings of lucerne. The browned areas in the galls contain a number of golden-brown spores, which function as resting sporangia and eventually give rise to zoospores. The spores are somewhat spherical, flattened slightly at one pole, having a brittle wall lined within by a thin, colorless wall. At germination, the spores become converted into sporangia containing a large number of biflagellate zoospores, which may either be passed into an extruded vesicle before dispersal or without forming an external zoosporangium they may escape into the soil through one or more cracks in the spore wall. The zoospores remain active for several hours and may or may not copulate before infecting the host. Buds of lucerne are infected during the early stages as they emerge from the crown. One, two or three zoospores may be found in a single epidermal cell. They appear as somewhat elongated, pear-shaped bodies. The narrow end of the body is drawn out into a thin hypha and the broader end enlarges to form a turbinate cell or bulb, while retaining the single nucleus and almost the whole of the protoplasm of the zoospore. The nucleus undergoes division, thereby 10 to 15 nuclei are formed in the turbinate cell. The top half of the bulb is divided into two to four cells by the formation of convex intersecting walls, enclosing one nucleus in each cell, while the other nuclei remain in the mother turbinate cell. This portion of the mulinucleate bulb gives rise to the spore or resting sporangium. At a point on the surface of this portion, a small vesicle is formed and all the contents of the basal portion of the bulb migrate into it. After a series of nuclear divisions, about a hundred nuclei are formed in the vesicle as it matures. On its equator, short, branched

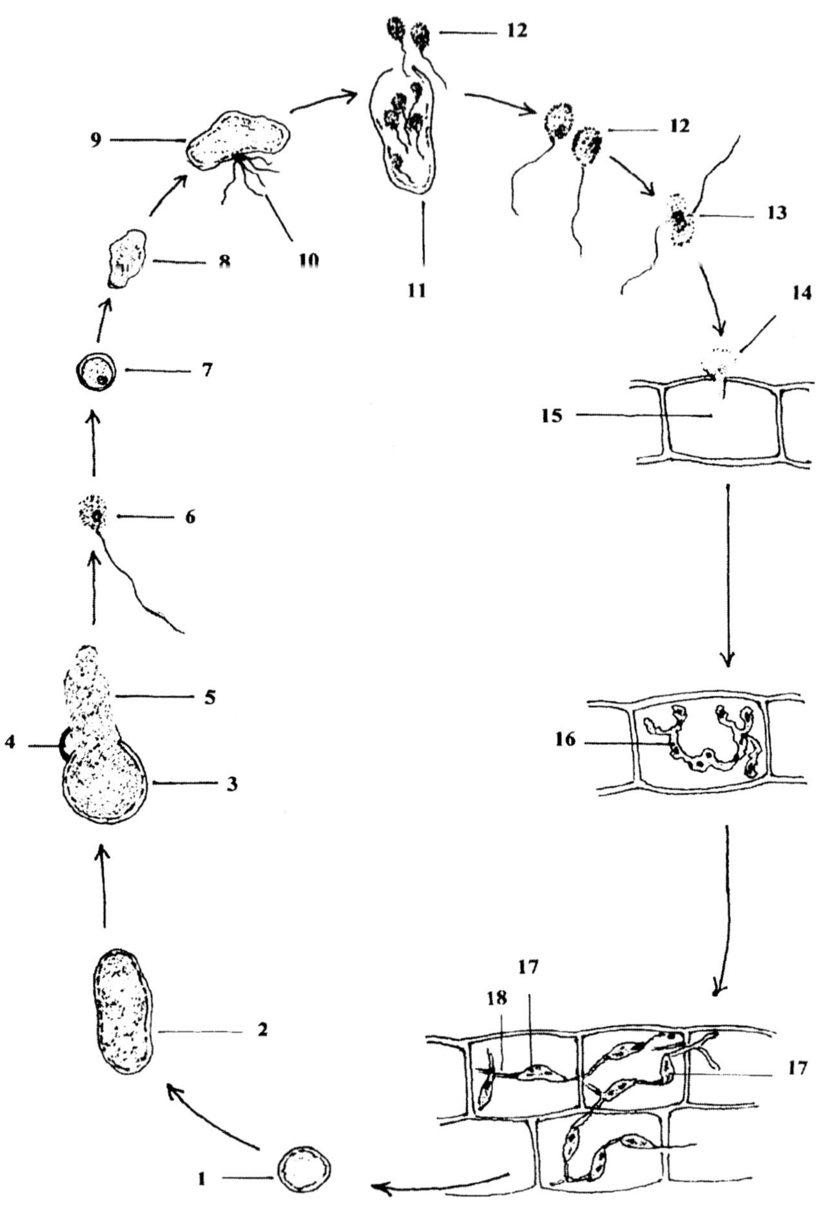

1.Resting spore 2,3. Germinating resting spore 4. Lid 5. Zoosporangium 6. Zoospore 7. Encysted zoospore 8. Formation of apophysis 9. Apophysis 10. Rhizoids 11. Thin-walled extramatrical zoosporangium 12. Zoospores 13. Copulation 14. Reinfestation of host epidermal cell 15. Endobiotic stage 16. Diploid, intramatrical mycelium 17. Turbinate cell 18. Rhizomycelium

Fig. 17 : Life cycle of *Physoderma zeae maydis*

hyphal processes, which serve as haustoria are produced at three or more points and the wall of the emergent vesicle thickens to become that of the spore. Side by side, from each of the two to four peripheral cells, narrow hyphae are produced, which penetrate the neighboring host cells and establish new points of internal infection. Each of these penetrating hyphae produces apical bulbs (turbinate cells) as described above and the sequence is repeated several times. Thus, these penetrating hyphae spread infection within the host **(Fig. 18)**

Operculate Chytridiales: The operculate series of the Chytridiales are characterized by the presence of a hinged lid-like or cap-like structure (operculum) on the sporangium wall. Just prior to spore release, the operculum opens like a hinged cap and the zoospores escape.

Family - Chytridiaceae. In this Family, the members produce sporangia, which are operculate; they are aquatic fungi, parasitic on algae, other aquatic fungi, pollen and protozoa or saprophytic. They are not of economic importance.

The other three Orders viz., Harpochytridiales, Blastocladiales and Monoblepharidales are also not of economic importance.

Class - 2. Hyphochytridiomycetes (Anteriorly uniflagellate fungi). The fungi belonging to this Class are aquatic, living in fresh water or marine habitats; the motile cells are anteriorly uniflagellate, with a tinsel-type flagellum; parasitic on algae and other fungi or saprophytic in nature. There is a single Order - Hyphochytridiales and the fungi of this Class are not of economic importance.

Class - 3. Plasmodiophoromycetes (Endoparasitic slime moulds). The Plasmodiophoromycetes are obligate endoparasites of vascular plants, algae and fungi, which usually cause an abnormal enlargement of the host cells (hypertrophy) and often abnormal multiplication of the host cells (hyperplasia), resulting in the enlargement of the infected parts leading to disruption of the vascular elements. The somatic phase of the Plasmodiophoromycetes is a plasmodium that develops within the host cell. Usually two types of plasmodia are formed; the 'sporangiogenous plasmodia' and the 'cystogenous plasmodia'. Sporangiogenous plasmodia give rise to thin-walled sporangia containing zoospores, whereas the cystogenous plasmodia cleave into thick - walled cysts each of which ultimately gives rise to a single zoospore. The zoospores from both zoosporangia and cysts are similar, each bearing two, unequal, anterior, whiplash flagella. The life cycle is initiated when the cysts germinate, each giving rise to a zoospore capable of infecting the host plant. There is only one Order in this Class, the Plasmodiophorales with a single Family, the Plasmodiophoraceae.

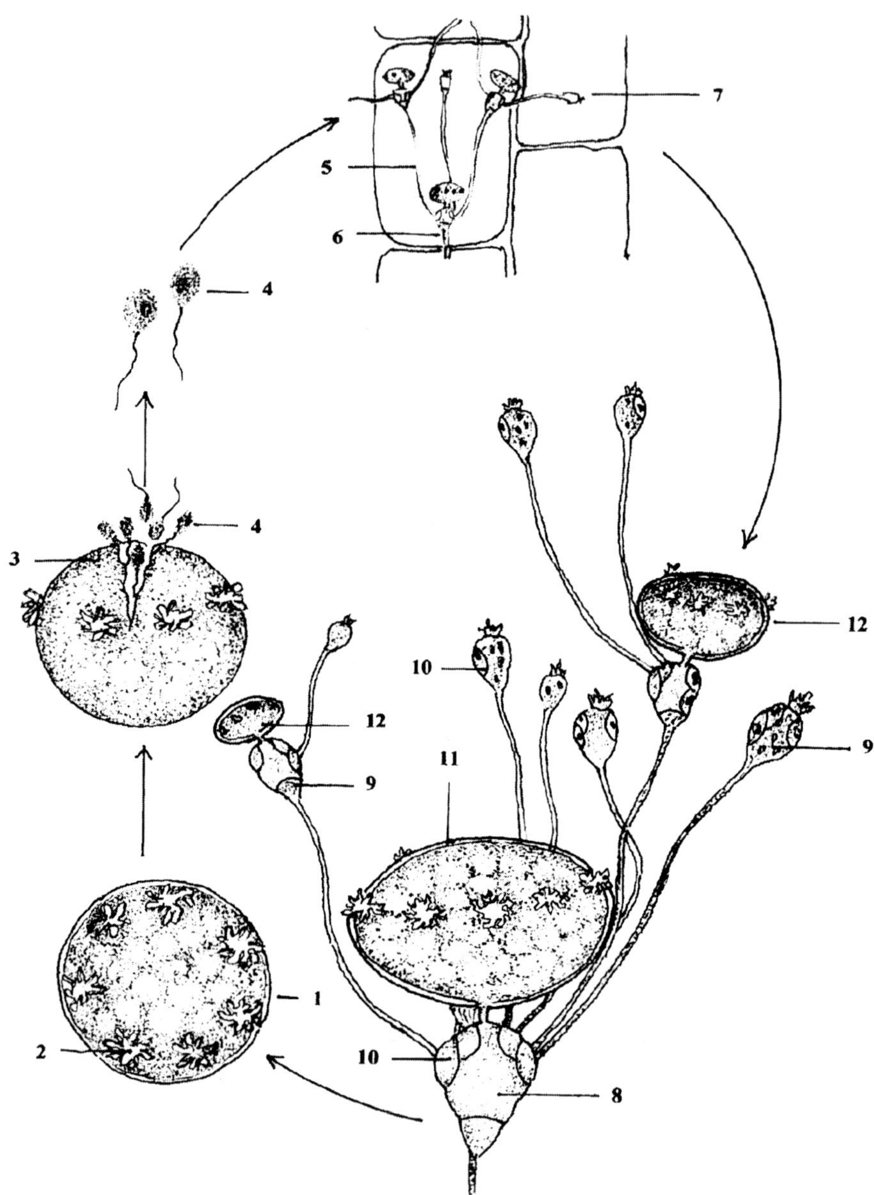

1.Mature resting spore showing a zone of haustoria 2. Haustoria 3. Zoospores released from cracked resting spore 4. Zoospores 5. Parasite inside host cell and spreading infection to adjacent cells 6. Primary infection 7. Secondary infection in adjacent cell 8. Primary turbinate cell (Bulb) 9. Secondary bulbs 10. Peripheral segments in the bulbs 11. Nearly mature resting spore 12. Young resting spore

Fig. 18 : Life cycle of *Urophlyctis alfalfae*

Genus - *Plasmodiophora* (e-g) *Plasmodiophora brassicae* - causes club root or finger and toe disease of cabbage and related plants, both cultivated and wild.

Life cycle of *Plasmodiophora brassicae*

Asexual cycle. The resting spores are dispersed in the soil by the disintegration of the host cells and they germinate to produce heterokont zoospores. The two, whiplash flagella are inserted anteriorly. The larger resting spores, which contain a greater number of nuclei, produce a correspondingly larger number of zoospores. The motility of the zoospore is interrupted by recurring periods during which their movement is amoeboid. The zoospore finally comes to rest on a root hair or on some other young epidermal cell of a suitable host plant, withdraws its flagella and surrounds itself with a delicate cell wall. The protoplast escapes from this wall and enters the cell of the host. If a resting spore comes in contact with a root hair, the naked protoplast may penetrate the host cell without first forming a swarm spore. The parasite develops into a multinucleate plasmodium within the host cell. The plasmodium may break up into daughter plasmodia by a budding-like process. This type of vegetative multiplication is called **'schizogamy'**. The mother plasmodium is called a **'schizont'** and the daughter plasmodia are called **'meronts'**. The young plasmodia finally dissolve the cell walls of the host tissue and migrate from cell to cell by amoeboid movement. The plasmodia extend to form pseudopodia of hyaline cytoplasm, into which flows the mass of remaining protoplasm. During unfavorable conditions, the plasmodium or its segments may secrete a firm, external, wall-like layer, thereby encysting itself. On the return of normal conditions, this protective layer is dissolved and the plasmodium resumes its normal activities.

When the plasmodium reaches maturity, it becomes segmented into uninucleate portions, which become rounded in form. The nucleus divides two or three times and then each rounded portion, now multinucleate becomes enclosed by a delicate hyaline wall. In this stage, the segment may be regarded as a summer sporangium. The cytoplasm within the summer sporangium segments into uninucleate portions, which become pyriform, heterokont zoospores. The zoospores are released into the soil by the disintegration of the host tissue and the cycle is repeated.

Sexual cycle. The zoospores function as planogametes and copulate in pairs. The young zygote then penetrates the epidermal cell of the host and develops into a multinucleate, diploid plasmodium. When reproductive maturity is attained, the nucleus undergoes meiosis by two nuclear divisions. The plasmodium then segments into uninucleate portions, which become rounded and are finally enclosed by a thin cell wall forming resting spores. Ultimately, the resting spore germinates to form a zoospore.

The life cycle of *Plasmodiophora* is similar to that of *Olpidium* and *Synchytrium* except that the vegetative body of *Plasmodiophora* is a plasmodium developing within the host cell. Since this plasmodium is multinucleate, a correspondingly large number of summer spores or resting spores are differentiated **(Fig. 19)**

Genus - *Spongospora* (e-g) *Spongospora subterranea* - causes powdery scab of potatoes.

Life cycle of *Spongospora subterranea*

Asexual cycle. The life cycle of *Spongospora subterranea* is similar to that of *Plasmodiophora brassicae* in some respects. But here, the resting spores adhere together forming more or less compact groups called **'cystosori'**.

The resting spore normally germinates to form one zoospore, which develops a cell wall. The naked protoplast escapes this wall and enters a root hair of the host plant. In the case of potato tuber, entry is effected through a lenticel. Within the host, the protoplast develops a multinucleate plasmodium, which may migrate through the tissues of the host and reach cells a considerable distance away from the original point of infection. The migrating plasmodium may become fragmented to form several plasmodia or it may fuse with adjacent plasmodia. The plasmodium may secrete an outer wall-like layer to protect itself against unfavorable conditions. When matured, it segments to form summer sporangia, which germinate to form zoospores.

Sexual cycle. Under some conditions, two zoospores function as planogametes and may copulate. The resulting zygote grows within the host cell and develops a diploid plasmodium. This nucleus undergoes reduction division (meiosis) as in *Plasmodiophora brassicae*. Finally the plasmodium segments to form a cystosorus composed of a spongy ball of resting spores **(Fig. 20)**

Subdivision - 2. Diplomastigomycotina. Fungi producing typical biflagellate zoospores; life cycle haplobiontic and diplobiontic; meiosis gametangial. There is only one Class - Oomycetes.

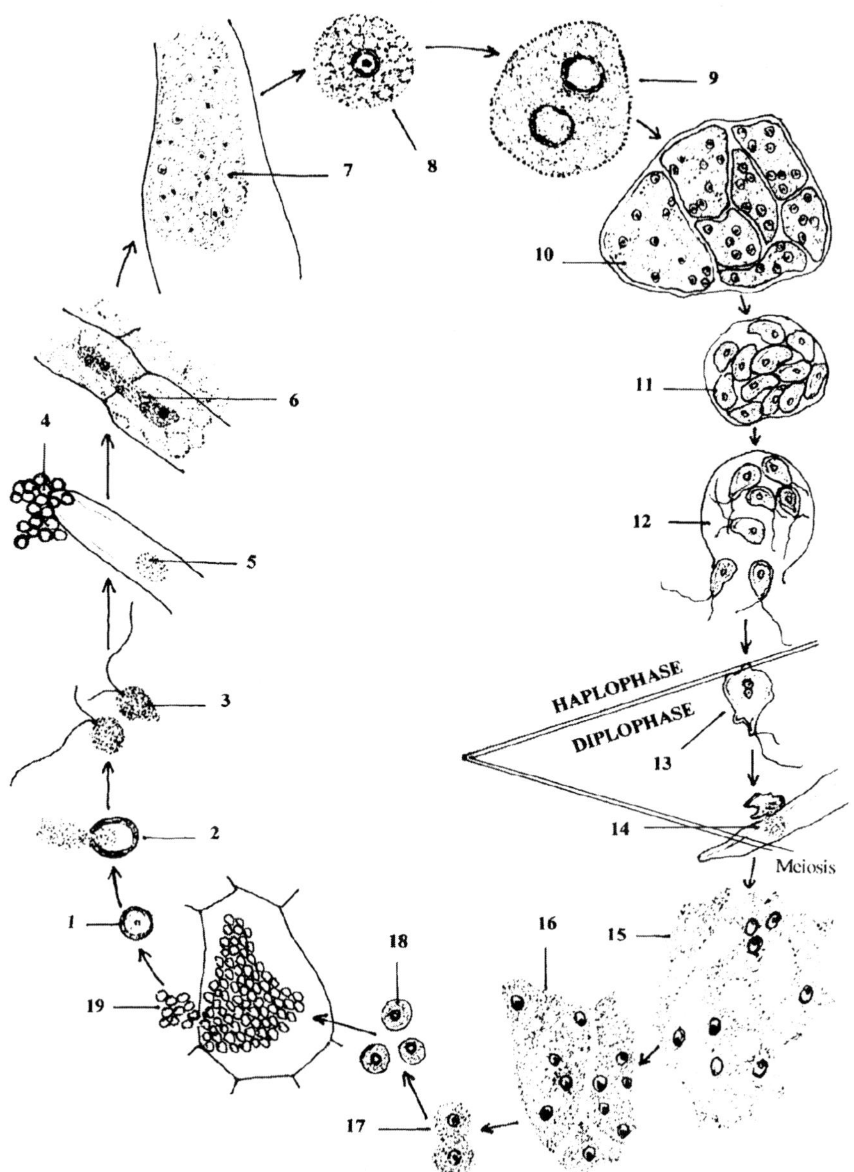

1. Resting spore 2. Resting spore germinating and releasing a zoospore 3. Heterokont zoospores 4. Empty zoospore membranes on a host root hair 5. Plasmodium has entered the root hair 6. Plasmodium migrating through the cell walls 7. Multinucleate plasmodium in a root hair 8. Uninucleate fragment of the plasmodium separated from the mother plasmodium and developing into a summer sporangium 9. Development of summer sporangium 10. Large segmented plasmodium whose segments are encysted 11. Encysted segment with zoospores 12. Zoospores released from zoosporangium 13. Zygote 14. Zygote infecting root 15. Vacuolated, diploid plasmodium after meiosis inside the host tissue 16. Segmentation of plasmodium 17. Rounding up of the immature resting spores 18. Mature resting spores 19. Resting spores released from disintegrated cell.

Fig. 19 : Life cycle of *Plasmodiophora brassicae*

Class - Oomycetes (The water moulds, white rusts and downy mildews). Mycelium usually elongated, filamentous, thin-walled, non-septate (coenocytic or aseptate) and multinucleate; hyphal wall contains mostly glucans and cellulose; zoospores biflagellate, one whiplash and one tinsel; flagella laterally inserted and extend in opposite directions; sexual reproduction oogamous; oospores (sexual spores or resting spores) produced by copulation of two morphologically different gametangia called **'antheridia'** (male) and **'oogonia'** (female). The Class - Oomycetes contains four Orders - Saprolegniales, Leptomitales, Lagenidiales and Peronosporales.

Order - Saprolegniales. Occur mostly in clear water; majority of the species are saprophytic, while a few are important parasites; mycelium well developed; zoospores produced in long, cylindrical sporangia arising from the mycelium; usually several oospores are present in an oogonium.

Family - Saprolegniaceae. Mostly aquatic, some live in brackish waters, while many live in moist soils; mycelium profusely branched and coenocytic; zoosporangia are long, cylindrical and produced terminally on the hyphae and septa are formed in the mycelium just below the reproductive organs; sexual reproduction by means of gametangial contact and several oospores are formed inside each oogonium.

Genus - *Aphanomyces* (e-g) *Aphanomyces euteiches* - causes root rot of peas.

Life cycle of *Aphanomyces euteiches*. The pathogen is soil-borne. The mycelium is hyaline and non-septate and branches moderately by producing short, stubby branches. The vegetative phase of the fungus within the host is very brief and the reproductive organs, followed by the oospores make their appearance as soon as the affected cortical tissues of the roots and basal part of the stem begin to show signs of collapse.

The contents of the entire thallus outside the host are cleaved into segments and form spores. These non-motile spores are eventually extruded one by one through a beak in the apex of the zoosporangium and immediately become spherical and encyst. From each cyst a reniform, flagellate zoospore emerges. This zoospore again encysts and soon germinates to form a germ tube, which becomes the mycelium. The mycelium penetrates the hypocotyl by putting forth one or more hyphae, which penetrate the epidermis at regions between two epidermal cells. However, the fungus can invade the stem only by travelling up from within the roots

The oogonia are formed on short stalk cells, usually at the ends of stubby branches of the wider mycelium. The oogonia are thin-walled, subglobose, densely granular and vacuolated. After fertilization, the oogonial

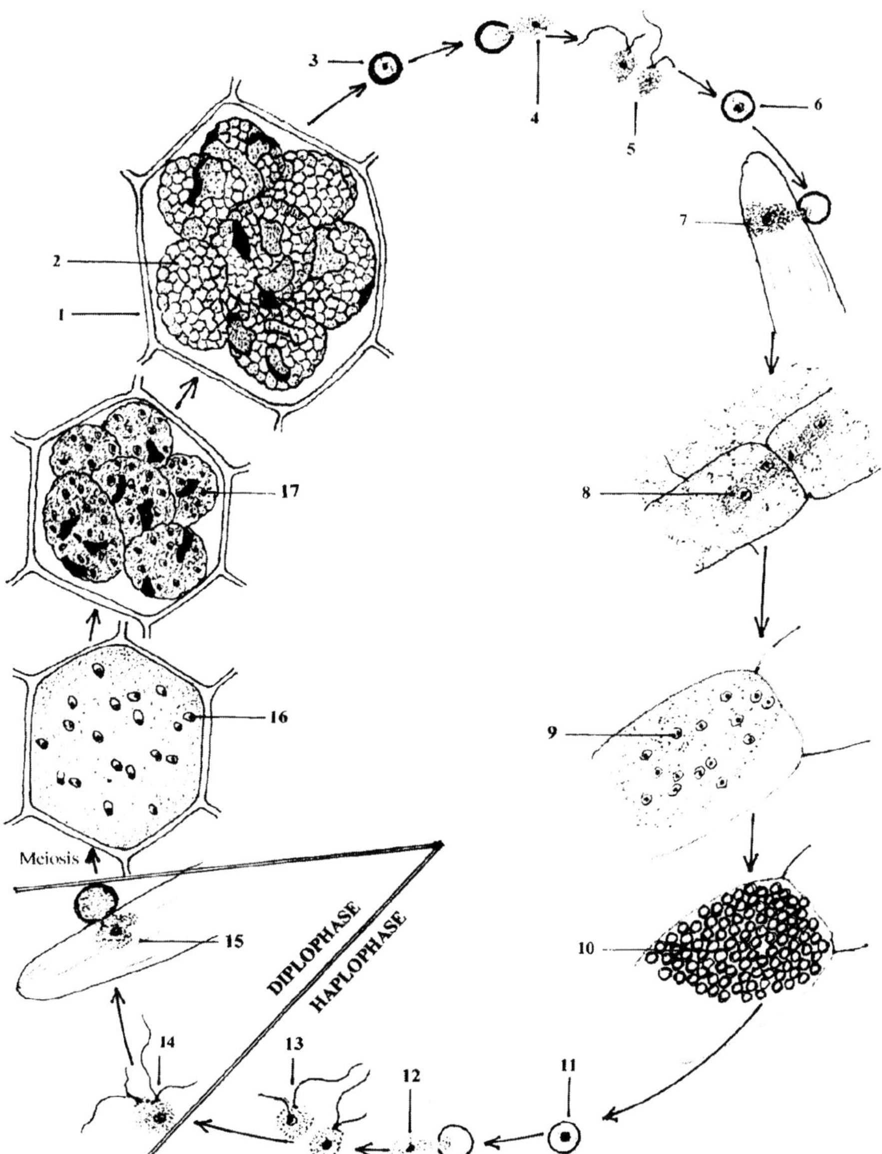

1.Resting spore balls inside the host cell 2.Spore ball (cystosorus) 3.Resting spore 4.Single primary zoospore emerging from the resting spore 5.Primary zoospores 6.Encysted primary zoospore 7.Plasmodium from encysted zoospore penetrating the host 8.Plasmodium migrating through the host cell walls and the nucleus undergoes division to form multinucleate plasmodium 9.The nuclei organize to form zoosporangia 10.Zoosporangia inside the host cell 11.Zoosporangium 12.Single secondary zoospore emerging from the zoosporangium 13.Secondary zoospores 14.Copulation of secondary zoospores 15.Secondary sporogenic plasmodium from the encysted spore entering the host 16.Nuclei after undergoing meiosis 17.Nuclei organizing to form spore balls

Fig. 20 : Life cycle of *Spongospora subterranea*

wall gets thicker and within an oogonium only one oospore is formed. The antheridia arise from less stout hyphae and one to five antheridial branches may invest an oogonium, putting forth fertilizing tubes. The oospores germinate, either directly by producing one or more germ tubes or indirectly by producing a long, tapering, slender hypha, which develops into a sporangium. The cytoplasm, which is passed into the sporangium from the oospore cleaves into several portions and are released through a small orifice formed at the tip of the sporangium. Before release, each portion rounds itself and becomes encysted to form a spore. On release, all the spores remain loosely attached to the sporangial tip for some time. After they separate, each spore releases a motile zoospore through a minute pore in the spore wall **(Fig. 21)**

Genus - *Saprolegnia* (e-g) *Saprolegnia parasitica* - causes diseases of fish and fish eggs

Order - Peronosporales. The Order consists of most specialized organisms of the Oomycetes. This large Order includes aquatic, amphibious and terrestrial species; most of the members are obligate parasites and many species cause destructive diseases in several economic plants and sometimes may result in epiphytotics; mycelium well developed, coenocytic and the hyphae branch freely; the hyphae of parasitic species may be intercellular or intracellular, the intercellular ones producing haustoria of various shapes; sporangia in most cases are oval or lemon-shaped and borne on ordinary hyphae or on sporangiophores; in majority of species, the sporangia liberate reniform, biflagellate zoospores, which swim in water for sometime, come to rest, encyst and then germinate by producing a germ tube that develops into the mycelium; in some species, the sporangium itself acts as a spore and germinates by producing a germ tube; sexual reproduction is by means of well differentiated oogonia and antheridia borne on the same or on different hyphae; these sexual organs fuse and produce an oospore; oospores germinate by giving rise to zoospores or to a germ tube with a terminal sporangium depending upon the species. The classification of Peronosporales is based mainly on the characters of the sporangiospores and sporangia.

Family - Pythiaceae. Aquatic, amphibious or terrestrial; many of them are serious crop pathogens; mycelium well developed; haustoria produced in some species; sporangia-bearing hyphae indistinguishable; in some species definite sporangiophores, which are usually indeterminate are formed; sporangia, usually zoosporangia are formed; the zoosporangia after maturity liberate a large number of zoospores; sometimes the sporangia germinate by means of germ tube; oogonia thin-walled; oospores thick-walled and smooth; under some conditions the oospores germinate by means of germ tube; mostly facultative parasites.

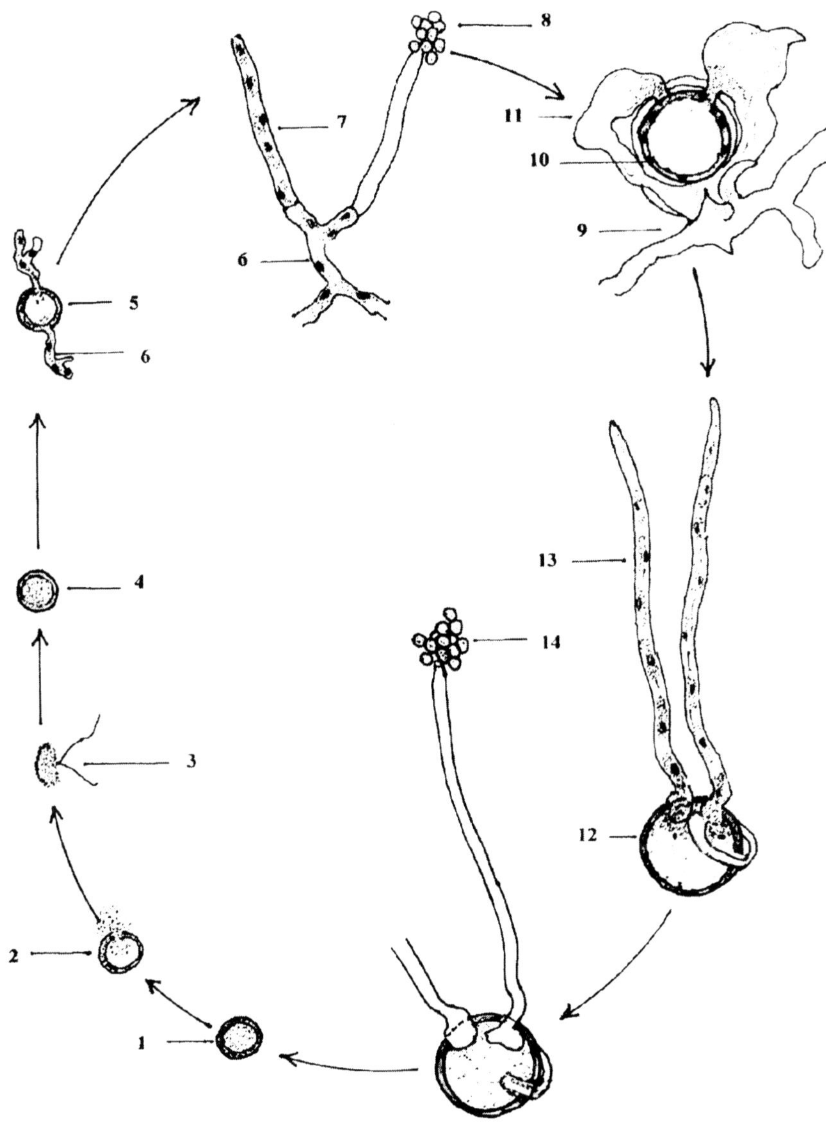

1.Spore 2.Germinating spore 3.Zoospore 4.Encysted zoospore 5.Germinating encysted zoospore 6.Mycelium 7.Sporangiophore forming spores 8.Spores released from sporangiophore 9.Formation of oospore 10.Oogonium 11.Antheridium 12.Germinating oospore producing sporangium 13.Formation of spores 14.Spores released from sporangiophore

Fig. 21 : Life cycle of *Aphanomyces euteiches*

Genus - *Pythium*. Causes damping off, seed decay, root rots etc.

(e-g) *Pythium debaryanum, P. aphanidermatum, P. ultimum* - causes damping off of seedlings of chillies, brinjal, tobacco, tomato, castor, potato and some other crops.

Life cycle of *Pythium debaryanum*. The fungus lives in the soil saprobically on dead organic matter or parasitically on the young seedlings of many susceptible species of seed plants. The mycelium consists of slender, coenocytic hyphae and is both inter- and intracellular. No haustoria are produced. In the sexual phase of reproduction, the fungus produces globose to oval sporangia from the somatic hyphae, which are either terminal or intercalary. The sporangia remain attached to the hyphae and germinate. Germination is either by zoospores or by germ tube. Zoospore production is preceded by the formation of a very thin, balloon-like vesicle at the tip of a long tube that develops from the sporangium. The protoplast from the sporangium flows into the vesicle through the tube and the zoospores are differentiated in the vesicle. The zoospores after formation move rapidly within the vesicle and the vesicular wall bursts and the zoospores are scattered in all directions. The zoospores are kidney-shaped and have two flagella inserted on the concave side. After swimming around for some time, the zoospore comes to rest, encysts and then germinates by a germ tube.

In the sexual reproduction process, oogonia and antheridia are produced in close proximity, mostly on the same hypha, with the antheridium just below the oogonium. The oogonium is globose with a multinucleate oosphere surrounded by a layer of periplasm. The antheridia are much smaller and somewhat elongated or club-shaped. After gametangial contact, a fertilization tube issues and penetrates the oogonial wall and the periplasm. In the mean time, mitosis takes place in both the gametangia and all the nuclei, except one functional nucleus in each of the gametangium disintegrate. The male nucleus passes through the fertilization tube into the oosphere, copulates with the female nucleus and forms the zygote. The oosphere develops into a thick-walled, smooth oospore, which germinates after undergoing a prolonged rest period. The oospore germinates by producing a germ tube, which develops into a mycelium. At lower temperatures, the oospore germinates by producing a vesicle at the tip of a tube. The protoplast gets transferred to this vesicle through the tube and zoospores are differentiated inside the vesicle **(Fig. 22)**

Genus - *Phytophthora*. Causes very serious diseases in many economically important crops.

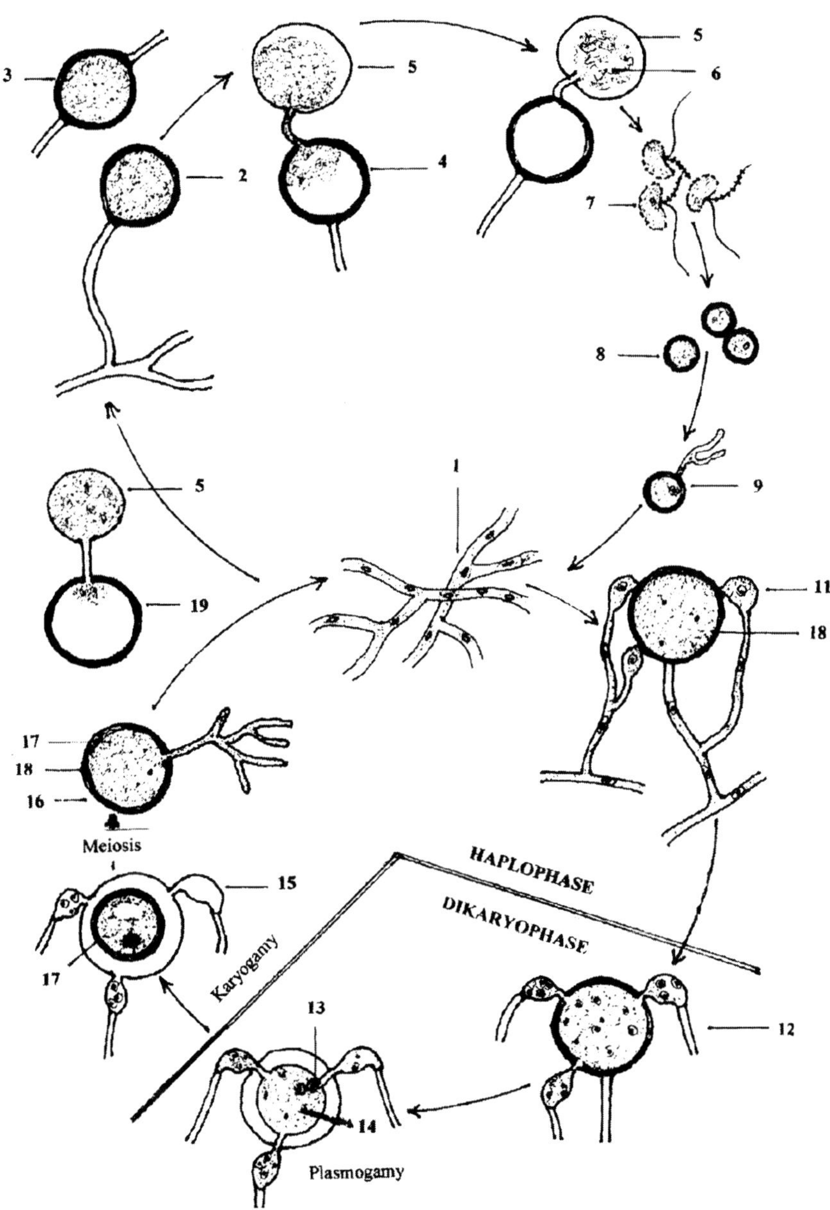

1.Somatic hyphae 2.Terminal sporangium with several nuclei 3.Intercalary zoosporangium 4.Germination of zoosporangium and formation of vesicle 5.Vesicle 6.Mature zoospores in the vesicle 7.Released zoospores 8.Encysted zoospores 9.Germinating zoospore 10.Oogonium 11.Antheridium 12.Female and male gametangia after meiosis 13.Entrance of male nucleus into the oosphere 14.Oosphere 15.Empty antheridium 16.Oospore germination by germ tube 17.Oospore 18.Oogonium 19.Oospore germination by zoospore production

Fig. 22 : Life cycle of *Pythium debaryanam*

(e-g) *Phytophthora infestans* - causes late blight of potato; *P. palmivora* - causes bud rot of coconut, black pod of cacao and leaf fall, canker and pod rot of rubber; *P. parasitica* - causes foot rot of tomato, black shank of tobacco and seedling rot of castor, cotton etc.; *P. citrophthora* - causes root rot in *Citrus* plants.

Life cycle of *Phytophthora infestans*. The fungus causes late blight of potato, which is a very serious and destructive disease of potato. In most temperate and cold regions, the fungus overwinters in the form of mycelium in infected potato tubers. With the arrival of favorable weather conditions in the spring, the mycelium resumes growth and production of sporangiophores and sporangia begins. Temperature and humidity are the two most important factors governing sporangial production. The disease spreads from infected tubers to young seedling sprouts and then to the leaves. The mycelium is profusely branched, intercellular, which sends long, curled haustoria into the plant cells.

Phytophthora infestans differs from many other members of the Family - Pythiaceae in that it produces sporangiophores distinguishable from the somatic hyphae. The sporangiophores are sympodially branched and are of indeterminate growth. They arise directly from the internal mycelium, emerging from the leaf through the stomata in small groups of about three to five. On the tubers, they mainly arise from lenticels or injuries in the rind. The lemon-shaped, papillate and hyaline sporangia are borne at the tips of the branches of the sporangiophores while they are still short, but the apical growth continues just below the spore, as a result the sporangium is pushed over to one side and usually falls off. At each point where growth has been renewed, there is a little nodular swelling in the stalk and up to 9 or 10 such nodes may be found on a single branch. When the fungus is active and the environmental conditions are favorable, normally sporangium germinates by liberating zoospores through the apical papilla. Otherwise the sporangium behaves like a conidium and germinates directly by giving out germ tubes. Nearly 100 % relative humidity is necessary for the germination of sporangia. The zoospores are reniform in shape and bear two flagella, which are inserted laterally on the concave side. One flagellum is whip-like, while the other is ciliated. After swimming in the film of water for some time, they come to rest, withdraw the flagella, encyst and germinate by a germ tube. The germ tube produces an appressorium, which helps to attach the spore on to the host surface. From the appressorium, a minute infection peg grows and enters the host epidermal cell.

Sexual reproduction takes place by means of antheridia and oogonia of opposite mating types. The antheridium is punctured by the oogonium, grows through it and develops into a globose structure above the antheridium. The antheridium forms a collar around the base of the oogonium. A fertilization tube from the antheridium penetrates the oogonium and fertilization takes place and eventually the oospore is formed.

In the absence of antheridium of the opposite type, oospore may be formed parthenogenetically. The oospore germinates by means of a germ tube that terminates in a sporangium. Sometimes the germ tube directly produces mycelium **(Fig. 23)**

Family - Peronosporaceae (The downy mildews). Highly specialized family in the Order - Peronosporales; all species obligate parasites of vascular plants; causes downy mildew disease on many widely cultivated commercial crops; sporangia borne on sporangiophores of determinate growth; sporangiophore on maturity produces a crop of sporangia on sterigmata at the apices of its branches; sporangia mostly of the same age; they are round, oval or lemon-shaped, deciduous and wind disseminated; sporangia germinate by zoospores or by germ tubes depending upon environmental conditions; only one oospore is present in each oogonium; oospores as a rule germinate by germ tubes. The morphology and branching characteristics of the sporangiophores are the main criterion for classification of the different Genera.

Genus - *Plasmopara*. Obligate parasites; sporangiophore appears like a panicle; the branches and sub-divisions of branches occur at almost right angles and are irregularly spaced and end in pointed tips on which sporangia are borne; sporangia normally produce zoospores

(e-g) *Plasmopara viticola* - causes downy mildew of grapevines; *Plasmopara halstedii* - causes downy mildew of sunflower.

Life cycle of *Plasmopara viticola*. Primary infection occurs from oospores present in the soil. The oospore germinates by issuing a short germ tube producing a single terminal sporangium, which gives rise to a number of zoospores. The zoospores are biciliate and pear-shaped. After swimming in a film of water for sometime, the zoospore comes to rest, withdraws the flagella, secretes a membrane around it and encysts. Subsequently, it puts forth a germ tube and enters the host through a stoma or a water pore in the leaf margin and by the elaboration of an intercellular mycelium invades the mesophyll tissues. The intercellular mycelium forms numerous globular haustoria that enter the cells.

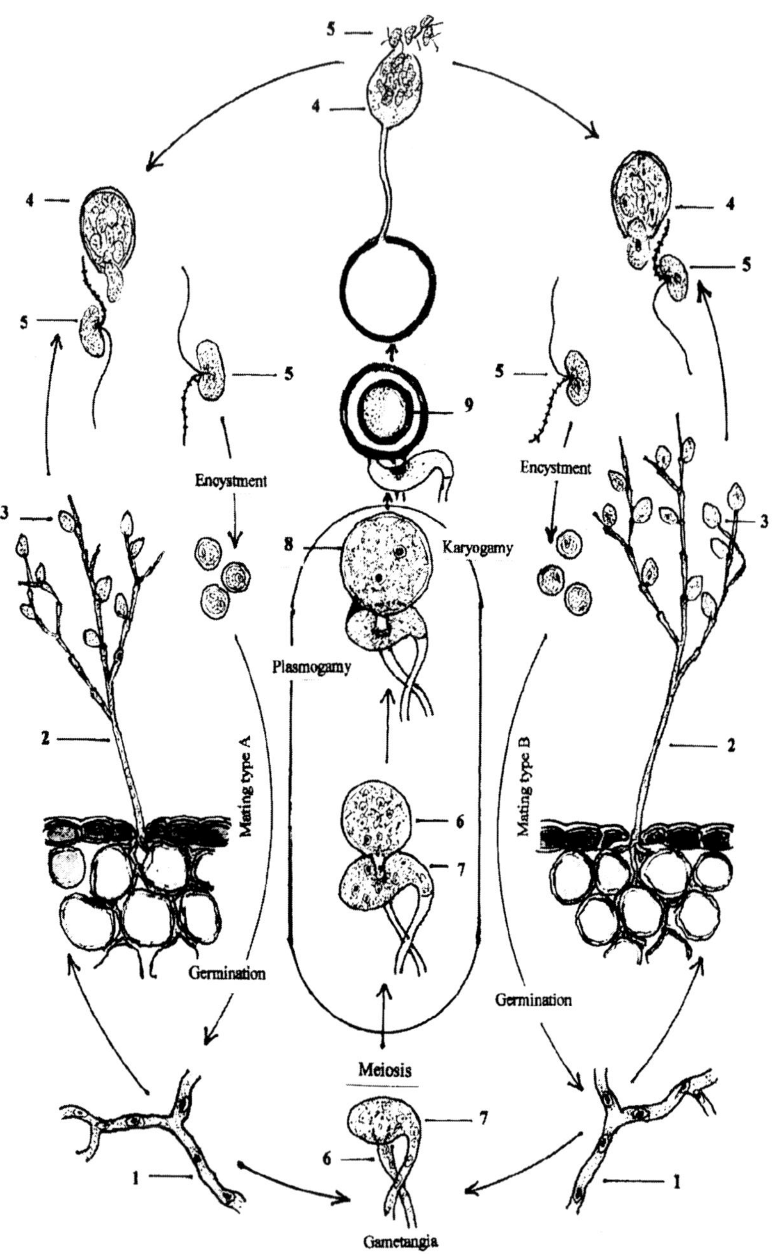

1.Somatic hyphae 2.Sporangiophore 3.Sporangium 4.Germinating sporangium 5.Zoospore 6.Oogonium 7.Antheridium 8.Developing oospore 9.Oospore

Fig. 23 : Life cycle of *Phytophthora infestans*

The mycelium produces sporangiophores, which emerge on the undersurface of the leaves and on the stems through stomata and through lenticels in young fruits. Four to six or more sporangiophores arise from a single stoma. Each sporangiophore produces four to six branches at nearly right angles to its main stem and each branch produces two or three secondary branches in a similar manner. At the tips of the short branches (sterigmata), single, lemon-shaped sporangia are produced. The sporangiophore with the sporangia appears like a panicle. The sporangia germinate by issuing zoospores. At the end of the growing season, the fungus develops oogonia, antheridia and oospores in the infected old leaves and sometimes in the shoots and berries. The formation of oospores is similar to that of *Pythium* and *Phytophthora*. Sometimes oospores may be formed apogamously without formation of antheridia **(Fig. 24)**

Genus - *Peronospora*. Obligate parasites; sporangiophore repeatedly dichotomously branched at acute angles and taper to curved, pointed tips on which sporangia are borne; the sporangia do not fall from the sporangiophore, but are actively discharged from it; germination of sporangia is regularly effected by a germination tube. The germ tube may arise from any part of the sporangial surface

(e-g) *Peronospora destructor* - causes downy mildew of onion; *P. manchurica* - causes downy mildew (blue mould) of soybean; *P. tabacina* - causes downy mildew of tobacco; *P. parasitica* - causes downy mildew of mustard, cabbage, cauliflower and other cruciferous crops **(Fig. 25)**

Genus - *Bremia*. Sporangiophore extensively branched and its apices end in concave cup-shaped enlargements called **'apophyses'**; the margin of each such enlargement bears four, minute, pointed sterigmata, each of which bears a single sporangium; germination is regularly effected by a germ tube and the germ tube develops from an apical papilla

(e-g) *Bremia lactucae* - causes downy mildew of lettuce **(Fig. 25)**

Genus - *Pseudoperonospora*. Sporangiophores exhibit pseudo (false) dichotomous branching at acute angles and ends in pointed tips on which sporangia are borne; germination usually accomplished by the production of zoospores; only under exceptional conditions, germination tube is produced

(e-g) *Pseudoperonospora cubensis* - causes downy mildew of cucurbits **(Fig. 25)**

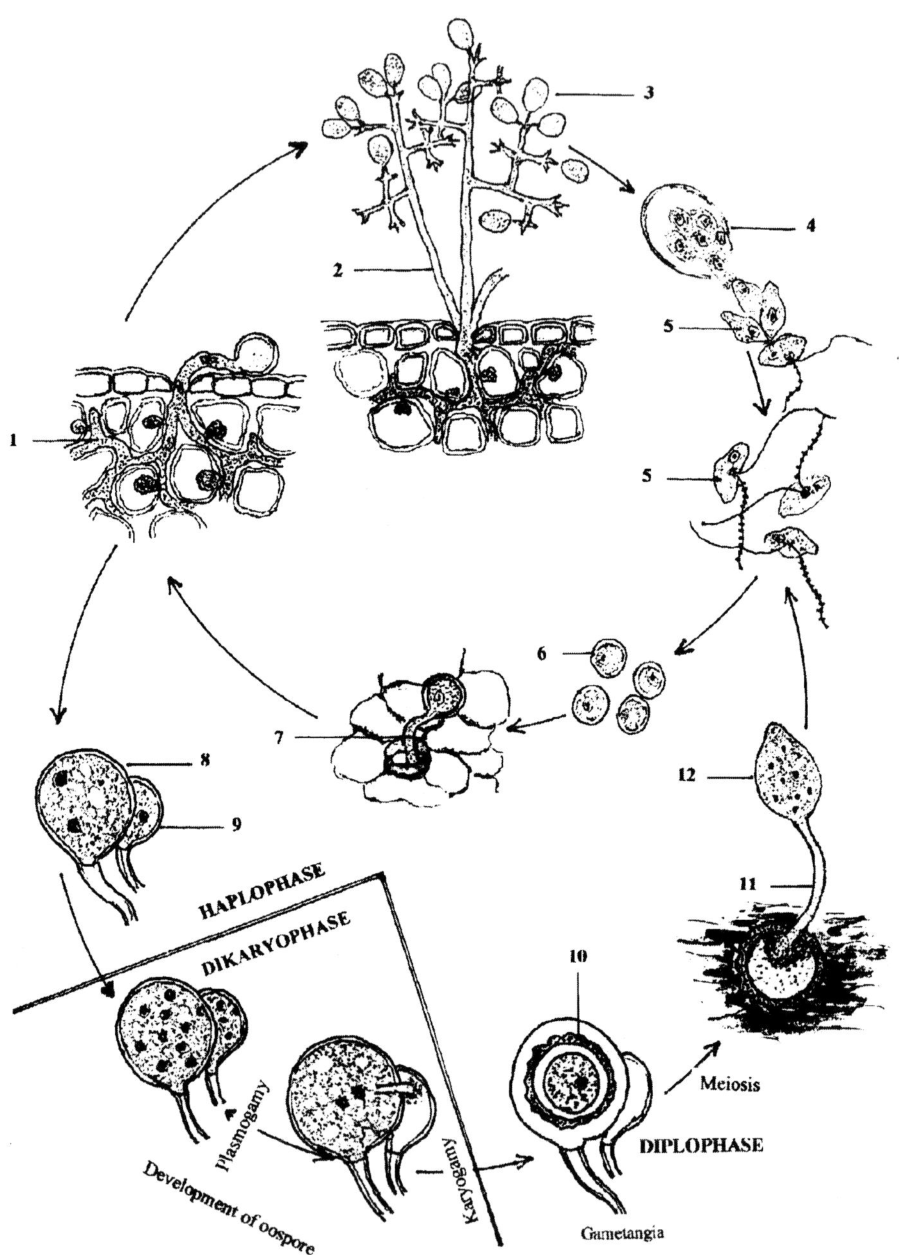

1.Mycelium in host 2.Sporangiophore with sporangia 3.Sporangium 4.Sporangial germination 5.Zoospores 6.Encysted zoospore 7.Spore germination and infestation 8.Oogonium 9.Antheridium 10.Oospore 11.Oospore germinating from soil 12.Sporangium

Fig. 24 : Life cycle of *Plasmopara viticola*

Genus - *Sclerospora*. Sporangiophore somewhat swollen and irregularly branched near the end; the short, pointed tips of each branch bears a sporangium; germination usually accomplished by the production of zoospores; rarely under exceptional conditions the sporangium germinates by germ tube

(e-g) *Sclerospora graminicola* - causes downy mildew of millets and grasses; *S. sorghi* - causes downy mildew of sorghum **(Fig. 25)**

Genus - *Basidiophora*. The sporangiophore is unbranched and is enlarged gradually towards its tip; the somewhat rounded apex bears several small projections called sterigmata, each of which bears a single sporangium; germination takes place by production of zoospores; under some specific conditions sporangium germinates by producing a germ tube

(e-g) *Basidiophora entospora* **(Fig. 25)**

Genus - *Peronosclerospora*. Sporangia show typical conidial germination

(e-g) *Peronosclerospora maydis* - causes downy mildew of corn.

Family - Albuginaceae (The white rusts). Obligate parasites causing diseases of vascular plants; mycelium intercellular; feeds by means of globose or knob-like haustoria; sporangia borne in chains from the tips of sporangiophores arranged in close proximity below the epidermis of the host; sporangia thin-walled and multinucleate; sporangia disseminated mostly by wind and water and they germinate usually by zoospores, but rarely by germ tubes; oogonium globose and contains a single oospore; oospore thick -walled and with ornamental markings characteristic of the species; oospore germinates to form zoospores.

Genus - *Albugo* (e-g) *Albugo candida (Cystopus candidus)* - causes white rust of mustard, cabbage, cauliflower and other cruciferous crops; *A. bliti* - causes white rust of eggplants.

Life cycle of *Albugo candida*. The mycelium of this fungus is intercellular and feeds by means of haustoria that penetrate the host cell walls through minute pores and expand on the inside of the cells into globose or knob-like structures. The mycelium grows and ramifies and at maturity produces short, club-shaped sporangiophores from the tips of a large number of hyphal branches in one locality. The sporangiophores are arranged in close proximity to one another in compact layers immediately below the host epidermis. The multinucleate sporangiophores produce a number of sporangia in succession from their tips in chains, one below the other. The oldest of the sporangium is at the top of the chain and the youngest at the base. Pads of gelatinous material called **'isthmus'** formed between each pair of

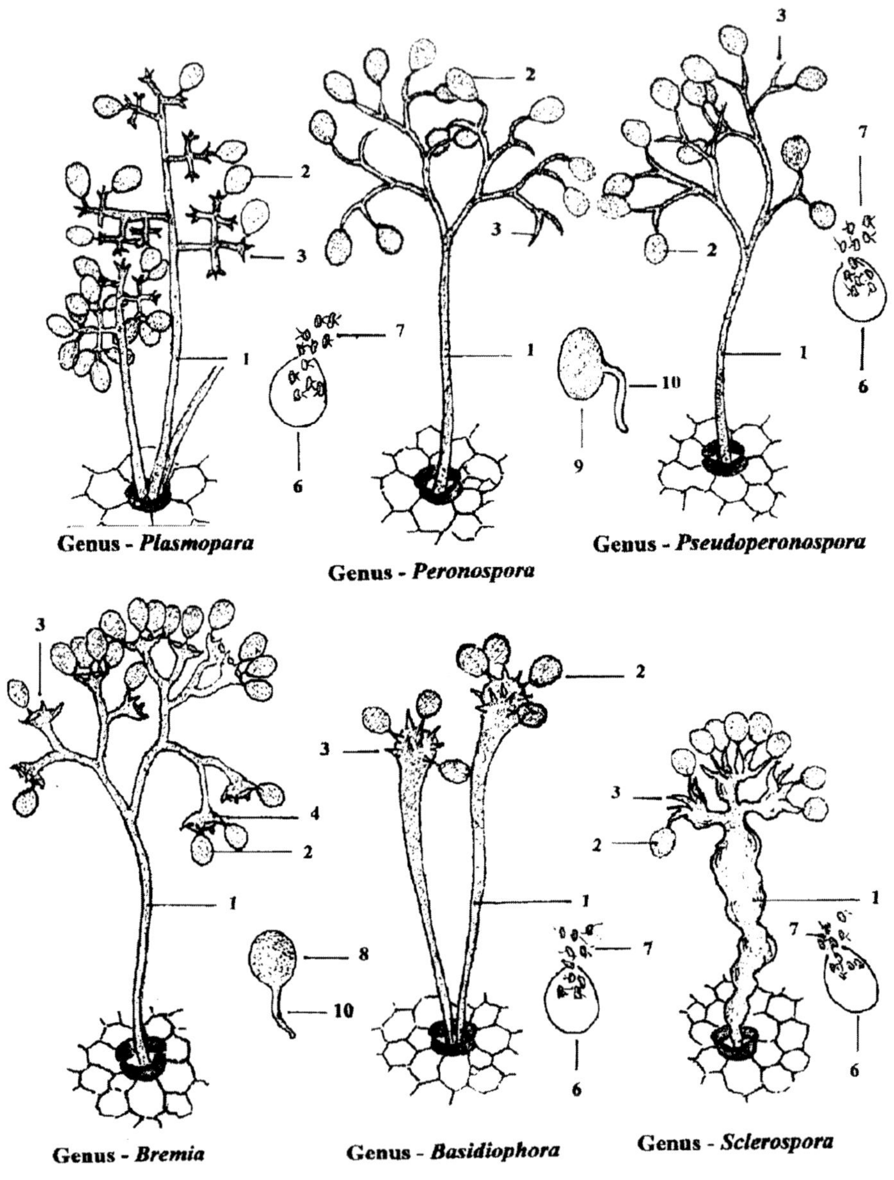

1.Sporangiophore 2.Sporangium 3.Sterigma 4.Apophyses 5.Stoma 6.Germination by zoospores 7.Zoospores 8.Germination by germ tube through apical papilla 9.Germination by germ tube from any part of the sporangial surface 10.Germ tube

Fig. 25 : Sporangiophore characteristics of the genera of Family Peronosporaceae and method of germination of sporangia.

sporangia, function as disjunctors, which on disintegration free the sporangia. As the sporangia mature, they become detached and accumulate in the space between the sporangiophores and the host epidermis. The growth of the mycelium and the continuous production of numerous sporangia exert pressure from below on the host epidermis, causing it to bulge and eventually rupture. Upon bursting of the epidermis, the sporangia are released and form a white crust on the surface of the host and so it is called **'white rust'**. Sporangia are normally globose, but due to pressure during their formation they may become flattened on the sides and some of them may be cuboid or polyhedral. They are somewhat thin-walled and multinucleate. Dissemination is by wind, water or other agents. The sporangia normally germinate by extruding four to twelve zoospores in a sessile vesicle, from which the zoospores emerge. Subsequently they encyst, germinate by putting forth germ tube and infect the host. Rarely the sporangia germinate by germ tube.

Sexual reproduction is almost similar to Peronosporaceae and occurs within the tissues of the host. A hyphal branch enlarges to form the multinucleate oogonium, while another branch nearby becomes a multinucleate antheridium and makes contact with the oogonium. The mature, globose oogonium contains a single oosphere surrounded by periplasm. The antheridium pushes a fertilization tube through a pore in the side of the oogonial wall and a single male nucleus passes through it along with some cytoplasm and fuses with a female nucleus. The resulting zygote (diploid) nucleus undergoes meiosis followed by several mitotic divisions, while the oospore develops a thick wall with ornamental markings. The oospore germinates to form zoospores that eventually encyst and germinate by germ tubes **(Fig. 26)**

Division-III. Amastigomycota

This Division comprises of a very large assemblage of fungi, many of which are quite common and familiar. The yeasts, moulds, mildews, cup fungi, rusts, smuts, bracket fungi, puffballs and mushrooms are included in this Division. All of them, except the yeasts produce a well-developed mycelium consisting of septate or aseptate hyphae. Motile cells are completely absent in these fungi. Asexual reproduction is by conidia, by budding or by fragmentation. Sexual reproduction, wherever present is by the production of zygospores, ascospores or basidiospores. The Division - Amastigomycota is divided into the Subdivisions - Zygomycotina, Ascomycotina, Basidiomycotina and Deuteromycotina. The Subdivision - Zygomycotina has two Classes - Zygomycetes and Trichomycetes, of which the former is of considerable importance.

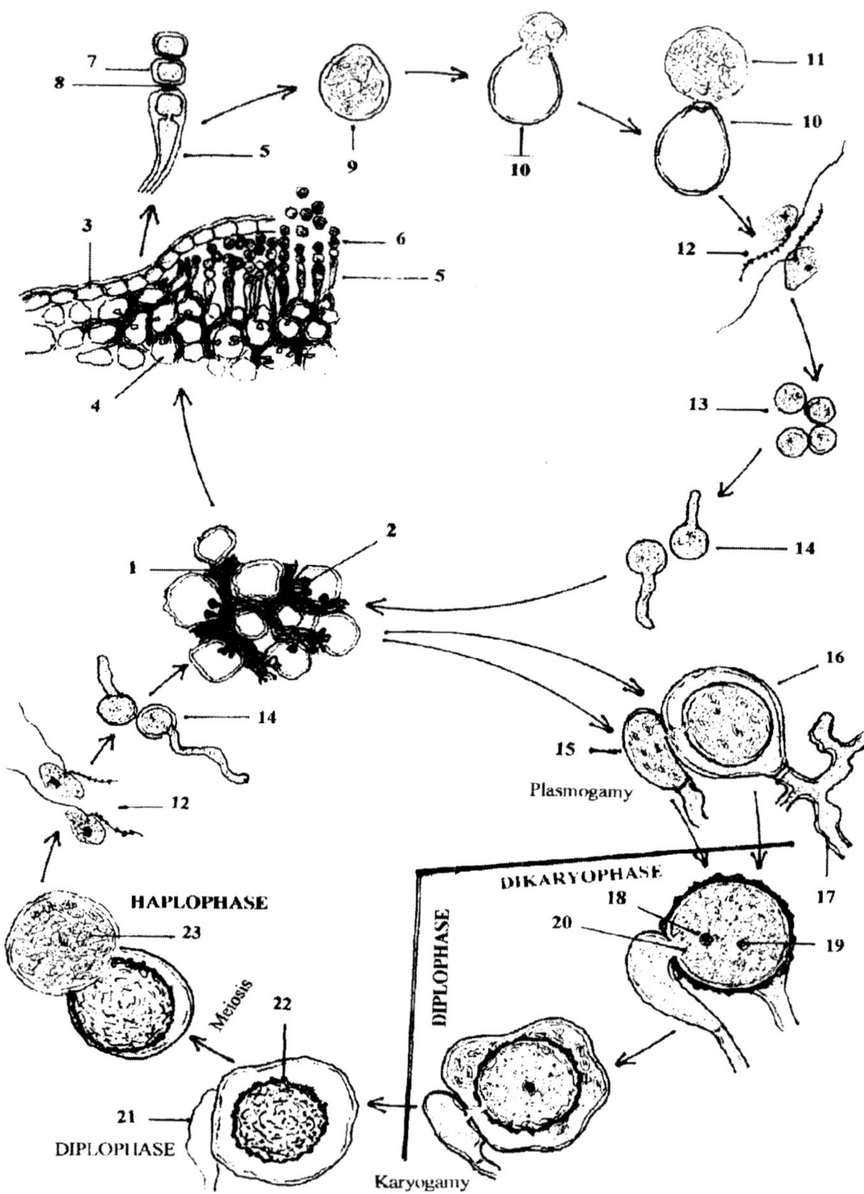

1.Intercellular mycelium 2.Haustorium 3.Host epidermis 4.Host mesophyll 5.Sporangiophore 6.Sporangia in chain 7.Sporangia 8.Abscission layer 9.Germinating sporangium 10.Release of zoospores 11.Zoospores inside sessile vesicle 12.Zoospores 13.Encysted zoospores 14.Germination of encysted zoospores 15.Antheridium 16.Oogonium 17.Somatic hypha 18.Male nucleus 19.Female nucleus 20.Fertilization tube 21.Empty antheridium 22.Oospore within oogonium 23.Zoospores inside vesicle

Fig. 26 : Life cycle of *Albugo candida*

Subdivision 1. Zygomycotina. Asexual reproduction by non-motile spores produced in sporangia; no zoospores produced; sexual reproduction by zygospore produced by the fusion of two morphologically identical gametes.

Class - Zygomycetes (The bread moulds). The term Zygomycetes refers to the production of a thick-walled resting spore called a **'zygospore'** that develops within a **'zygosporangium'** formed as a result of fusion of two equal or unequal gametangia. These gametangia may arise from the same mycelium or from different mycelia. Most Zygomycetes produce a well-developed mycelium consisting of coenocytic hyphae, but in some species mycelium is very much reduced and has septa at regular intervals. Motile cells are completely absent. Asexual reproduction is typically by sporangiophores, while some others produce chlamydospores or reproduce by budding. They are either saprophytes or weak parasites of plants or parasites of animals and humans. Some are used in the fermentation of food items, while some others are used commercially to produce certain enzymes, organic acids etc. Many saprophytic species attack and spoil human foods. Some are known to be important mycorrhizal fungi. Only the Orders - Mucorales and Entomophthorales of this Class are of economic importance.

Order - Mucorales. Mostly saprobes, living on substrates, such as dung, decaying plant or animal matter; many saprophytic species synthesize important industrial products, such as fumaric acid, alcohol, lactic acid, citric acid, oxalic acid etc.; few species parasitize fruits and other detached plant parts and cause rotting in transit and storage; in most of the species the hyphae are coenocytic, producing septa only at the bases of reproductive organs; the mycelium in some species produce rhizoids, which adhere to the substratum and anchor the fungus securely; asexual reproduction by sporangiospores (aplanospores) produced in sporangia, which are borne on simple or branched sporangiophores; sporangium formed at the tip of a sporangiophore; sporangium is globose and contains cytoplasm and many nuclei; a central columella delimits the outer sporiferous region from the sporangiophore; the sporangium contains many thousands of sporangiospores; sporangiospores differ in shape, size, markings and color; mostly they are globose to ovoid and in some species cylindrical; in many forms the spores are longitudinally striated; the spores often become multinucleate by nuclear division; the spores are liberated by the dissolution of the sporangial wall and dispersed by air currents; sexual reproduction takes place by the copulation of two multinucleate gametangia to form a prozygosporangium, which enlarges, develops a thick multilayered wall and becomes the zygosporangium in which a single zygospore is formed; the dormant zygospores usually take a long time to break the dormancy and germinate;

during the process of zygospore germination, meiosis takes place leading to segregation of strains; at the time of zygospore germination, the zygosporangium cracks open and a sporangiophore emerges and develops a germsporangium at the tip; the wall of the germsporangium on maturity ruptures and liberates the non-motile sporangiospores.

Family - Mucoraceae. Some members of this family are found only in soils, while others grow on dung and other organic refuse and a few others grow exclusively on mushrooms; zygospores are common in almost all the Mucoraceae and all members produce large, multispored sporangia with prominent columellae. Under unfavorable conditions, thick-walled **'gemmae'**, similar to chlamydospores, which are able to withstand adverse conditions, are produced by the hyphae, either singly or in chains.

Genus - *Mucor*. Thallus consists of highly branched mycelium that spreads evenly on the surface of the substratum; some of the hyphae penetrate into the substratum; they serve as fixative and absorptive organs; no rhizoids are formed; slender, upright sporangiophores develop from any part of the mycelium; sporangiophores typically unbranched with a single, terminal, globose sporangium; however, in a few species as in *Mucor racemosus* and *M. brunneus* sporangiophores are branched, each branch terminating in a sporangium; large, dome-shaped columella present; sporangiospores globose to oval in shape, thick-walled and dark-colored; columella not present in the zygosporangium.

(e-g) *Mucor racemosus; M. brunneus; M. pusillus; M. ramosissimus* - causes diseases in humans; *M. hiemalis* - parasitic on sugarcane leafhopper; *M. mucedo* - parasitic on common housefly **(Fig. 27)**

Genus - *Rhizopus*. Thallus does not spread evenly over the surface of the substratum; individual non-septate hyphae form stolons, which develop buds at intervals from which rhizoidal branches are produced and they penetrate the substratum; clusters of sporangiophores and sporangia are produced, usually at the points where the rhizoids are formed. The sporangia contain thousands of spherical sporangiospores. Columella is present. Adjacent hyphae produce short branches called progametangia, which grow toward one another. When they come in contact, the tip of each hypha is separated from the progametangium by a cross wall and the terminal cells become the gametangia. These two cells fuse and their nuclei pair. The cell thus formed by the fusion enlarges and develops a thick, black and wavy wall. This sexually produced spore is called a zygospore and is the overwintering or

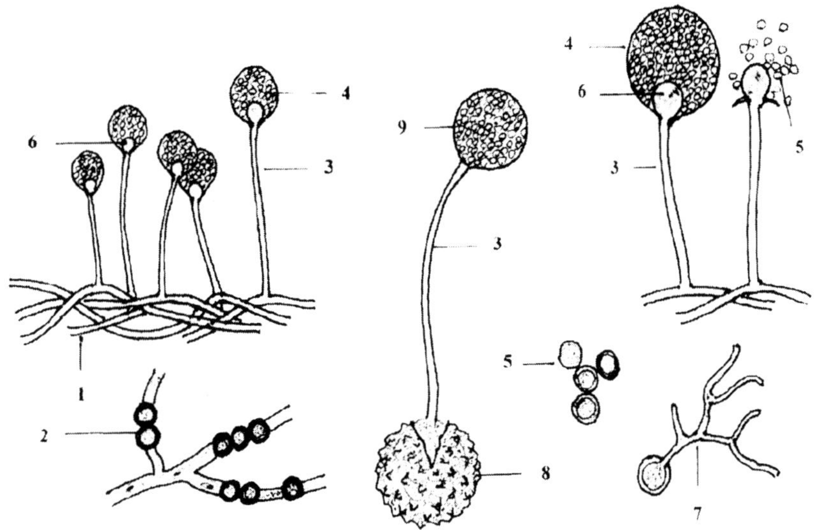

1.Hyphae 2.Gemmae 3.Sporangiophore 4.Sporangia 5.Sporangiospores 6.Columella 7.Germ tube 8.Germinating zygospore 9.Zygosporangium

Fig. 27 : Genus - *Mucor*

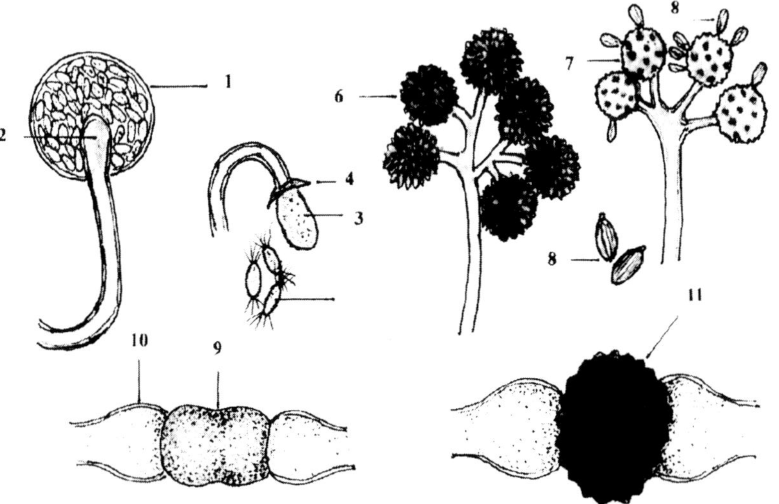

1.Large multispored sporangium 2.Columella 3.Remainder of the columella after shedding of the sporangiospores 4.Basal collar 5.Endogenous sporangiospores 6.Sporangioliferous head with sporangiola 7.Heads after shedding most of the spores 8.Sporangiola (Conidia) 9.Young zzgosporangium 10.Suspensor 11.Mature zygosporangium

Fig. 28 : Genus - *Choanephora*

resting stage of the fungus. When it germinates, it produces a long sporangiophore bearing a globular zygosporangium full of sporangiospores at its tip; columella present in the zygosporangium.

(e-g) *Rhizopus stolonifer (R. nigricans)* - the common black bread mould is used in the manufacture of fumaric acid, cortizone and lactic acid. It causes a serious transit disease of strawberries and soft rot of potatoes, sweet potatoes, cucurbits, peaches, cherries and many other fruits and vegetables during storage and marketing of these products. It also attacks and spoils our food **(Fig. 29)**; *R. oryzae* - it is used in the manufacture of alcohol; R. nodosus - it is used in the manufacture of lactic acid.

Family - Choanephoraceae. The fungi of this family produce large sporangia with prominent columella. The sporangial wall is persistent even after the sporangium has broken open. The sporangiospores are purple to brown in color, often striate and possess stiff, hair-like appendages at their ends. They also produce monosporous **'sporangiola'**, besides the larger sporangia. They are commonly found growing on flowers and fruits and may occasionally cause severe damage to cultivated crops. The zygospore formation is similar to that of *Mucor* and *Rhizopus.*

Genus - *Choanephora* (Fig. 28) (e-g) *Choanephora cucurbitarum* - attacks cucurbits and other plants. It attacks the withering floral parts and subsequently invades the fruit and causes a soft rot primarily of summer squash, but also pumpkin, pepper and bhendi.

Genus - *Gilbertella* (e-g) *Gilbertella persicaria* - it attacks peaches and tomatoes.

Order - Entomophthorales. The fungi of this Order are either parasitic on both plants and animals or saprobic. Many of them are able to parasitize insects. The mycelium of Entomophthorales is not so extensive as that of other Zygomycetes and is usually divided into uninucleate or multinucleate segments by septae, which are called **'hyphal bodies'**. Asexual reproduction is usually by means of single-celled sporangia, which are referred to as conidia. Sexual reproduction is by zygospores and their formation is quite similar to that of other Zygomycetes, with some differences characteristic of the group.

Family - Entomophthoraceae. Many fungi of this family are parasitic on insects.

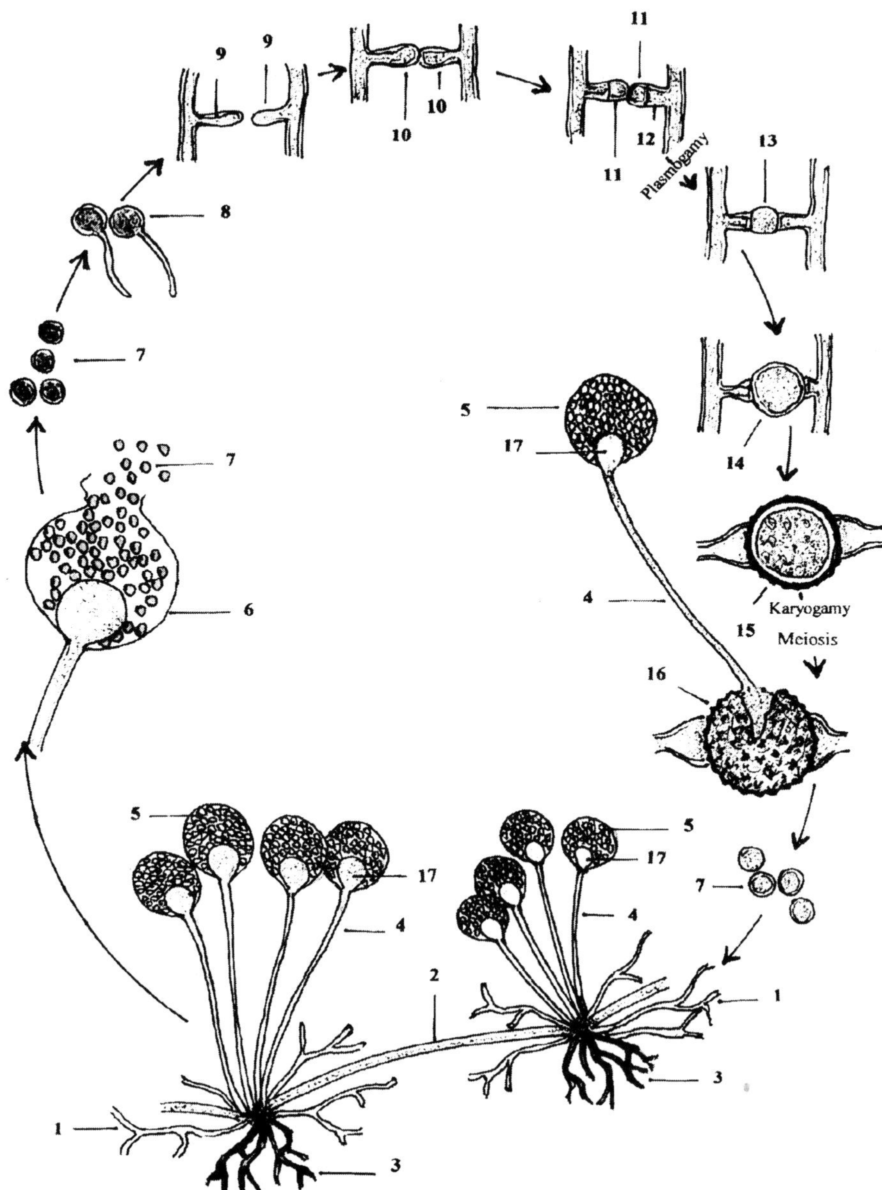

1.Hyphae 2.Stolon 3.Rhizoids 4.Sporangiophore 5.Sporangium 6.Germinating sporangium releasing sporangiospores 7.Sporangiospores 8.Germinating sporangiospores 9.Gamete 10.Progametangium 11.Gametangium 12.Suspensor 13.Zygote 14.Prozygosporangium 15.Zygospore 16.Germinating zygospore 17.Columella

Fig. 29 : Life cycle of *Rhizopus stolonifer*

Genus - *Entomophthora*. This is the largest and best-known Genus of Entomophthorales. Members of this Genus are parasitic on arthropods. The mycelium often breaks up into segments called **'hyphal bodies'**, each of that is capable of forming a sporogenous cell (sporangiophore) bearing one or more asexual spores called conidia. These spores at maturity are forcibly discharged. Each of these spores may again form sporogenous cells and form secondary spores. Sexual reproduction takes place, when hyphal bodies acting as gametangia copulate and develop a zygosporangium containing a zygospore. The zygosporangium develops from a lateral outgrowth or bud arising between the two fusing cells. In some species structures known as **'azygosporangia'** are formed parthenogenetically. They resemble zygosporangia and the zygospore within carries on the function of a resting spore just as true zygospores do **(Fig. 30)**

(e-g) *Entomophthora muscae* - this is called the **'fly fungus'**, which is often found on the dead bodies of houseflies. The spores, which are produced singly at the tips of unbranched sporogenous cells, are covered by a mucilaginous substance. The spore, which is catapulted when comes in contact with a fly, germinates immediately and penetrates the cuticle of the fly. Infected flies usually die within a week's time. The fungus produces azygosporangia parthenogenetically.

Subdivision 2. Ascomycotina. The fungi belonging to this Subdivision are called higher fungi and are considerably more complex in structure than the fungi dealt with so far. In Ascomycotina, a short-lived dikaryotic stage is present in-between plasmogamy and karyogamy. These fungi are believed to have been derived from the flagellate fungi by many Mycologists.

Class - Ascomycetes (The sac fungi). Most of the Ascomycetes have a sexual stage (teleomorph, ascigerous or perfect stage) and an asexual stage (anamorph or imperfect stage). The Ascomycetes are of much economic importance to humans. There are both harmful and beneficial members. Many of them are harmful in the sense that they produce worst diseases, such as powdery mildews, leaf spots, blights, rots etc., in several cultivated crops. However, many others are beneficial. The fermenting activities of certain yeasts are the basis of the baking and brewing industries. A few, such as *Morchella* are edible and are considered to be delicacies. The ergot is of great medicinal value.

The main distinguishing character of the Ascomycetes is the production of **'ascus'** (pl. - asci), a sac-like cell containing a definite number of ascospores, usually eight in each ascus, formed by free cell formation after karyogamy and meiosis. Most of the species produce a fruiting body, such as cleistothecium, perithecium or apothecium enclosing the asci. In Ascomycetes, flagellate cells are completely absent. Asexual spores called **'conidia'** are produced on free hyphae or in asexual fruiting bodies, such as pycnidia, acervuli etc. Generally ascospores act as the primary inoculum, which causes the primary infection. The conidial stage, which follows is responsible for the secondary infection and the rapid spread of diseases. The mycelium of the Ascomycetes is composed of septate and profusely branched hyphae, the walls of which contain mainly chitin. Some Ascomycetes, such as a few yeasts are unicellular. Asexual reproduction in the Ascomycetes may take place by fission, fragmentation or budding or by the formation of chlamydospores or conidia depending upon the species and the environmental conditions. Spores produced by budding are generally called **'blastospores'**. Conidia may arise, either directly from the somatic hyphae or from specialized conidiogenous cells called **'conidiophores'**. Conidiophores are produced free or they are cemented together in the form of complex structures called **'synnemata'**. In some other species, the imperfect stages of which are classified under Deuteromycetes, conidia are produced in definite fruiting bodies called **'pycnidia'** or **'acervuli'**.

The union of two compatible nuclei accomplishes sexual reproduction. In Ascomycetes, the two nuclei remain in close association and undergo successive divisions that result in a number of dikaryotic cells. Nuclear fusion then takes place in the ascus mother cell, which finally develops into the ascus. Meiosis of the diploid zygote nucleus occurs just after fusion and results in the production of four haploid nuclei after two meiotic divisions (meiosis 1 and meiosis 2). These four nuclei undergo one more mitotic division resulting in eight haploid nuclei. The eight nuclei ultimately become eight ascospores. The development of the asci and the ascospores are the main characteristic feature of the Ascomycetes. The two compatible nuclei are brought together in different ways viz., gametangial copulation, gametangial contact (gametangy), spermatization or somatogamy. In gametangial copulation, two identical gametangia touch at their tips or coil around each other and fuse and the fusion cell develops into an ascus as in the case of yeasts. In gametangial contact, which is the typical sexual reproductive process, morphologically differentiated uninucleate or multinucleate gametangia, known as **'antheridia'** and **'ascogonia'** are produced. The male nucleus passes from the antheridium into the ascogonium through a pore at the point of contact or through a trichogyne that is provided in the ascogonium to receive the male nucleus. The ascogonium,

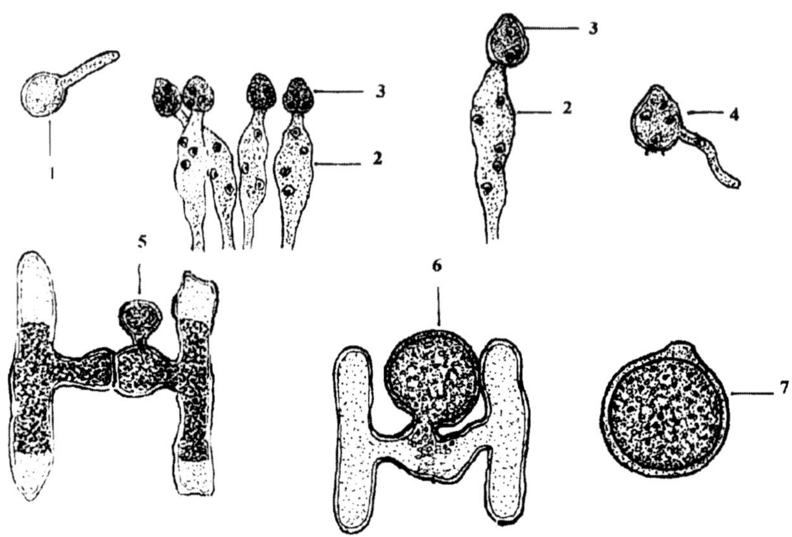

1.Germinating hyphal body 2.Sporangiophore 3.Sporangium 4.Germinating sporangium 5.Young zygosporangium 6.Developing zygosporangium 7.Mature zygosporangium

Fig. 30 : Genus - *Entomophthora*

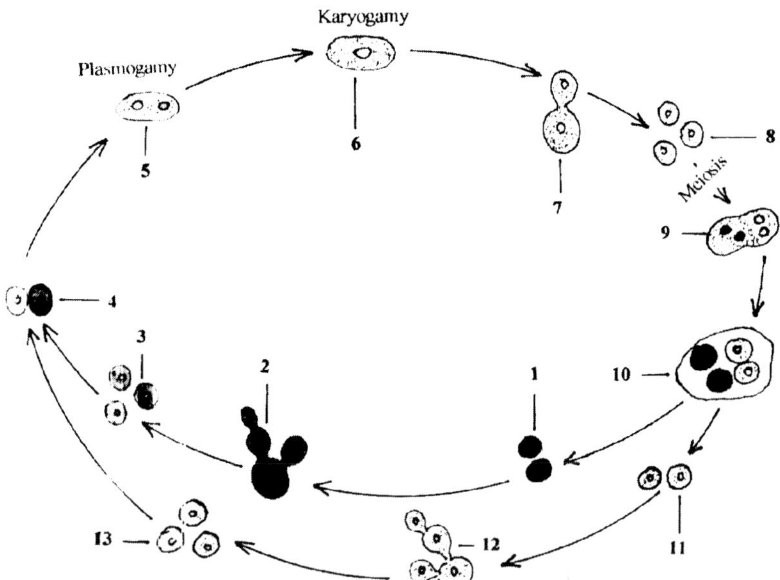

1.Ascospore (Type a) 2.Budding of ascospore (Type a) 3.Somatic cells (Type a) (n) 4.Copulation of type a and b 5.Union of nuclei 6.Zygote 7.Budding of zygote cell 8.Somatic cells (2n) 9.Young ascus after meiosis 10.Mature ascus 11.Ascospores (Type b) 12.Budding of ascospores (Type b) 13.Somatic cells (Type b) (n)

Fig. 31 : Life cycle of *Saccharomyces cereviciae*

which is the female organ ultimately develops into an ascus **(Fig. 32)**. In spermatization, no antheridia are formed and the male nuclei reach the ascogonia by means of minute, uninucleate, male sex cells called **'spermatia'** produced from specialized hyphae called **'spermatiophores'**. The spermatial nuclei migrate to the ascogonia and ultimately develop into asci. In somatogamy, fusion of somatic hyphae of two compatible mycelia takes place leading to the production of asci. The asci in some Ascomycetes, such as yeasts and leaf curl fungi are naked without any fruiting body. In all other Ascomycetes, the asci are produced singly or in groups in ascocarps (fruiting bodies), such as **'cleistothecium'** (a completely closed, spherical container) or **'perithecium'** (a flask-shaped structure with a small opening for the ascospores to escape) or **'pseudothecium'** / **'ascostroma'** (a cavity / locule within a stroma of mycelium, in which the stroma forms the wall of the ascocarp) or **'apothecium'** (an open cup- or saucer-shaped structure) **(Fig. 33)**. Ascocarps may be formed singly or in groups. They may be superficial, erumpent or deeply embedded in the substratum. The substratum may be composed entirely of host tissues or a hyphal stroma on or in which the ascocarps are formed.

In a vast majority of the Ascomycetes, the asci are elongated, either club-shaped or cylindrical. In some forms, the asci are globose or ovoid or rectangular. Asci may be stalked or sessile and may arise from a common fascicle and spread out like a fan or may arise singly at different levels within the ascocarp. A definite layer of asci, whether naked or enclosed in a fruiting body is called a **'hymenium'**. Three types of asci viz., **prototunicate asci**, **unitunicate asci** and **bitunicate asci** are found. The prototunicate asci have a delicate, thin wall and release the ascospores either by breaking apart or deliquescing. The wall of both unitunicate and bitunicate asci consist of two layers, exotunica (exoascus) and endotunica (endoascus). In the unitunicate ascus, the two layers are closely adherent all through their existence and the ascospores are released through a terminal pore, slit or hinged cap (operculum). In the bitunicate ascus, the endotunica extends up to twice or more of its original length and separates from the exotunica at the time of spore release and the spores are released through a pore at the tip of the endotunica **(Fig. 35)**

Ascospores vary greatly in size, shape and color, wall ornamentation and other characters. The size varies from minute to more than 1,000 mm in length, the shape from globose to thread-like, the color from hyaline to black and the number of cells from one to many, usually eight **(Fig. 34)**

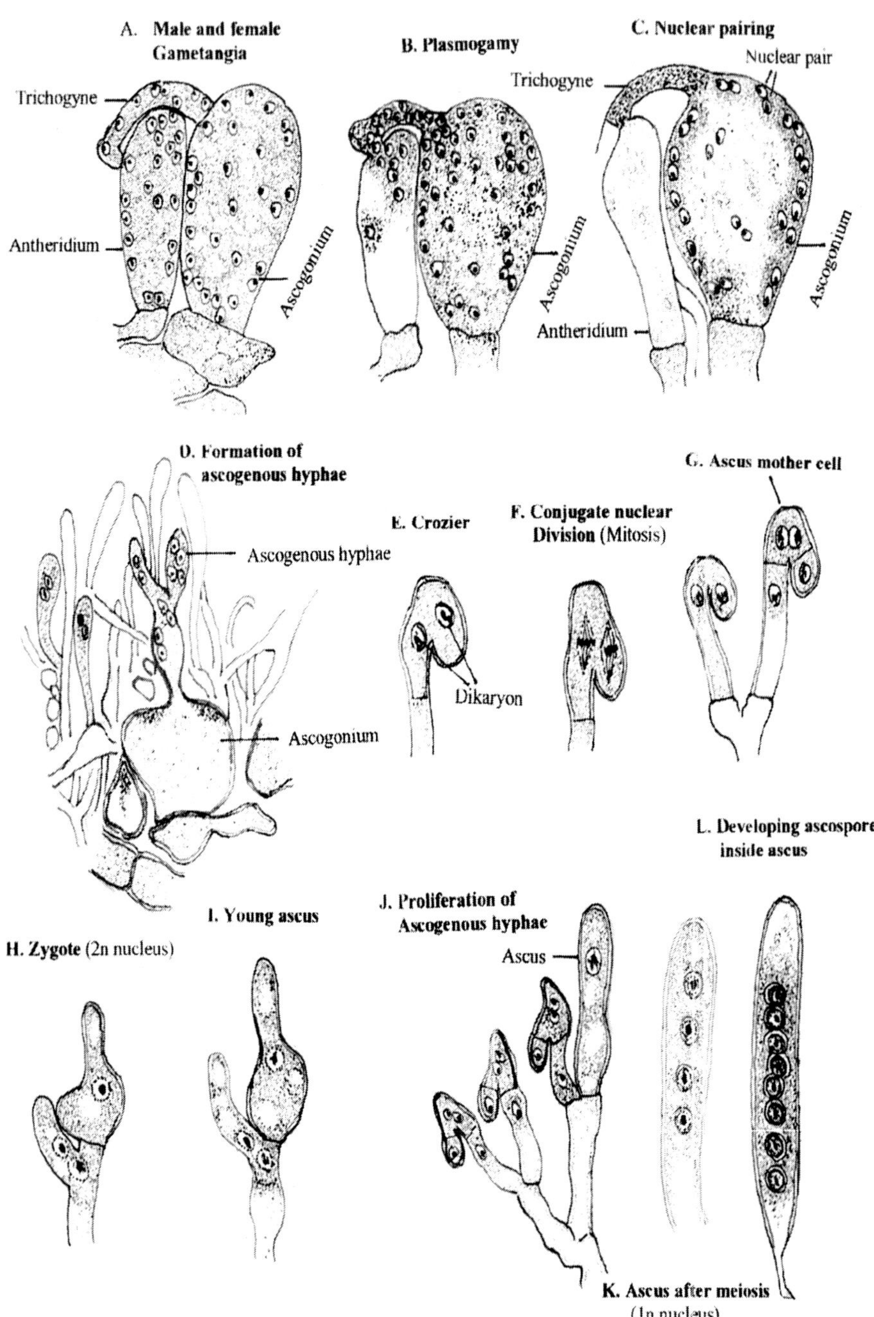

Fig. 32 : Sexual reproduction and ascus formation in the Ascomycetes

In addition to asci, many ascocarps contain various types of sterile thread-like structures, such as paraphyses, periphyses, periphysoids, apical paraphyses and pseudoparaphyses. These structures form part of the hymenium. **'Paraphyses'** are elongated, cylindrical or club-shaped, sometimes branched threads, usually non-septate, which originate at the base of the ascocarp and grow among the asci in the hymenium. They absorb water, expand and oscillate, thereby shaking the asci and thus aid in disseminating the ascospores. **'Periphyses'** are short, hair - like threads that form an internal fringe around the ostiole of the perithecium. They serve to direct the asci toward the ostiole prior to ascospore discharge. **'Periphysoids'** are lateral periphyses that originate all along the inner wall of the ascocarp and curve upward toward the apex. **'Apical paraphyses'** are paraphyses that originate at the tip of the perithecial centrum and grow downward. **'Pseudoparaphyses'** originate at the top of the centrum of an ascostroma. When the ascospores mature, some provision is usually made for their release and dissemination. The ascospores are released either by bursting of the ascus wall or through a pore or an operculum or a slit at the tip of the ascus. In the case of bitunicate asci, the spores are released through a pore at the tip of the endotunica. In a large number of Ascomycetes, the ascospores are forcibly ejected from the ascus by a puffing action. Wind, water, insects or other agencies then distribute the spores. When conditions are favorable, the ascospores germinate by producing one or more germ tubes, which develop into septate mycelium.

There are a number of Ascomycetes whose conidial stages, if they exist have not been found. Since classification of Ascomycetes is primarily based on the characteristics of the perfect stage there is no problem in classifying them. Many fungi that were classified as imperfect fungi earlier were found to produce ascospores and subsequently they were classified under Ascomycetes. However, Ascomycetes that rarely produce sexual spores are generally known by the name of their asexual stages.

Subclass - Hemiascomycetidae (Yeasts and other non-ascocarpic sac fungi). The Hemiascomycetidae are morphologically simple Ascomycetes. They produce rather scant mycelium or mycelium is totally lacking. They produce asci directly from zygotes or single cells without any intervention of a system of ascogenous hyphae. Ascocarps are completely absent. In the species that have no mycelium, no dikaryotic stage is present. In others, the mycelium itself may be dikaryotic. The asci are thin-walled and the ascospores are released either by bursting or deliquescing.

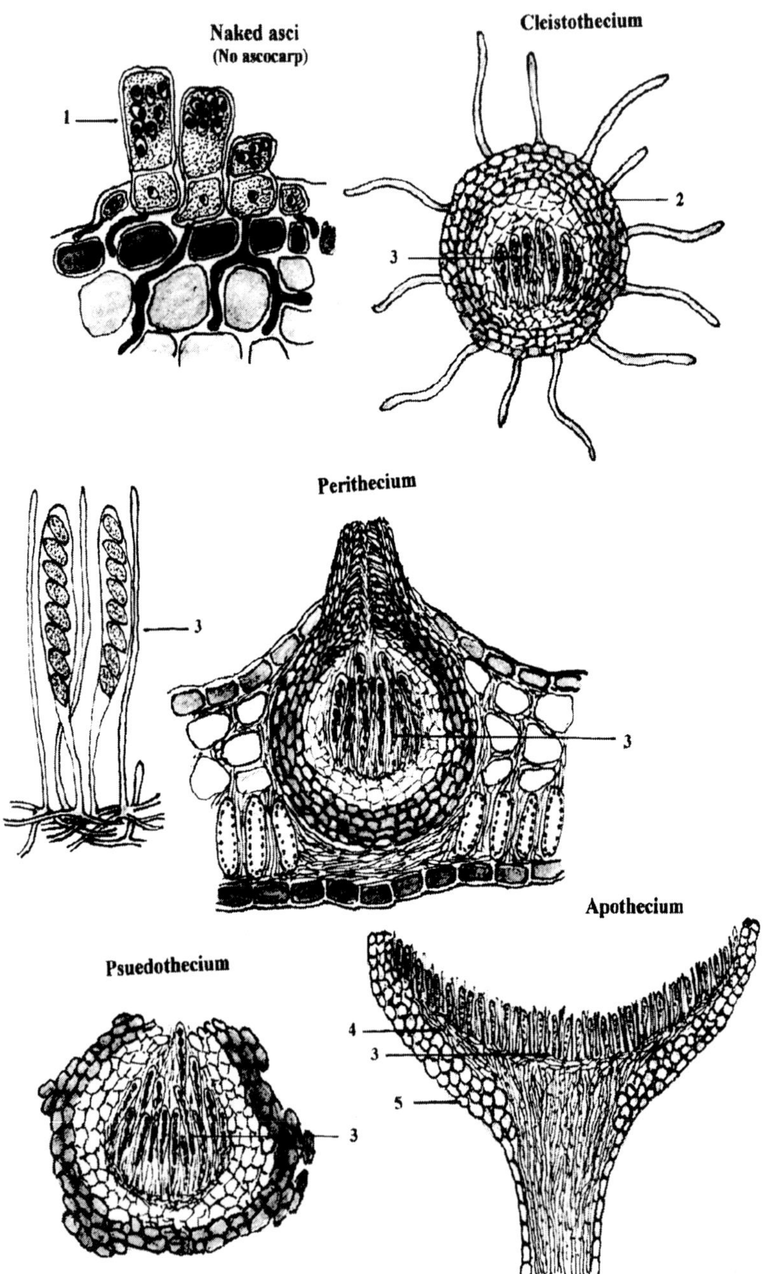

1.Ascus 2.Ascocarp wall 3.Asci and paraphyses (Hymenium) 4.Hypothecium 5.Excipulum

Fig. 33 : Different types of Ascocarps

Order - Protomycetales. The chief characteristic of fungi belonging to this Order is the presence of a compound spore sac or **'synascus'**. This compound ascus emerges from thick-walled resting cells formed by the mycelium. There is a single Family, the **Protomycetaceae**. This Family consists of a few Genera of which *Protomyces* and *Mixia* are important.

Genus - *Protomyces*. The mycelium of *Protomyces* is septate, endophytic, intercellular and diploid. Some of the cells of the hyphae become thick-walled resting cells or chlamydospores. A resting cell germinates and forms an elongated, more or less cylindrical sac, the compound ascus consisting of a large number of asci. The ascospores, which are four in numbers in each of the ascus, are forcibly ejected from the spore sac in a single mass. *Protomyces* are common parasites on members of the Umbelliferae. **(e-g)** *Protomyces macrosporus* - causes stem gall of coriander.

Order - Endomycetales. The zygote is derived from the copulation of two cells or parthenogenetically and is transformed directly into an ascus or produces an erect, septate ascospore. Some species are unicellular throughout their life cycle, except for the asci that may contain more than one ascospore. Some yeasts produce a number of buds in succession that remain joined for some time forming a pseudomycelium. In other species, septate hyphae are produced that bear asci at the tips of their branches. Fission, budding and formation of arthrospores are the three usual methods of asexual reproduction. Sexual reproduction occurs by the fusion of two single cells or two differentiated gametangia or two somatic hyphal cells. Karyogamy mostly occurs in the young ascus. There is no dikaryotic phase. Only the Families - **Endomycetaceae** and **Saccharomycetaceae** are of economic importance.

Family - Endomycetaceae. The mycelium of the fungi belonging to this Family is composed of well-developed, typical hyphae. Asexual reproduction is by means of arthrospores produced in chains. Plasmogamy is by fusion of two uninucleate gametangia or through somatogamy. A single zygote cell is transformed into an ascus. Each ascus contains 1 - 8 ascospores. The ascospores may be of different shape, but not needle-, spindle- or sickle-shaped.

Genus - *Endomyces*. (e-g) *Endomyces geotrichum*. Its imperfect stage is *Geotrichum candidum* of Family - Moniliaceae. It causes 'sour rot' of *Citrus* fruits, tomatoes, carrots and many other fruits and vegetables in storage.

Family-Saccharomycetaceae. The Family comprises of the Ascomycetous yeasts. The thallus is predominantly unicellular and reproduces asexually by budding, transverse fission or both. Ascospores are produced in a free ascus that originates either from a zygote or parthenogenetically from a single somatic cell. The ascosporogenous yeasts are abundant in substrata that contain sugars, such as nectar of flowers and the surface of fruits. They are also found in soil, milk etc. They are noted particularly for their ability to ferment carbohydrates. Because of this property and the resultant alcohol and carbon dioxide, the brewers and bakers use yeasts. In the brewery, alcohol is the industrial product whereas in the bakery carbon dioxide, which raises the dough is the important product and the alcohol is a waste. *Saccharomyces cereviciae* is universally employed in baking and brewing industries and also for the commercial production of yeast cake. The ascosporogenous species, such as *Candida utilis* are also used for the production of food yeast.

The Saccharomycetaceae are unicellular organisms that possess definite cell wall and a nucleus surrounded by cytoplasm. A large vacuole is often present inside the cell. The cells are hyaline and may be globose, ovoid, elongated or rectangular in shape according to the species. They sometimes adhere in chains forming a pseudomycelium. Asexual reproduction is by budding or by fission. Sexual reproduction is by means of union, either between two somatic cells or between two ascospores that function as gametangia. The zygote cell thus formed becomes an ascus, which contains ascospores. Usually 4 - 8 ascospores are found in each ascus. The ascospores are globose or ovoid. On release from the ascus, the ascospores start asexual reproduction by budding or by fission.

Genus - *Saccharomyces* (e-g) *Saccharomyces cereviciae* - the commercial yeast.

Life cycle of *Saccharomyces cereviciae*. In this yeast fungus, budding perpetuates both haplophase and diplophase. So, its life cycle is diplobiontic and the fungus is heterothallic with two of the ascospores in each ascus of one mating type **(a)** and the other two another mating type **(b)**. After fusion of cells of the two opposite mating types, the zygote begins to bud and several generations of diploid cells ensue before ascospore formation sets in. During this phase meiosis takes place and ascospores are formed **(Fig. 31)**

Family - Spermophthoraceae. The organisms belonging to this Family are considered to be intermediate between the Ascomycetes and Phycomycetes. The Spermophthoraceae are all plant parasites.

Spermophthora, Eremothecium, Ashbya and *Nematospora* are parasites on cotton. *Eremothecium* and *Ashbya* are very important in the commercial production of **'riboflavin'** and are the basis of that very important industry.

Genus - *Nematospora* (e-g) *Nematospora coryli* - parasitic on cotton.

Genus - *Eremothecium* (e-g) *Eremothecium ashbyi* - used in the commercial production of riboflavin (Vitamin B2)

Order - Taphrinales. The fungi included in the Order - Taphrinales are all parasitic on vascular plants and cause diseases, such as leaf curl, puckering, blisters, 'witches' broom' etc., as a result of malformation of the affected tissues. The Taphrinales resemble the yeasts in that the ascospores multiply by budding. However, the Taphrinales differ from the yeasts in that they produce definite, true mycelium. The ascus is not produced directly from a zygote cell by the copulation of two cells, but from a special binucleate, ascogenous cell derived from the mycelium. The Order has a single Family - **Taphrinaceae** and a single Genus - *Taphrina*.

Genus - *Taphrina*. The mycelium is composed of septate hyphae, consisting of binucleate cells or multinucleate cells, with the nuclei often occurring in pairs. The hyphae may be intercellular or subcuticular or may grow within the walls of the epidermal cells. Intercellular hyphae may penetrate deeply into the host tissues.

Asexual reproduction is by small, ovoid or spherical, uninucleate, haploid, yeast-like blastospores that bud directly from the ascospores, while they are still within the ascus or after they are released from the ascus. The blastospores may, either bud and produce more blastospores or may germinate and form mycelium that infects the host. Generally no plasmogamy takes place, but karyogamy occurs between sister nuclei. The dikaryotic condition arises through a mitotic division of the ascospore or blastospore nucleus at the time of germination.

(e-g) *Taphrina deformans* - causes leaf curl in peach and almond; *T. pruni* - causes plum pockets; *T. cerasi* - causes 'witches' broom' of cherries; *T. coerulescens* - causes curling and puckering of oak leaves; *T. maculans* - causes leaf spot of turmeric

Life cycle of *Taphrina deformans*. The ascospores soon after their formation produce small, round or ovoid blastospores by budding. This usually occurs within the asci and continues even after the spores have been released from the asci. The blastospores, like the ascospores are uninucleate and haploid. The blastospores may continue to bud and produce secondary

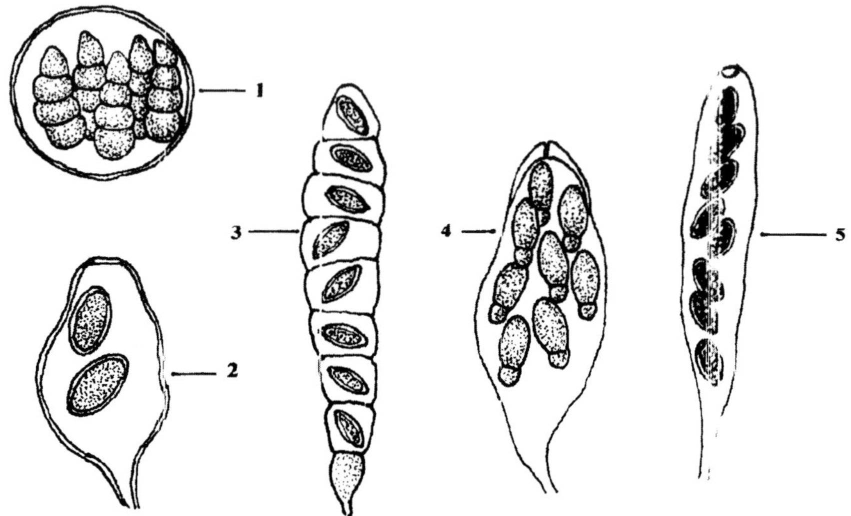

1.Globose 2.Broadly ovate with stalk 3.Septate 4.Clavate 5.Cylindrical

Fig. 34 : Various types of Asci

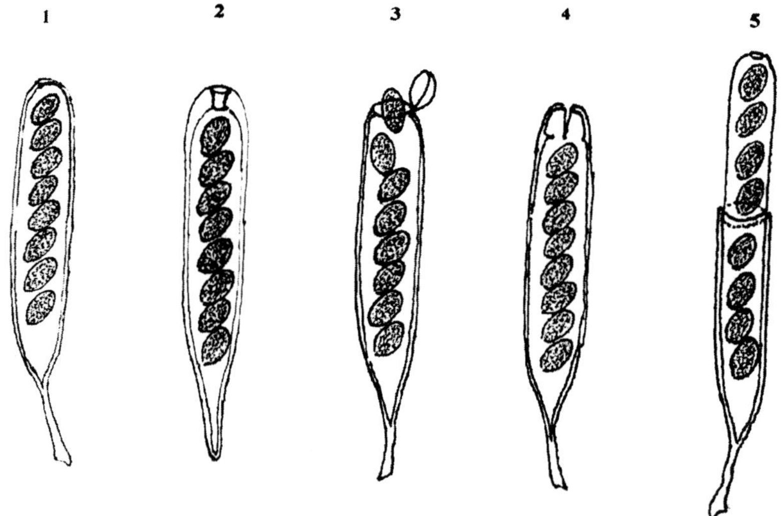

1,2,3 & 4. Unitunicate asci 5.Bitunicate ascus
1.No ascal opening 2.Ascal pore 3.Operculum 4.Split 5.Dehiscence of bitunicate ascus with pore at the tip of inner, expanding wall.

Fig. 35 : Various types of Ascal opening

blastospores on the host surface or may germinate by germ tubes that infect the host and produce mycelium. At the time of germination, the conidial nucleus divides and the resulting pair of nuclei migrates into the germ tube. As the hypha grows, conjugate division of the nuclei perpetuates the binucleate condition of the hyphal cells. The mycelium spreads between the cells and penetrates the tissues of the host. Hyphal strands eventually aggregate in the subcuticular region and break up into their component binucleate cells. These are the ascogenous cells. Karyogamy occurs within each ascogenous cell, which begin to elongate. At this stage the diploid nucleus divides mitotically. One daughter nucleus remains near the base of the cell, while the other moves toward the growing tip. Then a septum is formed between the two nuclei separating the cell into a basal stalk cell and an upper ascus mother cell, which is converted into an ascus. Meiotic and a subsequent mitotic division follows resulting in the formation of eight nuclei. Each nucleus along with a portion of cytoplasm is enveloped by a wall and develops into an ascospore. As the ascus mother cells enlarge, they exert pressure on the host cuticle and ultimately break through to form a compact surface layer of asci (hymenium) on the epidermis of the host. No fruiting bodies are formed, the asci remain naked and release the ascospores into the air by bursting at the tip. On release, the ascospores begin to bud and form numerous blastospores. The blastospores eventually produce germ tubes that infect the host **(Fig. 36)**

Subclass - Plectomycetidae (Black moulds, blue moulds and ringworms). The fungi classified under this Subclass produce globose to pyriform, evanescent, thin-walled asci. Asci do not form a regular hymenium, but are scattered at various levels, arising from ascogenous hyphae of various lengths ramifying through the ascocarp centrum. The ascospores are unicellular. The ascocarp is typically a cleistothecium, but in some cases are ostiolate. The ascocarp may not have a peridium and the asci are borne naked, or may have a definite pseudoparenchymatous wall. The Subclass consists of five Orders of which the Orders -Eurotiales and Microascales are of economic importance.

Order - Eurotiales. This Order includes seven Families of cleistothecial fungi with imperfect stages producing spores other than arthrospores. The cleistothecia vary from microscopic with one or two peridial layers, to small spheres with pseudoparenchymatous peridia bearing lines of dehiscence at maturity. The most important Family is **Eurotiaceae** in which the perfect stages of Form genera *Aspergillus* and *Penicillium* are included.

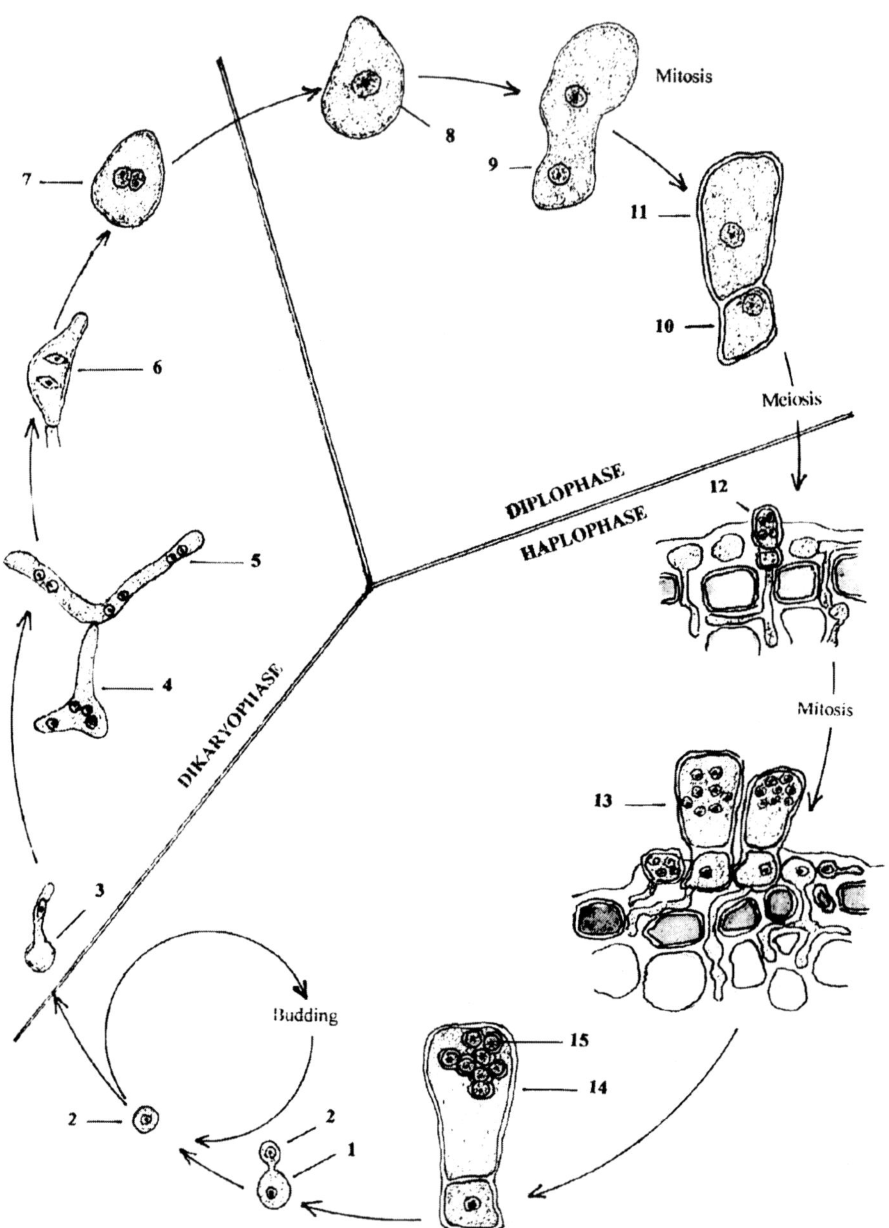

1.Budding ascospore 2.Sprout cell (Conidium) 3.Germinating sprout cell 4.Mycelium 5,6.Binucleate hyphal cells 7.Beginning of karyogamy 8.Karyogamy 9.Separation of the stalk cell 10.Stalk cell or foot cell 11.Developing ascus 12.Four nucleate stage of ascus 13.Young ascus (Eight nucleate stage) 14.Mature ascus 15.Ascospores

Fig. 36 : Life cycle of *Taphrina deformans*

Family - Eurotiaceae. The fungi classified under this Family are very widely distributed. The Family includes several Genera, most of these having well-formed cleistothecia with definite peridia. Many of them have their conidial stages in the Form genus *Aspergillus* and *Penicillium* of Subdivision - Deuteromycotina. So, this Family has often been called **'Aspergillaceae'**.

Form genus - *Aspergillus*. The Genus is very widely distributed, from the arctic to the tropics. The air everywhere contains the conidia of these organisms. The soil too, contains the spores of the fungi. *Aspergillus niger* group of fungi are commonly called as the **'black Aspergilli'** or **'black mould'**. The Aspergilli are capable of utilizing a large variety of substances for food, because of several enzymes they produce. *Aspergillus niger* and many other species are usually found on exposed foodstuffs and cause decay. During this process some species produce toxic substances called **'mycotoxins'**, the most important is **'aflatoxin'**. The organism *Aspergillus flavus* was found to produce a violent toxin in cattle and poultry feed and the toxin was hence named **'aflatoxin'** after the fungus. This name is now generally applied to many related toxins. The carcinogenic 'aflatoxin B1' may cause liver cancer in humans and animals. Several species of Aspergilli grow on leather and cloth fabrics and impart a musty odor when humid weather conditions prevail. *Aspergillus fumigatus, Aspergillus flavus, Aspergillus niger, Aspergillus terreus* and several other species are parasitic on animals, birds and humans and cause a disease complex of the lungs known as **'Aspergilloses'** (sing. - Aspergillosis), the symptoms of which closely resemble that of tuberculosis. Because of their great enzymatic activities, Aspergilli are employed in several industrial processes. Citric and gluconic acids are manufactured commercially by using *Aspergillus niger*. Several members of this Genus are used for the production of many other acids and various other organic chemicals. A number of enzyme preparations and antibiotics are also being produced from *Aspergillus* cultures.

The mycelium of the Aspergilli is well developed, profusely branched, septate, hyaline and the cells are multinucleate. The vigorously growing, young mycelium produces conidiophores in abundance, singly from the somatic hyphae. The hyphal cell that branches to give rise to the conidiophore is called the **'foot cell'**. The conidiophores are long, erect and terminate in a bulbous head called the **'vesicle'**. From the developing multinucleate vesicle, a large number of conidiogenous cells (mother cells) are produced over the entire surface, covering it completely. According to the species, one or two layers of such cells called **'sterigmata'** may be produced. When two layers, the primary and secondary layers are formed conidia arise from the secondary layer. The conidia bearing cells, whether primary or secondary are typical phialides. As the phialides reach maturity, they successively cut off daughter cells from their tips one below the other in chains, which become the conidia. The conidia are globose, unicellular and possess roughened walls. *Aspergillus* colonies appear black, brown, yellow, green and so on according to the color of the conidiophores and conidia, which cover the colonies. Sexual behavior of different species of *Aspergillus* varies considerably, from normal plasmogamy between two functional gametangia, to a complete absence of an antheridium and the development of the asci parhtenogenetically from the ascogonium alone. The cleistothecium develops as a single layer of cells around the sex organs and becomes a small, globose ascocarp with smooth walls. The asci are globose, ovoid or pear-shaped. They are evanescent and dissolve soon after ascospore formation leaving the ascospores free within the cleistothecium. The ascospores are shaped like pulley-wheels and the outside wall of the two halves is variously modified and sculptured. Usually eight ascospores are produced in each ascus. The ascospores give rise to germ tubes that develop into mycelium **(Fig. 37)**

Form genus - *Penicillium*. The Penicillia also are as common and as cosmopolitan as the Aspergilli. They are generally called as **'green moulds'** and **'blue moulds'** and are frequently found on *Citrus* and other fruits, on various foodstuffs, cheese, butter etc. The conidia of *Penicillium* are found everywhere. Some members are associated with animal and human diseases. They spoil silage and makes it unsuitable for cattle feed. Many species of *Penicillium* are capable of producing organic acids, such as citric, fumaric, oxalic, gluconic and gallic acids. Some are used for making cheese, antibiotics etc. *Penicillium digitatum* and *P. italicum* - common pathogens of *Citrus* fruits, cause green mould and blue mould respectively; *P. expansum* - causes decay of apples in storage. It also attacks and destroys leather articles and cloth fabrics; *P. roqueforti* and *P. camemberti* - are used for making Roquefort and Camembert cheese respectively; *P. notatum* and *P. chrysogenum* - are sources of the famous antibiotic - Penicillin; *P. griseofulvum* - produces an important antibiotic - Griseofulvin, which is very effective in the control of fungal skin diseases.

The mycelium of Penicillia produces simple, long, erect conidiophores that branch about two-thirds of the way to the tip in characteristic symmetrical or asymmetrical, broom-like fashion. The conidiophores commonly referred to as the **'brush'** is technically called **'Penicillium'**. The multiple branching of the conidiophore ends in a group of phialides that bear the long conidial chains. The conidia are globose to ovoid in shape. The enormous quantities of greenish, bluish or yellow conidia produced are chiefly responsible for the characteristic colony color of various species of *Penicillium*.

The conidia germinate into germ tubes from which mycelia develop. Sexual reproduction takes place by functional gametangia in some species, but in others antheridia no longer function but the ascogonial nuclei pair and pass into the ascogenous hyphae. From these hyphae asci develop within the cleistothecium in the usual manner. Karyogamy takes place in the ascus mother cells and the ascospores develop by meiosis and free cell formation. The asci are scattered at various levels within the cleistothecium. The ascocarp develops from somatic hyphae that envelop the sexual organs. The asci are globose or pear-shaped. The ascus wall dissolves soon after ascospore formation and the ascospores are released into the cleistothecium. The ascospores are pulley-wheel-shaped **(Fig. 38)**

Order - Microascales. Of the two Families classified under this Order, only Family - **Ophiostomataceae** is of economic importance, as many members of this Family cause serious plant diseases.

Family - Ophiostomataceae. The Aspergillaceae and related organisms produce closed fructifications called cleistothecia and the ascospores are released only by the disintegration of the peridium. This is known as **'cleistocarpy'** or **'angiocarpy'**. But the Family - Ophiostomataceae produces ostiolate fructifications called **'perithecia'** and the matured ascospores are extruded through this special opening or ostiole. This is known as **'gymnocarpy'**. The young fructifications or ascocarp is closed and globose in shape in the early stages. Later, the peridium differentiates into an outer, dark-colored, thick-walled layer and a more delicate inner layer. The more rapid growth and tangential expansion of the outer layer causes the apex of the peridium to arch upward. The hyphae then pull apart at the tip and a papilla-like ostiole is formed. The mouth of the ostiole is lined with short periphyses. The asci, which are formed inside the fructification, contain a compact group of eight ascospores in each ascus.

Genus - *Ceratocystis (Ophiostoma / Ceratostomella)*. The perithecia appear superficially or partially buried on the host surface. They have globose base and greatly elongated neck with shredded, feathery tip. The ascus wall gelatinizes early in the perithecial formation and the ascospores are extruded through the long neck of the perithecium and are embedded in mucus that forms a droplet at the ostiolar opening. Ascospores vary in shape from elongate to ovoid or crescent-shaped or hat-shaped. Different species exhibit entirely different conidial stages and some species produce two completely different types of conidia. Some species of *Ceratocystis* produce synnemata (sing. - Synnema). These fungi also produce conidia on short, simple conidiophores in chains one after the other and are held together by a drop of mucus

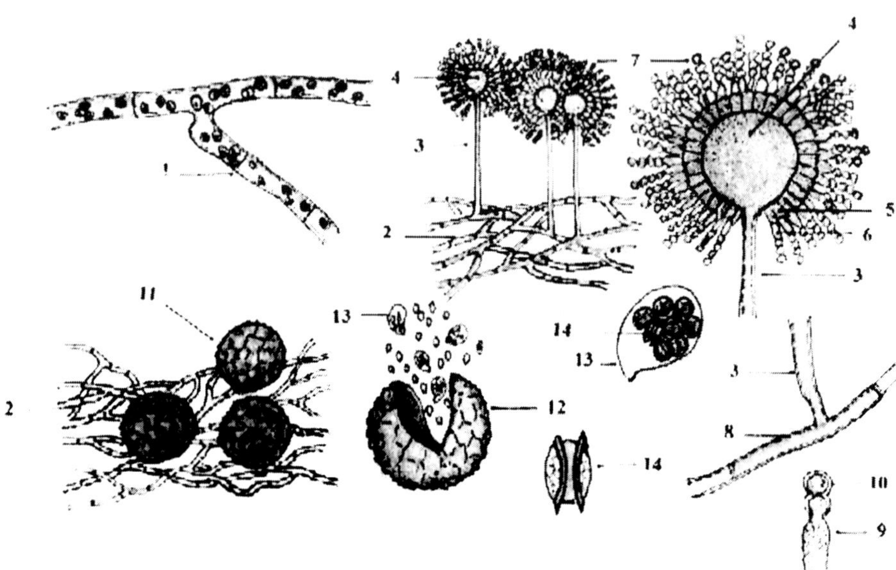

1.Multinucleate, septate hypha 2.Mycelium 3.Conidiophore 4.Vesicle 5.Primary sterigmata 6.Secondary sterigmata (Phialides) 7.Conidia 8.Foot cell 9.Phialide 10.Conidium 11.Cleistothecium 12.Asci and ascospores released from cleistothecium 13.Ascus 14.Ascospore

Fig. 37 : Form genus - *Aspergillus*

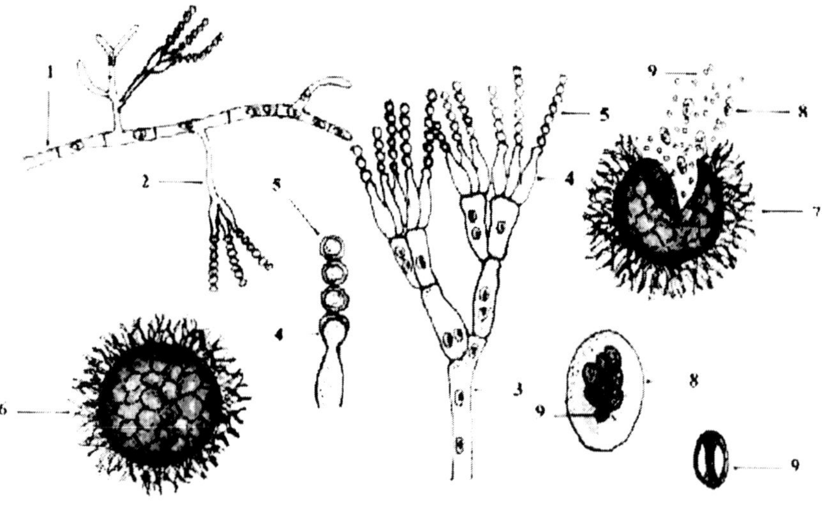

1.Somatic hyphae 2.Formation of conidiophore from somatic hypha 3.Conidiophore 4.Phialide 5.Conidial chain 6.Cleistothecium 7.Asci and ascospores released from the cleistothecium 8.Ascus 9.Ascospores.

Fig. 38 : Form genus - *Penicillium*

(e-g) *Ceratocystis fagacearum* - causes oak wilt; *C. ulmi* (=*Graphium ulmi* of Deuteromycotina) - causes 'Dutch elm' disease; *C. fimbriata* - causes sweet potato black rot; *C. pilifera* - causes 'blue stain' in lumber that greatly reduces their commercial value; *C. paradoxa* (=*Thielaviopsis paradoxa* of Deuteromycotina) - causes stem bleeding disease of coconut and 'pine apple' disease or sett rot of sugarcane.

Subclass - Hymenoascomycetidae

The fungi that are classified under this Subclass (the powdery mildews and the black mildews) produce unitunicate asci in hymenia, usually borne as basal layers in ascocarps, either cleistothecia, perithecia or apothecia. The structure and function of the ascal apex and the ascocarp centrum are quite important in classifying the fungi. The **'Pyrenomycetes'**, as they are usually called, produce typically club-shaped or cylindrical asci in a hymenium that lines the base and often the sides of the inner wall of the ascocarp. The asci are usually persistent, but may be evanescent. The ascocarp may or may not be embedded in a stroma, but in either case a definite perithecial wall (peridium) of its own is always present.

Four types of centra of the pyrenomycetous Hymenoascomycetidae are met with. They are the ***Phyllactinia* type centrum**, the ***Xylaria* type centrum**, the ***Diaporthe* type centrum** and the ***Nectria* type centrum**.

Subclass - Hymenoascomycetidae I

Pyrenomycetes I

The Phyllactinia Type Centrum. This type of centrum is characteristic of the Orders -Erysiphales and Meliolales. The centrum is characterized by a pseudoparenchymatous mass of cells filling the cleistocarpous or perithecial ascoma and this cellular mass disintegrates as the asci develop and cover the center of the matured ascocarp. The mycelium of both these Orders are mostly superficial and obtain nourishment from the host cells through haustoria. All species of these Orders are obligate parasites on vascular plants.

Order - Erysiphales. The fungi of this Order have a completely closed ascocarp (cleistothecium) and the asci form a characteristic basal layer. The asci are somewhat globose or pyriform and lack a definite apical structure to release the ascospores, but explode to release the ascospores. The mycelium is hyaline, but may be light-brown in color. The Order has a single Family - **Erysiphaceae.**

Family - Erysiphaceae (The powdery mildews). The Erysiphaceae produce enormous number of conidia on the surface of the host. These appear to the naked eye as a white, powdery coating and hence the group of diseases caused by them is commonly known as **'powdery mildews'**. In their parasitism, some of the Erysiphaceae are non-host specific as seen in *Erysiphe polygoni*, which is known to infect 352 host species. On the other hand, a few are host specific as is seen in *Podosphaera leucotricha*, which attacks only apple. Some species of powdery mildews consist of a number of physiological races, each with a limited host range.

The mycelium of the Erysiphaceae is completely superficial, except in *Leveillula taurica* and *Phyllactinia corylia*. It consists of a network of hyaline hyphae, abundantly present on the epidermis of the infected parts of the host and securely anchored to it by numerous haustoria that penetrate into the epidermal cells and obtain nourishment from the host protoplasm. Different members of the Family produce several types of haustoria **(Fig. 5)**. In *Leveillula taurica*, the hyphae penetrate into the leaf through the stomata and spread between the mesophyll cells. In *Phyllactinia corylia*, most of the hyphae are superficial, but these do not produce haustoria. However, some special hyphal branches enter into the host through the stomata and obtain nourishment from the mesophyll cells by sending in haustoria.

A few days after infection, the somatic hyphae produce a large number of long, hyaline, erect conidiophores. A mother cell at the apex of each conidiophore produces conidia. In some species, each newly formed conidium is abstricted before the next one matures. In other species, the conidia cling together in chains. In *Leveillula*, the conidiophores grow out of the stomata from the endophytic mycelium and produce conidia on the surface of the leaves. These conidiophores may be branched. In *Leveillula* and *Phyllactinia*, the conidia fall off, as soon as they are formed and no conidial chains are formed. The conidia of the Erysiphaceae are hyaline and single-celled. The shape of conidia varies according to the species, but generally they are ovoid or cylindrical with rounded edges. The conidia of *Phyllactinia* are clavate **(Fig. 39)**. Conidia of the Erysiphaceae do not require free water for germination and in some cases the conidia are capable of germination at very low or even at 0 % humidity level.

When conidial production slows down and finally ceases, young cleistothecia begin to appear on the white mycelium. They appear white in the beginning and the color gradually changes to orange, reddish-brown and finally black when fully matured. Some of the species are homothallic, while others are heterothallic. The Erysiphaceae are divided into three groups depending on whether the dikaryon originates by gametangy, somatogamy or apomixis.

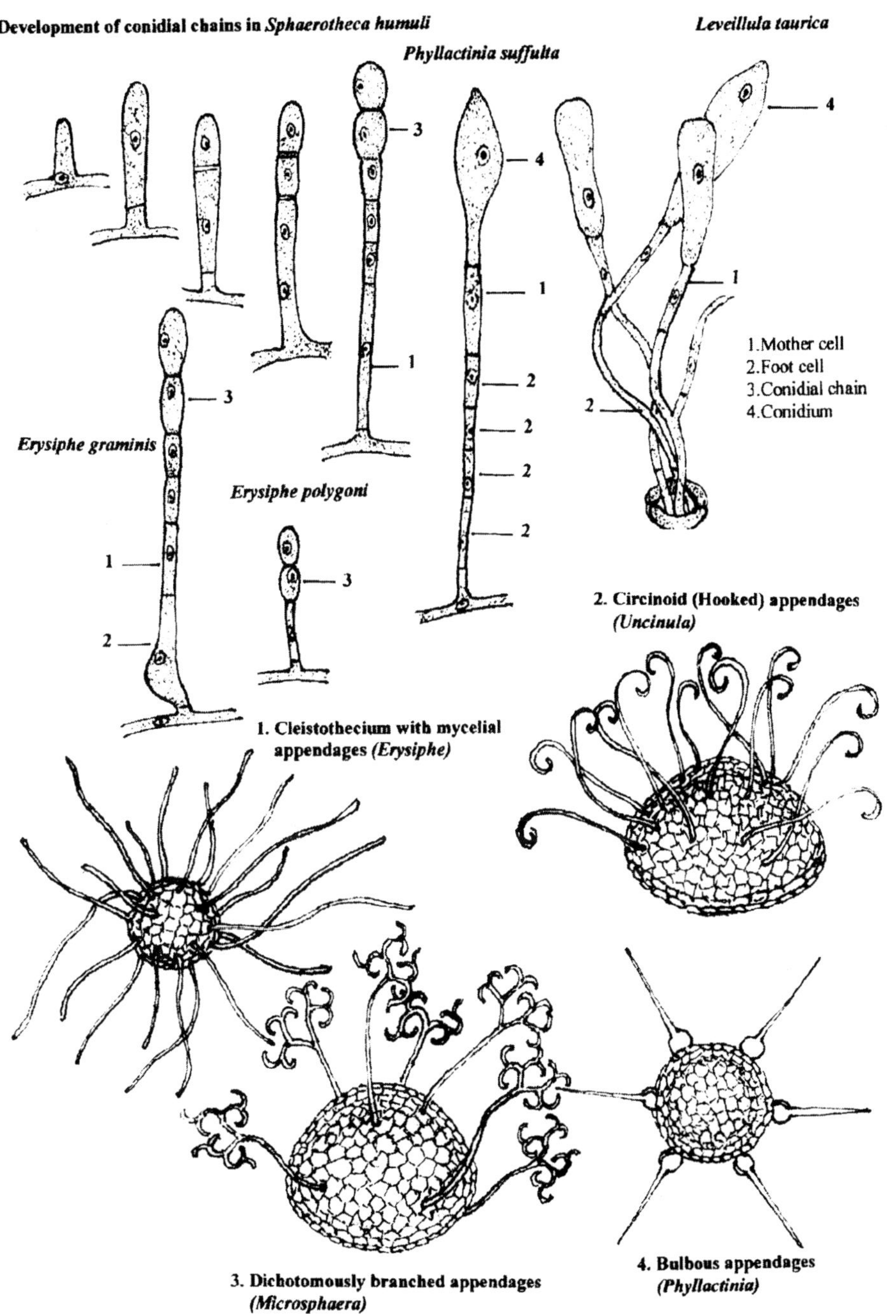

Fig. 39 : Conidiophore and conidial development in Erysiphaceae

In the species exhibiting normal gametangy, the male copulation branch consists of a stalk cell and a multinucleate antheridium. Similarly, the female copulation branch also has a uninucleate ascogenous cell and a stalk cell. Through an open union, the male nucleus migrates into the ascogonium. This type of union occurs in *Erysiphe martii, E. nitida, E. cichoracearum, Sphaerotheca humuli, S. fuliginea* and in many other related species. Numerous copulations may occur somatogamously between ordinary vegetative cells within a single developing fructification **(Fig. 41).**

In the forms exhibiting lower levels of sexuality, the male nucleus does not enter the ascogonium, but disintegrates within the antheridium and the ascogonium develops parthenogenetically. This type of development occurs in *Erysiphe lamprocarpa, Sphaerotheca mors-uvae, Microsphaera alphitoides, Uncinula salicis, Phyllactinia suffulta* and in many other species **(Fig. 41).**

The sexual apparatus of all types becomes enclosed with peripheral hyphae. These surrounding layers of hyphae become thickened and ultimately the fructification is differentiated into an outer brownish peridium and an inner hyaline ground tissue.

In the first group, which is represented by *Sphaerotheca fuliginea*, the ascogonium itself develops into an ascus. The dikaryon divides once and the four daughter nuclei become separated by cross walls into two uninucleate cells and one dikaryotic cell. The dikaryotic cell then becomes an eight-spored ascus **(Fig. 41).**

In the second group as exemplified by *Phyllactinia suffulta*, the binucleate subterminal cell of the ascogonium forms ascogenous hyphae from which asci are formed later. The ascogonium first becomes 8-nucleated by two nuclear divisions and then produces non-septate ascogenous hyphae in the ground tissue. The nuclei become arranged in a row and again divide. Finally side branches of these hyphae are cut off by cross walls. Each side branch forms a uninucleate terminal cell and binucleate intercalary cells, the uppermost of which produces the ascus. Ten to twenty asci are produced in this manner within a single ascocarp **(Fig. 41).**

In the third group as exemplified by *Erysiphe graminis*, no differentiated copulation branches are produced. Copulation may take place between intertwining lateral hyphal branches, which rise above the substratum.

The undifferentiated initial hyphae produce throughout their entire length dark-colored hyphae, which are multinucleate and form a thick ball of intertwined threads. The degeneration of the cells in the center of this ball produces a cavity, at the base of which the ascogenous hyphae develop into asci. A layer of about 20 asci of the hook-type is produced inside each ascocarp **(Fig. 41).**

Copulation between undifferentiated hyphal branches
(Eryslphe graminis)

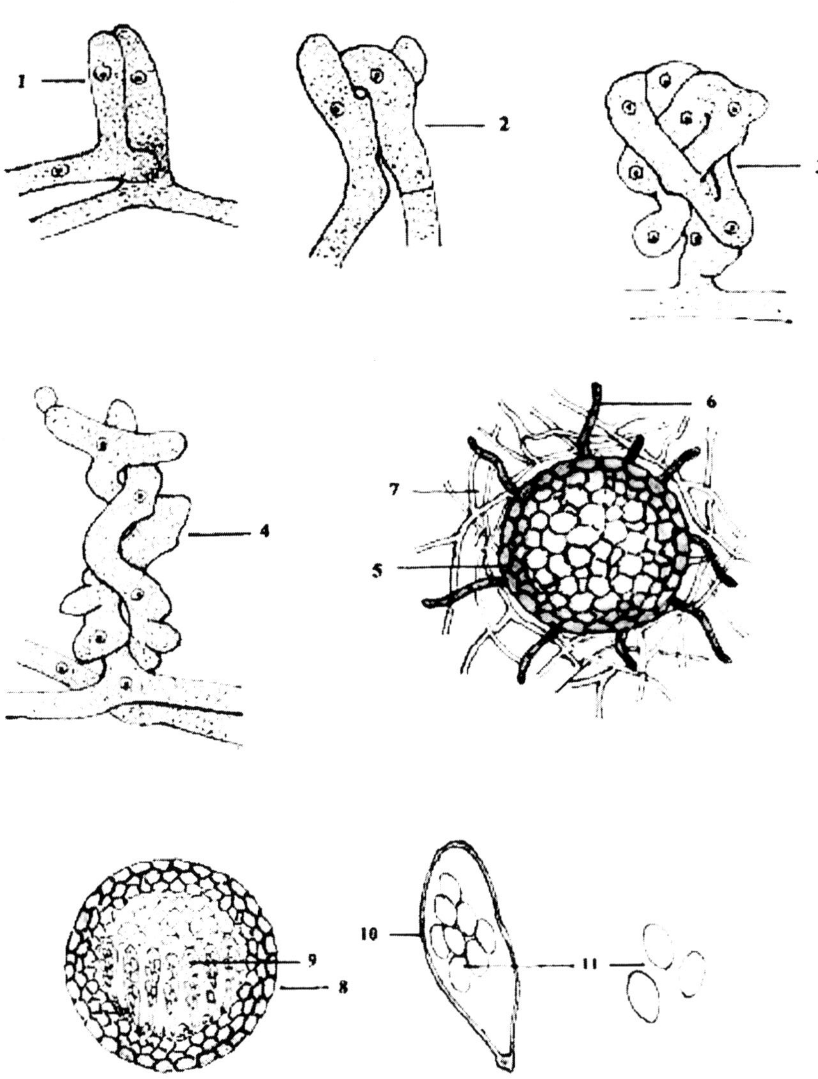

1 Undifferentiated lateral hyphal branches 2,3 Intertwinning hyphal branches rise above the substratum 4 Development of ascogenous hyphae and asci of the hook type 5 Cleistothecium 6 Appendages 7 Peripheral hyphae 8.Cleistothecium enclosing the asci 9.Asci 10 Ascus with ascospores 11.Ascospores

Fig. 40 : Types of Cleistothecia

The matured cleistothecia of *Sphaerotheca, Erysiphe* and *Leveillula* are usually symmetrical and spherical in shape and are similar in structure. But, the fructifications of *Microsphaera* and *Uncinula* are dorsiventral, as the basal cells are wider and thin-walled, while those toward the apex are narrower and thick-walled. The ascospores are released by the irregular rupture or crumbling of the peridium **(Fig.40)**. The cleistothecia aid the dissemination of the spores by means of hook-like appendages produced by the outer cells of the peridium. These appendages are so characteristic that they are useful for taxonomic identification. *Erysiphe, Sphaerotheca* and *Leveillula* have mycelioid appendages, which resemble somatic hyphae. In *Phyllactinia* the appendages are rigid, spear-like, with a bulbous base and pointed tips. In *Uncinula* and *Pleochaeta* the appendages are rigid with curled tips, while in *Microsphaera* and *Podosphaera* the appendages are rigid with dichotomously branched tips **(Fig. 40)**. The cleistothecium of *Phyllactinia* is extruded actively from the extramatrical hyphae by means of vesicles with thin-walled under surfaces and thick-walled upper surfaces that are inserted at the base of the pointed appendages. Dehydration of the vesicle modifies its shape thereby pushing the cleistothecium upward on the stalk and separating it from the surrounding mycelium. The attachment of the cleistothecium to a new surface is by special tufts of short hyphae, which form a drop of slime at its apex. The drying of the droplet firmly cements the cleistothecium to the new surface.

Genus - *Sphaerotheca*. Imperfect stage *Acrosporium (Oidium);* one ascus in each cleistothecium; appendages mycelioid, indefinite

(e-g) *Sphaerotheca macularis* - causes powdery mildew of strawberry; *S. mors-uvae* - causes powdery mildew of gooseberry and currant; *S. pannosa* - causes powdery mildew of rose.

Genus - *Podosphaera*. Imperfect stage *Acrosporium*; one ascus in each cleistothecium; appendages definite, their tips dichotomously branched

(e-g) *Podosphaera oxyacanthae* - causes powdery mildew of apricot, cherry, peach and plum; *P. leucotricha* - causes powdery mildew of apple.

Genus - *Erysiphe*. Imperfect stage *Acrosporium*; several asci found in each cleistothecium; appendages mycelioid, indefinite

(e-g) *Erysiphe cichoracearum* - causes powdery mildew of begonia, chrysanthemum, cucurbits, dahlia, zinnia etc. *E. polygoni* - causes powdery mildew of legumes, beets and crucifers; *E. graminis* f.sp. *tritici* - causes powdery mildew of cereals and grasses.

Gametangy and formation of the ascus
(Sphaerotheca)

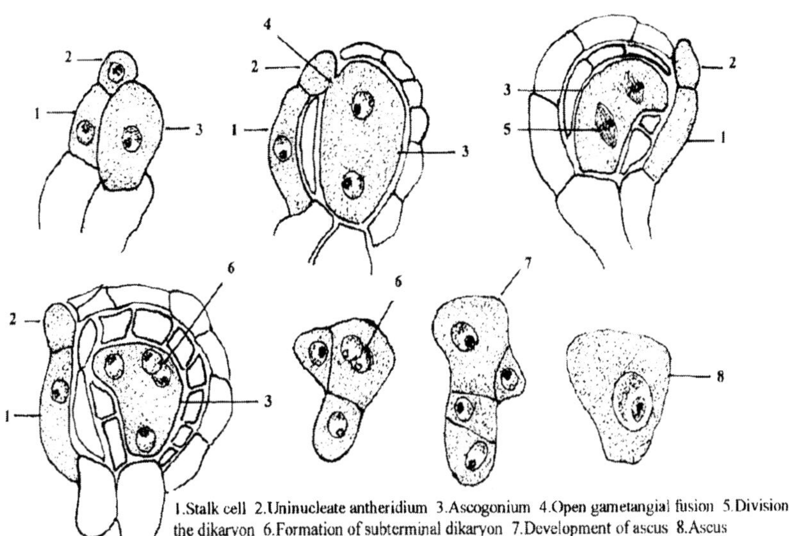

1.Stalk cell 2.Uninucleate antheridium 3.Ascogonium 4.Open gametangial fusion 5.Division of the dikaryon 6.Formation of subterminal dikaryon 7.Development of ascus 8.Ascus

Apomistic development of ascus
(Phyllactinia)

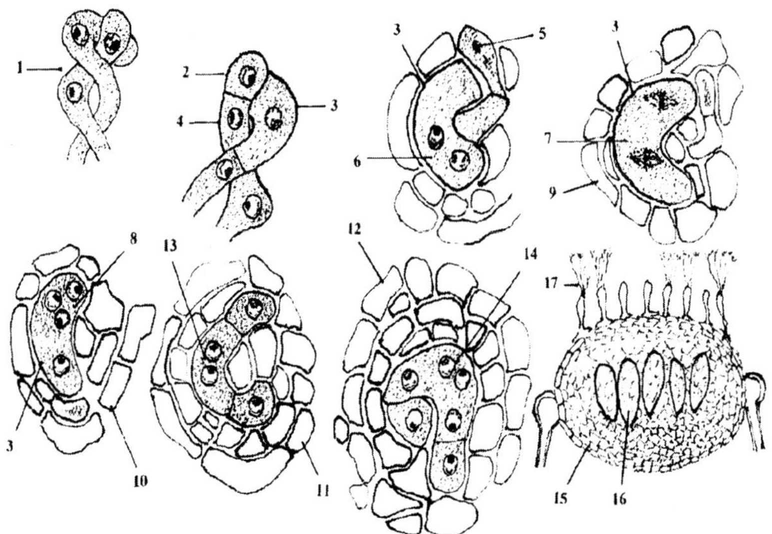

1.Young hyphal copulation branches not yet septate 2.Antheridium 3.Ascogonium 4.Stalk 5.Male nucleus degenerated 6,7,8. Development of the ascogonium into the multinucleate stage 9,10,11,12. The copulation branches are covered by vegetative hyphae 13.The ascogonium becomes septate, forms a subterminal dikaryon 14.The binucleate subterminal cell divides further 15.Mature cleistothecium 16.Asci 17.Special tufts of short hyphae which produce slime

Fig. 41 : Sexual reproduction in Erysiphaceae

Genus - *Microsphaera*. Imperfect stage *Acrosporium*; several asci in each cleistothecium; appendages definite, rigid; tips of appendages dichotomously branched

(e-g) *Microsphaera alni* - causes powdery mildew of lilac, many shade trees and woody ornamentals.

Genus - *Uncinula*. Imperfect stage *Acrosporium*; several asci in each cleistothecium; appendages definite, rigid; tips of appendages coiled

(e-g) *Uncinula necator* - causes powdery mildew of grapevine.

Genus - *Phyllactinia*. Imperfect stage *Ovulariopsis*; several asci in each cleistothecium; appendages with bulbous base; tips of appendages pointed

(e-g) *Phyllactinia suffulta(corylea)* - causes powdery mildew of shade and forest trees; *Phyllactinia guttata* - causes powdery mildew of mulberry.

Genus - *Leveillula*. Mycelium endophytic; several asci in each cleistothecium; appendages mycelioid, indefinite

(e-g) *Leveillula taurica* - causes powdery mildew of mulberry and tomato.

Subclass - Hymenoascomycetidae II

Pyrenomycetes II

The Xylaria Type Centrum. In the Xylaria type centrum, the ascogonia are formed either free on the mycelium or from the somatic hyphae within a stroma. Paraphyses grow upward from the base and inward from the sides of the ascocarp wall and the pressure exerted by them causes the ascocarp to expand and create a central cavity. The growth of the paraphyses in the upper part of the ascocarp toward the apex creates an ostiole. The club-shaped or cylindrical asci arising from the ascogonium from the base of the ascocarp grow among the paraphyses and form ascospores. The Orders - Xylariales and Clavicepitales are of economic importance.

Order - Xylariales (Sphaeriales). The Xylariales include all Pyrenomycetes with dark, leathery or carbonous, globose or pear-shaped, ostiolate or astomous ascocarps. Unitunicate asci mingled with paraphyses form a basal hymenial layer within the ascocarp. The hyphae are generally composed of uninucleate or multinucleate, elongated cells. Many of the Xylariales are saprobic, while a few are plant parasites. Only the Family - **Polystigmataceae** is economically important.

Family - Polystigmataceae. Stroma usually absent, but may be present in some cases; stroma if present are intramatrical; perithecia sunken within the stroma; the asci are persistent; asci without germ pores or slits. Most of the species of this Family are parasitic on vascular plants. When the spores germinate, they usually form appressoria, which are flattened hyphal structures that adhere on to the host surface, from which infection pegs grow and penetrate the host tissue. *Glomerella* is the only important Genus of this Family.

Genus - *Glomerella*. The perithecia are produced either in groups in a stroma or singly. The stroma may not be well-developed or may be lacking. The perithecia are beaked. The ostioles are well-developed and inlaid with periphyses. Paraphyses are present in some cases in large numbers. The asci possess short stalks and the ascospores are hyaline, uninucleate, oblong or elliptic or curved with pointed tips. All species of *Glomerella* have the same type of imperfect stage, *Colletotrichum* or *Gloeosporium* and produce hyaline, ovoid, cylindrical or dumb bell-shaped conidia in acervuli. The acervuli of *Colletotrichum* bear setae. There are many species of *Colletotrichum*, the perfect stage of which have not been discovered so far.

In the case of *Glomerella cingulata*, which is heterothallic, the perithecium arises from two uninucleate mycelial branches that develop into an inner and an outer coil. The outer coil develops into a peridium, while the inner coil becomes the ascogonium. The whole structure is a protoperithecium that develops into a perithecium after plasmogamy. When conidia of the opposite mating type germinate close to a protoperithecium, a conidial germ tube grows toward it, enters in and applies on to the ascogonial coil. The walls between them dissolve and plasmogamy occurs. One or more of the ascogonial cells become binucleate, but only one, usually the tip cell gives rise to the ascogenous hyphae that develop into asci from croziers along with paraphyses

(e-g) *Glomerella cingulata* (=*Colletotrichum gloeosporioides*, *Colletotrichum coffeanum* or *Gloeosporium cingulatum* of Deuteromycotina) - causes bitter rot of apples and anthracnose of several plants; *G. lindemuthiana* (=*Colletotrichum lindemuthianum* of Deuteromycotina) - causes bean anthracnose; *G. graminicola* (=*Colletotrichum graminicola* of Deuteromycotina) - infects maize; *Glomerella (Physalospora) tucumanensis* (=*Colletotrichum falcatum*) - causes red rot of sugarcane.

Order - Clavicepitales. The Clavicepitales produce perithecia within a well-developed stroma composed entirely of fungal tissue. The long, narrowly cylindrical ascus has a thick tip with a long cylindrical pore through which the ascospores escape. Paraphyses line the lateral inner walls and form part of the perithecial wall. Paraphyses are not present at the base of the perithecium. The ascospores are thread-like and in many species break into fragments after they are released, with each fragment serving as an individual spore.

All the species in this Order are grouped in the single Family - **Clavicepitaceae.**

Family - Clavicepitaceae. The Genus - *Claviceps* is the only important one in this Family.

Genus - *Claviceps*

(e-g) *Claviceps purpurea* - causes ergot of rye, oats, barley, wheat and many grasses; *C. microcephala* - causes ergot of pearl millet, many other lesser millets and grasses.

Life cycle of *Claviceps purpurea*. The thread-like ascospores produced by this pathogen are forcibly discharged from the perithecia. The wind-borne ascospores fall on the flowers of susceptible hosts, such as rye or other grasses, germinate, send germ tubes into the ovary and cause infection. The mycelium that develops destroys the ovary tissues and replaces them with a soft, white, cottony, mycelial mat that soon becomes covered by layers of short conidiophores, bearing minute, oval conidia at their tips. The conidia are mixed with a sticky, sweet, nectar-like secretion. Attracted by this nectar, insects visit the infected flowers and disseminate the conidia to uninfected flowers. The mycelial mat, which has produced the conidiophores continues to develop, begins to harden and is eventually transformed into a hard pink or purplish pseudoparenchymatous sclerotium. In shape the sclerotium resembles the grain of the host whose position it occupies, but it exceeds the host grain in length. The sclerotia are called **'secale cornutum'** or **'ergot'** in pharmacology. Thus, the matured heads bear sclerotia together with grains on the spikelets. The uninfected ovaries develop normally into grains and the infected ones are destroyed and replaced by the sclerotia of the fungus.

During the harvesting operations, many sclerotia are knocked off the spikelets and fall to the ground where they overwinter. The following spring, the sclerotia germinate and form several long-stalked, dark purple stromata with globose head. Within these stromatal heads and just below their surfaces, a number of minute cavities surrounded by the pseudoparenchymatous stromatic tissue arise. Each cavity contains a single, multinucleate ascogonium at the base of which one or more multinucleate antheridia are formed.

Plasmogamy takes place between one of the antheridia and the ascogonium. The male nuclei then migrate into the ascogonium. While the asci are forming as in other Ascomycetes, the perithecial walls develop around the sexual organs within the stromatal heads, producing definite perithecia that open out on the surface of the stroma through a long, neck-like ostiole. Each matured perithecium bears several elongated, cylindrical asci, each containing eight, thread-like ascospores **(Fig. 42)**

Ergot contains a number of powerful alkaloids, such as **'ergotamine'**, **'ergometrin'** and **'ergonovin'**, which are used medicinally to induce labor and prevent post partum haemorrhage during childbirth. The well-known hallucinogen, **'LSD'** is synthesized from **lysergic acid**, which occurs in ergot sclerotia. However, ergot contains a number of poisonous alkaloids, which are highly poisonous to humans and animals. Cattle are often poisoned by grazing on grasses that carry the sclerotia of the fungus or by consuming food contaminated with the ergot. Their legs, hoofs and tails become gangrenous and cows may abort their calves. This disease of animals is known as **'Ergotism'**.

Subclass - Hymenoascomycetidae III
Pyrenomycetes III
The Diaporthe and Nectria Type Centra

The Diaporthe Type Centrum. The centrum is pseudoparenchymatous at first and is destroyed partially or completely as the asci grow and extend into it. The asci are either evanescent at maturity, have short stalks that gelatinize in some Families or remain intact and expel their spores forcibly. The ascal apices are thickened with a narrow central canal through which the ascospores are expelled from persistent asci. There is only a single Order - **Diaporthales** with this type of centrum and only the Family - **Diaporthaceae** is of some economical importance.

Family - Diaporthaceae. The Diaporthaceae form their perithecia immersed in a stroma with only their long beaks protruding. The ascus has an apical refractive ring. The ascospores are hyaline and two-celled. The largest Genus in this Family is *Diaporthe*.

Genus - *Diaporthe*. The imperfect stages of all *Diaporthe* species belong to the Form genus - *Phomopsis* of Form class - Deuteromycetes. **(e-g)** *Diaporthe vexans* - causes fruit rot of egg plant; *D. phaseolorum* - attacks lima beans, soybean and a few other plants.

Genus - *Magnaporthe*. The perithecia are similar to that of *Diaporthe*. The ascospores are three-septate and the two middle cells pigmented. A toxin, **'pyricularin'** is produced by this fungus.

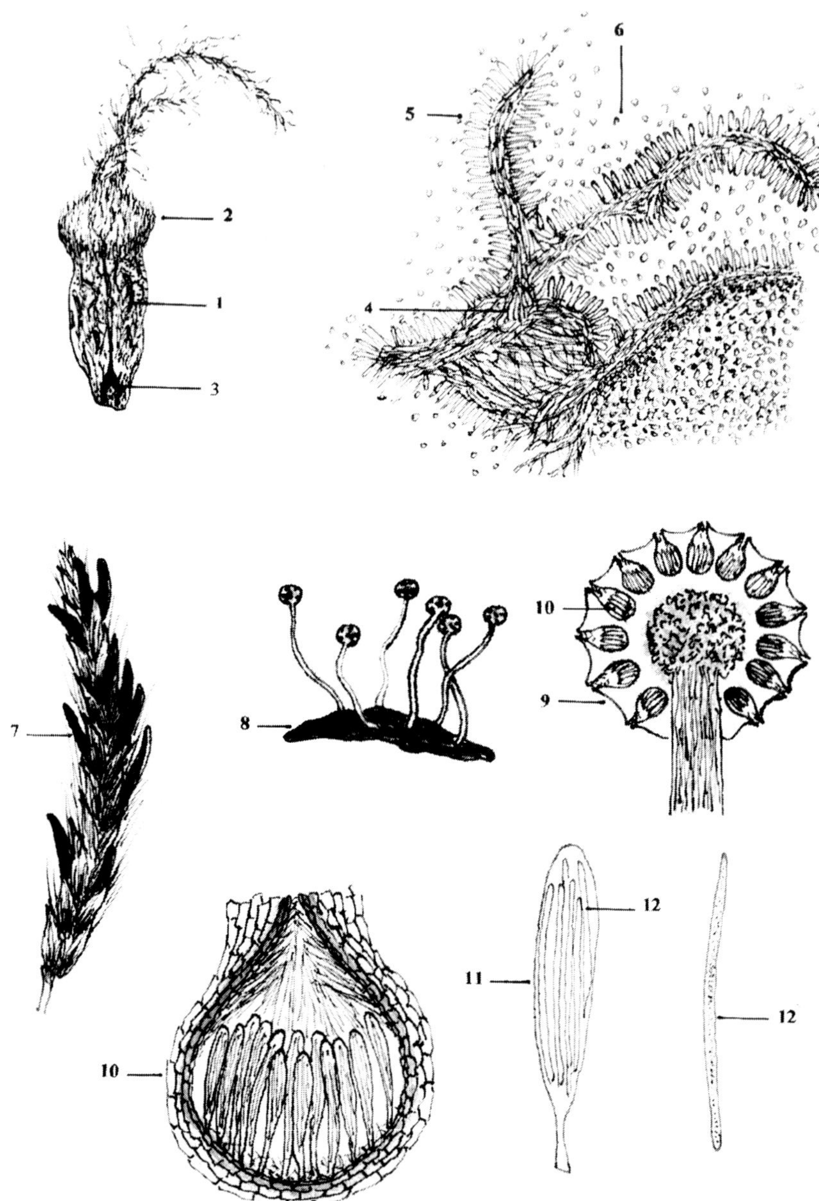

1.Infected young rye ovary covered except at the top with yellowish-white, branched, conidia producing mycelium 2.Hairs of the ovary at the apex 3.The primordium which later becomes the sclerotium 4.The enlarging pseudomorph covered with conidia producing hymenium 5.Conidiophore 6.Conidia 7.Sclerotia on the head of rye 8.Germinating sclerotium with perithecia producing heads 9.Perithecial head showing position of perithecia 10.Perithecium with asci 11.Ascus with ascospores 12. Asospore.

Fig. 42 : Life cycle of *Claviceps purpurea*

(e-g) *Magnaporthe grisea*. The imperfect stage of this belongs to Form genus - *Pyricularia oryzae* of Form class Deuteromycetes, Form subclass Hyphomycetes, Form order Moniliales, Form family Moniliaceae. This pathogen causes blast disease of rice.

Genus - *Leucostoma (Valsa)*. The fungus causes cankers in peach and other stone fruits, besides many other fruit, shade and forest trees. It is most commonly found in its anamorph stage, *Cytospora*. Small, pimple-like pycnidia are formed on the dead bark. The pycnidial stage *(Cytospora)* is mostly responsible for causing the cankers. The pycnidia consist of many connecting cavities with a single opening through which the spores are exuded in a gelatinous matrix. Perithecia and ascospores are not common. The perithecia are formed in a stroma side by side. The ascospores are hyaline and single-celled **(Fig. 43)**

(e-g) *Leucostoma(Valsa) eugeniae* - causes sudden death of clove

The Nectria Type Centrum. Here the centrum is parenchymatous. It is characterized by apical paraphyses originating from the perithecial apex just below the periphyses, growing down as a palisade layer and finally disintegrating as the asci grow up among them. The Nectria centrum is found in only the single Order - **Hypocreales** of which the Families - **Nectriaceae** and **Hypocreaceae** are of some economic importance. Several species of Hypocreales distributed over several Genera have a *Fusarium* type of imperfect stage.

Family - Nectriaceae. The Nectriaceae include species that produce their perithecia superficially, either on a well-developed stroma or without a stroma. The conidial stages resemble *Fusarium*, *Verticillium*, *Cephalosporium* etc. in the Deuteromycetes.

Genus - *Nectria*. The Genus - *Nectria* produces its perithecia on the surface of a cushion-shaped stroma. The ascocarps are brightly colored. The ascospores are generally two-celled, hyaline and often boat-shaped. The genus comprises of some important parasites of trees **(Fig. 44)**

(e-g) *Nectria cinnabarina* - attacks fruit trees, as well as shade trees and causes canker disease; *N. galligena* - causes apple and pear canker.

Genus - *Gibberella* **(e-g)** *Gibberella zeae* (= *Fusarium graminearum* or *F. roseum*) - causes stalk rot and red ear rot of maize; *G. xylarioides* - attacks coffee plants; *G. moniliforme (fujikuroi)* (= *Fusarium moniliforme*) - causes seedling disease of rice. This fungus is the source of **'gibberellic acid'**, which is extensively used as a growth promoting substance to promote flowering, growth and seed germination and to induce formation of seedless fruits.

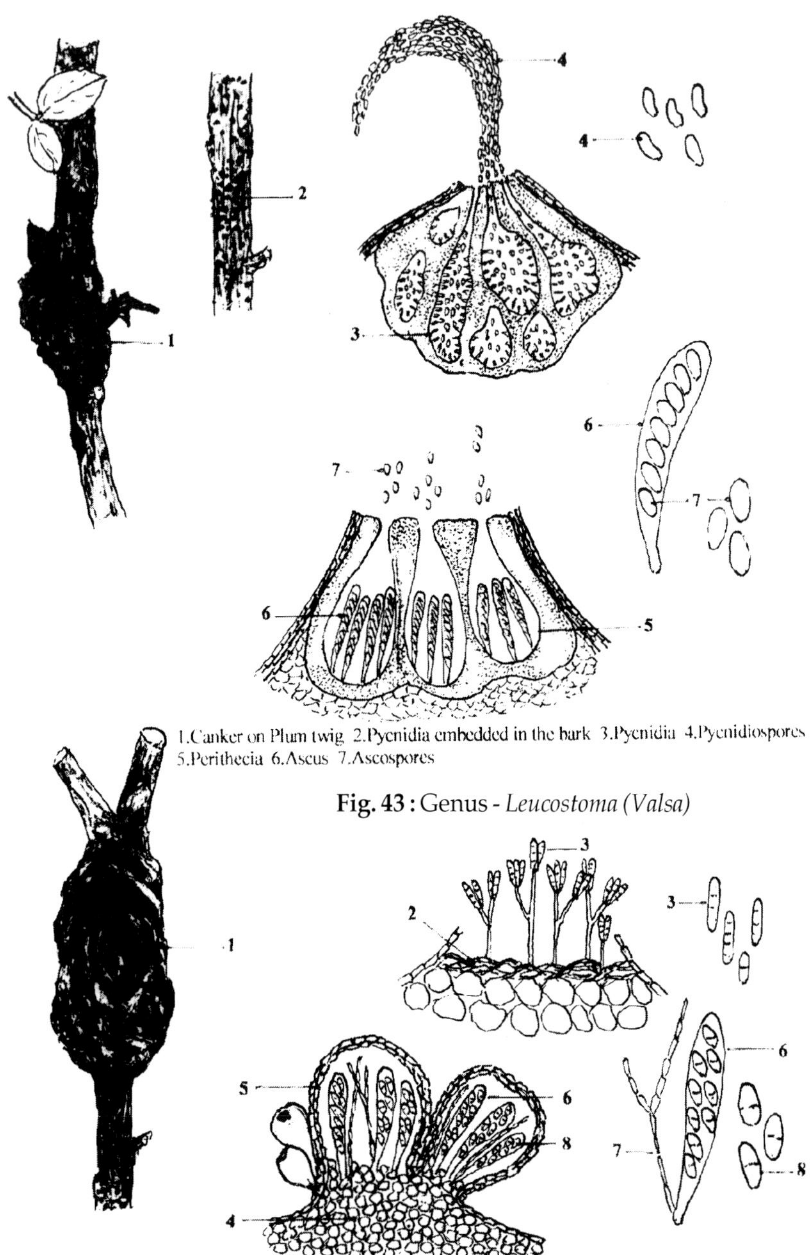

1.Canker on Plum twig 2.Pycnidia embedded in the bark 3.Pycnidia 4.Pycnidiospores 5.Perithecia 6.Ascus 7.Ascospores

Fig. 43 : Genus - *Leucostoma (Valsa)*

1.Canker caused by *Nectria galligena* on Apple branch 2.Sporodochium on canker 3.Conidia 4. Stroma on canker 5.Perithecia 6.Ascus 7.Paraphysis 8.Ascospores.

Fig. 44 : Genus - *Nectria*

Family - Hypocreaceae. The Hypocreaceae are stromatic with their perithecium immersed completely in a stroma consisting of hyphal elements. *Hypocrea* is the only important Genus in the Family.

Genus - *Hypocrea*. The asci in this Genus are narrowly cylindrical, containing eight, two-celled ascospores in each ascus and the spores are constricted at the septum and break at maturity thus, the ascus appears to have sixteen ascospores

(e-g) *Hypocrea rufa* - the culture of this fungus produces a conidial stage of *Trichoderma viride*; *Hypocrea citrina* - this fungus has a *Cephalosporium* imperfect stage, which is a Form genus in the Moniliaceae.

Subclass - Hymenoascomycetidae IV

Discomycetes. Discomycetes are Ascomycetes that produce their asci in apothecium, which is an open ascocarp. However, the truffles of the Order - Tuberales produce completely closed ascocarps. In the Discomycetes, the expanding hymenium ruptures the upper tissues of the developing ascocarp and forms an open apothecium instead of forming an ostiole as in the perithecium.

The Discomycetes include the cup fungi, the earth tongues, the morals and the truffles. Most of them can be recognized by their cup- or disc-shaped fruiting bodies that are produced on dead organic matter. The apothecia of some species are brilliantly colored red, yellow or orange, while some others are brown or black in color. Besides the typical cups or discs, there are sponges, bells, saddles, tongues and brain-like fruiting bodies. But the one common characteristic in these different kinds of apothecia is that they are open. They bear their asci on the surface or in large, open cavities and they puff their spores in clouds, in some cases with a hissing noise.

An '**apothecium**' consists of three parts: the hymenium, the hypothecium and the excipulum **(Fig. 33)**. The '**hymenium**' is the layer of asci that lines the surface or hollow of the disc or cup or other such structures. It is made up of club-shaped or cylindrical asci and paraphyses among them. In some apothecia the tips of the paraphyses may unite above the asci and form a layer called the '**epithecium**'. The '**hypothecium**' (pl. - hypothecia) is a thin layer of interwoven hyphae immediately below the hymenium. The fleshy part of the ascocarp, which is the apothecium proper is called the '**excipulum**' (pl. - excipula).

In classifying the fungi, more importance is given to the type of ascus and its dehiscence. The Discomycetes that form their ascocarps above the ground (epigean) are divided into two major groups depending upon their method of spore discharge. In the first group, **the inoperculate Discomycetes**, the asci release their spores through an apical pore. In the second group, **the operculate Discomycetes**, the asci release their spores through an operculum or a longitudinal slit. In the epigean inoperculate Discomycetes, Order - **Helotiales** is of economic importance

Order - Helotiales. The Helotiales have either cup- or disc-shaped apothecia with asci only slightly thickened at the apex and with ascospores that are round, elliptical or elongated, but rarely thread-like. Many of the Helotiales are saprobes, while a few are parasites of plants.

Family - Sclerotiniaceae. All the fungi of this Family are inoperculate. Many fungi classified under this Family live parasitically on plants. Apothecial initials arise from stromata or sclerotia. The apothecia are of medium size, generally brown and are most often borne on long stalks. The ascospores are mostly hyaline, single-celled, oval or somewhat elongated. The Genus - *Sclerotinia* is of much economic importance as it causes diseases in many plants. Another important Genus is *Monilinia*, which affects fruits of peach.

Genus - *Sclerotinia*. The fungus overwinters as sclerotia on or within infected tissues that have fallen on the ground and as mycelium in dead or living plants. In the spring or early summer, the sclerotia germinate and produce slender stalks terminating at a small disc- or cup-shaped apothecium in which asci and ascospores are produced. In some *Sclerotinia*, the sclerotia cause infection by producing mycelial strands that attack and infect young plant stems directly. No true conidia are produced

(e-g) *Sclerotinia sclerotiorum* - causes stem rot of many economic crops, such as mustard, chillies, gingelly, coriander, bhendi (lady's finger), gram, sunflower, tobacco, tomato, potato, brinjal (eggplant), soybean, peas etc., cottony rot of carrot, and watery soft rot of bean and cucumber; *S. fructicola* - attacks peach fruits; *S. minor* - causes destructive diseases of numerous succulent plants, particularly vegetables and flowers.

Life cycle of *Sclerotinia sclerotiorum*. The mycelium of the fungus is hyaline, much branched, consisting of large and closely septate hyphae, which are both inter- and intracellular and invade all the tissues of the affected region. No true conidia are produced, however microconidia are produced endogenously and exogenously on short stalks. The microconidia are cut off in loose chains and they lie embedded in a slimy substance. But, neither the microconidia, nor the hyphae serve as a means of vegetative propagation. Once the vegetative growth has ceased, the hyphae collect in small dense masses, which gradually become the sclerotia. The sclerotia are whitish or pinkish at first and later turn black and become smooth. The smaller sclerotia are usually round and the bigger ones are flattish and somewhat irregular in shape. The size of sclerotia depends on the substrate and the prevailing temperature. The sclerotia are composed of thin-walled, rectangular cells in the center and thick-walled cells at the periphery. The walls of the peripheral cells are impregnated with a dark, gelatinous material, which gives the sclerotia a black appearance. The sclerotia are the resting bodies of the fungus.

Under field conditions the sclerotia germinate during winter. Two to five stipes are produced by each sclerotium when it germinates. Sometimes the stipes are dichotomously branched and both the branches are fertile. The apothecial fundamentals are borne at the tips of the stipes as minute, brownish, funnel-shaped structures just above the soil surface. The fundamentals expand into shallow, flat or convex apothecia, which become darker in color with age. The asci are cylindrical and the ascospores are elliptical. Infection occurs only through the ascospores **(Fig. 45)**

Mycoparasites, such as *Trichoderma viride*, *Coniothyrium minitans*, *Gliocladium roseum*, *G. virens* and *Sporodesmium sclerotiorum* are capable of either destroying the sclerotia or inhibit the formation of new sclerotia, thereby markedly reduce the population of *Sclerotinia* in the soil.

Genus - *Monilinia* (e-g) *Monilinia (Sclerotinia) fructicola* - causes brown rot of peach and other stone fruits; *M. laxa* - attacks pome fruits and stone fruits.

Life cycle of *Monilinia fructicola*. The mycelium begins as a germ tube emerging from the ascospore or conidium in the spring and invades a susceptible host, causing twig blight or leaf blight. After the mycelium grows and reaches a certain stage, it produces long, branched conidiophores that break up into chains of oval or lemon-shaped conidia. This is the imperfect stage called the *Monilia* stage, which belongs to the Form genus - *Monilia* of the Deuteromycetes. The conidia break off easily from the chain and are disseminated by wind. On reaching a susceptible host, the conidia germinate in the presence of water, each conidium producing a germ tube and infecting the host, thus spreading the disease. Asexual reproduction is repeated several times, once in every few days in a single season.

The fungus invades peach fruits nearing maturity, which are less resistant, through hair sockets, insect punctures or other wounds and causes brown rot. The infected fruit first shows a soft brown spot. The mycelium spreads

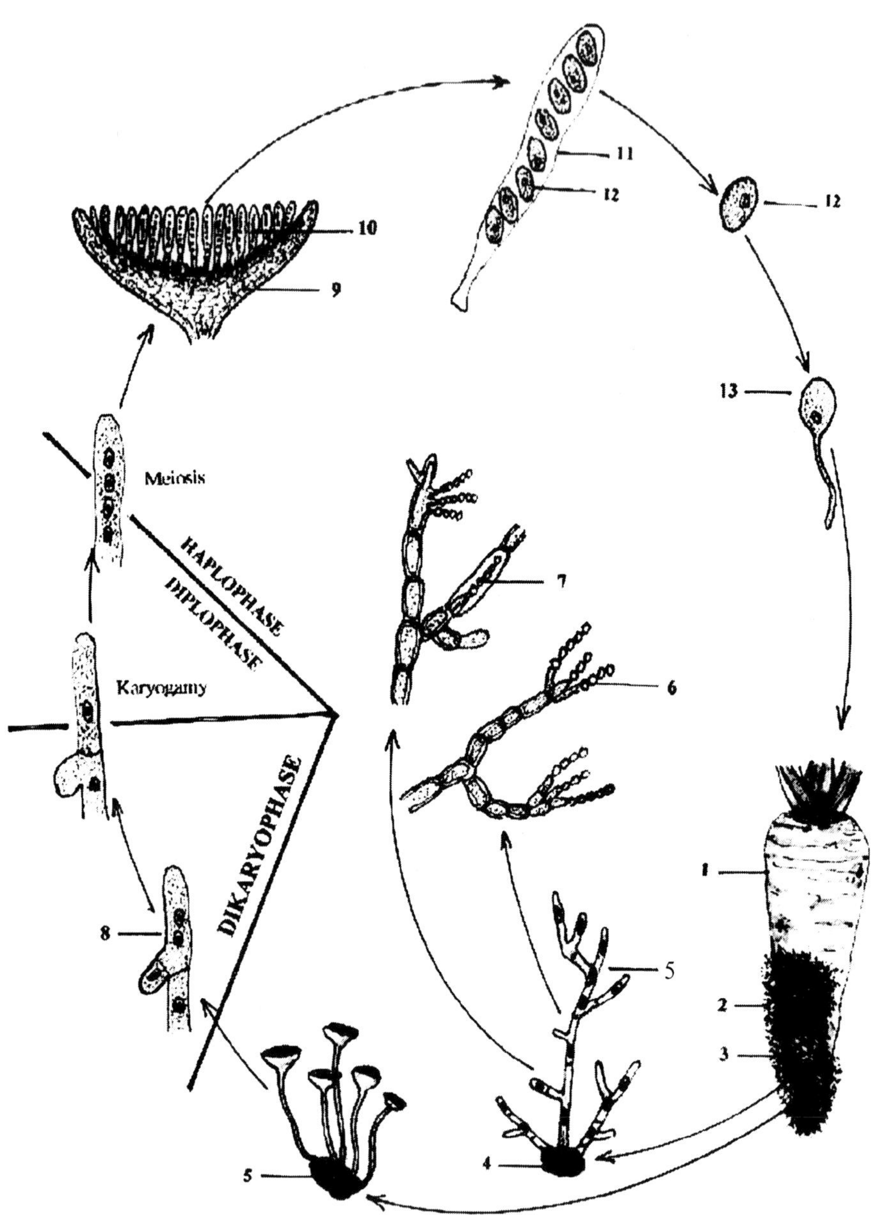

1,2,3. Cottony soft rot of carrot 4,5. Resting sclerotia germinating to produce mycelium or apothecial heads 6.7. Production of chains of exogenous and endogenous microconidia (Spermatia) 8. Ascogenous hypha 9. Mature apothecium 10. Asci 11. Ascus 12. Ascospore 13. Germinating ascospore.

Fig. 45 : Life cycle of *Scierotinia scierotiorum*

rapidly and secretes an enzyme that dissolves the middle lamellae of the host cells that renders the tissues soft. Gradually the mycelium penetrates the entire fruit, which shrivels and mummifies. The mummified fruits are covered with conidia. The mycelium that overwinters in the mummified fruits, which have either fallen to the ground or clinging to the twigs and in the twig cankers, as well as the conidia that may be present in a viable state initiate new infections in the spring.

The mummified fruits that have fallen on the ground and buried partially in the soil, in the course of one to three years produce the apothecia of the fungus from stroma on the mummies. *Monilinia fructicola* commonly produces spermatia or microconidia as they are called. But spermatization has not been definitely established. However, only some apothecial fundamentals alone develop into mature apothecia indicating that only those that are spermatized develop to form asci. Apothecia are long-stalked and are produced in large numbers on peach mummies, which contain asci and ascospores in the usual manner. By a puffing action, the asci forcibly expel their ascospores and air currents carry the spores to the blossoms, twigs and young leaves of peach trees and cause fresh infection.

Family - Dermataceae. The apothecia of the Dermataceae are erumpent or sessile, mostly fleshy and small, with an excipulum consisting of globose to angular cells and usually have dark walls. The Genus -*Diplocarpon* is an important plant pathogen in this family.

Genus - *Diplocarpon*. The apothecia produced by these fungi are well-developed and erumpent. The ascospores are more or less crescent-shaped

(e-g) *Diplocarpon rosae* - causes the very prevalent and serious black spot disease of roses. Its conidia are crescent-shaped and borne in acervuli. It is classified in the Form genus and species *Marssonina rosae* in Deuteromycotina; *D. maculatum* - causes black spot of apple and quince. Its conidial stage is Form genus - *Entomosporium* in Deuteromycotina.

Life cycle of *Diplocarpon rosae*. The black spot of rose pathogen attacks all varieties of roses. The symptoms appear as small, brownish-black spots on the upper, sometimes on the lower surface of the leaves. The spots are round and furnished with a radiating, fringe-like margin. Individual spots reach a diameter of 1.0 cm. The spots may coalesce to form large, irregular, black lesions. The leaf tissue around the lesions turns yellow. In case of heavy infections, extensive areas of the leaves may become blackened and are ultimately killed leading to extensive leaf fall. Tiny lesions also occur on the stems and on the flowers.

The pathogen produces *Marssonina*-type conidia in acervuli that are formed in-between the epidermis and cuticle and appear as tiny, raised specks within the blackened areas, mostly towards the margin of the spots. The conidiophores that arise form a thin, black stroma and the short conidiophores give rise to successive crops of conidia. The conidia are hyaline, oval to elliptical, bicellular and constricted at the septum, the cells sometimes breaking apart at the septum and function as separate spores. The fungus also produces spermatia from tiny spermagonia on older leaves similar to conidial formation. The spermatia are single-celled and bacilliform in shape. Sometimes typical bicellular conidia and spermatia are produced from the same acervulus. Ascospores are produced in tiny apothecia formed in old lesions on dead leaves. The apothecia are also formed in-between the epidermis and cuticle and are sunken deep in the tissue. They are spherical to disciform in shape. The asci are oblong or sub-clavate and are interspersed with septate, capitate paraphyses. The ascospores that resemble the conidia are hyaline and bicellular. All three types of spores can cause primary infection on leaves by direct penetration of the cuticle. On germination, the germ tubes form appressoria and after passing between the epidermal cells, the infection hyphae send haustoria inside the cells and ramify between the epidermis and the cuticle. Subsequently patches of the leaf epidermis turn brown from the death of its cells and then turn black.

The Operculate Discomycetes. In this group of Discomycetes, the ascus is provided with an operculum or slit at the apex or an operculum just below the apex. Only one Order - **Pezizales**, which consists of typical cup fungi, is of economic importance.

Order - Pezizales. The apothecia of the Pezizales may be open from the beginning or may be closed at first and open later on to release the ascospores. The majority of the Pezizales have only their ascogenous stages, the imperfect (conidial) stages of most are not known. Some are edible, while a few are poisonous.

The Order is divided into nine Families of which only the Family - **Morchellaceae** is of economic importance.

Family - Morchellaceae. The Morchellaceae is characterized by large, stalked apothecia, mostly with a sponge-like or bell-shaped pileus. The ascospores are always multinucleate.

Genus - *Morchella*. The morels belong to the Genus - *Morchella*. All the species have apothecia with a thick stalk in a pitted or ridged pileus that resembles a sponge. The color varies from a dirty grayish-white to dark-brown. It varies from 2.5 cm. or less in width and 10 - 14 cm. in height. The hymenial layer lines the pits of the pileus. It consists of long, cylindrical, operculate asci and elongated paraphyses interspersed among the asci. Each ascus contains eight, large, colorless, oval and multinucleate ascospores.

All morels are edible and delicious. The morels are actually the ascocarps of the fungus, the mycelium of which grows in the soil and obtains nourishment from organic matter in the soil

(e-g) *Morchella conica, M. semilibera (M. hybrida)* - the hybrid morel, *M. deliciosa, M. crassipes, M. esculenta* **(Fig. 50).**

Subclass - Loculoascomycetidae (The ascostromatic fungi). The most important distinguishing characters of this Subclass are that the asci are bitunicate and the ascocarps are ascostroma in which the asci are borne in locules. A large number of Loculoascomycetidae produce conidia, but many do not have the conidial stage and in such cases propagation is solely by means of their ascospores.

The conidial forms, when present are varied and sometimes the same fungus produces two or more types of conidia that bear no resemblance to each other. However, certain Genera produce typical conidial types. Sexual reproduction also varies greatly. Gametangial contact, spermatization, somatogamy or apogamous reproductions are found. The ascus in all cases is bitunicate and the inner wall (endotunica) extends two or three times the length of the outer wall (ectotunica) at the time the spores are expelled after dehiscence of the endotunica. The ascocarp is an ascostroma. The stroma in which the asci are located may be multilocular or unilocular, the structure of which is either prosenchymatous or pseudoparenchymatous. The unilocular ascostroma is generally called a **'pseudothecium'**. The Subclass is divided into five Orders of which the Orders - **Dothidiales** and **Pleosporales** are of economic importance.

Order - Dothideales. The chief characteristics of this Order are that the locules in the ascostroma are devoid of interascal threads of any sort and formation of bitunicate asci. The imperfect stages of many members of this Order are well known plant pathogens in Deuteromycotina. Usually only the conidial stage develops in the living host and causes damage. The Order has five Families, of which **Dothideaceae** and **Capnodiaceae** are important.

Family - Dothideaceae. The ostiolate, perithecioid pseudothecia are immersed in the host tissue or erumpent from a stroma. The ostioles are lined with periphyses. Fascicles of asci are produced in very small, ostiolate and spherical locules in a multilocular pseudothecia. Ascospores are usually single-septate.

Genus - *Mycosphaerella*. The pseudothecia in this Genus are small, separate and immersed in the host tissue, usually in dead leaves. The asci are eight-spored and the ascospores are hyaline or light-brown and have one septum near the middle. The conidial stages are varied and the conidia are produced in pycnidia, acervuli or on free conidiophores. Many species do not produce conidia at all. The conidial forms belong to such Genera as *Ramularia, Cercospora, Phoma, Ascochyta, Septoria* and *Phyllosticta* in Deuteromycotina **(e-g)** *Mycosphaerella musicola* (=*Cercospora musae* of Deuteromycotina) - causes the extremely destructive 'Sigatoka' disease of bananas; *M. fragariae* (=*Ramularia tulasnei* of Deuteromycotina) - causes leaf spot of strawberry; *M. citri* - causes *Citrus* greasy spot; *M. arachidicola* and *M. berkeleyii* (=*Cercospora arachidicola* and *C. personata* respectively of Deuteromycotina) - cause 'Tikka' disease of groundnut; *M. pinodes* (=*Ascochyta pinodella* of Deuteromycotina) - causes leaf, stem and pod spot, and foot rot of pea; *M. brassicola* - causes lesions on stem of young cabbage plant.

Life cycle of *Mycosphaerella musicola*. The pathogen produces spermatia in spermagonia, ascospores in perithecioid pseudothecia and conidia in sporodochia. The uninucleate microconidia or spermatia produced in the spermagonia possess a cell wall and resemble bacterial cells in appearance. They escape one after the other from the spermagonia. The spermatia are not able to infect the host and are incapable of germination. Successive crops of abundant conidia are produced on both surfaces of the leaf. The conidia, which are borne on short conidiophores, are olive-brown in color, long, slender and multiseptate. Wind and dripping or splashing of rainwater spread the conidia. Release and germination of the conidia depend on leaf wetness or high humidity.

Pseudothecia are produced during warm, humid weather and their ascospores are shot out violently in response to wetting of pseudothecia from bitunicate asci. The pseudothecia are small, separate and immersed in the host tissue, usually in dead leaves. The ascospores are hyaline or light-brown and have one septum near the middle. Ascospores are spread by air currents and are responsible for long-distance dissemination, while conidia are generally the most important means of local spread of the disease.

In the sexual cycle, a hyphal ball develops within the tissue of the host and several ascogonia are produced within this ball of hyphae. The ascogonium is a uninucleate single cell bearing a long, slender trichogyne. The trichogyne extends through the hyphal ball and ultimately reaches the outside through the host epidermis. No antheridia are produced. Copulation occurs between the ascogonium and a spermatium. The spermatium attaches itself to the trichogyne through which its nucleus migrates to the female nucleus of the ascogonium. Ascogenous hyphae develop from the ascogonium and produce asci of the hook type. The loose ground tissue in the area of the developing asci is absorbed and contributes to the nutrition of the asci. All asci within a fructification do not mature at the same time. Those in the center of the structure mature earlier than those near the periphery. The apex of the pseudothecium finally ruptures. Ascus after ascus elongates and reaches through the pore to the outside and then discharges its ascospores forcibly **(Fig. 46)**

Family - Capnodiaceae. The fungi belonging to this Family are found on many types of plants, especially on horticultural plants and are common in warm, humid weather. The organisms are epiphytic and live saprophytically on the honeydew exudates of certain sucking insects, such as leafhoppers, scale insects, aphids etc. The mycelium growing on the surface of leaves, stems etc. forms a dense, papery film or crust thereby reduces the amount of light reaching the plant surfaces. The mycelium is dark and consists of septate, tubular hyphae with a mucilaginous coating on the surfaces, which gives a black, sooty appearance to the plant parts, hence the name **'sooty mould'**.

There are considerable variations in the types of asexual reproduction. The ascocarps produced as a result of sexual reproduction are superficial. Periphyses are not present. The ovoid asci develop in a basal fascicle. Ascospores are hyaline to dark and usually one to several septate.

Genus - *Capnodium*. The fungi belonging to this Genus are all obligate saprophytes and live on the honeydew secretion of certain sucking insects, such as leafhoppers, aphids, scale insects etc. The mycelium covers the surfaces of leaves, stems and fruits and forms a dense, black, papery film or crust thereby reduces the light reaching the plant surfaces and leads to reduction of photosynthetic activity. The affected areas present a black, sooty appearance. The dark mycelium consists of tubular, septate hyphae with a mucilaginous coating on their surfaces.

Asexual reproduction takes place by different means. The fungi produce large number of conidia from the hyphal tips. Besides this, the organisms produce conidia from small, ostiolate, globose pycnidia. The conidia escape from the pycnidia through the small ostiole in large numbers. The conidia are dark-colored, globose or oblong in shape and single-celled. The conidia are embedded in the honeydew exudate of insects and are transmitted through insects, which are attracted to the sweet honeydew secretion. The conidia germinate by producing germ tubes.

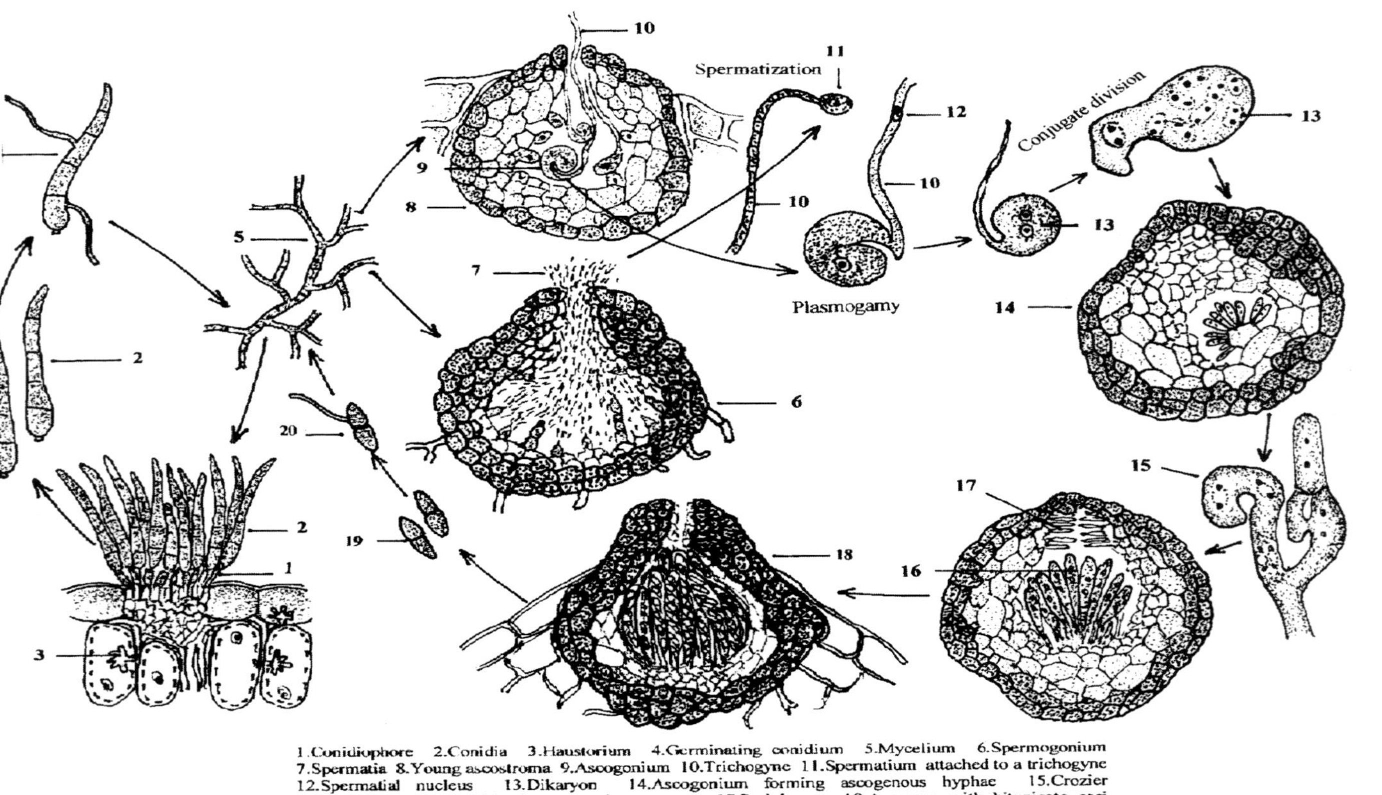

1.Conidiophore 2.Conidia 3.Haustorium 4.Germinating conidium 5.Mycelium 6.Spermogonium 7.Spermatia 8.Young ascostroma 9.Ascogonium 10.Trichogyne 11.Spermatium attached to a trichogyne 12.Spermatial nucleus 13.Dikaryon 14.Ascogonium forming ascogenous hyphae 15.Crozier 16.Developing asci within the stromatic ascocarp 17.Periphyses 18.Ascocarp with bitunicate asci 19.Ascospores 20.Germinating ascospore

Fig. 46 : Life cycle of *Mycosphaerella musicola*

As a result of sexual reproduction, these fungi produce superficial perithecioid pseudothecia. A mucilaginous substance also covers the pseudothecia. The asci, which are produced in a fascicle at the base of the pseudothecium, are ovoid and bitunicate. The ascospores are oblong, hyaline and one or more septate.

(e-g) *Capnodium ramosum* - causes sooty mould in several horticultural plants, such as mango, guava, cashew, sapota etc.; *C. braziliense* - causes sooty mould of coffee.

Order - Pleosporales. This Order is characterized by the Pleospora type centrum, in which the asci develop among pseudoparaphyses and grow upward between them. Pseudoparaphyses are attached both to the roof and the floor of the locule and they originate before the formation of the asci. They originate in the upper wall and grow downward. The ascostroma is either a pseudothecium or a multilocular cushion-shaped stroma. The Order is divided into eight Families of which **Pleosporaceae** and **Venturiaceae** are important.

Family - Pleosporaceae. The ascostroma of Genera of Pleosporaceae are perithecium-like or cleistothecium-like and are embedded in the substrate. Ascospores are one-septate to muriform. A number of important plant and animal parasites belong to this Family. In most members, the asexual conidial states are present throughout the growing season and sexual reproduction occurs after overwintering. Most of the members of the Family are known only by their conidial states than by their teleomorphs. The perithecia produced are medium to large in size and the ascospores are many septate. Only the Genus - *Cochliobolus* is of economic importance.

Genus - *Cochliobolus*. This Genus consists of a number of species parasitic on grain crops and has their asexual stage either in the Form genus *Bipolaris (Helminthosporium)* or *Curvularia*. Both these Genera are responsible for causing leaf spots and root rots on many grain crops and grasses

(e-g) *Cochliobolus heterostrophus* (= *Bipolaris / Drecshlera maydis*) - causes southern corn leaf blight; *Cochliobolus miyabeanus* (= *Helminthosporium / Bipolaris / Drechslera oryzae*) - causes brown spot of rice; *Cochliobolus setariae* (=*Helminthosporium setariae*) - causes leaf spot of finger millet.

Family - Venturiaceae. The Venturaceae are plant parasites that produce their ascostroma subepidermally or subcutaneously. Conidiophores are usually developed from such stromata and push through to the surface, where they produce conidia. The ascostromata may vary with the different Genera and species. Some are glabrous, but the majority bears hairs or setae, particularly around the pore over the matured asci. The ascospores are two-celled, ovoid or elliptical. At maturity they are olive-brown or grayish-green and rarely dark-brown in color. The ascospore characters constitute the main distinguishing feature of the Family. The Genus - *Venturia* is very important economically.

Genus - *Venturia*. *Venturia* is the most important Genus in this Family and includes a number of serious plant pathogens

(e-g) *Venturia inaequalis (=Spilocaea pomi)* - causes apple scab; *V. pyrina* - causes pear scab.

Subdivision 3 - Basidiomycotina

Class - Basidiomycetes (Smuts, rusts, jelly fungi, mushrooms, shelf fungi, puff balls, stink horns and bird's nest fungi). The Basidiomycetes differ from all other fungi in that they produce their spores, called **'basidiospores'**, on the outside of a specialized, microscopic, club-shaped spore producing structure called **'basidium'** and hence are called as **'club fungi'**. Basidiospores are generally uninucleate and haploid, but binucleate, homokaryotic basidiospores are also present in some cases. Like ascospores, basidiospores are the result of plasmogamy, karyogamy and meiosis, the last two occurring in the basidium. A definite number of basidiospores (usually four) are typically produced on each basidium.

The Basidiomycetes are an important group of fungi consisting of harmful species, as well as beneficial ones. Among the harmful parasites, the two groups viz., smuts and rusts cause tremendous loss every year. Many others attack a large number of food and other crops. Several Basidiomycetes cause diseases of forest and shade trees, resulting in considerable damage to lumber and wooden structures. On the other hand, many Basidiomycetes, especially the mushrooms are food delicacies. Some species of *Agaricus, Volvariella, Calocybe, Pleurotus* etc. are cultivated extensively for food all over the world. Many Basidiomycetes are also extremely valuable in nature because of their mycorrhizal relationship with many cultivated and non-cultivated plants.

Somatic structures. The mycelium of the Basidiomycetes consists of well-developed, septate, profusely branching hyphae that penetrate the substratum and absorb nourishment. Individually the hyphae are microscopic, but in mass the mycelium is visible to the naked eye and can be found in moist places in the woods, under barks of rotten logs, on wet, dead leaves or on other organic matter. The mycelium is usually white, bright yellow or orange in color and usually spreads out in a fan-shaped growth. In some forms, a number of hyphae lying parallel to one another are joined together to form a thick strand of mycelium called **'rhizomorph'**. These strands are enveloped in a sheath or cortex and behave as a unit **(Fig. 2)**

The mycelium of most Basidiomycetes passes through three distinct stages of development - the primary, the secondary and the tertiary before completion of its life cycle. The primary mycelium or homokaryon develops from the germination of a basidiospore and is septate and uninucleate. The secondary mycelium or dikaryon is developed by two different methods. In the first method, two uninucleate cells of the compatible homokaryotic mycelium fuse to form a binucleate cell. From this cell, secondary mycelium or dikaryon is formed by repeated conjugate divisions of the nuclei and development of septa, separating each pair of newly formed nuclei, thus forming new binucleate cells. In the other method, which is more common, special structures known as **'clamp connections'** are involved. When a binucleate cell is ready to divide, a short branch, the clamp connection arises between the two nuclei **(a)** and **(b).** The nucleus **(b)** now moves into the clamp. The two mother nuclei then divide simultaneously. While the mother nucleus **(b)** remains in the clamp, the daughter nucleus **(b1)** moves to the dividing cell. During the division of the mother nucleus **(a),** the daughter nucleus **(a1)** approaches the daughter nucleus **(b1),** while the mother nucleus **(a)** remains at the other end of the cell. In the mean time, the clamp bends over and its free end fuses with the cell, so that the clamp forms a bridge through which the mother nucleus **(b)** passes to the other end of the cell and pairs with **(a).** Then a septum is formed to seal the clamp at the point of origin and another septum is formed vertically under the bridge to divide the parent cell into two daughter cells, with nuclei **(a)** and **(b)** in one daughter cell and **(a1)** and **(b1)** in the other daughter cell **(Fig. 47)**

The organized, specialized tissues that form the sporophores (basidiocarps) represent the tertiary mycelium. Sporophores originate when the secondary mycelium forms complex tissues.

The basidiocarp. The more complex Basidiomycetes produce their basidia in highly organized fruiting bodies of various types called **'basidiocarps'**, which correspond to the ascocarps of the complex Ascomycetes. Basidiocarps may be thin and crust-like, gelatinous, cartilaginous, papery, fleshy, spongy, corky or woody and hard. Their size also varies greatly, from microscopic to a meter or more in diameter and these fruiting bodies attain greatest complexity in their structure and size. Most Basidiomycetes bear their basidia in basidiocarps. But the rusts and the smuts do not form any basidiocarps, with the exception of one or two species.

Fruiting bodies or basidiocarps of Basidiomycetes, such as mushrooms, bracket fungi, coral fungi, puffballs, bird's-nest fungi etc. are quite conspicuous, while the main body consisting of the extensive mycelium of the fungus that bears them usually goes unnoticed. Basidiocarps may be open from the beginning, exposing their basidia or they may open at a later stage when the spores are fully matured or they may remain closed till external forces disintegrate them.

The basidia are typically formed in definite hymenial layers, which are composed of basidia, as well as any other sterile structures, such as basidioles and cystidia. **'Basidioles'** appear to be either cells resembling basidia or basidia that have not yet produced spores and they provide support for the fertile basidia. **'Cystidia'** are larger and protrude beyond the other hymenial elements and may act as air traps besides aiding in the evaporation of moisture and other volatile compounds **(Fig. 48)**

The basidium. The basidium is a simple, club-shaped structure that bears on its surface a definite number of basidiospores (usually four) that are typically formed as a result of karyogamy and meiosis. The simple club-shaped basidium originates as a terminal cell of a binucleate hypha and is separated from the rest of the hypha by a septum over which a clamp connection is generally found. The narrow and elongated basidium soon enlarges and becomes broader. The nuclei within the young basidium fuse and the zygote nucleus undergoes meiosis, giving rise to four, haploid nuclei. Four, small outgrowths, the sterigmata push out at the top of the basidium and their tips enlarge forming the basidiospore initials. In many species, a mitotic division takes place following meiosis and thus eight, haploid nuclei are formed. One nucleus then passes into each of the four basidiospore initials, while the other four nuclei remain in the basidium itself and eventually four basidiospores are formed **(Fig. 49)**

Not all basidia bear four spores. Some may produce only two, while some others may produce more than four.

The basidiospore. The basidiospore is typically unicellular and haploid in nature, however in some cases two nuclei may move into the basidiospore from the basidium. Such spores may give rise to binucleate mycelium directly.

Basidiospores may be globose, oval, elongated or sausage-shaped, colorless or lightly pigmented green, yellow, orange, brown or black.

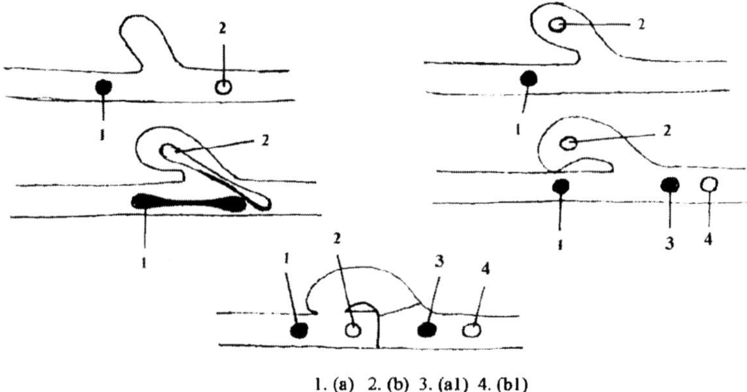

Fig. 47 : Mechanism of dikaryotic cell division by clamp connection

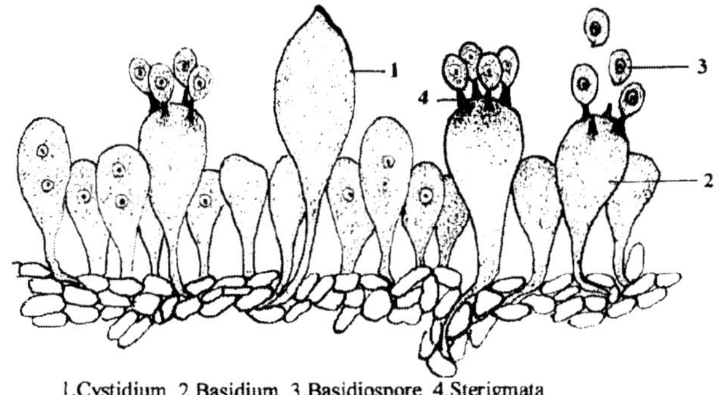

Fig. 48 : Typical hymenium of a Basidiomycete

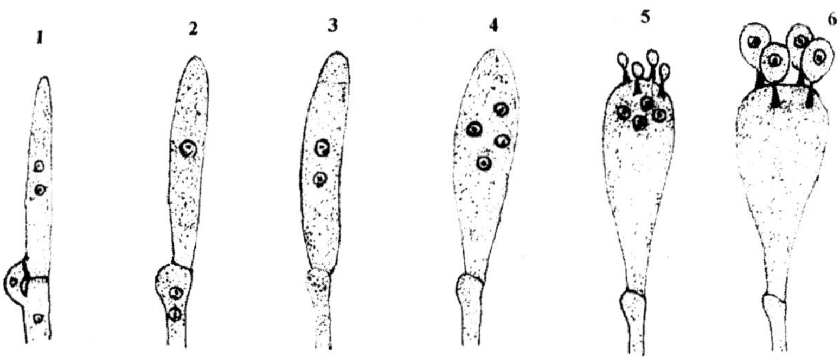

Fig. 49 : Stages in the development of a basidium

Asexual reproduction. In the Basidiomycetes, asexual reproduction takes place by means of budding, by fragmentation of mycelium and by the production of conidia, arthrospores or oidia. Conidia may be budded off from both the basidiospores and the mycelium. The urediniospores produced by the rusts are also conidial in nature. The arthrospores produced by the Basidiomycetes may be uninucleate or binucleate depending on whether they arise from primary or secondary mycelium. Oidia are produced from special, short, hyphal branches, the oidiophores, which cut off oidia in succession from the tip of the oidiophores.

Sexual reproduction. By sexual reproduction, Basidiomycetes produce basidium bearing haploid basidiospores. Within the basidium, karyogamy and meiosis take place. Sex organs, such as spermatia and receptive hyphae are produced only in the rust fungi. As in the Ascomycetes, in Basidiomycetes also, there are homothallic and heterothallic species.

In the Class - Basidiomycetes there are three Subclasses viz., Holobasidiomycetidae, Phragmobasidiomycetidae and Teliomycetidae

Subclass - Holobasidiomycetidae I (Hymenomycetes I). The Holobasidiomycetidae is divided into two groups viz., Gasteromycetes and hymenial Holobasidiomycetes. In **Gasteromycetes**, no distinct hymenium is visible by the time the spores are released from the basidiocarp (e-g) Puffballs, earthstars, stinkhorns and bird's - nest fungi. In the **Hymenial Holobasidiomycetes**, a well-developed hymenium exposing the basidia is formed even before the basidiospores are mature.

Order - Aphyllophorales (Pore fungi, tooth fungi, coral fungi and chanterelles). The Aphyllophorales, also known as **Polyporales** of the hymenial Holobasidiomycetes, produce single-celled, club-shaped basidia in well-defined, exposed hymenia, which are borne in various ways on sporophores.

The Aphyllophorales produce different types of basidiocarps. Many are fairly large and easily visible, while others are much smaller and less conspicuous. The hymenial layer, which may be smooth, ridged, warted, toothed, porous or lamellate may be typically found on one side of the sporophore (unilateral) or all over the surface (amphigenous). The texture of the sporocarp may be papery, leathery or hard and woody, but not soft or spongy as in the mushrooms. The basidiocarps may be sessile or stalked and may look like funnels, clubs, tooth-like, shelf-like or umbrella-like structures. The hymenium is borne on the undersurface of the umbrella or shelf-like portion. The porous forms commonly called polypores bear their hymenium inside the pores.

Some of the Aphyllophorales are serious parasites of forest and shade trees, causing root rots and heart rots, while many others are only saprophytes. Two types of decay are caused by the Aphyllophorales, the **'brown rot'**, in which only the cellulose part of the wood is decomposed and the **'white rot'**, in which both the cellulose and the brown, lignin components are decomposed, giving the rotten wood a white appearance.

The Order is divided into six Families, of which the Families - **Polyporaceae** and **Corticiaceae** are of much economic importance.

Family - Polyporaceae. Members of this Family are commonly called **'polypores'**, because of the porous nature of the hymenium in most species. Many cause diseases of trees, while others are important because they attack and destroy lumber. The fruiting bodies in the Polyporaceae may resemble crusts, shelves or mushrooms. The basidiocarps at maturity are generally tough, leathery, corky or woody. Although the hymenium is lamellate in a few species, in most of the species the basidia typically line the inner surface of pores or tubes. In this Family, the Genera - *Fomes* and *Polyporus* are important.

Genus - *Fomes*. The shelf-like or hoof-shaped fruiting bodies are hard, woody structures produced on the sides of dead or dying trees, as well as stumps. The fruiting bodies are perennial and may be found season after season. Each year perennial fruiting bodies increase in size and a new layer of pores is formed over the surface of the preceding layer. The pores are extremely small and can be seen only with the aid of a hand lens.

(e-g) *Fomes annosus (Heterobasidium annosum)* - causes root rot of pine trees. It invades and colonizes the surfaces of the freshly cut stumps, then spreads to nearby trees through natural root grafts. To prevent attack, the surfaces of the stumps are treated with chemicals, such as urea, creosote, liquid borate or dry borax granules.

Genus - *Polyporus*. The fruiting bodies of *Polyporus* are not so hard and they are annual, producing spores for only one season. The sporocarps are mostly bracket-like or shelf-like and may be produced in large number on trees, logs, stumps and branches. A few Polypores are edible

(e-g) *Polyporus versicolor* and *P. cinnabarinus* - cause rotting of wood.

Genus - *Ganoderma*. Several species of *Ganoderma* cause root and basal stem rots in conifers and hard woods, in palms and in other tropical plantation trees. The fruiting bodies are reddish-brown, whose lateral stem and upper surface are coated with a hard, shiny substance resembling shellac or sealing wax

(e-g) *Ganoderma lucidum (Polyporus lucidus)* - causes Thanjavur wilt disease or basal stem rot of coconut trees, 'Anabe roga' disease of arecanut and red root rot of pepper; *G. pseudoferarum* - causes red rot of rubber.

Family - Corticiaceae. The Corticiaceae includes both saprobic and parasitic forms. Their fruiting bodies are formed on wood, herbaceous plant parts and sometimes even in the soil. The sporophores are often little more than a thin layer of basidium bearing hyphae. The basidia develop on a webby or loose mycelium or on a thick, interwoven mycelium forming a crusty fructification overlying the substratum. Often there are sterile hyphae, which push upward between the basidia.

Genus - *Corticium*. In this Genus, the basidiocarps are often so thin, as to resemble a coating of drab paint on a fallen twig or branch.

(e-g) *Corticium salmonicolor (=Pellicularia salmonicolor)* - causes pink disease of rubber, tea, coffee etc.; *C. koleroga* - causes 'koleroga' or black rot on leaves and fruits of coffee plants; *C. fuciforme* - causes *Corticium* disease or red thread disease of turf grasses; *C. (Rhizoctonia) solani* - causes black scurf of potato; *C. theae* and *C. invisum* - cause black rot of tea.

Subclass - Holobasidiomycetidae II (Hymenomycetes II)

Order - Agaricales (Mushrooms and boletes). The Order, which also belongs to the hymenial Holobasidiomycetes, includes the fungi whose fruiting bodies are generally called **'mushrooms'**. Mushrooms are fleshy, sometimes tough, with umbrella-like sporophores that bear their basidia on the surface of gills or lamellae or line the inside of deep tubes or rarely shallow pits on the under surface of the fleshy basidiocarp. The Agaricales contain parasitic, saprobic and mycorrhizal forms.

In addition to mushrooms grown commercially, many wild species are also edible, while a number of other mushrooms are quite poisonous. The most severe type of mushroom poisoning is caused by species of *Amanita*. Only a relatively few species of mushrooms are grown commercially to any extent. *Lentinus edodes, Volvariella volvaceae* - the paddy straw mushroom and *Agaricus brunnescens (bisporus)* are grown widely.

Somatic structure. In the Agaricales, the mushroom (basidiocarp) is not the whole fungus, while the mycelium that covers a much greater area than the basidiocarp is actually the body of the fungus. The mycelium is typically Basidiomycetous, arising as a primary mycelium from a homokaryotic basidiospore, becoming dikaryotic and eventually becoming the tertiary mycelium, which form the mushroom proper. The primary mycelium in most species is of short duration, while the secondary, binucleate mycelium is more abundant, perennating and produces mushrooms year after year.

Many of the Agaricales are mycorrhizal fungi. **'Mycorrhizae'** are symbiotic associations between the hyphae of the fungi and the roots of higher plants. Without mycorrhizal associations most plants may not be able to survive in natural soil habitats. Mycorrhizal fungi increase the solubility of minerals in the soil, enhance the uptake of nutrients, such as nitrogen, potassium and phosphorus by the host plant, protect the host's roots against invasion by pathogens and produce plant growth hormones. In return, the mycorrhizal fungi obtain carbohydrates from their hosts.

Asexual reproduction. Only few members of the Agaricales are known to have a distinct asexual phase. Some species produce oidia that are thin-walled spores, resembling mycelial fragments. In some species the oidia perform a dual function, either germinating and giving rise to mycelium or behaving as spermatia and uniting with hyphae of the opposite mating type.

A few species, such as *Coprinus lagopus* and *Volvariella volvaceae* produce chlamydospores. Upon germination the chlamydospores give rise to mycelium.

Sexual reproduction. The majority of the species in the Agaricales exhibit heterothallism. Hyphal fusion (somatogamy) or oidization brings compatible nuclei together. Nuclear fusion and meiosis take place in the basidium and subsequently haploid basidiospores are produced.

The basidiocarp of most Agaricales is the well-developed mushroom. The most conspicuous parts of the mushroom are the **stipe** (stalk) and the **pileus** (cap). The tissues comprising the mushroom consist of closely packed dikaryotic hyphae that arise from the somatic hyphae growing within the substratum supporting the growth of the fungus. The initial events in the development of a fruiting body involve the formation of an intricate hyphal lattice by the interaction of dikaryotic hyphal branches, which results in the formation of fruit body primordium or button. Even when the primordium is about 1.0 mm. in diameter, a presumptive stipe, hymenium and pileus are already present. By the rapid enlargement of the cells, the elongation of the stalk and expansion of the pileus occur quite rapidly. During the early stages of basidiocarp development, tissues of the basidiocarp enclose the hymenium or fertile layer. The margin of the pileus is connected to the stipe by a membrane known as the **'inner veil'**. Shortly before the spores mature and are discharged from the basidia, the veil becomes severed from the margin of the pileus and remains attached to the stipe in the form of a ring or **'annulus'** (pl.-annuli).

In some other species, such as those belonging to the Genera - *Amanita* and *Volvariella,* a membranous structure called **'universal veil'** also covers the entire primordium. When the sporophore enlarges and the pileus finally expands the universal veil tears and leaves a cup-shaped body, the **'volva'** (pl.-volvae) around the bulbous base of the stipe. The remnants of that part of the universal veil that covered the pileus can often be seen in the form of scales on the cap.

The hymenium is found beneath the pileus. It may line the inside of tube-like structures as in boletes or more typically, it lines the lamellae (gills) that hang below the pileus. Gills are generally thin strips of tissue radiating from the margin of the pileus in toward the stalk. The inner tissue of the gills is called the **'trama'**. On the surface of the trama, covering both the sides of the gill and often the edge as well, is the hymenium, a closely packed layer of basidia interspersed with basidioles or cystidia. Each basidium usually bears four spores that are forcibly shot off and fall below the pileus. In still air, they get deposited below the pileus in a mass, forming a **spore print**.

Basidiospores vary in shape, size and color. They may come in a variety of colors and shades mostly white, pink, brown, purple or even black. In some species, they may be green or yellow. The external ornamentation of the spore wall also varies. The basidiospores are thick-walled and multi-layered. The Order is divided into a number of Families, but only a few are of much economic importance.

Family - Tricholomataceae. It is a very large Family comprising of white-spored species, with attached gills. Members of the Family are found on many different substrates and in many different habitats. Many species are edible and are of excellent flavor. The white oyster shell-like caps appear on logs or tree stumps in shelf-like layers. They are either sessile or have a very short, lateral stalk

Genus - *Armillaria*. Members of this Genus are distributed worldwide and some of them are known to cause severe root rot diseases of a wide variety of trees including conifers and hard woods. The rhizomorphs produced by *Armillaria* species serve as foraging structures and can resist long periods of drought and can even survive fire. The tips of the rhizomorphs represent the sources of inoculum that can penetrate and infect the roots of host plants

(e-g) *Armillaria mellea* - it is generally known as the **'honey mushroom'**. The sporocarps are quite variable in appearance. They are usually honey-colored with wet, smooth caps. Most of them have a prominent annulus especially when they are young and sometimes even have a double annulus. It is a serious root pathogen of oaks and produces sporocarps in clusters on the roots or on the dead stumps. It is one of the choicest of edible mushrooms. However, some species of *Armillaria* are poisonous. The mycelium and rhizomorphs of *Armillaria mellea* and several other species are **bioluminescent.**

(e-g) *Armillaria tubescens, A. ostoyae.*

Genus - *Pleurotus*. This Genus consists of mushrooms generally known as **'oyster mushrooms'**. They are very commonly found in nature and are choice edible species. The oyster shell-shaped caps appear on logs or tree stumps in shelf-like layers. The sporocarps are either sessile or have a very short, lateral stalk. However, *Pleurotus ulmarius*, a large, white, edible mushroom has a thick, almost central stalk

(e-g) *Pleurotus ostreatus, P. sapidus, P. ulmarius, P. sajor caju, P. citrinopileatus, P. florida* - all these species are edible mushrooms and are cultivated in various substrates, especially paddy straw **(Fig. 50)**

Genus - *Marasmius*. (e-g) *Marasmius oreades* - it is also a good edible mushroom.

Family - Volvariaceae. The members of this Family have free gills and produce pink-colored, elliptical basidiospores.

Genus - *Volvariella*. The mushrooms belonging to this Genus produce a volva and have a central stalk. The sporocarps have free gills and the basidiospores are elliptical in shape and are pink-colored. They are generally known as **'paddy straw mushrooms'** and are edible

(e-g) *Volvariella volvaceae, Volvariella diplasia* **(Fig. 50)**

Family - Agaricaceae. The Family contains the well-known Genera - *Agaricus* and *Calocybe*

Genus - *Agaricus*. Species of *Agaricus* generally produce a basidiocarp, with a white to brown or grayish-brown cap, free gills, an annulus, but no volva and a stalk that readily separates from the cap. In a young basidiocarp, the gills may be light in color, often pink or white, but eventually darken and become chocolate-brown. The spores are also chocolate-brown in color. The Genus contains some of the fine edible mushrooms

(e-g) *Agaricus campestris, A. brunnescens (bisporus)* - they are edible mushrooms **(Fig. 50)**

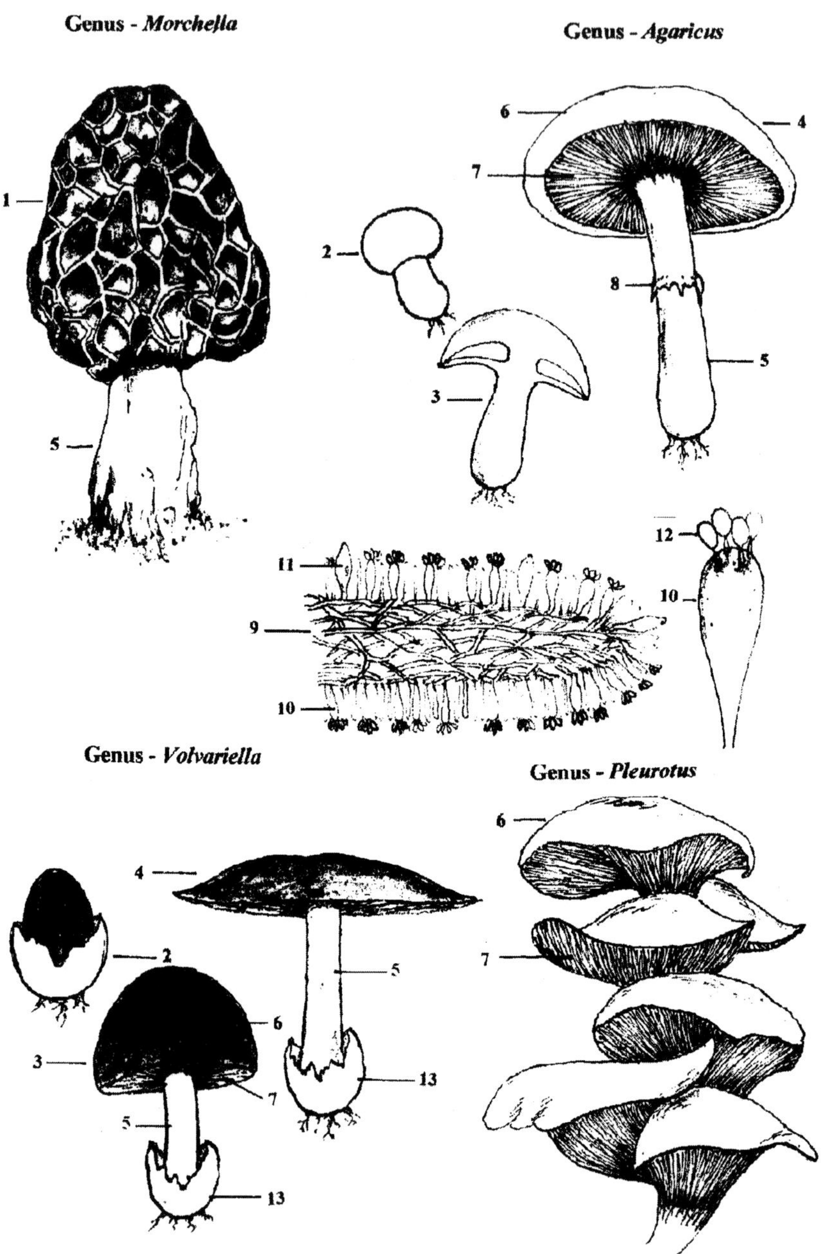

1.Ascocarp 2.Young basidiocarp 3.Developing basidiocarp 4.Fully developed sporocarp 5.Stipe 6.Pileus 7.Gills 8.Annulus 9.Hymenium 10.Basidium 11.Cystidium 12.Basidiospores 13.Volva.

Fig. 50 : Edible mushrooms

Genus - *Calocybe*. The mushrooms belonging to this Genus are generally known as **'milky mushrooms'** because of their white color and are edible

(e-g) *Calocybe indica.*

Subclass - Holobasidiomycetidae III (Hymenomycetes III)

Order - Exobasidiales. The Exobasidiales are basidial fungi, parasitic on several flowering plants and economic crop plants in addition to many wild plants. Infection of host often causes a swelling of the infected parts as a result of hypertrophy of the host cells. The diseased tissues usually become red. All the Exobasidiales are placed in the single Family - **Exobasidiaceae** and only the Genus - *Exobasidium* is economically important.

Family - Exobasidiaceae. All the fungi belonging to this Family are endoparasites. No fructifications are produced. The basidia are produced directly from dikaryotic hyphae between the epidermal cells of the host, push through the cuticle and form a layer on the surface of the host. The hymenium is discontinuous and the basidium is cylindrical. Four, six or eight basidiospores are produced following karyogamy and meiosis in the basidium. The basidiospores are uninucleate. They germinate either by budding, producing blastospores or by a germ tube that infects the host and develops into a haploid primary mycelium. By some means, which is not yet clearly understood, the dikaryotic mycelium originates. The dikaryotic mycelium, devoid of clamp connections is intercellular and feeds by means of haustoria.

Genus - *Exobasidium* (e-g) *Exobasidium vexans* - causes blister blight of tea.

Order - Tulasnellales. This Order includes saprophytic, as well as plant parasitic species. A number of species form mycorrhizal relationship with terrestrial orchids. The Tulasnellales produce effused basidiocarps that are often waxy in appearance. The web-like fructifications of some are very inconspicuous. In Tulasnellales, the sterigmata are either broad and finger-like or greatly inflated, often reaching considerable size. Order - Tulasnellalles is divided into two Families viz., **Tulasnellaceae** and **Ceratobasidiaceae**, of which the latter is important.

Family - Ceratobasidiaceae. This Family consists of four Genera of which the Genera - *Thanatephorus* and *Ceratobasidium* are economically important.

Genus - *Thanatephorus*. *Thanatephorus* is the perfect stage of the **'multinucleate *Rhizoctonia*'.**

(e-g) *Thanatephorus cucumeris* - it is the perfect stage of the important soil-inhabiting plant pathogen *Rhizoctonia solani. Rhizoctonia solani* attacks a wide range of host plants and causes a variety of diseases including root rots, cankers, damping off, fruit decay and even foliage diseases. Some of the most important diseases caused by *Rhizoctonia solani* are potato stem rot and tuber black scurf, tomato fruit rot, cabbage bottom rot, crater rot of carrot and bean pod rot. *Rhizoctonia solani* produces no spores, but only sclerotia.

Genus - *Ceratobasidium*. *Ceratobasidium* is the perfect stage of **'binucleate *Rhizoctonia*'**

(e-g) *Ceratobasidium anceps* - it is parasitic on *Pteridium.*

Subclass - Teliomycetidae (Rusts, smuts and basidiomycetous yeast). The rusts and smuts are characterized by the production of thick-walled, binucleate resting spores termed the **'teliospores'**. Karyogamy takes place within this structure. When the teliospore germinates, each of its cells gives rise to a short germ tube known as the **'promycelium'** into which the diploid nucleus moves. After meiosis, haploid basidiospores are formed. In the Teliomycetidae, the teliospore is considered to be the probasidium, because within this spore karyogamy takes place. This germ tube or promycelium is the **'metabasidium'**, because meiosis occurs within this structure. The small outgrowths from the metabasidium on which the basidiospores are formed are the **sterigmata**. In the smut fungi sterigmata are absent and the basidiospores are formed directly on the metabasidium.

Order - Uredinales. The Uredinales or rusts are very important economically. The mycelium is septate, uninucleate in its first phase and binucleate in its later stages.

When a rust spore lands on the surface of a susceptible host and germinates, a sequence of morphological and cytological events take place leading to the establishment of the mycelium within the host. Once established, the mycelium grows mostly in an intercellular fashion, obtaining nourishment from the host cells by means of haustoria. In nature the rust fungi are obligate parasites. Unlike most other Basidiomycetes, the rust fungi produce no basidiocarps. The life cycle of a rust fungus consists of a usually complex series of events. All the rusts except the so-called **'imperfect rusts'** produce teliospores. The teliospore is considered as the perfect stage of the Uredinales, since karyogamy and meiosis takes place in this structure. Besides the teliospore, most rusts produce other spores also. Typically there are either four or five distinct reproductive stages in the life cycle. The

Uredinales are divided into three categories based on the life cycle pattern: **(i)** the macrocyclic rusts **(ii)** the demicyclic rusts and **(iii)** the microcyclic rusts. **'Macrocyclic rusts'** typically exhibit all five reproductive stages, while in **'demicyclic rusts'**, the uredinial stage is absent and in **'microcyclic rusts'**, the teliospore is the only binucleate spore produced. The first spermagonial stage may be absent in some forms.

The five stages of typical macrocyclic rust are:

Stage - 0	Spermagonia bearing spermatia and receptive hyphae
Stage - I	Aecia bearing aeciospores
Stage - II	Uredinia bearing urediniospores
Stage - III	Telia bearing teliospores
Stage - IV	Basidia bearing basidiospores

Spermagonia and spermatia. The function of spermagonia (sing. - spermagonium), otherwise called **pycnia** (sing. - pycnium) in the life cycle of the rust fungi was discovered only in 1927, a long time after the other four stages were discovered and designated. However, since the spermagonial stage actually precedes the aecial stage, it was designated as **'Stage - 0'**, instead of changing the entire sequence. Spermagonium may be flask-shaped, conical or flat and sprawling, but mostly flask-shaped.

The wind-borne basidiospores (sporidia) falling on the leaves of the alternate hosts viz., barberry or mahonia germinate by producing a germ tube and penetrate the host epidermis directly and enter the host. If suitable alternate hosts are not available, then the sporidia are perished. About four days after infection of the host, a stromatic mass of hyphae consisting of uninucleate compartments develop below the host epidermis of the upper leaf surface. From this mycelial mat, a flask-shaped **'spermagonium'** develops with an outer spermogonial wall and a central cavity. Large number of closely packed, elongated, uninucleate **'sporogenous cells' (spermatiophores)** derived from the spermagonial wall-cells fill the spermagonial cavity. Each of these cells gives rise to a series of uninucleate **'spermatia' (pycniospores)** basipetally. As the spermatia are produced and released into spermagonial cavity, an ostiole is formed in the upper part of the spermagonium. A number of slender, stiff, tapering periphyses also develop from the upper edge of the spermagonial wall and curve upward. The tips of the periphyses push the host epidermis from below, rupture it and protrude through the opening (ostiole) they have made. Spermatia, which are produced in enormous numbers are also exuded through the ostiole in a droplet of nectar, a thick, sticky, fragrant, sweet liquid. From the upper part of the spermagonial wall, **'receptive hyphae'** are produced from just below or among the periphyses and grow out through the ostiole into the droplet of nectar containing the spermatia. The spermatia and receptive hyphae produced in spermogonia from the sporidia of one mating type **(+)** are all of the same mating type **(+)**, while the spermatia and receptive hyphae produced in spermogonia from the sporidia of the opposite mating type **(-)** are all of the opposite mating type **(-)**. Spermogonia producing either **(+)** or **(-)** spermatia and receptive hyphae may occur on the same leaf, but spermatia and receptive hyphae of the same mating type are not compatible and no copulation takes place between them. Sometimes, the periphyses also function as receptive hyphae.

Transfer of spermatia to compatible receptive hyphae are mostly effected by insects, which are attracted to the nectar or by rainsplash or wind. When a spermatium comes in contact with a receptive hypha of the opposite mating type, the receptive hypha acts as a trichogyne and receives the nucleus from the spermatium. The nucleus and the contents of the spermatium move into the receptive hypha through a pore dissolved at the point of contact and migrates down the receptive hypha, and thus initiates the **'dikaryotic phase' (stage - I)** of the rust life cycle.

Aecia and aeciospores. Aecia and aeciospores are next in line **(stage - I)** in the rust life cycle. As the spermogonia are being formed on the upper surface of the leaves, aecial primordia are formed from the uninucleate primary mycelium prior to dikaryiotization. The receptive hyphae are connected to the mycelium forming the aecial primordium. After fertilization, the nucleus from the spermatium migrates down the receptive hypha through the septal perforations and reaches the cells of the aecial primordium, making them binucleate. Conjugate division of the dikaryotic nuclei and formation of cross septa results in the formation of a large number of **binucleate (dikaryotic) cells.** Cell fusion in the aecial stroma also plays a vital role in the dikaryotization process. A number of such dikaryotic cells eventually form a well-defined group of **'sporogenous cells'** or **'aeciosporophores'**, all in close lateral contact at the base of the aecial primordium and develop into an aecium. The sporogenous cells that contain two nuclei in each cell, divide conjugately during the formation of the aeciospore initial. Two of the daughter nuclei remain in the sporogenous cell, while the other two move into the **'aeciospore initial'**. After the initial is delimited from the mother cell by a septum, the nuclei in the initial divide again and a transverse septum separates the initial into a **binucleate (dikaryotic) aeciospore** and a small wedge-shaped intercalary or **'disjunctor cell'**, which is also binucleate. The entire process is repeated many times, eventually resulting in the formation of a chain of aeciospores and disjunctor cells, with the oldest cell at the top and the youngest adjacent to the sporogenous cell at the aecial base. In most species of rusts, the peripheral cells of the aecial base by successive division give rise to a wall or **'peridium'** that surrounds the spore chains in the form of a cup. Initially the peridium surrounds the spore chains on all sides forming a dome over them. When the aecium matures, the spore chains push through the roof of the peridium and the spores are liberated. The torn peridium forms a lip around the aecial cup. As the aecium develops, the disjunctor cells disintegrate and the spores are separated from each other. The aecia are generally located on the under surface of leaves and break through the lower epidermis.

Uredinia and urediniospores. Urediniospores also known as **'uredospores'** or **'urediospores'** constitute the repeating stage **(stage - II)** of the rusts, since several generations of spores may be produced in one crop season. They are borne in structures similar to the acervuli, but are called **'uredinia'** (sing. - uredinium) or **'uredia'**, which are reddish in color. The uredinial cells are formed sub-epidermally from dikaryotic mycelium originating from an aeciospore or an urediniospore. Eventually, a palisade of hyphal tips appears in the uredinial initial and from this layer the urediniospores are produced. The spores are formed from buds originating from the sporogenous cells. A bud enlarges and is then divided by a septum into two cells. The upper cell enlarges into the spore proper, while the lower cell develops into a stalk-like structure called the **'pedicel'**. As the spores form, they exert pressure on the host epidermis from below and rupture it, pushing it out in the same manner as in the case of an acervulus. Matured urediniospores are dikaryotic and have a thick wall covered with minute spines. Mostly they are globose to oval in shape and pedicellate.

Telia and teliospores. 'Telia' (sing. - telium) are groups of binucleate cells that give rise to thick-walled cells called **'teliospores'** or **'teleutospores' (stage - III)**. In many rusts, the old uredinia are actually converted to telia. Teliospores may be unicellular or consist of two or more cells and are formed from the tips of binucleate cells of the telium. Each cell of the spore is at first dikaryotic, but later karyogamy takes place and each cell becomes diploid and uninucleate. Most rusts overwinter in the teliospore stage. In teliospores consisting of more than one cell, each cell germinates and gives rise to basidiospores. Classification of Uredinales into Families and Genera is based largely on teliospore characteristics. Teliospores may be sessile or stalked and they may be completely free from one another or united laterally to form small groups or columns or layers. Teliospores vary in color from colorless to dark reddish-brown, in size, in shape and in their external wall texture.

Basidia and basidiospores. When favorable conditions for germination of teliospore arises, the **'promycelium'** grows out from each of its cell. The diploid nucleus then migrates into the promycelium, undergoes meiosis and produces four haploid nuclei that distribute themselves at almost equal distances in the promycelium. Septa formed between the nuclei divide the promycelium into four uninucleate cells. Each cell then produces a sterigma, at the tip of which a basidiospore develops. The four nuclei then migrate into the basidiospores. Finally the basidiospores are forcibly discharged into the air. Upon germination, a basidiospore gives rise directly to a germ tube or produces another spore called **'sporidium'** from the tip of another sterigma, which is **'stage - IV'** of the rust life cycle.

Sexual compatibility in the rusts. All nuclei in the haploid mycelium arising from a germinating basidiospore are of the same mating type. Consequently the spermatia and receptive hyphae produced by the same mycelium carry the nuclei of the same mating type and are therefore not compatible. Therefore, a reciprocal exchange of nuclei between compatible mycelia is necessary. This is achieved by spermatization between receptive hypha having nucleus of one mating type and spermatium having nucleus of the opposite mating type. When karyogamy occurs in the teliospore, the nuclei of the two opposite mating types combine to form a single, diploid nucleus. When meiosis occurs in the promycelium, segregation takes place, as a result two of the four basidiospores produced are of one mating type and the other two of the opposite mating type.

Heteroecism. Many of the Uredinales are highly specialized in their parasitism, attacking only a single host, while others are capable of attacking several related hosts. In rusts, there is a peculiar condition known as heteroecism and fungi exhibiting such characteristics are called **'heteroecious rusts'**. Heteroecious rusts require two distinctly different hosts to complete their life cycle. They produce stages 0 and I on one host species and stages II, III and IV on another host species. In most of the heteroecious rusts the two hosts are very much unrelated. The host on which rusts produce the uredial and telial stages is called the **'primary host'** and the other is known as **'alternate host'** (e-g) *Puccinia graminis* produces its spermatia and aecia on the common barberry bush, which is a dicotyledon, while its uredinia and telia are formed on various grasses, which are monocotyledons and are the primary hosts.

Autoecism is the opposite of heteroecism. **'Autoecious rusts'** complete their entire life cycle in a single host species (e-g) *Uromyces fabae*, causing rust of peas and lentil, is autoecious and it does not require an alternate host to complete its life cycle.

Family - Pucciniaceae. The teliospores of the Pucciniaceae are mostly stalked. They may be borne free from each other or embedded in a common gelatinous matrix or united into groups of three or more on a common stalk, but not in the form of layers or crust. The spores may be one-celled or two-celled or many-celled. The teliospore produces a septate promycelium on germination. They are generally reddish-brown and thick-walled and the walls are either smooth or variously sculptured. Aecia possess a cylindrical pseudoperidium. Teliospore characteristics, aecial characteristics and type of life history form the basis for classification of Genera in this Family.

Genus - *Puccinia*. Teliospores are stalked, usually two-celled, thick-walled and are produced singly. They are borne free and their stalks do not deliquesce at maturity. Aecia usually have persistent pseudoperidia. Spermogonia are sub-epidermal. Some rust fungi are morphologically identical, but attack different host genera and these are regarded as **'special forms'** ***('formae specialis')***. Within these special forms there may be several **'physiological races'**, which attack only some varieties of the Genera **(Fig. 52)**

(e-g) *Puccinia graminis* f.sp. *tritici* - causes black rust or stem rust of wheat; *P. graminis* f.sp. *hordei* - causes rust of barley; *P. arachidis* - causes rust of groundnut; *P. striiformis* - causes yellow or stripe rust of wheat, barley and rye; *P. recondita* - causes leaf rust or brown rust of wheat and rye; *P. coronata* - causes crown rust of oats; *P. sorghi* - causes rust of corn; *P. purpurea* - causes sorghum rust.

Life cycle of *Puccinia graminis*. The two - celled teliospores produced on the leaves and stems of susceptible hosts, such as wheat, barley, rye, oats and some grasses, remain dormant during the winter season on the stubble in the fields. Overwintering takes place in the uninucleate, diploid stage after karyogamy has occurred. Early in the spring, each cell of the teliospore germinates and produces a promycelium into which the diploid nucleus migrates, undergoes meiosis and forms four, haploid nuclei. Septa are then formed, separating the nuclei from one another into four cells. Each cell of the promycelium produces a sterigma on which a minute basidiospore is formed. The nuclei squeeze through the sterigmata into the basidiospores. Two of the basidiospores are of one strain (mating type) and the other two are of the opposite strain. Soon, the basidiospores are forcibly ejected and carried away by wind. The basidiospores germinate in a drop of water on the leaves of the alternate host, the common barberry bush *(Berberis vulgaris)* and produce a slender germ tube. Most basidiospores never reach a barberry bush and therefore perish. The spores that happen to fall on barberries germinate and carry on the life cycle of the fungus. The germ tubes after penetrating the host tissues obtain nourishment from the host cell protoplasts through haustoria. Thus, a well-developed, haploid mycelium is formed. The nuclei carry either the (+) or (-) factor of the parent basidiospore. In nature, several basidiospores may reach and infect the same barberry leaf, as a result both (+) and (-) mycelium may develop side by side.

A few days after infection of the barberry leaf, the hyphae of the fungus nearest the upper epidermis develop flask-shaped spermagonia (pycnia). Each spermagonium contains numerous spermatiophores that cut off a succession of minute spermatia. These are exuded in small droplets of nectar through the opening (ostiole) at the tip of the spermagonium. Each spermatium contains a nucleus carrying (+) or (-) factor depending on the strain of mycelium that produced the spermagonium. All spermatia and receptive hyphae from a single spermagonium carry the same factor. The receptive hyphae arise in the spermagonium and protrude through the ostiole. The spermagonial periphyses may also change to receptive hyphae. Spermatia and receptive hyphae, some carrying the (+) factor and some the (-) factor are generally found on the same barberry leaves.

Spermatization takes place through the agency of insects. A fly or some other insect sips the sweet nectar exuding from the ostiole. During this process, spermatia in the nectar adhere to the mouthparts of the insect and are subsequently brushed off by the receptive hyphae and periphyses of the next spermagonium the insect visits. If (+) spermatia happen to be transferred to (-) receptive hyphae or (-) spermatia happen to be transferred to (+) receptive hyphae, spermatization is effected. The spermatial contents pass into the receptive hyphae through a pore dissolved in the wall at the point of contact, initiating the dikaryotic phase.

The dikaryotic mycelium penetrates the entire leaf and the hyphae form a number of aecial primordia near the lower epidermis. Each aecial primordium develops into an inverted cup-shaped aecium, which produces chains of hexagonal-shaped aeciospores attached to each other by disjunctor cells. The binucleate aeciospores formed in chains finally break through the lower epidermis of the barberry and the spores escape. Occasionally, aecia may develop on the upper surface of the leaves also. The aeciospores are disseminated by wind and if they happen to fall on susceptible grass hosts, germinate under favorable conditions to produce a binucleate mycelium. Thus, aeciospores can infect only the primary grass hosts and not barberry.

Soon after infection, the binucleate mycelium grows in the grass host and produces masses of cells to form uredinia. From the uredinia, binucleate urediniospores arise on fairly long stalks. The urediniospores are oval, yellowish and spiny. The pressure exerted by the developing spores causes the host epidermis to break into elongated, streak-like, reddish, rusty pustules. In masses, the urediniospores appear rust-red on the leaf surface and hence the disease is known as **'rust disease'**. The urediniospores are capable of germination immediately after maturity and can reinfect the grass host on which they were produced. As such, the urediniospores are the **'repeating spores'**, which perpetuate the fungus throughout the growing season and spread the disease from plant to plant and field to field.

As the host matures, the uredinia produce a few teliospores and gradually more and more teliospores and fewer urediniospores are produced and finally only teliospores are produced. Besides uredinia gradually changing into telia, telia are also produced directly from the dikaryotic mycelium produced by the germinating urediniospores. The telial pustules constitute the **'black stage'** of the rust. The masses of dark red teliospores appear black to the naked eye and hence the disease is also known as **'black rust' (Fig. 51)**

Biological specialization. A number of biological subspecies that differ only slightly in their morphology, but greatly in their ability to attack various grasses, constitute the species *Puccinia graminis*. To distinguish these subspecies, a third name has been added to the binomial of *Puccinia graminis*. Each subspecies is designated as follows:

Puccinia graminis tritici on wheat

Puccinia graminis hordei on barley

Puccinia graminis avenae on oats

Puccinia graminis secalis on rye

Further, each subspecies is composed of a large number of **'physiological races'** that differ in their parasitism on different varieties of the host species. *Puccinia graminis tritici* has more than 200 such races to wheat varieties, which are differentially susceptible. All these physiological races and subspecies have the common barberry bush as their alternate host.

The only practical method of controlling cereal rust is development of resistant varieties by continuous breeding for resistance. However, it is not possible to evolve varieties that are resistant to all the physiological races. New races of *Puccinia graminis* may originate by mutation, by hybridization or by introduction from other regions or countries.

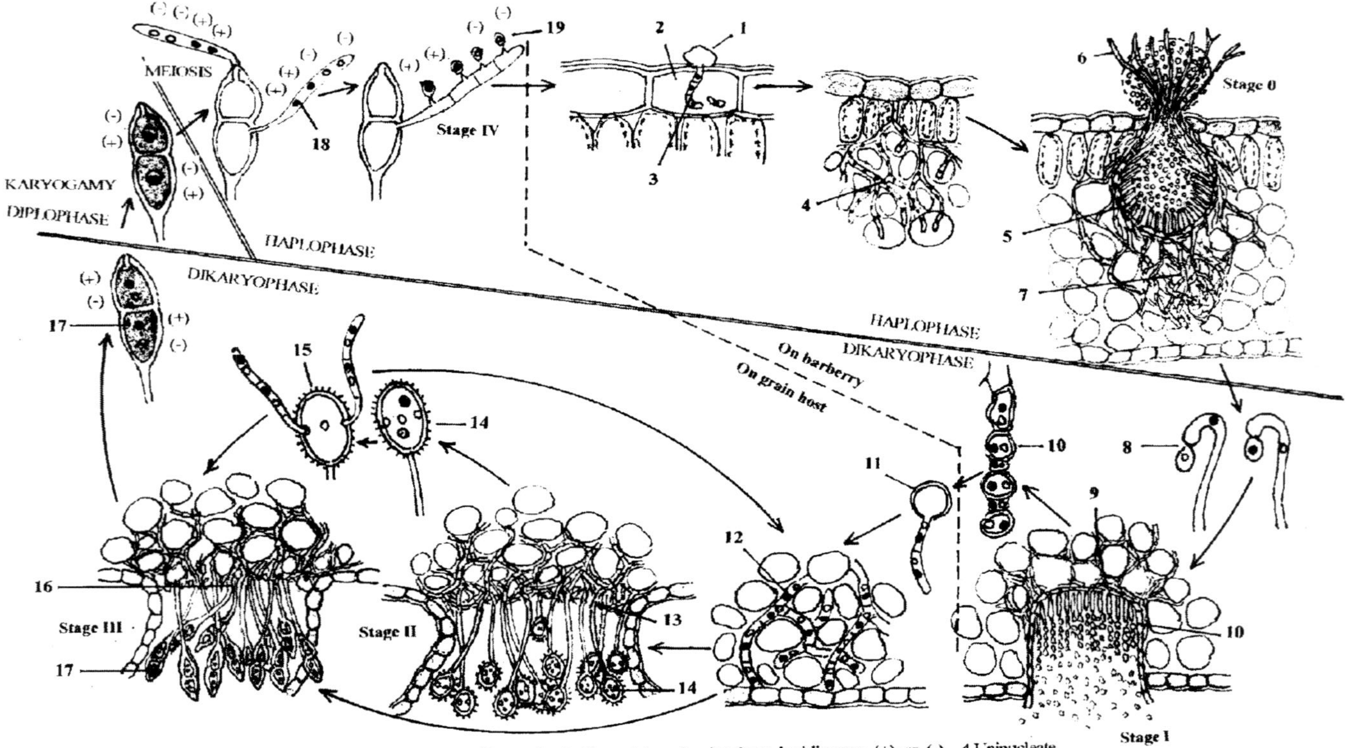

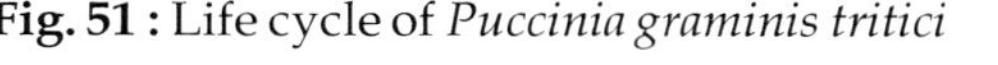

1.Basidiospore shell 2.Host cell 3. Germ tube of uninucleate basidiospore (+) or (-) 4.Uninucleate mycelium (+) or (-) 5. Spermogonia (+) or (-) 6.Receptive hyphae 7.Aecial primordium 8.Spermatization 9.Binucleate cells in aecial primordium 10.Aecium with chains of aeciospores 11.Germinating aeciospore 12.Binucleate mycelium in grain host 13.Uredium 14.Uredospore 15.Germinating uredospore 16.Telium 17.Teliospore 18.Promycelium 19.Basidiospore

Fig. 51 : Life cycle of *Puccinia graminis tritici*

Genus - *Uromyces*. Teliospores are stalked, one-celled and are produced singly. Aecia usually have persistent pseudoperidia. Spermogonia are sub-epidermal **(Fig. 52)**

(e-g) *Uromyces appendiculatus* - causes rust of bean; *U. caryophyllinus* - causes rust of carnation; *U. fabae* - causes rust of broad bean and lentil; *U. ciceris arietini* - causes rust of gram; *U. phaseoli typica* - causes rust of bean; *U. phaseoli vignae* - causes rust of cowpea; *U. pisi* - causes rust of pea.

Genus - *Phragmidium*. Teliospores long-stalked, several-celled, the cells arranged in a series and are surrounded by a gelatinous sheath, aecia without pseudoperidia of the uredo type, spermogonia sub-cuticular **(Fig. 52)**

(e-g) *Phragmidium rubi* - causes blackberry rust; *P. mucronatum* - causes rust of rose.

Genus - *Hemileia*. Teliospores usually produced in the same uredial sori. They are single-celled, pedicellate, round or turnip-shaped **(Fig. 52)**

(e-g) *Hemileia vastatrix* - causes coffee rust

Genus - *Phakopsora*. Telia sub-epidermal, but not erumpent. Teliospores sessile, irregularly arranged, not echinulated, single-celled and light-brown in color **(Fig. 52)**

(e-g) *Phakopsora gossypii (P. desmium)* - causes tropical rust of cotton; *P. pachyrrizi* - causes rust of soyabean.

Family - Melampsoraceae. The teliospores are sessile, single-celled and cylindrical and are borne in a layer. They germinate externally. Aecia do not possess a pseudoperidium

Genus - *Melampsora*. The teliospores are sessile, cylindrical, one-celled and reddish-brown in color and are arranged in a layer **(Fig. 52)**

(e-g) *Melampsora lini* - causes linseed rust; *M. ricini* - causes rust of castor.

Order - Ustilaginales (Smuts). The fungi included in the Order - Ustilaginales are commonly known as **'smuts'** because they produce black, dusty spore masses that resemble soot or smut. All smut fungi are plant pathogens. Economically, the smuts are very important because they cause severe grain losses. Farmers dread the smuts of cereals even more than the rusts, because many smuts attack the grain kernels and replace the kernel contents with smut spores. Thus, the reduction in yield is direct and the quality of the remaining produce is also drastically reduced because of the presence of the black smut spores on the surface of healthy grains. Besides various cereals, smuts also affect sugarcane, onions and some ornamental plants.

Most smut fungi attack the ovaries of grains and destroy the kernels completely, while several other smuts attack leaves, stems, roots, floral parts etc. Some smuts are systemic and grow internally in the seedling until they reach the inflorescence, while others cause only local infections on various plant parts. Black, smut spores replace cells in affected tissues. The spores are present in masses that may be held together temporarily by a thin, flimsy membrane or by a tougher membrane. Smut fungi rarely kill their hosts, but in some cases the infected plants may become severely stunted. Most smut fungi produce only two kinds of spores viz., teliospores or smut spores or chlamydospores and basidiospores. The teliospores are usually formed from mycelial cells along the length of the mycelium within the smut galls and the basidiospores, either bud off laterally from the basidium cells or are produced as a cluster at the tip of a non-septate basidium. The basidiospores of the smuts are not borne on sterigmata. When basidiospores germinate, the germ tubes either unite with compatible ones while still on the basidium and then infect the host or they penetrate the tissues directly. Their haploid mycelium, however, cannot invade tissues and does not cause typical infections until two compatible mycelia unite to produce dikaryotic mycelium. The dikaryotic mycelium then invades the host tissues inter- or intracellularly and produces the typical symptoms and teliospores or smut spores.

At the time of sporulation, the dikaryotic mycelium develops profusely in certain portions of the host, lays down many septa and forms masses of hyphae composed of short, dikaryotic cells. The protoplast of each hyphal cell rounds up and the hyphal wall gelatinizes. Each protoplast then secretes around itself a thick wall that eventually converts the protoplast into a round teliospore. Since smut teliospores are formed in the manner of chlamydospores, some authors refer to these spores as **'chlamydospores'**. Teliospore characteristics, especially the surface ornamentation is of much importance in the taxonomy of the Ustilaginales. The spores may be reticulate, spiny, tuberculate or smooth. Spore size, shape and color of individual teliospore are some of the other important characteristics in the taxonomy of the smuts. In many species, the spores are entirely free from each other, while in some species the spores cling together in groups or several spores may be cemented together and form specialized spore balls.

Asexual reproduction in the smuts is by means of conidia produced both from uninucleate and from binucleate mycelium. Budding is also a very common method of reproduction in the Ustilaginales. Both basidiospores and conidia may bud repeatedly.

Several races are present in the smut fungi however, the races of smut fungi are not as stable as the rust fungi because of repeated meiosis and gene recombination.

The Order - Ustilaginales is divided into two Families viz., **Ustilaginaceae** and **Tilletiaceae** based on teliospore germination. In the Ustilaginaceae the promycelium is transversely septate with lateral and terminal basidiospores. In the Tilletiaceae the promycelium is aseptate and basidiospores are produced only terminally.

Family - Ustilaginaceae. A number of economically important plant pathogens are present in the Family - Ustilaginaceae. The young teliospore is at first binucleate (dikaryotic). The nuclei fuse and the matured spore is uninucleate and diploid. The spores may germinate immediately or they may undergo a dormant, rest period before germination. At the time of germination, the spore wall cracks open and the promycelium emerges. The diploid nucleus migrates into the promycelium, undergoes meiosis and the four resulting haploid nuclei distribute themselves more or less uniformly in the promycelium. In some species, meiosis takes place in the teliospore itself and the nuclei migrate into the promycelium. Subsequently, transverse septa are formed separating each nucleus from one another. Each nucleus undergoes mitotic division and while one nucleus migrates into the bud that arises laterally from each of the promycelial cell, the other remains in the cell. These uninucleate buds develop into **'basidiospores'** or otherwise called **'sporidia'**. During meiosis, the genes controlling sexual compatibility segregate, as a result different strains of basidiospores are formed. After the release of the basidiospores, budding of basidiospores takes place. Dikaryotization takes place either by copulation of two secondary basidiospores of opposite strains or by the union of hypha arising from basidiospores of opposite strain. Without dikaryotization the uninucleate mycelium is short-lived and is not capable of teliospore production. Regardless of the manner in which dikaryotization occurs, it is only the binucleate mycelium that can carry on the life history of the fungus. This secondary binucleate (dikaryotic) mycelium in the host may produce branches that reach the surface of the host and produce binucleate conidia that are disseminated by wind and produce new infections. Later, the binucleate mycelium forms smut sori in which teliospores are produced in abundance. Generally, no symptoms appear on the infected host until smut balls are formed.

The smut fungi overwinter as teliospores in the soil or mycelium inside the host. In some species, the teliospores remain dormant for many years in the soil or in the host debris.

Genus - *Ustilago*. In the smut sori, the spore layer is without any sterile hyphae. The spores are borne singly

(e-g) *Ustilago maydis* - causes loose smut of corn; *U. avenae* - causes loose smut of oats; *U. nuda* - causes loose smut of barley; *U. hordei* - causes covered smut of barley; *U. tritici* - causes loose smut of wheat; *U. scitaminea* - causes smut of sugarcane; *Ustilaginoidea (Ustilago) virens* - causes false smut of rice.

Life cycle of *Ustilago maydis*. The pathogen causing corn smut disease, overwinters as teliospores in crop debris and in the soil, where it can remain viable for several years. The teliospore germinates under favorable conditions and produces four-celled basidium (promycelium) and from each cell, a basidiospore (sporidium) is formed. The basidiospores are carried by air currents or are splashed by water to young, developing tissues of corn plants. The basidiospore germinates and produces a fine hypha, which can enter the epidermal cells directly. However, after initial development in the host, its growth stops and the hypha usually dies, unless it contacts and fuses with a haploid hypha derived from a basidiospore of the compatible mating type. After fusion, the resulting hypha becomes dikaryotic, enlarges in diameter and grows into the plant tissues mostly intercellularly. The cells surrounding the hypha are stimulated to enlarge and divide and galls are formed. Galls are formed on the stem region, as well as in the earheads replacing the kernels. Later, the cells of the dikaryotic mycelium are transformed into black, spherical or ellipsoidal teliospores inside the galls **(Fig. 53)**

Genus - *Sphacelotheca*. The spore layer of the smut sori is covered with a hull of sterile hyphae. The chlamydospores or smut spores are borne singly

(e-g) *Sphacelotheca sorghi* - causes grain smut or short smut or covered smut of sorghum; *S. cruenta* - causes loose smut of sorghum; *S. reiliana* - causes head smut of sorghum.

Genus - *Tolyposporium*. The smut spores are cemented together in the form of a spore ball.

(e-g) *Tolyposporium penicillariae* - causes grain smut of pearl millet; *T. ehrenbergii* - causes long smut of sorghum.

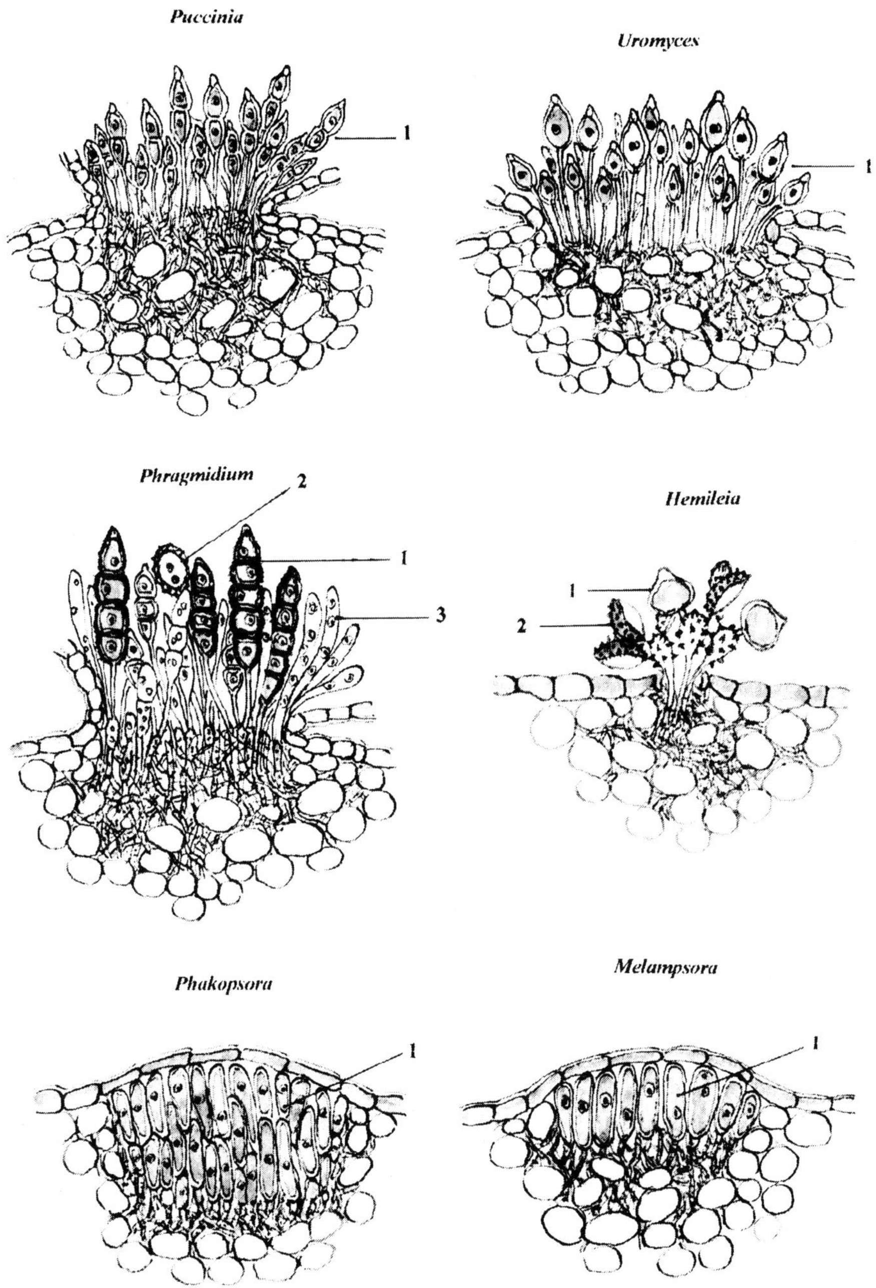

1.Teliospores 2.Uredospores 3.Periphyses

Fig. 52 : Morphological variations of telial stages of Uredinales

Family - Tilletiaceae. The Tilletiaceae differ from Ustilaginaceae in the method of teliospore germination. The diploid nucleus of the matured teliospore undergoes meiosis at the time of germination, followed by a mitotic division resulting in the formation of eight haploid nuclei, which migrate into a promycelium arising from the teliospore. The promycelium does not become septate as in the case of Ustilaginaceae. Instead a number of basidiospores, typically eight, are formed at the tip of the promycelium. One nucleus migrates into each of the basidiospore. The basidiospores are of opposite strains, half of them carrying the **(+)** and the other half **(-)** factor for sexual compatibility. Copulation takes place between two compatible basidiospores, uniting them in the form of **'H'**, even when they are still attached to the promycelium. Plasmogamy takes place between the two, with the protoplast from one basidiospore migrating through the connecting tube into the other basidiospore. Crescent-shaped secondary sporidia (conidia) generally develop on sterigmata produced by the binucleate member of the **'H'** piece. The nuclei divide conjugately and one pair migrates into the conidium. After discharge, the conidia germinate and develop binucleate mycelium that infects the host. Ultimately, this mycelium produces smut sori and teliospores. Binucleate mycelium arising from the binucleate member of the **'H'** piece may also infect the host directly.

Genus - *Tilletia*. Smut spores produced singly and dispersed like dust

(e-g) *Tilletia caries* and *T. foetida* - cause bunt or covered smut of wheat; *T. indica* - causes kernel bunt of wheat; *Tilletia (Neovossia) barclayana* - causes bunt or kernel smut or black smut of rice.

Life cycle of *Tilletia caries*. The pathogen causing bunt disease of wheat, overwinters as teliospores on contaminated wheat kernels and to some extent in the soil. As the young seedling emerges from the kernel, the teliospores on the kernel or near the seedling in the soil also germinate. Just prior to germination, the diploid nucleus in the teliospore undergoes meiosis followed by zero to two simultaneous mitotic divisions. The haploid nuclei migrate into the promycelium, which is produced from the teliospore. From the tip of the promycelium, a number of basidiospore initials, typically eight are formed. Each nucleus in the promycelium moves into a basidiospore initial. After the nucleus has entered a spore initial, it divides mitotically and one of the daughter nucleus moves back into the promycelium. These nuclei may either enter anucleate basidiospore initials (basidiospore initials without nucleus) or eventually disintegrate. *Tilletia caries* being heterothallic, some basidiospores are of one mating type, while the others are of the opposite mating type. Copulation tubes develop between two compatible basidiospores,

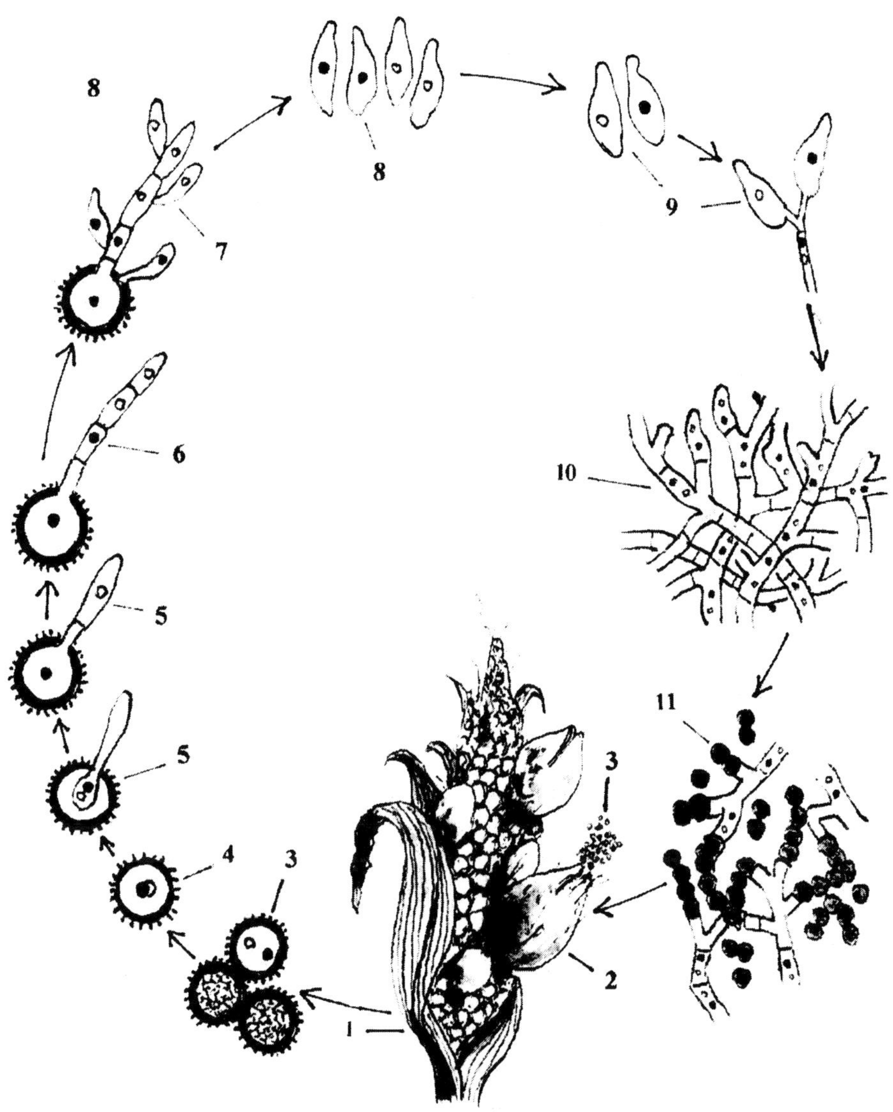

1.Galls on ear of corn 2.Galls full of teliospores 3.Teliospores 4.Zygote 5.Germinating teliospore 6.Formation of basidium 7.Formation of basidiospores 8.Basidiospores 9.Compatible basidiospores 10.Dikaryotic mycelium in galls 11.Formation of teliospores inside the galls

Fig. 53 : Life cycle of *Ustilago maydis*

uniting them into '**H**' pieces, while they are still attached to the promycelium. Plasmogamy then takes place between the members of the '**H**' piece and crescent-shaped secondary sporidia develop on sterigmata produced by one member of the '**H**' piece. The nuclei pass into the secondary sporidium and divide conjugately and one pair passes back into the parent member of the '**H**' piece. After discharge and germination, the secondary sporidia develop a binucleate mycelium that infects the host. Binucleate mycelium issuing from the '**H**' piece may also infect the host directly. Ultimately, the binucleate mycelium in the host tissues produces teliospores that are mixed with sterile cells **(Fig. 54)**

Genus - *Entyloma*. Smut spores produced singly and the spores remain within the host tissue

(e-g) *Entyloma oryzae* - causes leaf smut of rice; *E. dahliae* - causes leaf smut of dahlia.

Genus - *Urocystis*. Smut spores clustered in balls. Spore balls possess sterile marginal spores and are dispersed freely

(e-g) *Urocystis cepulae* - causes smut of onion.

Subdivision 4 - Deuteromycotina

Form class - Deuteromycetes (The imperfect fungi). Several fungi that have septate hyphae reproduce only asexually by producing conidia. These fungi are included in the Form class - Deuteromycetes. Since these fungi do not have a sexual phase (perfect stage), they are commonly called **'imperfect fungi'** or **'Fungi Imperfecti'**. The conidial stages of most of these fungi are quite similar to the conidial stages of some of the common Ascomycetes. However, except a few, the ascigerous stages of these fungi are formed very rarely in nature or have not been discovered yet or have been dropped from their life cycle during the evolution of these organisms. Some of these fungi, which were named and described as imperfect fungi, have been found to produce their sexual stages in nature or in culture several years later. These organisms have subsequently been classified under Ascomycetes based on the characteristics of their ascigerous stages. In a few cases, the perfect stages discovered at a later period have been found to belong to Basidiomycetes. However, there are still many Deuteromycetes that do not produce sexual stages at all.

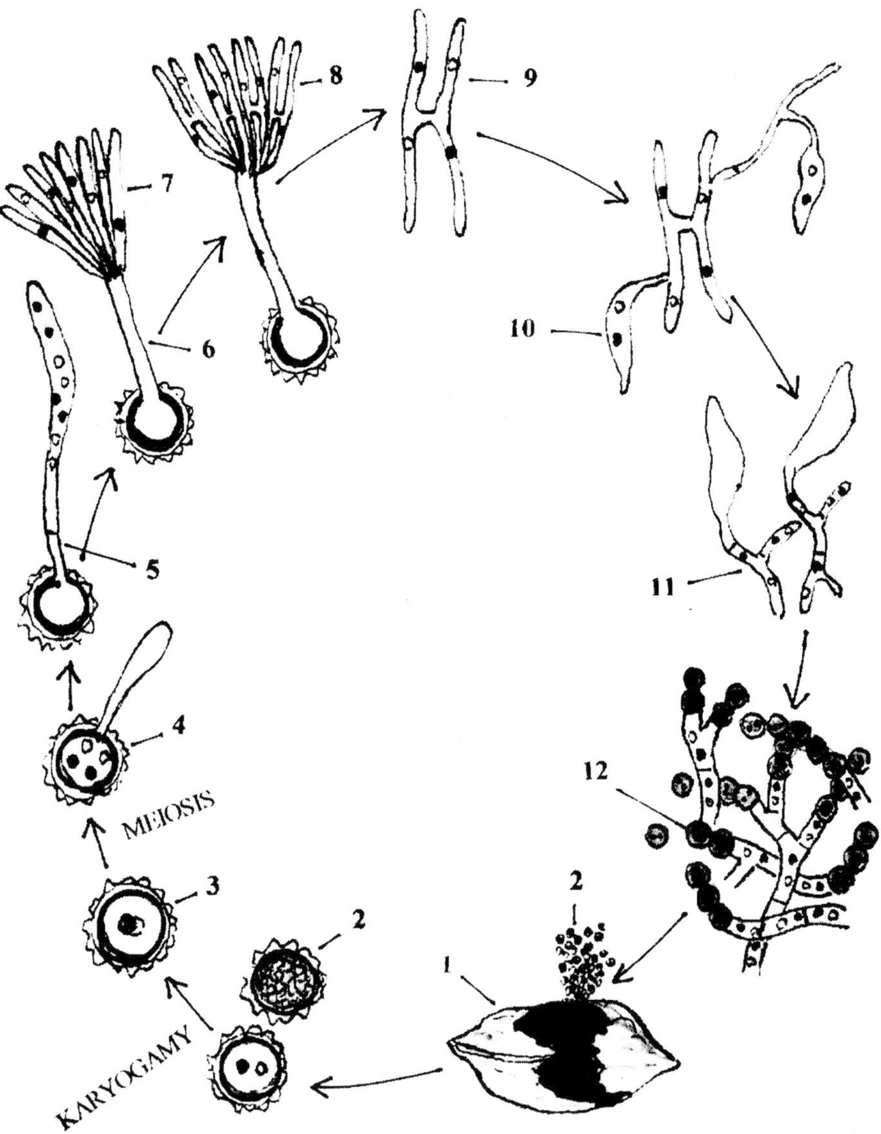

1.Smutted wheat kernel full of teliospores 2.Teliospores 3.Zygote 4.Germinating spore 5.Formation of basidium 6.Formation of basidiospores 7.Basidiospore 8.Fusion between basidiospores of opposite mating type 9.Plasmogamy 10.Binucleate conidium 11.Binucleate mycelium 12.Mycelium inside kernels producing teliospores

Fig. 54 : Life cycle of *Tilletia caries*

About 15,000 fungal species have been included in the Subdivision - Deuteromycotina. While most of them are terrestrial, a few are found to have marine or fresh water habitats. Most of them are either saprophytic or weak parasites on plants. Some are parasitic on other fungi or nematodes. The parasitic forms are of great importance, because they cause some of the serious diseases of plants, animals and humans. Further, some are used in the commercial production of certain chemicals and antibiotics. Species of Form genus - *Penicillium* are capable of producing **'Penicillin'**, one of the most important antibiotics.

Somatic structure. The Deuteromycetes typically produce well-developed, septate, branched hyphae, except in the case of Blastomycetidae, which produce yeast-like cells. The cells in the hyphae are multinucleate.

Reproduction. All Deuteromycetes, except one group under Form order - Agonomycetales reproduce by means of spores known as conidia. A **'conidium'** is a non-motile, asexual spore produced directly from the somatic hyphae or produced in more complex structures, such as **'pycnidia'**, **'acervuli'**, **'sporodochia'** or **'synnemata'** (coremia). However, species that produce conidia on conidiophores arising directly from the somatic hyphae reproduce more rapidly. Sporulation in the Deuteromycetes is strictly an asexual process.

There are considerable variations in the morphology of conidia. They may be spherical, ovoid, elongated, cylindrical, thread-like, spirally curved, star-shaped etc. They may be single-celled to many-celled, with either transverse or both transverse and longitudinal septa. Further, conidia may be hyaline or colored.

Conidia may be produced singly or in groups. In some species, the conidia are embedded in a droplet of mucus. Conidia produced in chains are called **'catenulate'**. New conidia may be added to the chain at random or in a regular manner, either at the tip or at the base of the chain depending on the species. If the oldest conidium is at the tip and the youngest at the base of the chain, then the conidial succession is said to be **'basipetal'**. On the other hand, if the oldest conidium is at the base and the youngest at the tip, then the conidial succession is said to be **'acropetal'**.

The hyphal cell from which or in which a conidium is directly formed is called the **'conidiogenous cell'** or **'conidiophore'**. A conidiophore is a simple or branched hypha arising from a somatic hypha. Conidiophores may be formed singly or grouped together to form specialized structures, such as **'sporodochia' (Fig. 12)** or **'synnemata'** (coremia) **(Fig. 13)** or produced in fruiting bodies known as pycnidia or acervuli. A **'synnema'** (pl.-synnemata) consists of a group of conidiophores, usually united at the basal portion and the conidiophores are often branched at the top. Conidia arise from the tips of numerous conidiophores. In some synnemata the stalk-like portion is longer than the branched top portion. In a **'sporodochium'** (pl.-sporodochia) the conidiophores arise from a central, cushion-shaped stroma. The conidiophores are packed tightly together and conidia are produced from the tips of the conidiophores.

A **'pycnidium'** (pl. pycnidia) is a globose or flask-shaped structure, the inside of which is lined with numerous conidiophores bearing conidia at their tips. Pycnidia may be completely closed or may possess an ostiole. They may have a small papilla or a long neck leading to the ostiole. Pycnidia vary greatly in size, shape and color, as well as in the consistency of the pseudoparenchymatous wall. They may be superficial or sunken in the substratum **(Fig. 55)**

An **'acervulus'** (pl. acervuli) is a flat or saucer-shaped stromatic hyphal bed, from which short conidiophores grow side by side in large numbers. They are mostly produced sub-epidermally or sub-cutaneously and ultimately become erumpent. They lack definite wall structure and possess no ostiole **(Fig. 56)**

Classification. The classification of imperfect fungi is based on the morphological characteristics of their asexual reproductive structure. However, this is only an arbitrary system of classification in the absence of the formation of the perfect stage.

Whenever the perfect stage of the so-called 'imperfect fungi' is observed, it is then classified under the respective Class based on the characteristics of the sexual stage. Here also anomalies do occur sometimes. Two fungi classified earlier in a particular Form genus under imperfect fungi, because they produce almost identical conidia might have quite different sexual stages to place them in different ascomycetous Genera. For example, the two imperfect fungi, *Septoria rubi* and *S. avenae*, classified under the Form genus -*Septoria* based on their conidial characters were found to have different perfect stages viz., *Mycosphaerella rubi* and *Leptosphaerella avenaria* respectively of Ascomycetes.

Similarly, two Ascomycetes belonging to the same Genera may have quite different conidial stages. For example, *Mycosphaerella fragariae* on strawberry produces conidia on unorganized conidiophores whereas, *Mycosphaerella rubi* on raspberry bears conidia inside a pycnidium. So, in the arbitrary system of classification of the Deuteromycetes, the conidial stage of *Mycosphaerella fragariae* should be placed in the Form genus *Ramularia* of Form order - Moniliales, while that of *Mycosphaerella rubi* should be placed in the Form genus *Septoria* of the Form order - Sphaeropsidales.

Further, many Ascomycetes, especially the parasitic ones form their sexual stages only once in a year or more than one year, though they produce their conidial stages quite often. By including these fungi under Deuteromycetes in the arbitrary system of classification, it is quite easy to identify a fungus by its conidial stage in the absence of its ascigerous stage.

According to the **'International Rules of Nomenclature'**, a fungus is to have only one valid name. Consequently if a fungus, which had been classified earlier under Deuteromycetes, is found to produce an ascigerous stage, then the name of the imperfect stage is given in parentheses after the valid name, that is the name of the ascigerous stage. Thus *Spilocea pomi* the conidial stage of apple scab, the ascigerous stage of which was later found as *Venturia inaequalis* is referred to as *Venturia inaequalis* (= *Spilocea pomi*). However, when dealing only with the conidial stage of this fungus and not the apple scab disease as a whole, it is quite valid to refer the fungus as *Spilocea pomi*. In case, the ascigerous stage of imperfect fungi is not yet found, it is quite valid to name it according to its conidial characteristics in the arbitrary system of classification under the Class - Deuteromycetes.

The Form class - Deuteromycetes is divided into three Form subclasses viz., Coelomycetidae, Hyphomycetidae and Blastomycetidae of which the former two are of economic importance.

Form subclass - Coelomycetidae. The fungi belonging to this Form subclass have well-developed, septate hyphae. Reproduction is by means of conidia borne in specific fruiting bodies, such as pycnidia or acervuli. Those forms producing pycnidia are placed in the Form order - Sphaeropsidales, while those producing acervuli are placed in the Form order - Melanconiales.

Form order - Sphaeropsidales. The chief characteristic of the Sphaeropsidales is the formation of **'pycnidium'**. The appearance of the pycnidia differs from one form species to the other. They may be superficial or immersed in the substratum; globose, elongate, papillate, beaked or cup-shaped; unilocular or multilocular; light-colored or dark-colored; produced individually or may be aggregated. Based on the pycnidial characters, the Form order is divided into four Form families viz., **Sphaeropsidaceae** - pycnidia dark-colored, globose, leathery to carbonous; **Nectrioidaceae** - pycnidia similar to those of Sphaeropsidaceae, but light-colored and soft or waxy; **Leptostromataceae** - pycnidia shield-shaped, elongate or flattened and **Exipulaceae (Discellaceae)** - pycnidia cup-shaped or saucer-shaped, of which the former two are of economic importance.

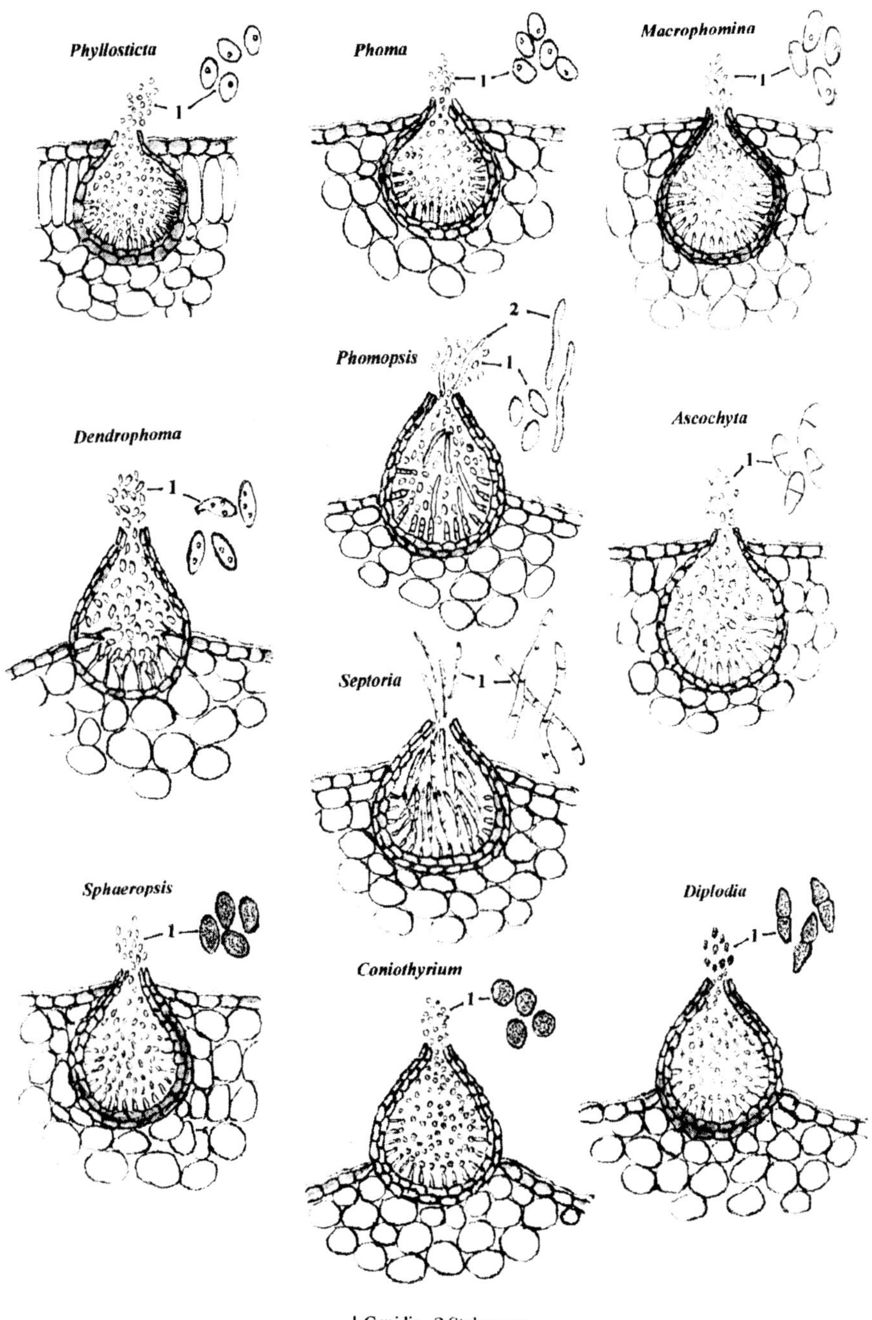

Fig. 55 : Pycnidia and conidia of important form genera of Sphaeropsidales

Form family - Sphaeropsidaceae. This Form family consists of saprophytic, as well as plant parasitic species. The pycnidia are dark-colored, globose, leathery to carbonous, stromatic or non-stromatic, generally provided with a circular ostiole. Of the **'hyalosporae'** (fungi that produce hyaline spores) the following Form genera are important.

Form genus - *Phyllosticta*. Pycnidia small, ostiolate, sunken in the substratum, phialides very short, conidia hyaline, spherical or oval and measure upto 15μ in size, attacks mostly leaves **(Fig. 55)**

(e-g) *Phyllosticta maydis* - causes yellow leaf blight of corn; *P. solitaria* - causes apple blotch; *P. citricarpa* - causes black spots of *Citrus*; *P. cajani* - causes leaf spot or frog's eye spot of red gram; *P. phaseoli* - causes leaf spot of cowpea; *P. arachidis* - causes leaf spot of groundnut; *P. bosanis* - causes leaf spot of castor; *P. nicotianae* - causes leaf spot of tobacco; *P. brassicola* - causes ring spot of cabbage and cauliflower. Its perfect stage is *Mycosphaerella brassicola*; *P. sulata* - causes leaf spot of papaya; *P. pirina* - causes leaf spot of apple.

Form genus - *Phoma*. Pycnidia small, ostiolate, sunken in the substratum, phialides very short, conidia hyaline, spherical or oval, conidia measure upto 15μ in size, attacks mostly the stem region **(Fig. 55)**

(e-g) *Phoma betae* - causes leaf blight, seedling and root rot (black leg) and tuber rot (heart rot) of sugarbeet. Its perfect stage is *Pleospora bjorlingii*; *P. cajani* - causes stem canker of redgram; *P. medicaginis* - causes blight of redgram; *P. exigua* - causes leaf blight of sunflower and canker of lady's finger; *P. piperina-betle* - causes leaf spot of betelvine; *P. destructiva* - causes fruit rot of tomato; *P. arachidicola* - causes web blotch or leaf spot of groundnut. Its perfect stage is *Didymella (Mycosphaerella) arachidicola*; *P. lingam* - causes black leg of crucifers. Its perfect stage is *Leptosphaeria maculans*; *P. glomerata* - causes blight of mango.

Form genus - *Macrophomina*. Pycnidia similar to those of *Phyllosticta*, conidia cylindrical, more than 15μ in length **(Fig. 55)**

(e-g) *Macrophomina phaseolina* - causes charcoal rot (root rot) of soybean, potato and sorghum, dry root rot and stem blight of cowpea, root rot of cotton, dry root rot of groundnut, root and stem rot of gingelly, root rot of betelvine, leaf blight of lady's finger, rot of onion etc. Its sterile stage is *Rhizoctonia bataticola*, which causes diseases in several different hosts.

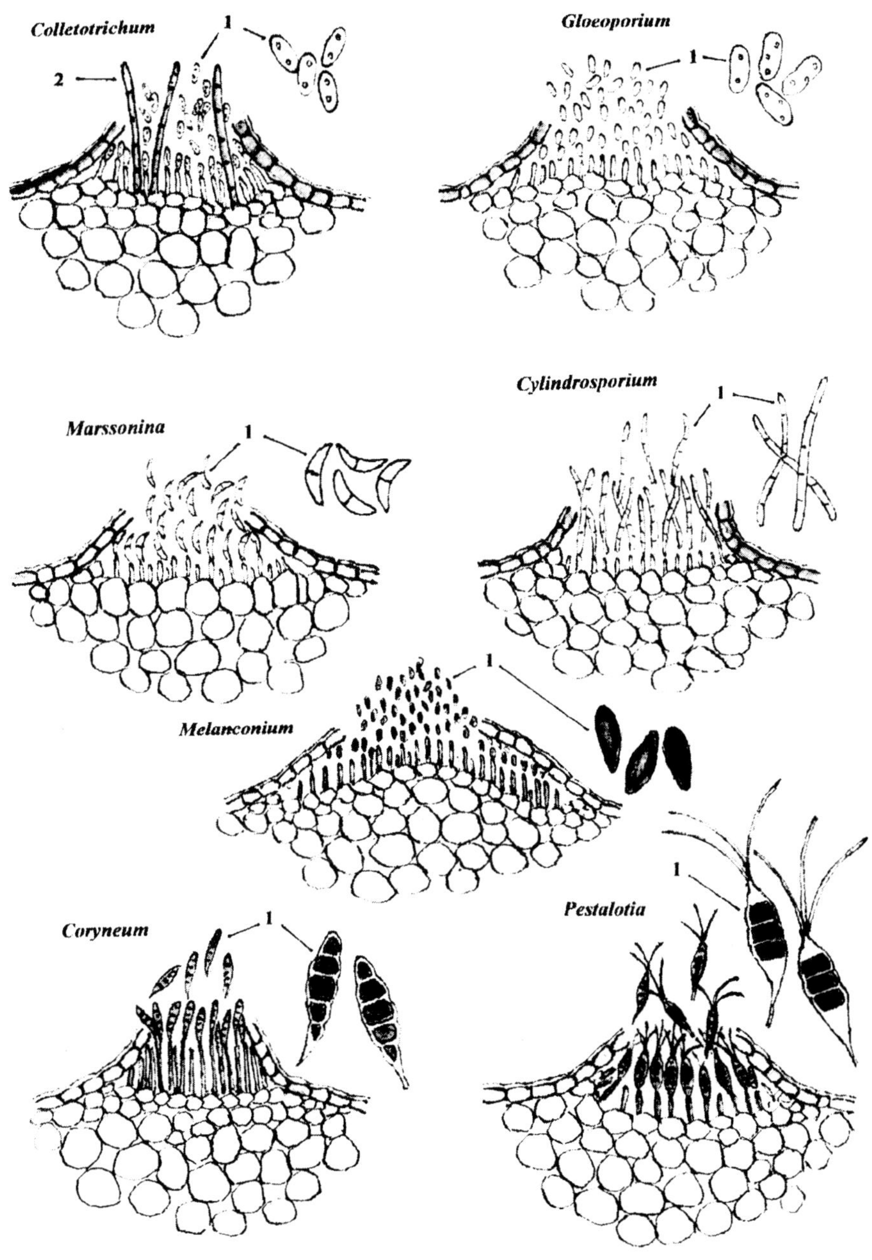

1.Conidia 2.Seta

Fig. 56 : Acervuli and conidia of important genera of Melanconiales

Form genus - *Dendrophoma*. Pycnidia similar to those of *Phyllosticta*, conidiophores long and branched, conidia hyaline, oval and upto 15μ in size **(Fig. 55)**

(e-g) *Dendrophoma obscurans* - causes strawberry leaf blight.

Form genus - *Phomopsis*. Pycnidia similar to those of *Phyllosticta*, produces two types of conidia referred to as **alpha** and **beta** conidia. Alpha conidia are similar to those of *Phoma*, while beta conidia, also called as **'stylospores'** are elongated and in some species bent at the top like a walking stick. The perfect stage of all *Phomopsis* found till now, belongs to the Genus - *Diaporthe* of the Ascomycetes **(Fig. 55)**

(e-g) *Phomopsis cinerescens* - causes fig canker; *P. sojae* - causes stem and pod blight of soybean and leaf spot of groundnut. Its perfect stage is *Diaporthe sojae*; *P. citri* - causes stem blight of *Citrus* plants. Its perfect stage is *Diaporthe citri*; *P. viticola* - causes soft rot of grapevine; *P. psidii* - causes fruit rot of guava; *P. vexans* - causes blight and fruit rot of eggplant. Its perfect stage is *Diaporthe vexans*; *P. manihotis* - causes leaf spot of tapioca; *P. palmicola* - causes leaf blight of arecanut.

Among the single-celled **'phaeosporae'** (fungi producing dark-colored spores), the following Form genera are important.

Form genus - *Sphaeropsis*. Pycnidia similar to those of *Phyllosticta*. Conidia dark-colored, spherical or oval **(Fig. 55)**

(e-g) *Sphaeropsis malorum* - causes black rot of apple and frog's eye spot on leaves. Its perfect stage is *Physalospora obtusa*, which belongs to the Ascomycetes.

Form genus - *Coniothyrium*. Pycnidia similar to those of *Phyllosticta*. Conidia are spherical and dark colored, slightly smaller than the conidia of *Sphaeropsis* **(fig. 55)**

(e-g) *Coniothyrium fuckelii* - causes stem canker of rose and raspberry. Its perfect stage is *Leptosphaeria coniothyrium* that belongs to the Ascomycetes; *C. diplodiella* - causes blight of grapevine; *C. pyrinum* - causes leaf spot of apple; *C. jasmini* - causes leaf spot of jasmine.

Among the fungi which produce two-celled or more than two-celled conidia, the following Form genera are important.

Form genus - *Ascochyta*. Pycnidia similar to those of *Phyllosticta*. Conidia two-celled and hyaline **(Fig. 55)**

(e-g) *Ascochyta rabiei* - causes blight of gram. Its perfect stage is *Mycosphaerella rabiei*; *A. pisi* - causes blight of pea and lentil; *A. carthami* - causes leaf spot of safflower; *A. nicotianae* - causes leaf spot of tobacco; *A. pinodella* - causes leaf, stem and pod spot and foot rot of pea. Its ascigerous stage is *Mycosphaerella pinodes*; *A. fabae* - causes brown spot of broad bean; *A. chrysantheme* - causes black rot of *Chrysanthemum*. Its perfect stage is *Mycosphaerella (Didymella) ligulicola*; *A. lycopersici* - causes leaf spot of tomato. Its perfect stage is *Didymella lycopersici*.

Form genus - *Diplodia*. Pycnidia similar to those of *Phyllosticta*. Conidia two-celled and brown in color **(Fig. 55)**

(e-g) *Diplodia natalensis* - causes sett rot of tapioca, gummosis disease in *Citrus*, cucurbits and a number of other cultivated crops. Its perfect stage is *Physalospora rhodina*; *D. cajani* - causes stem canker of redgram; *D. gossypina* - causes boll rot of cotton. Its perfect stage is *Glomerella gossypii (Thanatephorus cucumeris)*; *D. rosarum* - causes die-back of rose.

Form genus - *Septoria*. Pycnidia almost similar to those of *Phyllosticta*. Conidia long, slender and filiform (scolecospore), with one to several septa, generally slightly curved and hyaline or greenish in color. When the pycnidia get wet, they swell and the conidia are exuded in long tendrils **(Fig. 55)**

(e-g) *Septoria lycopersici* - causes leaf spot of tomato; *S. chrysanthemella* - causes leaf spot of *Chrysanthemum*; *S. apii* - causes blight of celery; *S. tritici* - causes leaf blotch of wheat and other cereals; *S. vignicola* - causes leaf spot of cowpea; *S. glycines* - causes brown spot of soybean; *S. helianthi* - causes leaf spot of sunflower; *S. pisi* - causes blight of pea; *S. bataticola* - causes concentric leaf spot of sweet potato; *S. rosae* - causes leaf spot of rose. Its perfect stage is *Sphaerulina rehmiana*. Many Ascomycetes have *Septoria* imperfect stages. *Mycosphaerella rubi* (= *Septoria rubi*) - causes leaf spot of blackberry and *Mycosphaerella sentina* (= *Septoria pyricola*) - causes pear leaf spot.

Form family - Nectrioidaceae. The pycnidia are generally light- or bright colored and their walls are soft or waxy. They may be either stromatic or nonstromatic; conidia single-celled.

Form genus - *Zythia*. Conidia single-celled

(e-g) *Zythia fragraiae*. Its perfect stage is *Gnomonia fragariae (fructicola)* of the Ascomycetes.

Form order - Melanconiales. The Melanconiales are assembled in a single Form family - the **Melanconiaceae**. Many are plant parasites and cause anthracnose disease in many crops. The characteristic **'acervuli'** produced may be sub-epidermal or sub-cuticular. When the conidia mature, the acervuli become erumpent and the conidia are released in droplets of white, cream, pink, orange or black fluid.

Form genus - *Colletotrichum*. Acervuli are sub-epidermal and become erumpent on maturity. Conidia are hyaline, single-celled, ovoid or elongated, sometimes slightly curved with rounded ends and are slightly narrower in the middle than at the ends. They are produced from the tips of phialides. Masses of conidia appear pink. Dark, long, sterile, hair-like setae are often formed in the acervuli **(Fig. 56)**

(e-g) *Colletotrichum falcatum* - causes red rot disease of sugarcane. Its perfect stage is *Glomerella (Physalospora) tucumanensis* of the Ascomycetes; *C. gloeosporioides* - causes anthracnose of mango fruits, leaf spots, blossom blight and wither tip of mango, anthracnose of *Citrus*, leaf spot of pomegranate, anthracnose of cashew, leaf blight of jasmine, anthracnose or pollu disease of pepper, inflorescence die- back of arecanut etc. Its perfect stage is *Glomerella cingulata*; *C. manihotis* - causes anthracnose and die- back of tapioca. Its perfect stage is *Glomerella manihotis*; *C. capsici* - causes ripe fruit rot and die-back of chillies, leaf spot of turmeric, leaf spot of rose etc. *C. zingiberis* - causes leaf spot of ginger. Its perfect stage is *Glomerella cingulata*; *C. circinans* - causes smudge disease of onion; *C. graminicola* - causes red spots on leaves and stem rot of sorghum, maize and several other cereals and grasses; *C. lindemuthianum* - causes anthracnose of beans and lab lab and leaf spot of onion. Its perfect stage is *Glomerella lindemuthianum*; *C. lagenarium* - causes anthracnose of cucurbits; *C. phomoides* - causes anthracnose or fruit rot of eggplant and tomato; *C. coffeanum* - causes anthracnose of coffee; *C. pisi* - causes anthracnose of pea; *C. papayae* - causes anthracnose of papaya; *C. melongenae* - causes anthracnose of eggplant.

Form genus - *Gloeosporium*. Acervuli and conidia are similar to that of *Colletotrichum*, except that in *Gloeosporium* no setae are produced in the acervuli **(Fig. 56)**

(e-g) *Gloeosporium album* - causes bitter rot of apple. Its perfect stage is *Glomerella cingulata*; *G. ampelophagum* - causes anthracnose of grapevine. Its perfect stage is *Elsinoe ampelina*; *G. musarum* - causes banana anthracnose or fruit rot of banana; *G. mangiferae* - causes shoot and inflorescence blight of mango; *G. psidii* - causes anthracnose of guava; *G. concentricum* - causes anthracnose of cabbage and cauliflower.

Form genus - *Melanconium*. Acervuli, as well as conidia are dark-colored, conidia single-celled and are produced at the tips of annelides (a type of conidiogenous cells).

(e-g) *Melanconium fuligineum* - causes bitter rot of grapes.

Form genus - *Marssonina*. Conidia two-celled and hyaline **(Fig. 56)**

(e-g) *Marssonina rosae* - causes black spot of rose. Its perfect stage is *Diplocarpon rosae*.

Form genus - *Pestalotiopsis (Pestalotia)*. The conidia are several-celled and dark with hyaline, pointed cells at both the ends. Two or more hyaline apical appendages are present on each spore **(Fig. 56)**

(e-g) *Pestalotia theae* - causes gray blight of tea; *P. guepini* - attacks camellias and other ornamentals; *P. palmarum* - causes gray leaf spot or blight of coconut and oil palm; *P. psidii* - causes fruit canker of guava; *P. versicolor* - causes leaf spot of sapota and fruit spot of pomegranate; *P. mangiferae* - causes gray blight or leaf spot of mango; *P. microspora* - causes gray blight of cashew; *P. meñezesiana* - causes fruit rot of grapes.

Form genus - *Cylindrosporium*. The conidia are long and slender **(Fig. 56)**

(e-g) *Cylindrosporium hiemale* - causes leaf spot of cherries. Its perfect stage is *Higginsia hiemalis*; *C. pomi* - attacks apple; *C. chrisanthemi* - attacks *Chrysanthemum*; *C. mali* - causes stem canker of apple.

Form genus - *Coryneum*. The acervuli produced by members of this Form genus appear like coremia of conidiophores arising on leaf spots and in black, sunken lesions on stem and fruit. The conidia are elongate-fusoid, pale-olivaceous, 3 - 5, mostly 3 septate, with slight constriction at the septa **(Fig. 56)**

(e-g) *Coryneum beijerinckii* - causes shot hole or gum spot of stone fruit trees, such as almond, apricot, peach, plum, cherry and allied trees.

Form subclass - Hyphomycetideae. The Hyphomycetideae produces neither pycnidia nor acervuli. The Subclass has been divided into two Form orders based on spore types, color of conidia, arrangement of conidia and type of conidiophores. The Form order - **Moniliales** includes all hyphomycetous fungi that produce conidia, while the Form order - **Agonomycetales** include those forms lacking conidia and reproducing only by fragmentation of mycelium.

Form order - Moniliales. This Order includes a large number of species, many of which are of much importance as plant, animal or human pathogens or industrial fungi. Among the industrial fungi there are species of *Penicillium* and *Aspergillus,* which form their sexual stages rarely. The Form order - Moniliales is divided into four Families viz., **Moniliaceae**, **Dematiaceae**, **Tuberculariaceae** and **Stilbellaceae**.

Form family - Moniliaceae. This Form family includes all imperfect fungi that produce conidia, which are hyaline on unorganized, hyaline conidiophores or directly from hyaline hyphae. Many of them are well known plant parasites. Some are animal and human pathogens. A number of species are known to destroy nematodes.

Form genus - *Pyricularia*. The mycelium, conidiophores and conidia produced by members of the Form genus are hyaline. The conidiophores are long, erect and unbranched and arise singly or in groups. Conidia are produced in succession, one at a time at the tips of the conidiophores. Conidia are narrowly pyriform to obclavate, with rounded base and narrow towards the tip. Conidia are usually two-septate **(Fig. 57)**

(e-g) *Pyricularia oryzae* - causes blast disease of rice. Its perfect stage is *Magnaporthe grisea* belonging to Order - Diaporthales of Ascomycetes; *P. grisea* - causes blast disease of ragi and some grasses.

Form genus - *Cephalosporium*. Conidia are produced at the tips of short, aseptate, simple or often verticillate stalks, on which they form mucilaginous white or yellow heads. Conidia are hyaline, ovoid or oblong-ellipsoid and without septation **(Fig. 57)**

(e-g) *Cephalosporium ulmi* - causes Dutch elm disease. Its perfect stage is *Ceratostomella (Ophiostoma) ulmi*; *C. sacchari* - causes wilt disease of sugarcane. Its perfect stage is *Ceratocystis (Ceratostomella) paradoxa.*

Form genus - *Verticillium*. Members of this Form genus are commonly isolated from soil. Some species cause *Verticillium* wilt disease in many crops. Slender conidiophores are formed abundantly from the mycelium. They are more or less erect, hyaline, verticillately branched, 3 - 4 phialides arising at each node of the conidiophore. Conidia are produced singly at the apices of phialides. Conidia are elliptical to sub-cylindrical, hyaline, mostly single-celled, rarely two-celled **(Fig. 57)**

(e-g) *Verticillium albo-atrum* - causes *Verticillium* wilt disease of castor, cotton, tomato, brinjal, chillies, tobacco etc.; *V. theobromae* - causes cigar-end rot of banana; *V. dahliae* - causes *Verticillium* wilt of potato, *Chrysanthemum* etc.

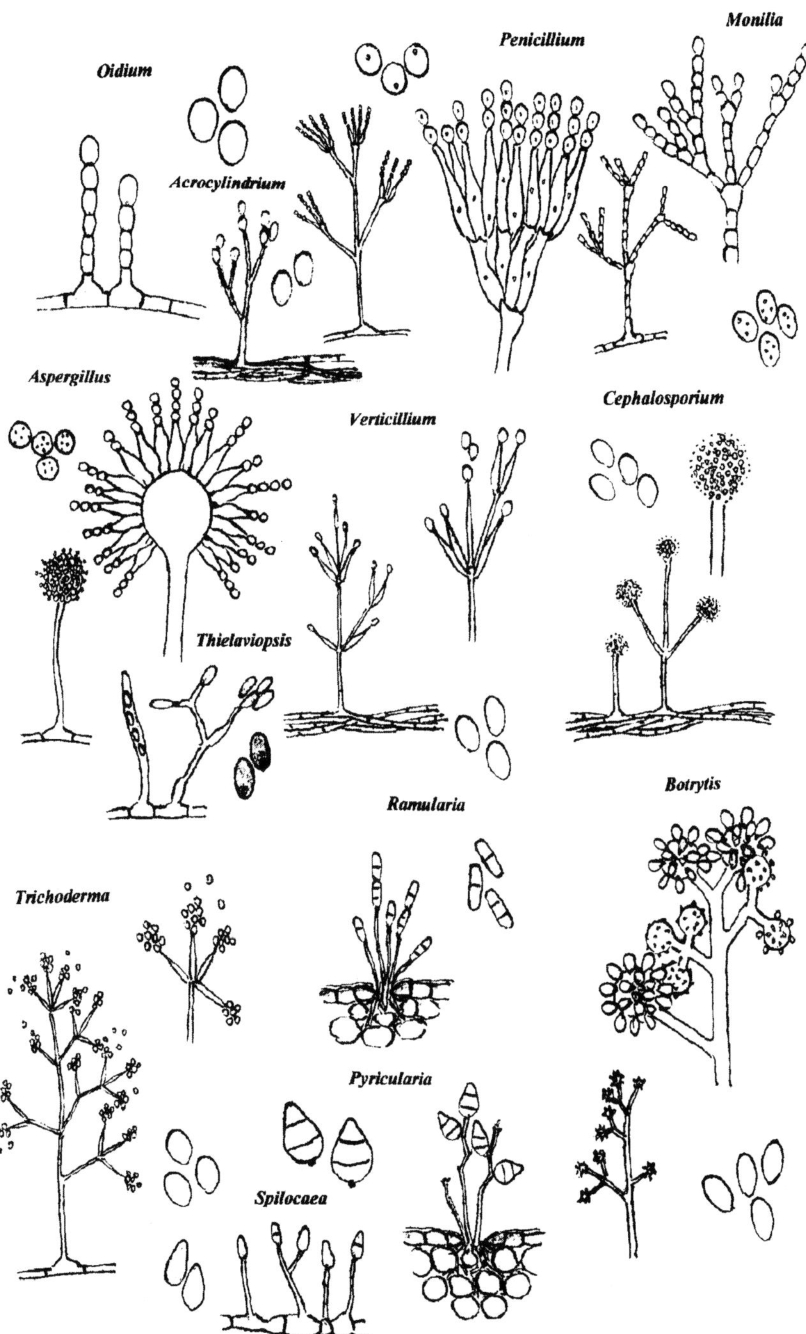

Fig. 57 : Conidiophores and conidia of important genera of moniliaceae.

Form genus - *Trichoderma*. *Trichoderma* is an ubiquitous soil fungus and produces white, yellowish or green colonies, when cultured. The highly branched conidiophores bear phialides singly or in groups that bear small, hyaline, single-celled conidia. A number of species of the Genus - *Hypocrea* of Ascomycetes have *Trichoderma* imperfect stage **(Fig. 57)**

(e-g) *Trichoderma viride* - It is used in the commercial production of the enzyme **'cellulase'**. It is also widely used in the biological control of a number of plant pathogens, especially the sclerotial pathogens. Its perfect stage is *Hypocrea rufa*; *T. harzianum* - parasitizes mycelia of *Rhizoctonia* and *Sclerotium* and inhibits growth of many fungi, such as *Pythium, Phytophthora, Fusarium* etc.; *T. lignorum* - causes fruit rot of *Citrus*.

Form genus - *Aspergillus*. The **'black Aspergilli'** or *Aspergillus niger* group are commonly called the **'black mould'**. Some species produce toxins called mycotoxins, such as **'aflatoxins'**. The conidiophores produced are not organized in any particular way, but arise singly from the somatic hyphae. The hyphal cell that branches to give rise to the conidiophore is called the **'foot cell'**. The conidiophores are long, erect hyphae, each terminating in a bulbous head or vesicle. Over the entire surface of the vesicle, a large number of conidiogenous cells or sterigmata are produced. When two layers of sterigmata (the primary and secondary sterigmata) are formed according to the species, conidia arise only from the secondary sterigmata, which are typical phialides. Conidia are formed from the tips of the phialides, one below the other in chains. Conidia are globose and unicellular with roughened walls. *Aspergillus* colonies may appear black, brown, yellow, green and so on depending on the species.

The perfect stages of most species of *Aspergillus* are not known. If at all the sexual stages are known, they belong to the ascomycetous Genera - *Eurotium, Sartorya* and *Emericella* of Family - Eurotiaceae **(Fig. 57) (e-g)** *Aspergillus fumigatus, A. flavus, A. niger, A. terreus* - these species are animal and human pathogens that cause a group of diseases known as **'Aspergilloses'** of the lungs; *A. niger* - causes black mould rot of mango, collar rot or crown rot of groundnut etc.

Form genus - *Penicillium*. Many species of *Penicillium* cause the **'blue mould rots'** and the **'green mould rots'**, also known as **'*Penicillium* rots'**. They affect many kinds of fruits and vegetables. These organisms also produce several mycotoxins, such as **'patulin'**, which contaminate juices and sauces. The mycelium produces simple, long, erect conidiophores that branch at the top portion in symmetrical or asymmetrical fashion. The conidiophore is known as the **'penicillus'**. The multiple branching of the conidiophore ends in a group of phialides that bear the long conidial chains. The conidia are globose to ovoid. The mass of greenish, bluish or yellowish conidia produced are responsible for the characteristic colony color of different species of *Penicillium*. The ascigerous stages of most of the species of *Penicillium* are not known. However, the perfect stages of a few species of *Penicillium* that have been found so far belong to the Genera - *Talaromyces* and *Carpenteles* of Family - Eurotiaceae **(Fig. 57)**

(e-g) *Penicillium roqueforti* - gives the characteristic flavor to the highly prized Roquefort cheese; *P. camemberti* - gives flavor to Camembert cheese; *P. griseofulvum* - produces one of the best antibiotic - **'Griseofulvin'**, which is effective in the control of fungal skin diseases; *P. notatum* and *P. chrysogenum* - sources of the most famous antibiotic - **'Penicillin'**; *P. digitatum* and *P. italicum* - cause blue mould rot of *Citrus* and grape fruits; *P. expansum* - causes decay of apples in storage.

Form genus - *Botrytis*. Conidia rather large, oval or spherical, formed at the tips of erect conidiophores that are simple or branched. The conidiogenous cell itself is swollen and many conidia are produced simultaneously over its surface. Conidia are hyaline or brightly colored and the term **'gray mould'** is often used in reference to *Botrytis*. The perfect stages of many form species of *Botrytis* are unknown. If at all the perfect stage is found, it belongs to the ascomycetous Genus - *Botryotinia* **(Fig. 57)**

(e-g) *Botrytis tulipae* - causes blight of tulip; *B. peoniae* - causes blight of peonies; *B. cinerea* - attacks mature or nearly mature grapes.

Form genus - *Monilia*. The members of this Genus produce conidiophores, which are often branched. Conidia are produced in chains and they are spherical, oval or lemon-shaped, hyaline or pinkish in color. The ascigerous stage is rarely found and whenever found belongs to the Genera - *Neurospora* and *Monilinia* **(Fig. 57)**

(e-g) *Monilia sitophila* - it is commonly referred to as the **'bakery mould'** or **'red bread mould'**, because it frequently infests bakeries and causes considerable damage. Its perfect stage is *Neurospora sitophila*; *M. fructicola* - causes brown rot of peach and other fruits. Its perfect stage is *Monilinia fructicola*.

Form genus - *Oidium (Acrosporium)*. The conidiophores produced from the superficial mycelium are simple, erect and septate. Conidia are hyaline, unicellular, elliptical and are borne singly or in short chains **(Fig. 57)**

(e-g) *Oidium mangiferae* - causes powdery mildew of mango; *O. tingitaninum* - causes powdery mildew of *Citrus*; *O. caricae* - causes powdery mildew of papaya; *O. anacardii* - causes powdery mildew of pomegranate; *O. jasmini* - causes powdery mildew of jasmine; *O. heveae* - causes powdery mildew of rubber.

Form genus - *Ramularia*. The conidiophores that come out through the stomata in bunches are short, septate and branched from the lower portion. Conidia are borne singly or in chains at the tips of conidiophores. Conidia are irregularly cylindrical with rounded or flat ends, non-septate or with 1 - 3 septa **(Fig. 57)**

(e-g) *Ramularia areola* - causes areolate mildew or gray mildew of cotton. Its perfect stage is *Mycosphaerella areola.*

Form genus - *Acrocylindrium (Sarocladium)*. The mycelium is very slender, well-branched, hyaline and septate. From the mycelium, conidiophores that are slightly thicker than the normal hyphae arise. Conidiophores are hyaline, erect, branched once or twice to form 3 or 4 branches in a whorl. The branches taper towards the tips. Conidia are produced consecutively at the tips of the branches. Conidia are hyaline, single-celled, cylindrical and smooth **(Fig. 57)**

(e-g) *Acrocylindrium (Sarocladium) oryzae* - causes sheath rot of rice.

Form genus - *Spilocaea*. Conidiophores are short, light-brown in color, waxy and are closely ringed. Conidia are one or two-celled, hyaline to light-brown in color **(Fig. 57)**

(e-g) *Spilocaea pomi* - causes apple scab. Its perfect stage is *Venturia inaequalis*.

Form genus - *Thielaviopsis*. Conidiophores are slender, arising laterally from hyphae and produce cylindrical to oval endoconidia. Conidia are pale brown and smooth-walled **(Fig. 57)**

(e-g) *Thielaviopsis basicola* - causes black root of tobacco; *T. paradoxa* - causes stem bleeding disease of coconut. Its perfect stage is *Ceratostomella (Ceratocystis) paradoxa.*

Form genus - *Geotrichum*. The fungi reproduce by means of holothallic conidia formed in random chains. The conidia are cuboid or short cylindrical in shape with somewhat flattened ends

(e-g) *Geotrichum candidum* - causes sour rot of citrus fruits, tomatoes, carrots and many other fruits and vegetables. Its perfect stage is *Endomyces geotrichum* of Endomycetales.

Form family - Dematiaceae. Both hyphae and conidia produced by members of this Form family are typically dark-colored, but sometimes either the hyphae or the conidia alone are dark-colored. As in the Moniliaceae, no definite fruiting body is produced. Most of the species are saprobic, while some are serious plant pathogens and a few are parasitic on animals and humans.

Form genus - *Alternaria*. Species of this Form genus produce conidia referred to as **'dictyospores'**. The conidia are rather large and multinucleate with both transverse and longitudinal septa that are typical of this Form genus. The conidia are usually borne acropetally in chains or are formed singly at the tips of conidiophores that are indistinguishable from the somatic hyphae. It is a large universally occurring Genus. A number of form species are parasitic on plants causing serious diseases **(Fig. 58)**

(e-g) *Alternaria solani* - causes early blight of tomato and potato, and fruit rot of chillies; *A. alternata* - causes fruit rot of apple, leaf spot of sugarbeet etc.; *A. tenuissima* - causes rot of mango; *A. citri* - causes fruit rot of *Citrus*; *A. brassicae* and *A. raphani* - cause leaf spots of crucifers; *A. jasmini* - causes leaf blight of jasmine; *A. macrospora* - causes *Alternaria* blight of cotton; *A. ricini* - causes blight of castor.

Form genus - *Helminthosporium*. Species of this Form genus also produce rather large **'phragmospores'**, multicellular conidia possessing transverse septa only. Conidia are thick-walled and typically obclavate. Many species of *Helminthosporium* are serious plant pathogens **(Fig. 58)**

(e-g) *Helminthosporium solani* - causes silver scurf disease of potato; *H. oryzae* - causes brown leaf spot of rice. Its perfect stage is *Cochliobolus miyabeanus*.

Form genus - *Drechslera*. The conidia produced by species of this Form genus are rather large phragmospores, multicellular with only transverse septa. The septa produced are not typical septa, but are called pseudosepta. They are thick-walled, curved, clavate or cylindrical. This Form genus was formerly included under *Helminthosporium*. A number of ascomycetous Genera, such as *Trichometasphaeria*, *Pyrenophora* and *Cochliobolus* have imperfect stages of *Drechslera*. Several species are plant parasitic and are commonly found on Gramineae including oats, sugarcane, wheat, corn, sorghum etc. Conidia, which germinate from the two terminal cells, are usually referred to as Form genus - *Bipolaris* **(Fig. 58)**

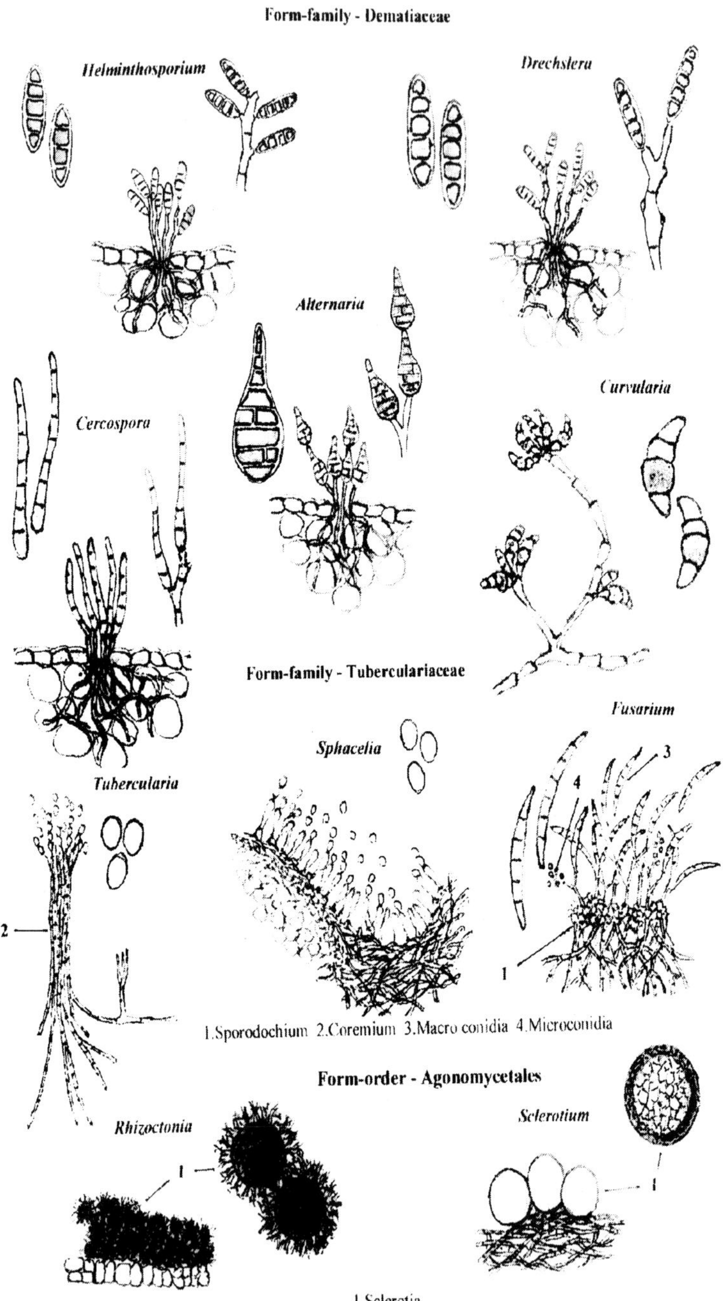

Fig. 58 : Coniophores and conidia of important form genera of Deinatiaceae and Tuberculariaceae.

(e-g) *Drechslera (Helminthosporium) maydis* - causes leaf blight of corn; *Drechslera (Bipolaris) rostrata* - causes leaf spot of grapevine, pomegranate, tapioca etc.; *Drechslera (Helminthosporium) sacchari* - causes eye spot of sugarcane; *D. heveae* - causes bird's eye spot of rubber; *Drechslera (Bipolaris) spicifera* - causes leaf spot of cotton. Its perfect stage is *Cochliobolus spicifer*; *D. rostratum* and *D. tetramera* - cause grain discoloration in rice.

Form genus - *Curvularia*. Species of *Curvularia* produce somewhat large phragmospores. The multicellular conidia possess only transverse septa. They are usually curved, the central cells broader and darker and the end cells paler in color than the other cells **(Fig. 58)**

(e-g) *Curvularia maculans* - causes leaf spot of coconut; *C. lunata* - causes leaf spot of oil palm and grain discoloration of rice; *C. verrucolosa* - causes leaf spot of tobacco and betelvine

Form genus - *Cercospora*. Many species of Form genus - *Cercospora* are plant parasites and some of them cause very serious diseases of cultivated crops. Conidia are long, slender and multicellular. They are produced sympodially on clustered, dark conidiophores that usually emerge through the stomata of infected leaves **(Fig. 58)**

(e-g) *Cercospora personata* - causes late leaf spot or late 'tikka' leaf spot of groundnut. Its perfect stage is *Mycosphaerella berkeleyi*; *C. arachidicola* - causes early leaf spot, also called early 'tikka' leaf spot of groundnut. Its perfect stage is *Mycosphaerella arachidicola*; *C. oryzae* - causes narrow brown leaf spot of rice. Its perfect stage is *Sphaerulina oryzina*; *C. nicotianae* - causes frog's eye leaf spot of tobacco; *C. viticola* - causes brown leaf spot of grapevine; *C. melongenae* and *C. deightonii* - cause leaf spot of brinjal (eggplant); *C. capsici* - causes leaf spot of chillies; *C. manihotae* - causes leaf spot of tapioca.

Form family - Tuberculariaceae. The Tuberculariaceae are characterized by the fact that they all produce sporodochia. The conidia are produced from phialides.

Form genus - *Tubercularia*. The sporodochia produced by members of this Form genus are brightly colored and shaped somewhat like a mushroom with a wavy stalk. The surface layer of the sporodochium consists of innumerable simple or sparingly branched conidiophores on the tips of which single-celled, elongated, oval conidia are borne. Members of this Form genus are mostly saprobic on wood, but a few are pathogenic on plants **(Fig. 58).**

(e-g) *Tubercularia vulgaris* - causes canker in the trunk, branches and twigs and die-back of maple. Its perfect stage is *Nectria cinnabarina*.

Form genus - *Fusarium*. Members of this Form genus typically produce two types of conidia viz., macro- and microconidia. Both types of conidia are produced from phialides. Macroconidia are long, multiseptate, crescent-shaped that are generally borne in sporodochia or may occur separately. The micro conidia are small, usually single-celled, spherical or oval in shape.

A number of Fusaria are parasitic and cause wilting of the host plant. Mycelium of the fungus invades the vascular tissues and along with microconidia physically blocks xylem vessels, thus preventing the translocation of water. In addition, *Fusarium* is known to produce toxins that may affect the permeability of cell membranes and disrupting cell metabolism, thereby induce wilting **(Fig. 58)**

(e-g) *Fusarium oxysporum* f.sp. *cubense* - causes *Fusarium* wilt of banana; *F. oxysporum* f.sp. *lycopersici* - causes wilt of tomato; *F. oxysporum* f. sp. *vasinfectum* - causes wilt of cotton; *F. oxysporum* f.sp. *sesami* - causes wilt of gingelly; *F. oxysporum* f.sp. *melonis* - causes wilt of cucurbits; *F. oxysporum* f.sp. *pisi* - causes wilt of pea; *F. moniliforme* - causes foot rot of rice, 'Pokkah boeng' or twisted top disease of sugarcane, fruit rot of banana and *Citrus*, sett rot or stem rot of sugarcane etc. Its perfect stage is *Gibberella moniliformis* or *G. fujikuroi* var. *subglutinans*; *F. solani* - causes wilt of brinjal, potato, chillies etc.; *F. equiseti* - causes fruit rot of pomegranate.

Form genus - *Sphacelia*. The mycelium producing conidia consists of intricately woven hyphae, which emerge from the surface of the ovary in slimy masses. They give rise to a more or less close palisade of short conidiophores, which abstrict minute conidia in abundance. These conidia accumulate on the surface of the inflorescence like drops of honeydew **(Fig. 58)**

(e-g) *Sphacelia sorghi* - causes sugary disease of sorghum. Its perfect stage is *Claviceps sorghi*; *Sphacelia segetum* - causes sugary disease or 'ergot' of rye. Its perfect stage is *Claviceps purpurea*.

Form order - Agonomycetales. The fungi included in this Form order are referred to as the **'Mycelia sterilia'**, since they are not known to produce conidia. Reproduction is by random fragmentation of the hyphae. However, these fungi produce sclerotia, which consist of dense mass of mycelium.

Form genus - *Rhizoctonia*. The fungi belonging to *Rhizoctonia* are soil inhabitants and cause serious diseases on many hosts by affecting the roots, stems, tubers, corms and other plant parts that develop in or on the ground. They do not produce spores of any kind, either sexual or asexual. However, it has been found now that at least some species produce spores very rarely. The sclerotia are small, white in the beginning and become brown or black at maturity. The sclerotia are covered by septate hyphae **(Fig. 58)**

(e-g) *Rhizoctonia solani* - causes sheath blight of rice, root rot of tomato, pea, sugarbeet etc., damping off of brinjal, crucifers, cotton etc., root and stem rot of cluster beans, crown rot of carrot, black scurf and stem canker of potato, foot rot of *Chrysanthemum*, collar rot of coffee etc. Its perfect stage is *Thanatephorus cucumeris*, which belongs to Basidiomycetes; *R. bataticola* - causes charcoal rot or stem rot of gingelly, sugarbeet, cotton etc. Its conidial stage is *Macrophomina phaseolina.*

Form genus - *Sclerotium*. Like *Rhizoctinia*, members of this Form genus are also soil inhabitants and cause serious diseases, such as damping off, root rots, stem rots etc. of various crops. They produce small, sclerotia on white, cottony hyphae. The sclerotia are spherical, smooth and black at maturity **(Fig. 58)** **(e-g)** *Sclerotium rolfsii* - causes root rot of tomato, chillies, tapioca, sugarbeet etc., seedling blight of apple, damping off of brinjal, *Sclerotium* rot of potato, basal wilt of pepper, stem and pod rot of groundnut, stem and root rot of tobacco, wilt of betelvine etc. Its perfect stage is *Pellicularia rolfsii*; *S. bataticola* - causes stem blight of bean. Its conidial stage is *Macrophomina phaseolicola*; *S. cepivorum* - causes white rot of onion and garlic; *S. oryzae* - causes stem rot of rice. Its perfect stage is *Leptosphaeria salvinii.*

Survival of Plant Pathogens

Pathogens that infect perennial plants have enough food supplies throughout the year for continuing and completing their life cycle. Though they may have to face unfavorable seasonal conditions, their life process may continue unabated and they are capable of existence as long as the hosts are alive. However, in the case of most of the common pathogens living on annuals, biennials or seasonal crops, they have to face a period of the year, when their usual hosts are not available and so, even in the absence of their regular hosts, they have to survive for several months. In order to overcome such crisis, the pathogens have evolved various mechanisms for survival and perpetuation of the diseases in the ensuing seasons.

Fungi have evolved many types of mechanisms for persistence between crops. Of these, the most important one is that several fungi produce asexual or sexual resting spores for their survival. The pathogens which survive such a crisis and infect the host for the first time is known as **'primary infection'** and the inoculum, which is either the mycelium, spores, resting spores or sporocarps is called the **'primary inoculum'**. In the case of diseases, which are seed-borne or soil-borne, the inoculum is present in the seed or soil respectively and serves as primary inoculum. Once the host is infected by the primary inoculum, the pathogen grows in the host and produces spores, resting spores or sporocarps, which are known as **'secondary inoculum'** and the infection caused by the secondary inoculum is called **'secondary infection'**. So, the primary inoculum is responsible for the initial infection and the secondary inoculum is responsible for the rapid spread of the disease. In the case of pathogens, such as viruses, viroids, mycoplasma-like organisms etc., the primary inoculum that is present in the alternate or collateral hosts are transmitted to the primary hosts to cause primary infection through vectors or by other means.

Survival of plant pathogens during the off-season and subsequent development of primary infection is achieved in three different ways:

1. The infected host itself may serve as a storehouse of spores and other fructifications. This is made possible in two different ways :

(i) The primary host itself serves as a storehouse of inoculum and this is common in many annuals and perennials. The bacteria, *Xanthomonas campestris* pv. *citri*, causing canker in *Citrus* plants survive in the cankers produced in the leaves, stems, fruits etc. all through the year. The potato ring disease bacterium *(Pseudomonas solanacearum)* can survive in the left over tubers for very long periods. Sugarcane red rot pathogen, *Glomerella tucumanensis (Colletotrichum falcatum),* though capable of producing resting spores, the mycelium and chlamydospores present in the setts obtained from infected canes serve as primary inoculum. Rice brown leaf spot pathogen, *Cochliobolus miyabeanus (Helminthosporium oryzae)* and rice bacterial leaf blight pathogen, *Xanthomonas campestris* pv. *oryzae* live inside the seeds in a dormant state and cause primary infection, when such seeds are sown. Similarly, the pathogen of angular leaf spot or black arm of cotton, *Xanthomonas campestris* pv. *malvacearum* can also persist as internal infection in cotton seeds. The bean anthracnose pathogen, *Glomerella lindemuthianum (Colletotrichum lindemuthianum)* also infects the seeds and the pathogen persists in the seeds as dormant mycelium. The perithecia produced by the coconut stem bleeding pathogen *(Ceratocystis paradoxa)* are found embedded in the disintegrated tissues of the stem beneath the bark. In potato late blight *(Phytophthora infestans)* the fungus, which remains dormant in the tubers, sometimes gives rise to diseased plants when they sprout. *Peronospora destructor,* the onion downy mildew pathogen persists as dormant mycelium within the bulbs. In peach leaf curl, the pathogen, *Taphrina deformans* survives as perennial mycelium in the branches. Several powdery mildew pathogens, such as *Podosphaera leucotricha* causing powdery mildew of apple, *Uncinula necator* causing powdery mildew of grapevine etc. persist as dormant mycelium in the bud scales and on perennial parts of the plant. *Venturia inaequalis* causing apple scab is able to persist and overwinter in the mycelial state on the shoots and bud scales. In the case of loose smuts of cereals, the pathogen can persist in the mycelial stage in the grains of the host. Plants, plant parts or seeds harboring internal mycelium are said to be **'systemically infected'**. In bunt of wheat *(Tilletia caries)*, the spores which are somewhat greasy, adhere to the seed grains at the time of harvest, threshing or storing and remain viable for many years and serve as primary inoculum.

(ii) Susceptible hosts belonging to the same family as that of the primary host or any other alternate cultivated or wild hosts may serve as a storehouse of inoculum in the absence of the primary host. This trait enables them to remain active throughout the year. *Sclerotinia sclerotiorum, Botrytis cinerea,* species of *Pythium, Rhizoctonia* etc. are capable of living on more than one host plant. *Sclerotinia sclerotiorum* has a wide host range, consisting of numerous succulent plants, especially vegetables and flowers. The gray mould rot fungus, *Botrytis cinerea* attacks several plants of botanically different families, such as lettuce, gooseberry, broad bean etc. *Pythium de baryanum* and *P. aphanidermatum* infect a number of vegetable crops, field crops, ornamental plants, forest plants etc. In the nurseries of crops, such as chillies, brinjal, tobacco, tomato etc., they cause damping off of seedlings. Soft, fleshy organs of vegetables, such as cucurbits, fruits, green beans, potatoes, ginger etc., which are in contact with the soil are also attacked by these fungi. Several species of *Rhizoctonia* and *Fusarium* are also capable of attacking many diverse hosts belonging to different families. *Rhizoctonia solani* attacks rice, cotton and many other crops, while *Fusarium oxysporum* attacks pea, tomato, banana, cumin, cotton etc. and *Fusarium udum* attacks sunnhemp, red gram etc. *Rhizoctonia bataticola,* which causes collar rot of groundnut, attacks several other crops, such as green gram, black gram, red gram etc. The rice blast pathogen *(Pyricularia oryzae)* survives in several collateral grass hosts during the off-season, when rice crop is not available. The root knot nematode (*Meloidogyne* sp.) has several alternate hosts, such as potato, brinjal, tomato, lady's finger, cucurbits, crosandra etc., in which they continue their lives and survive throughout the year.

In the case of some fungal pathogens, which attack different plants of widely separated botanical families, the fungus assumes different forms on the separate hosts and passes one phase of its life cycle on the first host and another phase on the second host, thus requiring both the hosts for the completion of its life cycle. The fungi, which live such a dual life, are called **'heteroecious fungi'** and this type of adaptation is termed **'heteroecism'**. Heteroecious fungi are seen almost entirely among the rusts. The most familiar case of heteroecism occurs in the black rust of wheat *(Puccinia graminis tritici)*. This destructive pathogen passes the first part of its life cycle consisting of the pycnial and aecidial stages on the common barberry plants and the second part of its life cycle consisting of the uredial and telial stages on wheat and other cereals and grasses. In the absence of wheat crop the fungus survives in the perennial barberry plants and in the advent of wheat crop, it is infected by the aeciospores produced in the barberry plants. So, for this pathogen, wheat is the primary host and barberry is the secondary host. Another example of heteroecism is the cumbu rust, where the primary host is cumbu and the secondary host is eggplant. In the absence of the primary host i.e., cumbu, the pathogen lives in the secondary host i.e., eggplant and survives all through the year.

2. In the absence of the regular host, the pathogen may lead a saprophytic life in the soil. Many facultative parasites are capable of continuing their lives as saprophytes in plant remains, plant refuse and such organic matter present in the soil. Several fungal organisms belonging to species of *Pythium* and *Phytophthora*, which cause seedling rot, *Fusarium* and *Verticillium*, which cause wilt, *Rhizoctonia, Sclerotium, Sclerotinia* etc., which cause stem rot and root rot are capable of living as saprophytes for prolonged periods. The potato ring disease bacterium *(Pseudomonas solanacearum)* remains in the soil in a living state for about two and a half years.

3. Fungal pathogens produce various types of reproductive structures, such as conidia, resting spores, spore bodies, sporocarps etc. for their survival, which are found on the surface of the host parts or inside the tissues of the hosts or outside the hosts. These reproductive structures, which are mostly very hardy, remain viable for very long periods when their natural hosts are not present or when conditions are not conducive for their perpetuation. Such reproductive structures include chlamydospores, sclerotia, sporodochia etc., which are asexually formed, and oospores, zygospores, cleistothecia, perithecia, apothecia, sporocarps etc., which are sexually formed. These spores and spore

bodies, which remain in the soil or in the plant debris for prolonged periods in a viable state, germinate in the advent of suitable hosts and favorable climatic conditions and cause primary infection.

The oospores of the pathogens *Pythium de baryanum* and *P. aphanidermatum*, causing damping off of seedlings of tobacco, brinjal, chillies etc.; *Peronospora pisi*, causing downy mildew of peas; *Albugo candida*, causing white rust of cauliflower, mustard etc.; *Sclerospora graminicola*, causing downy mildew of pearl millet, as well as several other such pathogens; chlamydospores and mycelium of *Phytophthora parasitica*, causing seedling blight of castor; mycelium of *Phytophthora infestans*, causing late blight of potato; sclerotia of *Rhizoctonia bataticola*, causing collar rot and root rot of groundnut; ergot or sclerotia of *Claviceps microcephala*, causing sugary disease of pearl millet; cleistothecia of *Erysiphe cichoracearum*, causing powdery mildew of cucurbits; conidia of *Cercospora personata*, causing 'tikka' leaf spot of groundnut and *Alternaria melangenae*, causing leaf spot of brinjal etc., reach the soil and perpetuate the disease. In several fungi belonging to Ascomycetes, the ascospores are protected by the wall of the spore bodies, such as perithecia, cleistothecia etc. and preserve their power of germination for many months even under extremes of climatic conditions, when cast into the soil. Many rusts and smuts, and most of the Peronosporaceae are carried over by the prolonged germinative capacity of their sexual or other durable spore forms. In the onion mildew *(Peronospora destructor)*, the spores usually do not germinate until several years after their formation. The spores of wheat bunt *(Tilletia caries)* can survive in the soil for a considerably long time, produce sporidia under favorable conditions and infect the host.

Some bacteria are also capable of surviving in the soil along with plant refuse consisting of various affected and dead parts of the host plants and remain in the soil in a viable state for long periods and cause primary infection. *Pseudomonas solanacearum*, which causes ring disease of potato, survives in the soil for as long as two and a half years.

Viruses, viroids, mollicutes and protozoa can survive only in living plant tissues and they cannot exist in a free, dormant state. So, they are transmitted from one living host to another living host by insect vectors or by other means for their survival.

Nematodes can survive in the soil as cysts or inside galls produced by them in a dormant state. The cysts of the potato golden nematode *(Heterodera rostochiensis)* and the sugar beet cyst nematode *(Heterodera schachtii)* can remain in the soil in a dormant state for several years.

Many of the phanerogamic parasites also can survive in the soil for several months or years. The seeds of *Cuscuta gronovii*, *Orobanche cernua* etc. remain in the soil in a viable state for years together.

Mode of Entry of Fungal Pathogens

For successful initiation of a disease in the host, first of all the pathogen has to gain entry into the host tissue and then it has to establish itself in the host and then cause infection. Once the infection process is successfully accomplished, then a period of incubation follows and eventually visible signs of the presence of the pathogen inside the host are exposed as disease symptoms. Most often, this phase is followed by the production of reproductive structures usually on the surface of the diseased parts. Entry into the host is mostly effected by germ tubes from spores, but sometimes by the mycelial hyphae also.

The entry of the pathogen into the host may be effected through natural openings on the host surface, such as stomata, lenticels, hydathodes, nectaries, cracks between main and lateral roots or by direct penetration of the cuticle and outer walls of the epidermal cells or through wounds. In some cases, even in a single pathogen different spore forms may infect in different ways. In the rust fungi, infection by the uredospores is through the stomata, while the sporidial infection is by direct penetration of the cuticle and epidermis **(Fig. 59)**

Entry through stomata and other natural openings. In order to attack a leaf, a spore has to remain long enough on the surface of the leaf to germinate and produce an infection hypha. Usually this requires high moisture and in some cases free water droplets or a thin film of water on the leaf surface. Besides these, temperature, light or combinations of these factors, as well as the microclimatic factors influence the germination process of the spores. Further, spore germination is often stimulated by nutrients, such as sugars and amino acids, which are exuded from the plant surface. In the case of pathogens, which attack roots of the host plants, the root exudates favor germination of the spores. After spore germination, the growth of germ tubes towards successful penetration sites is also influenced by the above physical factors and the chemical stimuli associated with the openings, such as stomata, lenticels, wounds etc. The thigmotropic (contact) responses of the host direct the germ tubes to grow at right angles to the circular ridges that surround the stomata and eventually to enter into the host tissues. The direction of movement of motile spores (zoospores) towards successful penetration points is also regulated in a similar way by these factors.

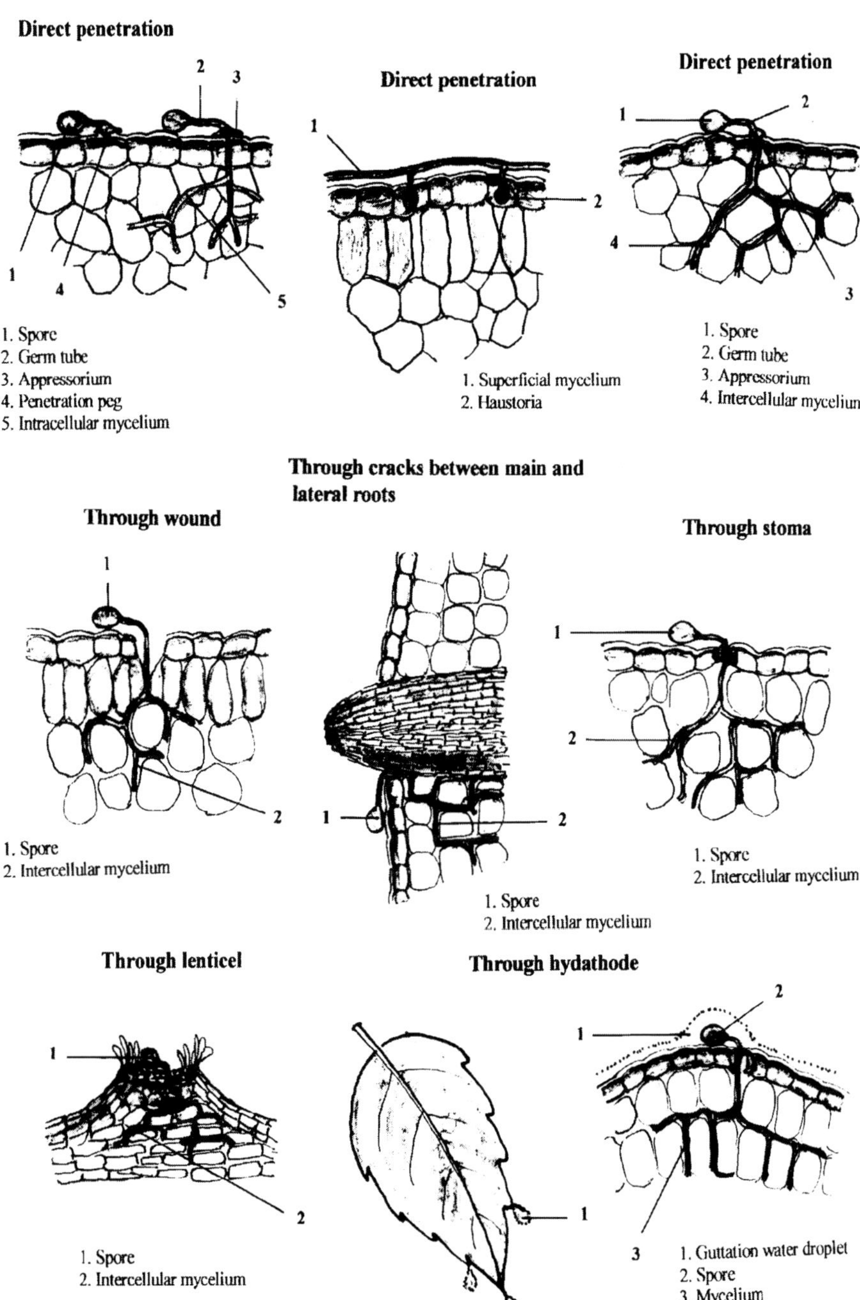

Fig. 59 : Mode of entry of fungal pathogen into the host

The formation of an appressorium, which may be a small swelling or a mere thickening at the tip of the germ tube usually precedes actual entry of the pathogen into the host. The appressorium is fixed to the surface of the guard cells of the stoma by some kind of sticky secretion and the entry through the stoma is effected by a slender infection hypha arising from the underside of the appressorium as in the case of *Puccinia hordei,* causing barley rust. In some other fungi, the tips of the germ tubes, which later form the mycelium, enter the stomata directly without forming an appressorium. The infection hypha swells up into a small vesicle in the sub-stomatal cavity and puts forth branches from it. In a susceptible host, these branches send haustoria into the neighboring host cells and through these haustoria food is taken from the host cells as in the case of *Cladosporium fulvum,* which causes leaf mould of tomato. The common green mould of citrus fruits, *Penicillium digitatum* causing decay in oranges and grapefruits enters through wounded oil vesicles in the rind of the fruit. In the case of uncongenial host, the spore may germinate and enter into the host, but the haustorium is not formed and even if the haustorium is formed it is unable to obtain food from the host cells, as a result the hypha dies **(Fig. 59)**

In the case of *Penicillium expansum* and a few other fungi causing rotting of apples, penetration occurs mostly through lenticels in the fruit. The apple canker-inducing fungus *Nectria galligena* enters through lenticels, as well as by penetrating the leaf scars or galls produced by insects, such as wooly aphids and also through wounds. Lenticel infection also occurs in common scab of potatoes caused by *Actinomyces scabies* **(Fig. 59)**

Entry through surface cells. Most parasitic fungi penetrate into their hosts by boring through the outer wall of the surface layer of cells, while in some cases the germ tube passes down in-between two adjoining epidermal cells without actually entering into the cells. The germ tube arising from the spore or the infection hypha arising from the appressorium, which is usually anchored to the host surface, then penetrates the host cell by mechanical pressure. If the outer wall of the epidermal cell is more cutinised then penetration becomes difficult. Such thick, cuticularised epidermal wall may be a cause of resistance of a particular host plant to parasitic invasion and such resistance to parasitic attack may increase with the age of the plant part concerned. Once the cutin layer of the outer wall is penetrated, further penetration into the deeper layers may be made possible by enzyme action. Older barley leaves with thick cuticle resist attack by *Erysiphe graminis,* the powdery mildew pathogen. *Ascochyta pinodella,* causing foot rot of peas attacks chiefly the base of the epicotyl, where the cuticle is rather thin. The plants grown under too moist conditions are comparatively more susceptible to attack by pathogens because of the natural softening of the cuticle. Accumulation of silica in the outer epidermal wall under high moisture and nitrogen nutrition induces resistance to rice plants against rice blast fungus, *Pyricularia oryzae* **(Fig. 59)**

Mycelial infection. In many fungi, infection can be caused by mycelial hyphae as readily as by the germ tubes from spores. The most extreme example is found in fungi that infect by means of rhizomorphs as in *Armillaria mellia,* a root parasite of tea and several other trees. When a rhizomorph comes into contact with a healthy root of a susceptible host, it penetrates the root by its tip or by a newly formed hyphal branch at the point of contact. The tip of the rhizomorph enters the root and spreads inside the root tissues. The entry of the rhizomorph is effected by mechanical pressure and also by chemical action. Behind the growing tip inside the tissues, individual hyphae, which are the feeding hyphae, grow out from the sides. In other cases, such as *Ophiobolus graminis*, causing 'take-all' and 'whitehead' of wheat, long runner hyphae are produced on the surface of the root from which several feeding hyphae arise and penetrate the roots. In the case of *Corticium koleroga*, causing black rot of apple twigs, the greater part of the fungus remains on the surface as a white covering, while penetration is effected by hyphae arising from the lower side of this layer. In such of these pathogens infection from spores play a very minor part **(Fig. 59)**

Wound infection. Infection through wounds is common in several fungi, especially the pathogens of woody plants in which wounds are sometimes the only path of entry **(e-g)** 'Dutch elm' disease caused by *Ceratostomella ulmi*. In *Stereum purpureum,* the cause of 'silver leaf' of plums and other trees, infection from spores occurs through pruning and other wounds. In the canker and die-back diseases of roses *(Griphosphaeria corticolax)* infection from spores occurs through pruning cuts **(Fig. 59)**

Entry through specific parts or organs. In a large number of parasitic diseases of plants, entry of the pathogen is possible only through certain parts or organs of the plant. After entry, the pathogen may infect the whole or part of the plant or may remain strictly localized in particular organs. The powdery mildew of mango *(Odium mangiferae)* is often confined to the inflorescence, but can attack young leaves in moist areas. Many leaf-spotting fungi are restricted to the leaves. Some of these may be found only on young leaves, while many others are usually confined to the outer, older leaves. Ring spot of cabbage *(Mycosphaerella brassicola),* black rot of grapevine *(Guignardia bidwellii),* leaf spot of beet *(Cercospora beticola)* etc. are usually confined to older leaves.

Many root parasites are restricted to the underground parts of plants and some around the collar region only.

Pathogens that enter only through stomata or other such natural openings are naturally limited to parts having these natural openings.

In some plants, there are weak spots through which some pathogens get access. *Colletotrichum falcatum,* the red rot pathogen of sugarcane and *Sclerospora sacchari,* the downy mildew pathogen of sugarcane, enter near the nodes where it is broken by shoot eyes and by the gaps caused by the ring of adventitious root organs, as these pathogens are incapable of entering through the tough rind of the cane. *Phytophthora erythroseptica,* causing pink rot of potatoes infects potato tubers after harvest only through the eyes.

The smuts show many cases of restricted points of entry. In the loose smut of wheat *(Ustilago tritici)* and loose smut of barley *(Ustilago nuda),* infection occurs through the flowers. Smut spores blown on to the feathery stigma germinate there and penetrate down the style and finally reach the ovary. The bunt of wheat fungus *(Tilletia caries)* enters through any part of the coleoptile by penetrating the epidermis. The sugarcane smut *(Ustilago scitaminea)* enters through the hairs on the scale leaves of young shoot buds, which develop at every node. The maize smut *(Ustilago zeae)* can cause infection in any growing part. The fungus penetrates not only the epidermal cells but also the underlying parenchyma.

Some fungi infect the host plants at the seedling stage only. *Helminthosporium avenae* and *Helminthosporium gramineum* causing stripe disease of oat and barley respectively, enter through the coleoptile only and infect the seedlings.

Dissemination or Spread of Plant Pathogens

A vast majority of fungal pathogenic organisms are disseminated, either by spores formed asexually or various fruiting bodies, which are mostly formed by sexual reproduction, while some others are disseminated by vegetative propagants, such as mycelium, sclerotia, rhizomorphs etc. In order to initiate diseases in succession, the propagants produced by the pathogens have to be carried to the hosts. When a pathogen infects a particular host, grows and multiplies in the host, a stage is reached when there is lack of sufficient food and space for continuing its life. Under such circumstances, the pathogen has to seek other host plants, which are not affected by the pathogen, or in the absence of its regular host has to seek alternate or collateral hosts for its survival. In case it is not possible to find a new host, the pathogen may die or may not be able to complete its life cycle.

So, to disseminate the pathogen from one host to another suitable host, nature has provided several ways. Under favorable environmental conditions, the pathogens produce large number of asexual and sexual spores, either on the surface of the infected portions or inside the tissues of infected parts. The spores thus produced are disseminated through many natural agencies. Propagants may be disseminated either by active or autonomous methods or by passive or indirect methods. Knowledge about the dissemination of pathogens helps to a large extent in preventing disease occurrence and to control them effectively.

Active or Autonomous Dispersal

1. Dispersal through Soil

(i) **Dispersal in soil.** Some pathogenic fungi, especially the facultative parasites are capable of living in soil as soil inhabitants. Some species of *Pythium, Phytophthora, Rhizoctonia* etc. are soil inhabitants. Generally, these unspecialized soil-borne pathogens have a wide range of hosts. Species of *Pythium* and *Phytophthora* living in the soil may produce zoospores or swarming spores. These spores may traverse very short distances in water sediments present in the soil and come in contact with susceptible host plants and infect them. The mycelium of some fungi, such as *Fusarium solani, Pythium aphanidermatum, Sclerotium rolfsii, Fusarium oxysporum* etc. can grow and spread in the soil over very short distances and reach the roots of adjacent plants and cause infection. Several wood rotting fungi, such as *Armillaria, Fomes, Ganoderma* and *Polyporus* spp. belonging to Hypomycetes can migrate independently through the soil from plant to plant or even from field to field by their active mycelial growth. However, in the absence of suitable hosts and congenial environmental conditions, the existence and survival of these pathogens in the soil become very remote. Some bacterial and nematode pathogens can also move short distances on their own power and can thus move from one host to another one very close to it.

(ii) **Dispersal by soil.** In this case, the pathogen gets access to disease-free fields from soils of infected fields and then causes fresh infection. Diseases may be introduced to disease-free soils through soils from disease affected fields, dead parts of disease affected plants, which may contain the pathogen, infected seeds etc. and may cause fresh disease infection. During cultural operations in the fields, pathogens, which may be present only in certain pockets of the field, may be spread over the entire field. Soils from infected fields, which contain propagants of pathogens may stick on to the legs of cattle, agricultural implements etc. and are liable to be carried to disease-free fields. Live plants or plant parts used for propagation, taken from infected fields may carry soils along with the propagants of the pathogens, which may find their way to disease-free fields. In papaya nursery, if *Pythium de baryanum*, the seedling collar rot pathogen is present, the soil, which may adhere to the roots of the seedlings, may contain the pathogen and serves as a source for introducing the disease in the newly transplanted fields. Seedlings of fruit trees when transported from one place to other distant places may carry many pathogens along with the soil adhering to the roots and may effect dispersion of the pathogens in the newly introduced places. Similarly, the banana wilt pathogen, *Fusarium oxysporum cubense* may also be dispersed by soil. In such ways diseases may spread from one state to another or from one country to another.

(iii) **Persistence in the soil.** The spores, resting spores, resting spore bodies etc. produced by several fungi can remain in the soil without losing their viability for long periods, sometimes even for many years, as they possess extraordinary powers of resistance to adverse environmental conditions, such as extremes of heat and cold. Further, spores and spore bodies are mostly less sensitive than hyphae to adverse environmental conditions. These resistant properties of the spores of many fungi facilitate their survival in the soil in the face of such adversities. The conidia of some species of *Helminthosporium, Alternaria* etc. can remain in the soil along with infected plant debris in a viable state for long periods. The oospores produced by Oomycetes, the zygospores of Zygomycetes, the cleistothecia produced by the powdery mildew fungi, the teliospores and sporocarps produced by some Basidiomycetes, the sclerotia produced by *Rhizoctonia, Sclerotinia* etc. and sometimes even the mycelium of some of the fungi persist in the soil in a viable state for prolonged periods and cause diseases on the arrival of suitable hosts and favorable environmental conditions. Many bacterial and nematode pathogens can survive in the soil for prolonged periods of time in the crop debris and transmit the diseases.

2. Dispersal through Seed

Many plant diseases are disseminated through seeds. Some diseases are spread by means of mycelium present in the seeds in a dormant state. Many others are spread by means of spores or spore bodies present on the seeds. The oospores of the downy mildew pathogen of pearl millet, sclerotia of ergot pathogen of pearl millet and rye, bunted grains containing spore balls of the wheat bunt pathogen etc. are found mixed with the seeds as contaminants and cause diseases when the seeds germinate. These disease causing agents are called **'seed contaminants'**.

Teliospores of the smut pathogen of cholam and barley, sclerotia of groundnut root rot pathogen etc, stick on to the surface of the seed, either during the crop growth or at the time of harvesting and threshing and are disseminated through these seeds. Many other fungi, such as *Synchytrium endobioticum* causing potato wart disease and *Rhizoctonia solani* causing potato black scurf are also transmitted in a similar manner. The diseases caused by such **'externally seed-borne pathogens'** are known as **'externally seed-borne diseases'**.

Some other pathogens, during the growth of the host plants, enter the seeds, remain dormant inside the endosperm and embryo of the seeds in the mycelial state and cause diseases when such seeds are sown. The diseases caused by such **'internally seed-borne pathogens'** are known as **'internally seed - borne diseases'**. Rice brown leaf spot, *Helminthosporium* blight and loose smut of wheat and barley, *Alternaria* blight of wheat etc. are examples of internally seed-borne diseases. Bacterial blight of rice caused by *Xanthomonas campestris* pv. *oryzae* is also internally seed-borne. Bacterial blight or 'black arm' of cotton caused by *Xanthomonas malvacearum* is both externally and internally seed-borne. Some virus diseases are also transmitted through seeds. Bean mosaic is transmitted through pollen grains and carried in the seeds.

3. Dispersal through Plants and Vegetative Plant Propagants

Besides seeds, various plant parts used for vegetative propagation, such as transplants, setts, stem cuttings, tubers, bulbs, rhizomes etc. may also serve in the process of dissemination of several pathogens. Potato late blight, sugarcane red rot, sugarcane whip smut, *Fusarium* wilt of banana and many other diseases are spread in this way. Several bacterial diseases are carried through infected seed materials. Bacterial wilt of banana, caused by *Pseudomonas solanacearum*, ring rot of potato, caused by *Clavibacter michiganense* sub sp. *sepidonicum* and bacterial brown rot of potato, caused by *Pseudomonas solanacearum* are carried through infected potato tubers. A large number of viruses are transmitted through vegetatively propagated plant parts. 'Bunchy top' of banana, sugarcane mosaic etc. are transmitted in this way.

Passive or Indirect Dispersal

1. Dispersal through Humans

Disease-causing pathogens of many crop plants propagated by vegetative propagation, such as fruit trees, tuber crops, bulb crops, commercial crops, ornamentals etc. are disseminated from one state to another state or from one country to another country mainly by humans in the form of planting materials.

Similarly humans are responsible for dissemination of many pathogenic organisms through seeds. During exports and imports of seeds either for seed purposes or for consumption, sufficient care is not taken to avoid entry of pathogenic organisms through seeds and this results in introduction of new pathogenic organisms from one place to another. Many destructive diseases of important crops have been introduced to our country through such propagants.

Spores and spore bodies of fungal pathogens may stick on to the legs, hands or other parts of the body and on clothes worn by farm laborers and may be dispersed from one plant to another or from one field to the neighboring fields inadvertently. Humans are also responsible for disseminating pathogenic organisms from one field to another field through soil along with the pathogenic organisms sticking on to various agricultural implements. Some pathogens may be disseminated through knives used in the fields for cutting purposes.

Humans disseminate several bacterial pathogens and sap transmissible viruses inadvertently. Angular leaf spot of cucumber, caused by *Pseudomonas lachrymans* and bacterial canker of tomatoes, caused by *Clavibacter michiganense* pv. *michiganense* are transmitted by humans while handling the plants. *Tobacco mosaic virus* is transmitted from diseased tobacco plants to healthy plants while handling the plants.

2. Dispersal through Insects

Insects are also responsible for dispersal of many diseases. The sweet honeydew secreted by some plants in which the spores of some pathogens are extruded attracts several insects. In the case of ergot of rye and pearl millet, caused by *Claviceps purpurea* and *Claviceps microcephala* respectively, the insects, which are attracted by the honeydew at the sphacelia stage, visit the sphacelia and carry the conidia and disseminate them on to other flowers. Many soil-inhabiting insects disperse several root rot pathogens. The sporangia of *Phytophthora palmivora,* the causal organism of coconut bud rot are disseminated to other trees by the rhinoceros beetle as well as by other insects, which come to gather nectar from the flowers. In many cases insects that cause wounds, such as the plant borers carry the fungal pathogens on their bodies and serve to spread infection. The 'Dutch elm' disease *(Ceratostomella ulmi)* is mainly disseminated by bark boring insects. Aphids are known to carry mycelial fragments and sporangia of *Phytophthora infestans* on their bodies and leg bristles and disseminate the pathogen over short distances.

Various insects spread many bacterial diseases. *Erwinia tracheiphila,* the cucumber wilt pathogen is transmitted by cucumber beetles. Insects attacking potato tubers also disseminate black leg of potato, caused by *Erwinia carotovora.* Flies and ants transmit *Erwinia amylovora,* causing fire blight of apples and pears. *Xanthomonas campestris* pv. *citri,* causing *Citrus* canker is carried by the leaf miner from diseased to healthy plants.

Insects are responsible for spreading a vast majority of virus diseases. Besides transmitting and disseminating the diseases, they are agents of inoculation also. Aphids, leafhoppers, white flies, mealy bugs and treehoppers are mostly responsible for transmitting virus diseases. Most of the insect-transmitted virus diseases are disseminated by insects with sucking mouthparts, while some others are disseminated by insects with biting and chewing type of mouthparts. Some viruses are transmitted by the vectors in a persistent manner, while others are transmitted in a non-persistent manner. Some insect species can transmit only one or a few virus diseases, while some others are capable of transmitting several virus diseases. More than 50 species of aphids are known to transmit plant viruses. *Myzus persicae* is a vector of over 50 different viruses. 'Katte' or 'Marble mosaic' disease of cardamom, mosaic and grassy shoot of sugarcane, mosaic streak of wheat, mosaic of maize, barley, bean, cowpea, banana, papaya, tomato, chillies etc., bunchy top of banana, leaf roll of potato etc. are transmitted by aphid vectors. Tobacco leaf curl, mung yellow mosaic, yellow mosaic of lab lab, double bean etc. are transmitted by white fly vectors. 'Tungro' disease of rice is transmitted by the green leafhoppers. The red pumpkin beetle transmits bottlegourd mosaic. The *Spotted wilt virus* is transmitted by thrips. Mites are also known to transmit some virus diseases.

3. Dispersal through Nematodes

Most of the soil-borne pathogens, such as species of *Pythium*, *Phytophthora*, *Fusarium*, *Rhizoctonia*, *Verticillium*, *Sclerotium* etc., causing root rot, collar rot, wilt etc. are disseminated by some soil-inhabiting nematodes, which are motile and can traverse short distances in the soil. The small wounds caused by the nematodes in the roots of host plants pave the way for the entry of the fungal pathogens into the host.

Similarly, some bacterial pathogens and viruses are also transmitted by nematodes. Leaf gall of various herbaceous plants, caused by *Corynebacterium fascians* is carried by the nematode *Aphelenchoides*, an ectoparasite. Yellow ear rot of wheat, caused by *Clavibacter tritici* is dispersed by the ear cockle nematode *(Anguina tritici)*. The *Grapevine fan leaf virus* is disseminated by *Xiphenema index*.

4. Dispersal Through Birds and Animals

Birds can fly and cover much longer distances than insects. The spores of fungal pathogens may adhere to the feathers of birds and may be carried to distant places and dispersed.

Animals having a dense coating of fur or hairs, such as rabbits, squirrels etc., as well as rats may carry the spores of fungal pathogens sticking on to the hairs and disperse them in the nearby fields.

The oospores of the fungal pathogen, *Sclerospora graminicola* causing green ear disease of pearl millet, which are present inside the tissues of the green ear, when fed to cattle is not killed during the process of digestion, but come out through the dung in a viable condition. When this dung is applied to the fields as farmyard manure, the oospores present in the manure may cause infection. The oospores are not killed even after the manure is composted.

5. Dispersal through Water

Mostly dispersal of pathogens through water is possible only for short distances. However, cyclones, storms, floods etc. may carry pathogens over very long distances. Some pathogens may be carried by irrigation water or rainwater and dispersed. Sugarcane red rot, *Fusarium* wilt, root rots and other soil-borne diseases are liable to be dispersed by running water. Bacteria, nematodes and spores and mycelial fragments of fungi present in the soil are disseminated by rain or irrigation water that flows on the surface of the soil from one field to another. *Xanthomonas campestris* pv. *oryzae* is carried through irrigation water and attacks the leaves touching the water. The nematode parasite attacking potato and *Piper longum* is carried through surface water from field to field. Water is an important agency of dissemination for the seeds of phanerogamic parasites, such as *Orobanche*, *Cuscuta* and *Striga*.

During rainfall also many pathogens are dispersed. Splashing of raindrops, which fall on diseased parts may carry the pathogen also along with the droplets to the nearby plants or else the raindroplets are carried by blowing wind to host plants, which may be far away. Rice blast disease, late blight of potato, downy mildews etc. are dispersed in this way. Several fungal pathogens belonging to *Septoria, Fusarium, Colletotrichum, Gloeosporium, Phyllosticta* etc. are splash dispersed. The spores of *Hemileia vastatrix* causing coffee rust is also largely dispersed by rainsplash. The fire blight bacteria, *Erwinia amylovora* is mostly disseminated by washing down or splashing of raindrops. *Xanthomonas campestris*, causing black rot of cabbage and other crucifers, *X. campestris* pv. *oryzae*, causing bacterial blight of rice, *X. translucens*, which attacks cereals and grasses, *X. campestris* pv. *malvacearum*, causing angular leaf spot of cotton and *Corynebacterium michiganensis*, causing bacterial blight of tomato are examples of other bacterial diseases that are disseminated largely by splashing of raindrops. Large drops of rain or overhead irrigation are more effective in the dispersal of pathogenic organisms

Fungal spores of some pathogens, because of their very light weight just float in the air in a suspended state. During rainfall, these spores are also brought down, fall on the host plants and cause infection.

Further, in many fungi, for the production of spores and for spore discharge water is required. So, when such conditions conducive for the pathogens occur, spores are produced in very large numbers and discharged effectively and are dispersed over long distances. Some fungal pathogens, such as *Phytophthora arecae, P. palmivora, P. infestans, P. parasitica* and several other downy mildew pathogens produce large number of zoospores in the presence of water that is supplied by rain or heavy dew. These zoospores are then dispersed to adjacent leaves or plants by splashing raindrops.

6. Dispersal through Wind

In a vast majority of fungal pathogens, the spores are the main organs of dissemination. Most of the pathogens, such as the powdery mildews, downy mildews, rusts, smuts, polypores, sooty moulds, leaf spot fungi etc. produce numerous spores in succession. Further, the spores are so small and so light, they are very easily caught up in the air currents and effectively dispersed by wind. The spores also show high degree of resistance to extremes of heat and cold and other aerial adverse conditions. These resistant properties of spores facilitate the aerial dissemination of many fungi, otherwise under such extremes of environmental conditions, which they have to face in the air, the spores may be desiccated and killed.

Several fungi possess special adaptations, which may be useful for effective dissemination through air. The vertical elongation of the sporophores of many fungi is often a powerful aid to exposing the spores freely to air currents. Means of propagation of spores from the stalks into the air current, as is common in many ascomycetous fungi, is yet another such adaptation.

Long distance dissemination of spores of pathogens, such as *Synchytrium endobioticum,* causing potato wart disease, *Exobasidium vexans,* causing blister blight of tea and several other fungal pathogens are carried over long distances by wind. Long distance dissemination of rusts is very common. It is an established fact that urediniospores of *Puccinia graminis tritici* are transported from the Nilgiris and Palney hills to Central India by wind. Spores of *Phytophthora infestans* have been reported to travel distances of 200m. to more than 60 km. along with the wind. Besides spores, bits of mycelium are also disseminated by wind to distant places. Although fungal spores are found abundantly near the earth surface, spores of many fungal pathogens, moulds and some bacteria are found at very high altitudes. This may facilitate easy and unobstructed transport of such spores over very long distances, possibly from one state to another or from one country to another or even from one continent to another.

Urediniospores of *Puccinia graminis tritici* have been trapped from altitudes as high as 14,000 feet above infected fields. Similarly, living spores of several fungi have been trapped at very high altitudes and from far off places from their source of origin in aeroscope studies.

Spores of some fungal pathogens are provided with some special provisions, which may aid in wind dissemination. Conidia of *Pestalotia* species possess two or three long appendages at their tips, which aid the spores to fly in the air.

Some plant pathogenic bacteria are carried over short distances by wind. The bacterium of fire blight of apple and pear produces fine strands of dried bacterial exudate containing bacteria and these strands are broken off and disseminated by wind. Some of the bacterial pathogens may survive in the dead plant tissues lying in the soil for a fairly long time and these may be blown off by wind and carried to distant places. Nematodes may also be disseminated by wind. The cysts of *Heterodera major*, causing 'Molya' disease of wheat and barley is believed to have been introduced to Haryana from Rajasthan through cysts carried by wind.

Physiological Specialization of Pathogenic Fungi

Many fungal pathogens causing rusts, smuts, powdery mildews, downy mildews etc., which are capable of attacking several different species of plants have developed distinct races. These races, which are morphologically identical and indistinguishable from one another, differ in their ability in infecting and living on certain hosts. Further, a single species of host plant may have a large number of varieties, particularly in the case of economically important crops, which are subjected to continuous breeding and selection. In such cases, there may be several different races of the pathogen on the particular species of the host, which may be differentiated only by their capacity to attack particular varieties more or less severely than others. The differences between these races are in their physiological characters and not in their morphological characters. The splitting of a particular parasite into specialized races on different host plants is called **'specialization of parasitism'** and the races are known as **'physiological races'** or **'pathogenic races'**.

Most of the rust fungi are specialized parasites and attack only certain Genera or only certain varieties of a species. Rust fungi, that are morphologically identical but attack different host Genera are regarded as **'special forms'** or ***'formae specialis'*** (f. sp.). Thus the black rust of wheat is called *Puccinia graminis* f. sp. *tritici* and the barley rust *Puccinia graminis* f. sp. *hordei*. Within each special form of rust, there are several physiological (pathogenic) races. These so called **'physiological races'** can attack only certain varieties within the host species and can be detected and identified only by a set of **'differential varieties'** (differential hosts), which they can infect. Each physiologic race is differentiated based on the intensity of spotting, which they produce when inoculated with the uredospores on the differential hosts. Twelve standard wheat varieties have been selected as differential hosts for differentiating the races of wheat rust *Puccinia graminis*; five oat varieties for differentiating the races of *Puccinia graminis* on oat and five rye varieties for the rust of rye. Based on the differences in the intensity of spotting on the differential hosts, separate race numbers are allotted.

Thus, *Puccinia graminis tritici* '50' indicates physiological race No.50 of special form *tritici* of *Puccinia graminis.* Sexual reproduction process of fungi may play an important part in the production of new races. But, even in many fungi, where the sexual process is very rare, new races appear periodically. Chemical heterogeneity of the host or the defense mechanisms posed by the host may also induce the pathogen to produce new races. In a general way, the physiological races of a specialized pathogen are quite stable. Wheat, barley and oats have many distinct races of rust and mildews. The black rust of wheat, *Puccinia graminis tritici* cannot attack oats and the oat rust *Puccinia graminis avenae* cannot attack wheat under natural conditions. However, the races, which occur on the cereals may attack various grass hosts. This is quite commonly seen in the case of *Puccinia graminis tritici* and *Puccinia graminis avenae,* which have many grass hosts. These grass hosts in all possibilities may be responsible for the persistence of the races of cereal rusts from one season to the next.

The cereal rusts are heteroecious and the pycnial and aecidial stages are produced on alternate hosts, such as barberry, buckthorn etc. The sexual stage of the fungus develops on the alternate hosts and union of spermatia produced from different spermagonia takes place. Thus, different physiologic races can hybridize resulting in the development of a **'form'**, which is heterozygous in respect of the character of physiologic specialization. If a heterozygous physiologic race on a cereal or grass host passes again on to its alternate host, seggregation may occur and forms that are new to the locality may arise. In countries like Australia, India and Kenya, where barberries play a very little part in the life history of *Puccinia graminis,* the number of physiologic races is much less than in Europe and North America, where barberries play a major role in the life history of the fungus. While there are over 200 physiologic races of *Puccinia graminis tritici* in Europe and North America, only a few races are found in India. In the other rusts, such as *Puccinia triticina, P. coronata, P. glumarum* and *P. hordei,* the number of physiologic races identified so far are much less than in *P. graminis.*

The cereal powdery mildew, *Erysiphe graminis* is more strongly specialized than *Puccinia graminis.* The physiologic races of *Erysiphe.graminis* on wheat infect only species of *Triticum,* those on barley, species of *Hordeum* and those on oats, species of *Avena.* Thus, physiologic races differ from one another in their powers of attack and the different species and varieties of the host are not all equally attacked by any one 'form' of the pathogen under natural conditions. Specialization of parasitism is strongly evident in some smuts of cereals also, as in the rusts and mildews. Here also, as in the case of rusts, the sexual process plays a vital part in the production of new races.

The infective powers of a specialized race are identical in all its spore forms, such as aecidial, uredial and telial stages in the rusts and in the conidial and ascospore stages in the mildews. In the case of *Claviceps purpurea* also the same trend is evident.

In general, the physiologic races of a specialized pathogen are quite stable. However, the range and degree of their parasitism may vary under varying conditions of environment and nutrition, but these variations are only temporary and not permanent.

Because of the continuous emergence of new physiological races, many popular varieties of important commercial crops, especially cereals, which are supposed to be resistant to the existing races of the pathogens, become vulnerable to attack by the new races and become susceptible hosts to the new races. So, evolution of new resistant varieties by breeding, selection and other techniques by humans, and breaking down of resistance of the new varieties by new races of the pathogens occurring in nature is a never ending ordeal.

Epidemiology of Crop Diseases

'**Epidemiology**' has been defined as a 'Science that deals with the increase or decrease of a plant disease in a host population in time and space' (Kranz, 1974). In other words the study of epidemics and the factors that influence them is called '**epidemiology**'. When a disease affects single or isolated plants within a population, its occurrence is called '**sporadic**'. On the other hand, when the disease occurrence increases rapidly in time or space, it is known as '**epidemic**' or '**epiphytotic**'. When a disease is established in a particular area, district or state over a long period of time, it is termed '**endemic**'. When an epidemic extends over most of the continent and causes mass destruction, it is termed '**pandemic**'.

Epidemic diseases of plants can be divided into two main groups: **(i)** the disease may be endemic normally, but may assume epidemic proportions under certain conditions. Here, the host, the pathogen, the environmental conditions and human activities are involved in the development of epidemics **(ii)** the disease may be a newly introduced one.

Epidemics occur as a result of a conflict between the host and the pathogen, aided by environmental factors favorable for the pathogen or unfavorable for the host. In this conflict, if the pathogen takes the upper hand that may lead to occurrence of an epidemic. The factors responsible for the development of disease epidemics are presented in **Fig. 60**.

Epidemics of various plant diseases had been reported from various countries the worldover in the past. Epidemics of potato late blight *(Phytophthora infestans)*, which caused famine in Ireland during 1845-'46, coffee rust *(Hemileia vastatrix)*, which devastated coffee plantation industry in Sri Lanka during 1869, the grapevine mildew *(Plasmopara viticola)*, which swept through France and threatened the wine industry during 1878, brown leaf spot of rice *(Helminthosporium oryzae)*, which caused the Bengal famine in India during 1943 are some of the examples. India had also witnessed many other disease epidemics in several crops during the past, such as stem rust of wheat *(Puccinia graminis tritici)*, brown spot of rice *(Helminthosporium oryzae)*, red rot of sugarcane *(Colletotrichum falcatum)*, downy mildew of pearl millet *(Sclerospora graminicola)*, apple scab *(Venturia inaequalis)*, rice blast *(Pyricularia oryzae)*, bacterial blight of rice *(Xanthomonas campestris* pv. *oryzae)*, rice 'tungro' virus disease etc.

Factors Responsible for the Development of Epidemics

(i) Host. Presence of a susceptible host over extensive areas at its most vulnerable growth stage for the attack by the pathogen is a prerequisite for the occurrence of an epidemic. When the distance between the host and the source of primary infection is more, then the chances of infection and disease occurrence are much less. So, an epidemic would rarely break out in the host plants, which are isolated and far apart even if the host is highly susceptible to a particular disease. On the other hand, the accumulation of a susceptible host over extensive areas distributed in different localities, paves the way for large-scale production of spores and rapid spread of the disease leading to an epidemic. Epidemics of rice blast, groundnut 'tikka' disease, red rot of sugarcane, bacterial blight of rice, 'tungro' virus disease of rice etc. had occurred in Tamil Nadu and other states of India, because of large-scale cultivation of susceptible varieties in large areas.

The growth stage of the crop is also important in the development of epidemics. In the case of downy mildew of grapevine, the foliage and first buds are more prone to attack by the pathogen, while the older leaves may remain free from attack. In late blight of potato, the foliage and tubers are attacked to a larger extent during the later stages of growth, as the resistance of potato plants decreases with age. In the case of brown spot of rice, late 'tikka' disease of groundnut etc., the older plants are more susceptible, while the young plants have some amount of resistance to the disease. On the contrary, rice blast is more severe during the early growth stages. 'Tungro' virus disease is also more severe during the young growth stages of the crop. During the vulnerable stages of crop growth, if there is abundant supply of inoculum, an epidemic may result. In the case of heteroecious rusts and many other diseases, the presence of alternate and collateral hosts may play a vital role in the development of epidemics by providing enough initial inoculum during the vulnerable growth stage of the host. In the case of vector transmitted virus diseases, the presence of large population of vectors may cause epidemics.

(ii) Pathogen. Epidemics of many crop diseases depend on the presence of a more virulent pathogen capable of infecting the host rapidly. In each and every disease causing organisms, there may be highly virulent or less virulent strains. Virulence in the strain of a pathogen may be altered due to changes in the habitat, hybridization or mutation. Sometimes, under ideal environmental conditions favorable for the development of the pathogen, even a less virulent pathogen may become highly virulent. A highly virulent strain may attack even some of the resistant varieties of the crop, besides the susceptible varieties and cause epidemics.

The reproductive capacity of a pathogen is another factor that is responsible for the development of epidemics. Pathogens causing rice blast, wheat rust, powdery and downy mildews, potato late blight etc. have high reproductive capacity. A typical leaf spot of rice blast produces 2,000 to 6,000 conidia each night for about two weeks. In the black rust of wheat about 50,000 to 4,00,000 uredospores are produced from a single uredosorus. In most of the bacterial diseases also reproduction is very rapid and millions of bacteria are produced from each lesion. When spores are produced in such large numbers, more spores get easy access to the host and the infection spreads quite rapidly leading to development of epidemics.

Production of inoculum in succession may also promote the chances of development of epidemics. Many fungi, bacteria and viruses are known to have short reproductive cycles. The uredospore of the wheat black rust fungus can infect a plant and in about 10-15 days time produce uredosori and another crop of uredospores. The rice blast fungus continues to release conidia for about two weeks from each spot. Most of the powdery mildews produce conidia in chains and continue to release them one by one for many days. The downy mildews also release the sporangia continuously for many days. Such polycyclic pathogens viz., rusts, mildews, leaf spots etc. that produce many generations in a single growing season are responsible for most of the disease epidemics throughout the world. When environmental conditions favorable for the pathogen and spore release coincide, large-scale infection and spread may occur leading to epidemics. On the other hand, in the case of soil fungi, such as *Fusarium*, *Verticillium*, many nematodes etc., which have only a few reproductive cycles in each growing season, as well as smuts, short-cycle rusts etc., which are monocyclic are rather slow in developing epidemics.

Many of the fungal pathogens produce their spores on the surface of the aerial parts of the host and the spores are dislodged and dispersed with ease and can cause widespread epidemics. In the case of many bacterial diseases, the bacteria exude out from the surface of the aerial parts as bacterial ooze. These bacteria can also be dispersed easily by wind and rainsplash. The seeds of parasitic higher plants are also disseminated by wind easily. Pathogens, such as vascular fungi and bacteria, viruses, mollicutes etc. reproduce inside the plants. In such cases spread of the inoculum is mostly done by vectors and as such, the chances of development of widespread epidemics are very much limited.

The mode of spread of inoculum is another important factor that favors the development of epidemics. The spores of many pathogenic fungi causing rusts, mildews, leaf spots, smuts etc. are released into the air directly. Being minute and very light in weight, they can be dispersed by air currents or strong winds over distances ranging from a few centimeters up to several kilometers. These kinds of pathogens are often responsible for causing widespread epidemics. Many of the pathogenic viruses are transmitted by vectors, such as aphids, white flies, thrips, mites etc., which may be carried by air currents over long distances.

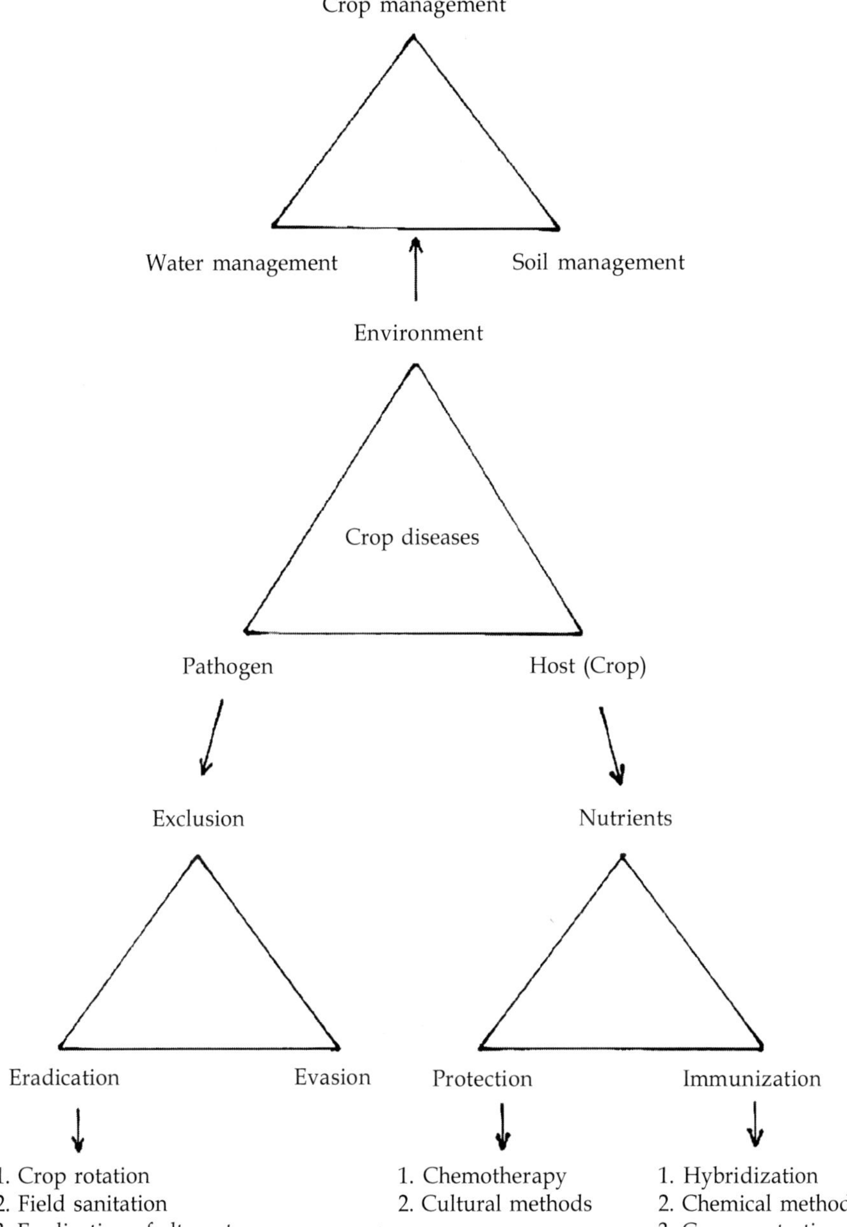

Fig. 60 : Manipulation of the Disease Triangle
(Factors responsible for disease epidemics)

Further, many pathogenic organisms can adopt themselves to environmental conditions and survive through adverse conditions. Reproductive organs, such as oospores, cleistothecia, sporocarps etc. remain dormant during the off season or when conditions are unfavorable for the development of the pathogens. On return of favorable environmental conditions, they produce spores in large numbers and initiate epidemics.

(iii) Environmental conditions. Even when a susceptible host is present and enough inoculum is available, a disease may not attain epidemic proportion, unless environmental factors are favorable for the pathogen. Environmental conditions play a vital role in the dispersal of inoculum, infection of the host, establishment of the pathogen in the host and subsequent production of spores. Even the microclimatic factors play an important part in the development of epidemics. Further, environment may affect the multiplication, intensity and spread of vectors transmitting several diseases, especially virus diseases. The most important environmental factors that influence the occurrence of plant disease epidemics are moisture, temperature and human activities.

Moisture. Moisture is available in the form of rain or irrigation water on the plant surface and around the roots of plants or as atmospheric humidity or as dew, mist etc. Moisture is essential for the germination of fungal spores, penetration of the host by the germ tube, establishment of the pathogen inside the host and reproduction of the pathogen. Moisture is also indispensable for the activation of bacterial, fungal and nematode pathogens before they can infect the host. Late blight of potato, downy mildew of grapes, apple scab, rice blast, bacterial blight of rice and several other diseases are more severe only in areas with high rainfall or high relative humidity during the growing season. Most fungal and bacterial pathogens require free water to infect the host successfully. Free water is also required for the production of zoospores from sporangia in the case of downy mildews, potato late blight and in the case of many pathogenic bacteria, which swim about in the free water on the host surface for some time before infecting the host. However, in the case of some diseases like powdery mildews, rice brown spot etc., spore production and spore germination are high in bright sunlight than in rainy, damp weather. Some diseases caused by soil-borne pathogens, such as *Fusarium* and *Streptomyces* are more severe in dry than in wet weather, but such diseases rarely develop into epidemics.

Temperature. Temperature has an important bearing on the development of epidemics. The influence of temperature on plant epidemics is mainly due to its effect on the pathogen during the different stages of growth viz., spore germination, host penetration, establishment in the host, sporulation and dissemination of the spores. At temperatures optimum for the pathogen, a polycyclic pathogen can complete its life cycle within a very short period and may produce several generations in a single growing season. With each infection cycle, the amount of inoculum is multiplied several times. This may result in more and more plants getting infected and thus may lead to development of epidemics. In most cases, the mycelium of the pathogen ceases to grow at temperatures below 10°-12°C and above 30°C. The optimum atmospheric temperature for the development of epidemics lies between 22°- 24°C. However, in a vast majority of plant diseases, optimum moisture and temperature combined together influence the development of epidemics

Environmental factors unfavorable for the hosts may also predispose them to infection and may lead to development of epidemics.

(iv) Human activities. Many activities of humans in the name of cultural practices have a direct or indirect bearing on plant disease epidemics. The use of seed, nursery stock and other propagative materials that carry pathogens, increase the amount of initial inoculum, thereby may favor development of epidemics. Continuous monoculture, cultivating a single variety, especially a susceptible one over extensive areas, application of high levels of nitrogen fertilizer, thick planting, overhead irrigation, poor sanitation etc increase the chances of development of epidemics.

Introduction of new diseases. Although endemic diseases may assume epidemic proportions under certain conditions, sometimes a new pathogen introduced from another country may cause epidemics. If the foreign pathogen acclimatizes itself to the new environmental conditions prevailing in the country of introduction, it may cause epidemics. There are several instances wherein introduced pathogens have been responsible for causing serious epidemics. Late blight of potato, leaf rust of coffee, downy mildew of grapes, black shank of tobacco, powdery mildew of rubber, onion smut, bacterial blight of rice, 'tungro' virus disease of rice, bunchy top of banana and golden nematode of potato are some of the examples of introduced pathogens that had caused widespread epidemics in the past. Sometimes, introduction of a new host variety from a foreign country may also lead to development of epidemics. The new varieties having no resistance against the indigenous pathogens may lead to widespread occurrence of diseases and epidemics.

Plant Disease Surveillance

Objectives of Plant Disease Surveillance

1. To identify the diseases attacking various crops.
2. To assess the intensity of diseases.
3. To study the relationship between environmental conditions and disease development.
4. To find out the occurrence of new diseases.
5. To study the influence of various improved cultural practices on the development of diseases.
6. To identify areas or localities where certain diseases of particular crops occur epidemically.
7. To recommend adoption of appropriate control measures to combat the diseases.
8. To forecast widespread occurrence and possible epidemics of diseases

Methods of Conducting Disease Surveillance

Disease surveillance is conducted in two ways viz., **1. Fixed plot survey** and **2. Roving survey.** In both these methods, the surveillance has to be undertaken on a particular day of every week all through the growing season of the crop, from the time of sowing to harvest of the crop depending upon the disease. During each survey, the intensity of the disease affecting the crop is assessed. At the time of survey, the amount of inoculum (number of spores) present in the atmosphere just above the crop canopy is also recorded by means of spore traps and the number of spores present in each cubic centimeter of air is estimated. Further, the day to day weather parameters, such as maximum and minimum temperature, relative humidity during the morning and evening hours, intensity and duration of sunlight, wind velocity, rainfall, number of rainy days and amount of dew for the week prior to each and every survey are also recorded. From these data, correlation between weather parameters and disease occurrence and spread is established. Such data, collected continuously over several years is of help in determining the chances of widespread occurrence of the disease and timely forecasting the chances of appearance of the disease on a wide scale. This will also help in taking up appropriate control measures at the proper time, thereby the disease can be controlled effectively without incurring appreciable loss in yield. Such surveys also help in locating the presence of alternate and collateral hosts of the pathogen, the destruction of which may also help to reduce the development of the disease to a large extent.

1. Fixed Plot Survey

This survey is conducted in one and the same field throughout the growing season of the crop. In one acre of the field, five one-meter square plots are selected, four plots at the four corners of the field and one plot at the centre of the field. The corner plots should be three meters inside from the field bunds. The plots are marked with pegs of suitable height. The disease intensity should be collected from these five plots every week.

In the case of closely planted crops with large number of plants per unit area, such as rice, groundnut, finger millet etc., 20 plants are selected at random from each plot. Each plant as a whole is taken as a unit and the disease intensity assessed. In all, the disease intensity is assessed in 100 plants from the five plots. In the case of bigger plants, such as sorghum, sugarcane, castor, cotton, sesame etc., which are planted with wider spacing, there will be less than 20 plants in a one-meter square area. In such cases, from each plot five plants are selected at random and the disease intensity assessed from four leaves viz., 3rd, 5th, 7th and 9th leaves from the tip of the main branch of each plant. Here, each leaf selected is taken as a unit and from five plants 20 leaves are selected for disease assessment. Thus, in all 100 leaves are selected from 25 plants in the five plots. In the case of bigger plants, such assessment may be done in another way also. Here instead selecting plants from five different plots, 25 plants are selected at random from the two diagonal lines connecting the opposite corners of the field and the disease intensity assessed from four leaves from each plant as mentioned before. In this method selection of plants from the first three meters from the corners should be avoided. The disease intensity is assessed every week from these marked plants.

2. Roving Survey

This survey is also conducted on a particular day every week, but the survey is carried out in different fields situated in different villages every week and not in the same field as in the case of fixed plot survey. However, the survey should be carried out in crops of almost the same age

The surveys are conducted by the Agricultural Officers in fields of villages under their control, Managers of State Seed Farms and Scientists of the Agricultural University and the survey reports sent to the concerned authorities in the prescribed proforma every week.

Scrutiny of the Survey Reports

On a particular day of every week, the survey reports received from the various Agricultural Officers of the Agricultural Department and Scientists of the Agricultural University are scrutinized by the representatives of the Agricultural University and the Deputy Director of Agriculture (Plant Protection) in each and every district separately. The particulars from the survey reports received are consolidated and the findings recorded. This consolidated report provides a specific idea about the intensity of various diseases in the different locations, variations in the occurrence and spread of diseases over the previous weeks, possible influence of environmental factors on further development of diseases, appearance of new diseases etc. After detailed deliberations on these facts, consensus is reached as to whether it is necessary to take up plant protection measures immediately and if necessary, the kind of plant protection measures, as well as the plant protection chemicals to be applied are decided. The necessity to take up mass ground sprayings or aerial sprayings to control certain epidemic diseases are also taken by this committee. The recommendations are sent to all the offices concerned in the district and if necessary to the adjacent districts and states, so that they can be in a state of preparedness in case of any eventuality. The final consolidated reports from all the districts are sent to the Plant Protection Center of the Agricultural University and the Directors of Agriculture, Oilseeds and Horticulture every week. If there is any possibility of widespread occurrence of any disease based on such scientific monitoring, the farmers in the district, as well as in the neighboring districts are forewarned through various media, such as television, radio, newspapers etc., so that the farmers can take up timely and appropriate control measures to combat the disease. Plant disease surveillance helps in the prevention of use of plant protection chemicals unnecessarily and whenever necessary recommends the use of appropriate plant protection chemicals at the correct dosage and at the proper time. Thus use of plant protection chemicals on a large scale unnecessarily is avoided, thereby the beneficial insects, such as parasites and predators are protected and environmental pollution is also minimized to a considerable extent. This leads to savings on unwarranted expenditure on plant protection also.

Assessment of Disease Intensity

The object of assessment of disease intensity is to convert the extent or severity of a disease from a qualitative state into a quantitative state for better expression. Expression of disease intensity in a quantitative state helps in estimating the yield loss more or less accurately. According to **Chester** (1950), when the yield loss is less than 1 %, it is ignored; when the disease occurrence is spread to a small extent over large areas and the yield loss is up to 1 %, it is negligible; when the disease occurrence is spread to a slightly larger extent uniformly and the yield loss is 1 - 5 %, it is considered to be low; when the disease occurrence is fairly widespread and the yield loss is 5 - 10 %, it is considered to be moderate; when the disease occurrence is somewhat severe, affecting many plants and the yield loss is 10 - 20 %, it is considered to be high and when the disease occurrence is very severe, affecting almost all the plants uniformly and the yield loss is more than 20 %, it is considered to be very high and such conditions result in high economic loss and cultivation of a crop is unremunerative and becomes a waste.

Granger (1959) has classified the occurrence and severity of diseases under three categories: - **(i)** As a result of the disease infection, the entire plant may die or the plant may become very weak and famished and unproductive **(ii)** One or a few parts of the plant may be affected or only a few plants in a field may be affected **(iii)** The ultimate effect of the occurrence of the disease may persist for several years to come.

Depending upon the type and symptoms of the disease, the method of disease assessment and yield loss varies. The occurrence of some diseases may cause total destruction and death of the affected plants. Several other diseases may affect portions of leaves or some parts of the foliage or parts of fruits, tubers, bulbs, rhizomes etc. or parts or the entire earheads.

In the case of diseases, which may cause death of the entire plant or destruction of the affected parts completely, such as wilts, seedling rots, sorghum head smut, red rot of sugarcane etc. and a few virus diseases, the disease intensity is expressed directly as percentage of affected plants or affected parts. In such cases no yield is obtained from the affected plants or affected plant parts. Hence the yield loss is directly proportional to the disease intensity.

Example 1. In a cotton field affected by *Fusarium* wilt, the extent of disease intensity was assessed. From the two diagonal lines connecting the two opposite corners, 100 plants were selected at random, 50 plants from each line. Of the 100 plants selected, 12 plants were found to be affected. Calculate the disease intensity and yield loss.

Number of plants affected out of 100 plants selected = 12

$$\text{So percentage of disease intensity} = \frac{12}{100} \times 100 = 12\ \%$$

From the affected plants no yield is obtained, because the plants are killed completely and as such, the loss in yield is directly proportional to the disease intensity.

Hence the loss in yield = 12 %

If there had been no disease incidence, the yield would have been 12 % more.

Calculation of yield loss. After assessing the intensity of the disease in terms of percentage, the yield loss can be calculated by applying the formula:

$$\text{Yield loss} = \frac{Pr}{(100 - Pr)} \times Pa$$

Where, 'Pr' is the intensity of the disease expressed as percentage and 'Pa' the actual yield obtained.

Using this formula, the yield loss can be calculated only in the case of diseases, which cause death of the affected plants or destroy the affected parts, such as earheads or grains completely. This formula cannot be applied to calculate the yield loss in case of diseases that affect only parts of leaves or other plant parts, such as leaf spots, leaf blights, rusts etc., as in such cases the yield loss is not directly proportional to the disease intensity.

In the above example, if the actual yield obtained is 1,200 kg /acre, the yield loss can be calculated directly by using the formula :

$$\text{Yield loss} = \frac{Pr}{(100 - Pr)} \times Pa$$

Pr = 12 % and Pa = 1,200 kg.

$$\text{Yield loss} = \frac{12}{(100 - 12)} \times 1{,}200 = 163.6 \text{ kg. /acre}$$

If there had been no disease incidence, the yield would have been 1,200 + 163.6 = 1,363.6 kg. /acre

Example 2. In a sorghum crop affected by head smut, the extent of disease incidence was assessed. In the 100 plants selected at random from the two diagonal lines connecting the opposite corners, there were 240 earheads in all. Out of these, 36 earheads were affected by the disease. The actual yield obtained was 1,500 kg. /acre. Calculate the disease intensity and yield loss.

The affected earheads are destroyed completely as a result of infection and no yield is obtained from such affected earheads. Here the yield loss is directly proportional to disease intensity.

Out of 240 earheads, 36 earheads are affected.

$$\text{So percentage of disease intensity} = \frac{36}{240} \times 100 = 15\ \%$$

From these 15 % of earheads no yield is obtained. The loss in yield is directly proportional to the disease intensity. Hence yield loss = 15 %.

Pr = 15 % and Pa = 1,500 kg.

$$\text{Yield loss} = \frac{Pr}{(100 - Pr)} \times Pa = \frac{15}{(100 - 15)} \times 1{,}500 = 264.7 \text{ kg. /acre}$$

If there had been no disease incidence, the yield would have been 1,500 + 264.7 = 1,764.7 kg. /acre.

In the case of diseases, such as pearl millet smut, sorghum grain smut etc., only some grains in the earheads are affected and destroyed completely, while the rest of the grains are normal. In such cases, the assessment of disease intensity is somewhat different. Here 100 earheads are collected at random from the affected field. Then the average percentage of infected to the total number of grains is ascertained. As the affected grains are completely destroyed, the yield loss is directly proportional to the percentage of grains affected. However, in such cases, to assess the disease intensity, the number of affected grains and the total number of grains in the 100 earheads collected have to be counted, so as to ascertain the disease intensity. But, this is practically a very tedious or almost impossible task.

Example 1. In a pearl millet crop affected by grain smut, the disease intensity was assessed. In 100 earheads collected at random, the percentage of grains affected was estimated as 9.0 %. The actual yield obtained was 500 kg. /acre. Calculate the yield loss.

Percentage of grains affected = 9.0 %

From this 9 % of grains, no yield is obtained.

$$\text{So yield loss} = \frac{Pr}{(100 - Pr)} \times Pa = \frac{9.0}{(100 - 9.0)} \times 500 = 49.5 \text{ kg. /acre.}$$

If there had been no disease incidence, the yield would have been 500 + 49.5 = 549.5 kg. /acre.

Assessment of disease intensity in the case of other diseases

Assessment of disease intensity quantitatively as mentioned above is not feasible in the case of many other diseases, where the affected plants or plant parts are not completely destroyed. Occurrence of these diseases, their symptoms, extent of damage etc. may vary to a considerable extent from plant to plant and from leaf to leaf. In several diseases, such as leaf spots, leaf blights, rusts, powdery mildews etc., there is much variation in the disease incidence and symptoms. Further, in such diseases the affected plants or affected plant parts are not completely destroyed. So, to evaluate the disease intensity from a qualitative state to a quantitative state, different methods have to be adopted. The method, which is commonly followed, is to adopt a scale with different grade values based on the extent of symptoms expressed. **Nathan Cobb** first introduced the scale for the quantitative expression of diseases for evaluating the disease intensity of wheat rust and the scale was called as **'Cobb scale'**. Subsequently **Parker** (1922), **Peterson** (1948), **Chester** (1950) and other Pathologists made several changes in the

Cobb scale to suite other diseases. A scale to assess the disease intensity of potato late blight was evolved by **James** (1971-'72). Although different scales were adopted to determine the disease intensity of many important crop diseases, grade values ranging from 0 - 9 were used in most of the cases. The disease intensity is expressed as **'percentage disease index'** (PDI).

Table 1. Scale for evaluating the disease intensity of potato late blight

Grade	Extent of disease incidence (%)	Symptoms expressed
0	0.0	No disease symptoms
1	0.1	Symptoms in only a few plants scattered in the field. In an area covering 10 meter radius, one or two small leaf spots are seen.
2	1.0	In each plant, up to 10 small leaf spots are seen.
3	5.0	In each plant, up to 50 small leaf spots are seen. Out of 10 leaves, one leaf shows typical leaf symptom.
4	10.0	In each plant, up to 100 small leaf spots are seen. Out of 10 leaves, two leaves show typical leaf symptoms.
5	25.0	Symptoms appear in all the leaves, but no blighted areas are seen. The plants appear normal and the field looks green.
6	50.0	All the plants are affected. Symptoms appear in all the leaves and 50% of the leaf areas are blighted. The field looks brownish in scattered areas.
7	75.0	All the plants are affected to a larger extent and 75% of the leaf areas are blighted. The field looks brownish in most of the areas.
8	95.0	Most of the leaves are blighted completely and have fallen down. Only a few leaves remain in the plants. The stem portions remain green. The field looks completely brownish.
9	100.0	All the leaves are completely blighted and have dropped off. The stem portions are also rotten completely and dead.

Table 2. Scale for evaluating the disease intensity of rice blast

Grade	Extent of disease incidence (%)	Symptoms expressed
0	0.0	No disease symptoms
1	1.0	Pin head-like small, brownish spots seen scattered on some leaves in the field
3	10.0	Small, brownish isolated spots appear on many leaves in the field. Few typical eye-spots are also seen on some leaves. Up to 10% of the leaf areas are affected.
5	25.0	Eye-like spots slightly enlarged. Spots found in more number of leaves. A few spots coalesce to form bigger spots. Up to 25% of the leaf areas are affected.
7	50.0	Eye-like spots greatly enlarged and many spots coalesce to form bigger patches. The leaf tips are dried. Up to 50% of the leaf areas are affected. The affected areas are dried and the leaves present a burnt-up appearance.
9	< 50	Almost all the spots have coalesced and most of the leaf areas have dried and the entire crop presents a burnt-up appearance.

The diagram representing the scale for assessing rice blast intensity is presented in **Fig. 61**.

In the case of grain diseases, such as pearl millet smut, sorghum grain smut etc., where only the affected grains are completely destroyed, the disease intensity may also be assessed using grade charts and the yield loss calculated from the percentage disease index. The grade chart for assessing the disease intensity of pearl millet smut is given in **Fig. 62**.

Sampling for Assessment of Disease Intensity

'**Sampling**' is very important in the assessment of disease intensity. The plants selected are supposed to represent the entire population of plants in the field. So, the plants have to be selected strictly at random in an unbiased manner and intentional selection of plants with either more disease infection or less disease infection is to be avoided. The growth stage of the crop at which the disease intensity is assessed is also very important. Generally in cereal crops, such as wheat, barley, maize etc., there are 10 quite distinct growth stages from the germinating stage to the harvesting stage.

Among such different growth stages, the disease intensity is assessed at the stage at which there will be maximum loss in yield as a result of disease occurrence. In the case of rice, there are four distinct growth stages. The first stage is in the nursery, about 10 days after sowing. The second, third and fourth stages are the maximum tillering stage, booting stage and the flowering stage respectively. Amongst these stages, if the disease occurs during the maximum tillering stage and booting stage, then the loss in yield is considerably more.

Calculation of Disease Intensity

Example 1. In a TKM.9 rice crop, blast disease intensity was assessed using a scale with grade values ranging from 0 - 9. Of the 100 pants selected from five one-meter square plots in a one-acre field, 25 plants were found to be free from blast. Of the remaining 75 plants, 15 plants were placed in grade 1, 15 plants in grade 3, 25 plants in grade 5, 15 plants in grade 7 and 5 plants in grade 9 respectively. The disease intensity is calculated as below:

The disease intensity is calculated using the following formula:

$$\text{Disease intensity (Percentage disease index)} = \frac{\Sigma (n \times v)}{N \times V} \times 100$$

Where,

n = the number of plants under each grade

v = grade values

N = total number of plants evaluated

V = the highest grade value in the scale

Σ (n x v) = summation of the multiplied values of the grade value and the number of plants under each grade.

The above example may be tabulated as below:

Percentage of leaf area affected in the selected plants (%)	Grade value (v)	Number of plants under each grade (n)	Disease ratings (n x v)
0	0	25	0
1	1	15	15
10	3	15	45
25	5	25	125
50	7	15	105
< 50	9	5	45
	Total	100	335

Total number of plants evaluated (N) = 100

Total disease rating [Σ (n x v)] = 335

Highest grade value in the scale used (V) = 9

$$\text{Percentage disease index (PDI)} = \frac{\Sigma (n \times v)}{N \times V} \times 100 \text{ i.e. } \frac{335}{100 \times 9} \times 100 = 37.2\ \%$$

Example 2. In a cotton crop affected by bacterial blight, the disease intensity was assessed at the boll forming stage using a scale with grade values 0 - 9. In an acre of field 25 plants were selected at random from the two diagonal lines connecting the corners of the field. The disease incidence was assessed from the 3rd, 5th, 7th and 9th leaves from the tip of the primary branch of each plant. In the 100 leaves thus selected, 10 leaves were free from any disease symptoms. Of the remaining 90 leaves, 30 leaves were placed in grade 1, 15 leaves in grade 3, 25 leaves in grade 5, 15 leaves in grade 7 and 5 leaves in grade 9 respectively. Calculate the disease intensity.

Percentage of leaf area affected in the selected plants (%)	Grade value (v)	Number of plants under each grade (n)	Disease ratings (n x v)
0	0	10	0
1	1	30	30
10	3	15	45
25	5	25	125
50	7	15	105
< 50	9	5	45
	Total	100	350

Total number of leaves evaluated (N) = 100

Total disease rating [Σ(n x v)] = 350

Highest grade value in the scale used = 9

$$\text{Percentage disease index} = \frac{\Sigma (n \times v)}{N \times V} \times 100 \text{ i.e. } \frac{350}{100 \times 9} \times 100 = 38.9\ \%$$

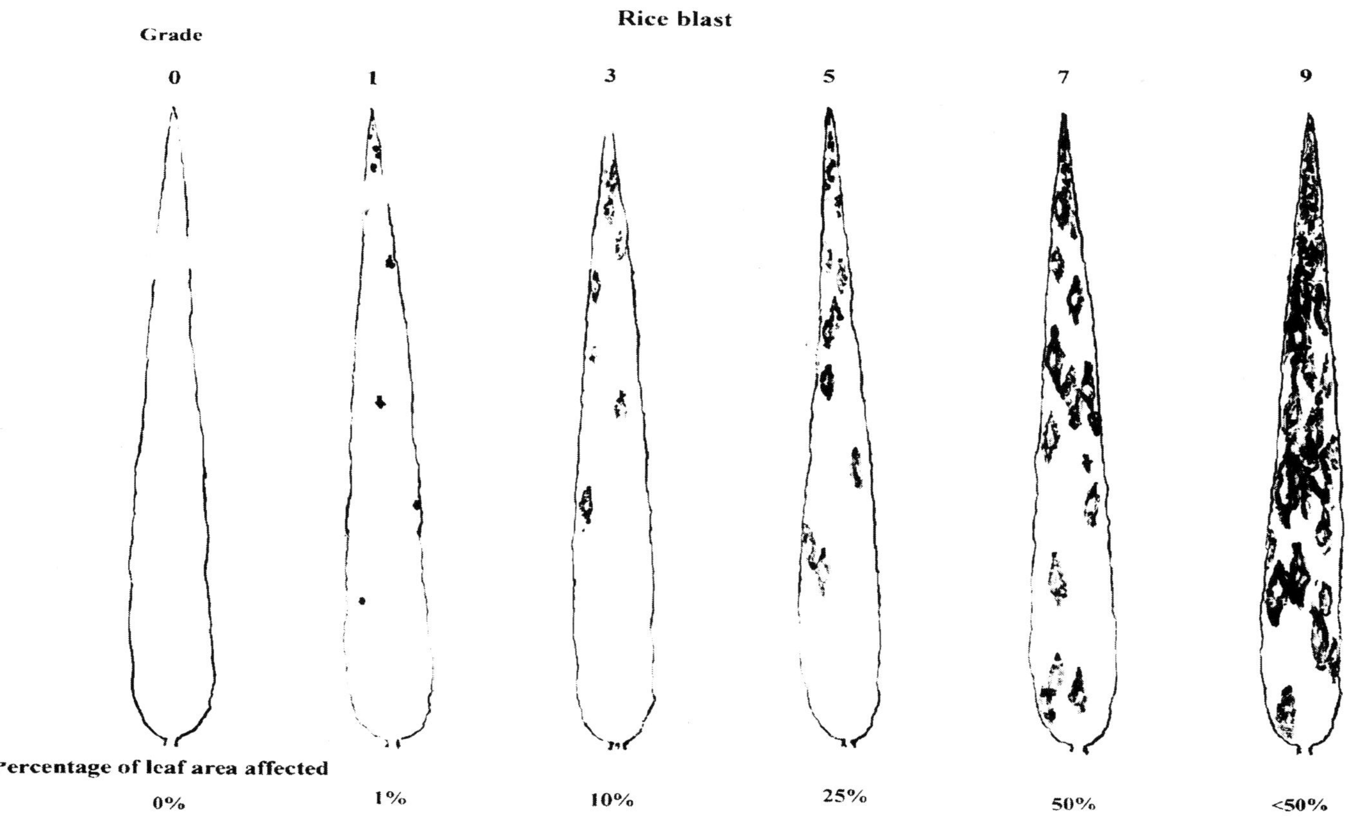

Fig. 61 : Grade chart for Disease Assessment (Rice Blast)

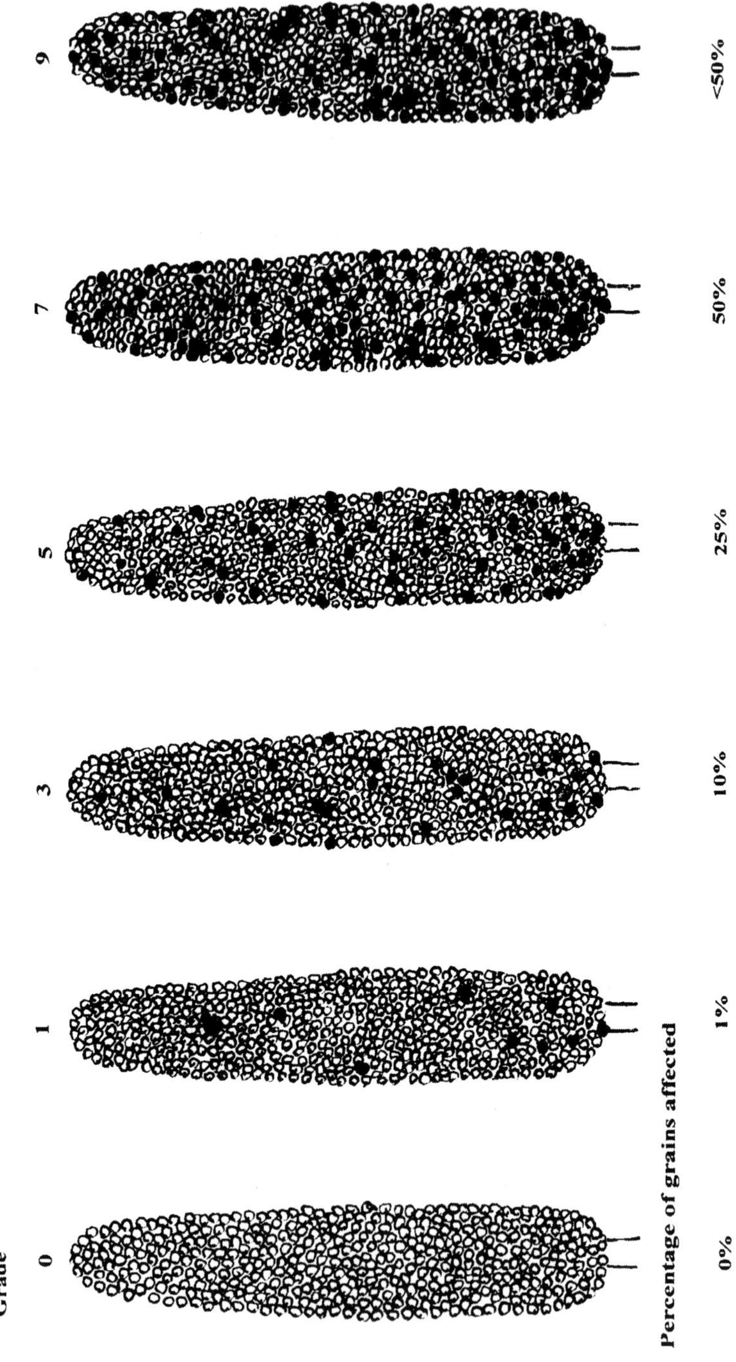

Fig. 62 : Grade chart for disease assessment (Pear millet)

In the case of leaf diseases, such as blast, rust, leaf blight etc., the yield loss is not directly proportional to the disease intensity, as the leaves affected are not completely destroyed. Further, the extent of disease infection may vary to a considerable extent from leaf to leaf and from plant to plant. Hence the yield loss due to the disease infection cannot be calculated directly. In such cases, the disease intensity is assessed first. Then specific number of plants under each disease grade is selected and the yield recorded separately. From this data the yield loss can be evaluated for the particular percentage disease index.

Such data collected over several years with different varieties and under different conditions are correlated, which may serve as a scale to determine the yield loss under different disease grades.

Disease Forecasting

Diseases do not occur uniformly in all the seasons or in all the years. A disease that occurs in an epidemic form in a particular season may occur in a much lesser proportion in some other seasons or the disease may not occur at all in some other seasons. Occurrence of diseases and their severity is governed by several factors, such as susceptibility of the host, weather parameters, environmental conditions around the crop, abundance of inoculum at the vulnerable growth stage of the crop etc. If all the conditions are favorable for the growth, reproduction, spread, infection and establishment of the pathogen, then the disease may attain epidemic proportions.

Based on certain changes in the weather conditions and initial appearance of some diseases, it may be possible to forecast the chances of large-scale occurrence of such diseases. Diseases, such as potato late blight, rice blast, grapevine downy mildew, apple scab, downy mildew of cucurbits, rice 'tungro' etc. can appear and spread very rapidly within a short span of time. However, these diseases can be effectively controlled if timely plant protection measures are taken up. Disease forecasting may give the growers a chance to mobilize all available resources and keep them in a state of readiness to take up appropriate control measures at the proper time, so as to avoid extensive damage to the crop and prevent yield loss.

Forecasting of large scale occurrence of certain diseases is possible based on different observations :

(i) Forecasting based on the presence of abundant inoculum in the atmosphere. Severity of a disease depends to a large extent on the availability of inoculum in the atmosphere. Aeroscope studies and observations at the crop level and in the atmosphere at higher altitudes, using different types of aeroscopes provide information on the presence of inoculum in the atmosphere. Based on the presence of unusually large number of spores in the atmosphere, it is possible to forecast the occurrence of certain diseases, especially in the case of diseases, such as wheat rusts, coffee rust, smuts, blast etc. Similarly, the occurrence of some of the insect transmitted virus diseases may also be predicted by ascertaining the vector population using light traps, sticky traps, pheromone traps etc. The occurrence of rice 'tungro' and some other virus diseases may be forecasted in this way.

(ii) Forecasting based on the weather parameters. There is always a close correlation between disease incidence and the prevailing weather conditions. In the disease surveillance programs, details of weather factors, such as maximum and minimum temperature, relative humidity, rainfall, number of rainy days, wind velocity, dew fall etc. are recorded regularly every week. Certain changes in the weather conditions may be either favorable or unfavorable for the infection, growth, reproduction and spread of some pathogens. The weather parameters and disease incidence are correlated statistically and the possibility of further widespread occurrence of the diseases with the changes in the weather conditions may be predicted. Very low night temperature, high atmospheric humidity, heavy dew and low wind velocity are highly favorable for the rapid spread of rice blast. Low temperature, very high relative humidity and rainfall favor rapid spread of potato late blight, downy mildews etc. Warm and humid weather favor powdery mildews and many leaf spot diseases.

(iii) Forecasting based on disease surveillance in cultivated crops. Information collected regularly on the occurrence and intensity of diseases from the fixed plot surveys and roving surveys of the surveillance programs in the cultivated fields provide valuable information on the occurrence, spread and severity of diseases. Based on the increasing trends in the severity of diseases, the possibility of flare up of certain diseases may be forecasted. Such forecasting may be possible in the case of bacterial leaf blight of rice, rice 'tungro', bacterial blight of cotton, groundnut 'tikka' etc

(iv) Forecasting based on the locality. Some diseases occur regularly in some localities in an endemic form in all the seasons and in all the years. Based on this fact, the possibility of some of these diseases assuming serious proportions may be predicted and forecasted, whenever susceptible varieties of the host are cultivated on a large scale. In South Arcot and Chinglepet districts of Tamil Nadu, 'tikka' disease occurs in a severe form when susceptible groundnut varieties are grown extensively in the rainfed season.

Similarly, brown leaf spot of rice occurs in some localities regularly and in some seasons may become very serious.

Whenever there is a possibility of widespread occurrence of a disease, such information is given wide publicity through various media, such as news papers, radio, television etc., so that precautionary measures may be taken by the growers in advance to combat the disease.

Based on any of the above mentioned forecasting methods, the possibility of occurrence and spread of a particular disease is forecasted just prior to cultivation of a crop or in the growing stage of a crop. Such forecasting is known as **'short-term forecasting'**.

However, in the case of some of the most devastating diseases, long term forecasting may be possible. Some diseases may occur in an epidemic form in cycles once in a few years or once in several years. Such severe incidence of the disease may be entirely due to the climatic conditions during that particular season or year. If such identical conditions recur after a cyclic period of a few or many years, there is every possibility that the disease may also occur in an epidemic form. Such kind of prediction may be possible by maintaining a record of weather parameters continuously for several years and studying the cyclic changes. Rice 'tungro', rice blast, potato late blight, wheat rusts etc. appear in an epidemic form mostly once in every 5 or 6 years in a cycle, when identical climatic conditions recur. Such recurrence of climatic conditions may be predicted as a result of meteorological studies and based on such studies forecasting the recurrence of certain diseases may be possible. This is known as **'long-term forecasting'**.

Forewarning of Crop Disease Occurrence

When a serious crop disease appears in an epidemic form in a village, district or state, there is every possibility that the same may spread to the adjoining villages, districts or states respectively. In such cases, the information about the possibility of the disease spreading to the neighboring villages, districts or states is conveyed through various media, such as news papers, radio, television etc. as a warning, so that the growers become aware of the impending menace and take appropriate precautionary measures to prevent large scale occurrence and spread of the disease. This is known as **'forewarning'**. Such forewarnings are given when serious crop diseases, such as rice blast, rice 'tungro' etc. occur in an epidemic form in one locality.

Bacteria

General Characters of Bacteria

Bacteria and mollicutes are Prokaryotes. They are microscopic, single-celled and are the simplest of all living organisms. Bacteria are mostly hyaline or yellowish-white in color. Just as fungi, the bacteria are also either parasitic or saprophytic in nature. Some bacteria are photosynthetic and are capable of manufacturing their own requirements of food. After the discovery of bacteria by **Anton van Leewenhock** (1676), they were considered to belong to the Animal Kingdom based on their motile nature. But several other characters of bacteria, such as unicellular nature, lack of well-defined nucleus, lack of sexual reproduction etc., resemble the blue-green algae of the Plant Kingdom and hence were included along with the fungi. It was **Ehrenberg** (1829), who created a separate division **'Bacterium'** to accommodate these organisms. The German Scientist, **Anton de bary** (1831-`88) separated bacteria from fungi and made it into a separate division and they were subsequently classified under Class - Schizomycetes in the Plant Kingdom. **Louis Pasteur** (1822 -`55) had conducted elaborate studies on bacterial fermentation of many substances and various diseases caused by bacteria. Now bacteria have been classified under the Class -Proteobacteria of Kingdom - Prokaryotae.

Bacteria are ubiquitous and are present in the air, soil and water, and in all living beings including plants, animals and humans. They possess high degree of resistance to very high temperature (75°C), very low temperature (-19°C), severe drought and other such unfavorable environmental conditions. Certain bacteria and mollicutes (Mycoplasma-like organisms) cause diseases in plants, animals, humans etc. Plant pathogenic bacteria form the largest group of pathogenic Prokaryotes and produce a variety of disease symptoms.

Both motile and non-motile forms of bacteria occur in nature. The motile bacteria possess long, delicate, whip-like appendages called **'flagella'** (sing. - flagellum). Spherical, cylindrical (rod), falcate (curved), spiral, as well as filamentous bacterial forms are present. They reproduce by binary fission. However, the falcate and spiral bacterial forms reproduce by longitudinal fission. The exceptional forms of bacteria, such as *Streptomyces*, *Actinomyces* etc., which are filamentous, produce spores (conidia) in chains from the tips of the branches. In *Rhizobium* the bacterial cells are **`Y'** - or **`X`** - shaped.

Some bacterial forms are found to form endospores, which can resist adverse environmental conditions and survive for long periods of time **(Fig. 65).** All bacteria can be grown in artificial solid or liquid culture media.

When bacteria are grown in solid culture media, they produce visible, viscous masses called **'colonies'**. Colonies of different species vary and may be circular, oval or irregular, with smooth, wavy or angular margins. The colonies appear flat, slightly raised or wrinkled and are whitish to yellowish in color. Some may produce diffusible pigments in the agar media. The spherical bacteria are called **'cocci'** (sing. - coccus). They occur either singly **(monococci** or **micrococci)** or in pairs **(diplococci)** or in chains **(streptococci)** or in groups **(staphylococci)**. The cylindrical or rod shaped bacteria are called **'bacilli'** (sing. - bacillus). They also occur either singly **(monobacilli** or **microbacilli)** or in pairs **(diplobacilli)** or in chains **(streptobacilli)**. The falcate or curved bacterial forms are called **'vibrio'** and the spiral bacteria **'spirillum'**. These two types are mostly found singly **(Fig. 66)**

The cocci types of bacteria have no flagella and so, they are non-motile. The bacilli types of bacteria invariably have flagella. A single polar flagellum at one long end of the bacterium **(monotrichous)** or a bunch of polar flagella at one long end of the bacterium **(cephalotrichous)** or a single flagellum at each of the two polar ends of the bacterium **(amphitrichous)** or few or several flagella distributed all over the surface of the bacterium **(peritrichous)** may be found. The bacilli type of bacteria are capable of quick movement by the lashing action of the flagella. In the vibrio type, a single flagellum is found at one long end of the bacterium **(monotrichous)**. In the spirillum type of bacteria, a single flagellum at each of the two long ends **(amphitrichous)** or many flagella at the two long ends **(lophotrichous)** are found **(Fig. 67)**

A bacterial species is really a group of bacterial strains, which share certain phenotypic and genotypic characteristics. One of these strains is actually the **'type strain'**, while the other strains differ from the type strain to a certain extent. These strains may differ from one another in their morphological, cultural, physiological, biochemical or pathological characteristics. When a host is infected by one strain or a group of strains, but not infected by the other strains of the same species, the particular strain or group of strains comprises a **'pathovar'** (pv.) of the species.

Morphology of Bacteria

Bacteria are generally unicellular organisms. Their genetic material (DNA) is not bound by a membrane and is not organized into a well-defined nucleus. The bacterial cell consists of a cell wall, which envelops the protoplasm. The protoplasm consists of cytoplasm, nucleus, granules and vacuoles. The cytoplasm is a complex mixture of proteins, lipids, carbohydrates and many other organic compounds, minerals and water.

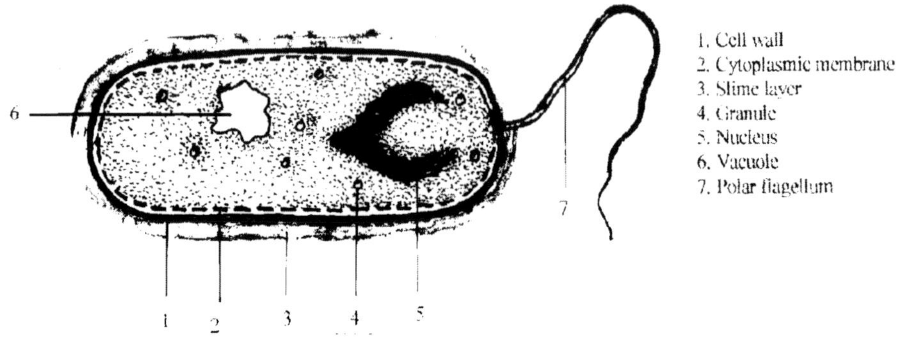
1. Cell wall
2. Cytoplasmic membrane
3. Slime layer
4. Granule
5. Nucleus
6. Vacuole
7. Polar flagellum
6
7
1
2
3
4
5

Fig. 63 : Structure of a typical bacterium

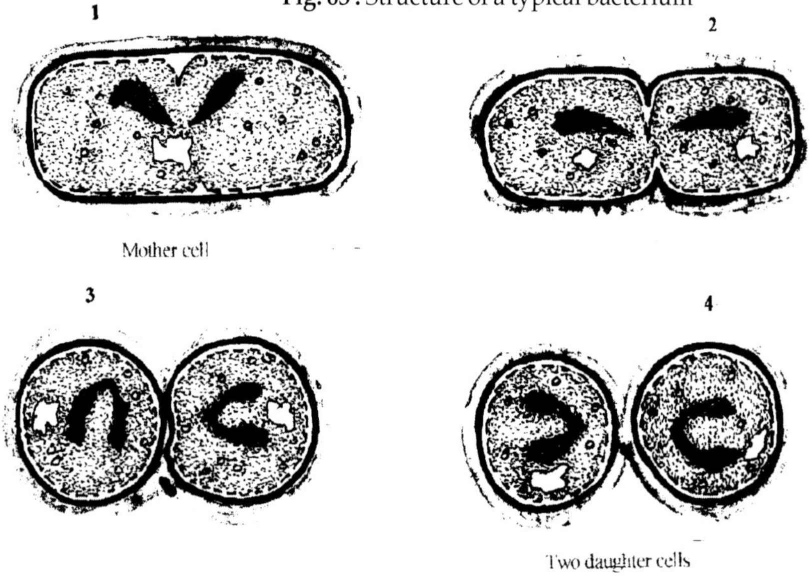
1
2
Mother cell
3
4
Two daughter cells

Fig. 64 : Reproduction of bacteria by binary fission

Spore within the bacterial cell
Spore released from the bacterial cell

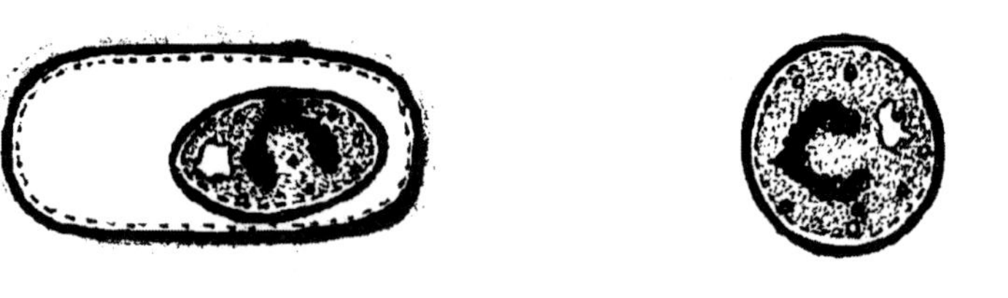

Fig. 65 : Endospore formation in bacteria

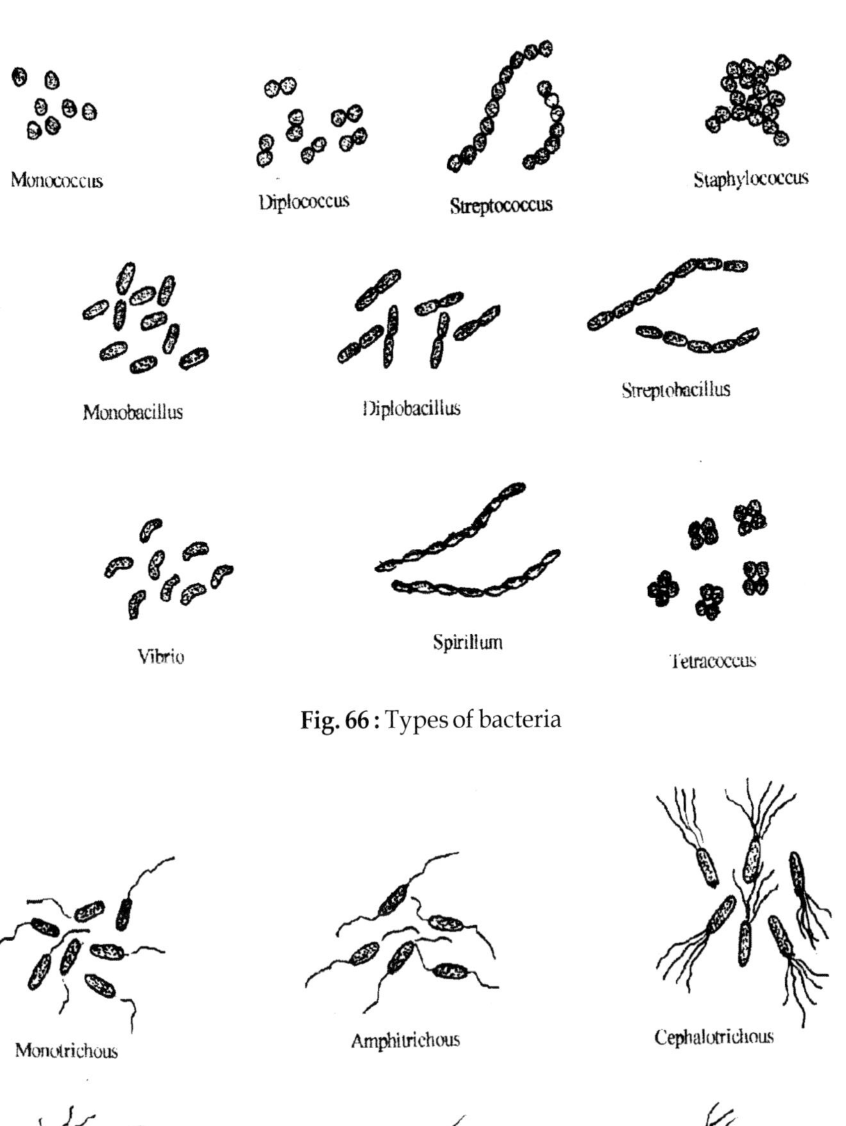

Fig. 66 : Types of bacteria

Monotrichous
Amphitrichous
Cephalotrichous
Peritrichous
Monotrichous
Lophotrichous

Fig. 67 : Types of flagella in bacteria

Often, bacteria also have smaller, spherical, genetic material called **'plasmids'**, which may contain several non-essential genes. In addition, the photosynthetic bacteria contain **'chromatophores'**. A thin slime layer covers the cell wall. In some types, the slime layer is thick and forms a definitive mass around the cell, which is called a **'capsule'**. The cytoplasm is enclosed within a cytoplasmic membrane. The cytoplasmic membrane determines the selective permeability of various substances into and out of the cell. Gram-negative bacteria have another outer membrane that appears to merge with the capsule.

The bacterial cell wall, besides giving protection to the protoplasm, gives shape to the bacterial cell. The cell wall is tough and rigid, at the same time it is elastic in consistency. No nuclear membrane is present around the nucleus. The vacuoles, which are present in the protoplasm, contain cell sap. Food materials rich in protein are stored in the granules.

In the absence of sufficient water and food materials or when there is a sudden change in the prevailing temperature or other changes in the environmental conditions, which are detrimental to their existence and survival, a vast majority of the bacteria die. However, a few Bacillus-type bacteria can withstand extremes of temperature, pH, oxygen tension, osmotic and atmospheric pressure and other such adverse conditions and survive. In the event of such adverse conditions, some bacteria form **'endospores'** within the bacterial cell. Endospores are formed, either in the middle or at the end portion of the bacterial cell. Before formation of the endospore, the protoplasm concentrates in a spherical form and a thick wall is synthesized around this. These resting spores, which are spherical and thick-walled, are capable of survival under extreme conditions. When favorable conditions return, the endospore comes out of the bacterial cell and subsequently the protoplast inside the spore germinates and comes out as a bacterium. However, this is not a means of reproduction in bacteria **(Fig. 63)**

Reproduction in Bacteria

In general, there is no sexual reproduction in bacteria and as such, they reproduce only asexually, either by fission or budding. Only a few forms of bacteria are capable of producing spores. Of all living organisms, bacteria can multiply very rapidly. The mode of reproduction in bacteria is also very simple

(i) Binary fission. This is the common method of reproduction in most of the bacteria. First of all, the nucleus of the bacteria undergoes mitotic division and is divided into two separate nuclei. Simultaneously, a transverse cross septum is formed in the middle of the bacterial cell by the inward growth of the cytoplasmic membrane, enclosing one nucleus in each of the section. The cytoplasm is also divided into two approximately equal parts. The granules and the other components also duplicate themselves and are distributed equally in the two sections. Between the two layers of the cytoplasmic membrane, two layers of cell wall material, continuous with the outer cell wall are also synthesized. After the formation of these cell walls is completed, the two layers separate, splitting the two cells apart into two daughter cells. In this way, fission takes place usually once in every 20 - 50 minutes, as a result bacteria multiply very rapidly. At this rate, a single bacterium may be able to produce more than one million progenies in one day under favorable conditions. In the case of *Vibrio* and *Spirillum* types of bacteria, fission takes place longitudinally (Fig. 64)

(ii) Budding. A few species of bacteria resort to this type of reproduction under certain conditions. Here also, first of all the nucleus undergoes mitotic division and divides into two separate nuclei. Then a small bud-like outgrowth is formed from the mother cell. One of the nuclei then passes into the bud along with a portion of the cytoplasm and its contents through a minute opening. Subsequently the bud detaches from the mother cell and becomes a new bacterium.

(iii) Spore formation. Some bacterial types, such as *Actinomyces* and *Streptomyces* produce spores called **'conidia'** from the tips of the branches of the filamentous hyphae by abstriction of spores in chains **(Fig. 68)**

Classification of Bacteria

The bacteria and mollicutes are classified under the Kingdom - Prokaryotae

KINGDOM - PROKARYOTAE. Microorganisms included in this Kingdom are generally single-celled, where the genetic material is not bound by a membrane and is not organized into a well-defined nucleus.

DIVISION - GRACILICUTES: Only gram-negative bacteria are included in this Division.

CLASS - PROTEOBACTERIA: Mostly single-celled, gram-negative bacteria are classified under this Class.

Order - Eubacteriales. The bacteria under this Order are single-celled and rod or cylindrical in shape. Some of them are motile, while others are non-motile. The motile bacteria possess peritrichous flagella. This Order has three important Families.

Family 1 - Azotobacteriaceae. The bacteria belonging to this Family do not have any direct association with any plants, but live independently in the soil. At the same time they inhabit the rhizosphere around the root zones of plants and fix atmospheric nitrogen in the soil, which is called 'non-symbiotic nitrogen fixation'. These bacteria live as saprophytes and utilize the food obtained from organic matter in the soil. The most important Genus in the Family is Azotobacter.

Genus - *Azotobacter*. The bacteria of this Genus are naturally soil-inhabitants and live entirely on organic matter and carbohydrates present in the soil. The energy generated from the food taken is utilized to fix atmospheric nitrogen in the soil as amino acids, which are later converted into proteins. When the bacteria die in the soil, the cells disintegrate and the protein and other organic matter present inside the cells are added to the soil, which in turn are utilized by plants. These bacteria, when applied to the soil as culture, help many crops, such as wheat, cotton, rice, sugarcane, cholam etc. to get additional supply of atmospheric nitrogen through nitrogen fixation. These bacteria, when present in the soil at the rate of 10,000 numbers per gram of soil can fix up to 5 - 10 kg. of atmospheric nitrogen in one acre of land under conditions favorable for their growth and reproduction. *Azotobacter* bacteria are slightly bigger than many other bacteria found in the soil and are rod-shaped. They are aerobic in nature and can form endospores under unfavorable conditions.

(e-g) *Azotobacter chroococcum*
Azotobacter agilis

Family 2 - Rhizobiaceae. The bacteria of this Family do not form endospores. They are not found free in the soil, but most of them have symbiotic relationship with some particular plant species, especially the legumes and a few oilseed crops. They are either rods or club-shaped or **'X'** - or **'Y'**- shaped and may have one or more flagella. Most of these bacteria form root nodules in the roots of plants. The most important Genus of this Family is *Rhizobium*.

Genus - *Rhizobium*. The bacteria of this Genus are mostly **'X'** - or **'Y'** - shaped. They do not form endospores and cannot live in the soil independently for long. They live symbiotically along with certain plant species, especially legumes and a few oilseed crops. The bacteria enter into the host roots through the root hairs and multiply inside the root tissues in large numbers. Because of the presence of these bacteria inside the roots and the physiological reactions caused by them, the roots become enlarged to form nodules at some points due to hyperplasia and hypertrophy of the cells of the root tissues. The bacteria live within the protective covering of the nodules, take sugars and carbohydrates from the host roots for their requirements of food and with the help of the enzyme **'nitrogenase'** produced by them, fix atmospheric nitrogen as nitrogenous compounds in the nodules, which are utilized by the host plants. This process is known as **'symbiotic nitrogen fixation' (Fig. 68)**

(e-g) *Rhizobium leguminosarum* - lives symbiotically with peas, groundnut etc.

Rhizobium phaseoli - lives symbiotically with bean

Rhizobium saponicum - lives symbiotically with soybean

Genus - *Agrobacterium.* The bacteria of this Genus are rod-shaped. They are motile by means of one to four peritrichous flagella. When only one flagellum is present, it is often lateral than polar. They are rhizosphere- and soil-inhabitants. A few species cause plant diseases.

(e-g) *Agrobacterium tumefaciens* - causes crown gall tumors on rose stems.

Family 3 - Enterobacteriaceae. The bacteria of this Family are anaerobic in nature. Many of them are capable of fermenting sugars. Several of them are found inside the intestine and respiratory tracts of animals. Some species infect plants and cause diseases. They are mostly rod-shaped and possess one or more flagella, mostly on the lateral sides. The most important Genus in this Family is *Erwinia.*

Genus - *Erwinia*. The bacteria belonging to this Genus are rod-shaped and are motile by means of several peritrichous flagella. They occur singly or in pairs or in chains. They do not form endospores and are facultative anaerobes. Some species possess strong pectic enzyme activity and cause soft rots, while those, which do not have pectolytic activity cause wilts, galls etc. They do not require proteins for their living.

(e-g) *Erwinia carotovora* - causes soft rot of carrots and potatoes.

Erwinia amylovora - causes fire blight of pear and apple.

Erwinia tracheiphila - causes wilt disease of cucurbits.

Order - Pseudomonadales. Many bacteria belonging to this Order are rod-shaped or spiral-shaped. The flagella are mostly found at the polar ends. They do not form endospores and reproduce by binary fission. This Order has only one Family **Pseudomonadaceae.**

Family - Pseudomonadaceae. Many bacteria of this Family are capable of causing diseases in several crops. They are mostly facultative parasites. There are two important Genera in this Family.

Genus - *Pseudomonas*. The bacteria of this Genus are straight or slightly curved rods. They are motile by one or many polar flagella and are aerobic in nature. Many species are common inhabitants of soil or fresh water or marine environments, but several other species are found to infect plants and cause leaf spots, leaf necrosis etc., while a few can infect animals and humans. Some species, such as *Pseudomonas syringae* are called **'fluorescent bacteria'** because of the presence of yellowish-green fluorescent pigments. There are many pathovars in several of the species.

(e-g) *Pseudomonas campestris* pv. *mangiferae indicae* - causes bacterial leaf spot of mango.

Pseudomonas campestris pv. *sesami* - causes leaf necrosis of gingelly.

Pseudomonas solanacearum - causes bacterial wilt disease of banana.

Genus - *Xanthomonas*. The bacteria of this Genus are all straight rods and are motile by means of a single polar flagellum. They are aerobic in nature. All species are plant pathogens and are found only in association with plants or plant materials. Several of them attack the vascular region of plants and cause wilt or blight diseases. There are many pathovars in several species.

(e-g) *Xanthomonas malvacearum* - causes black arm of cotton.

Xanthomonas campestris pv. *citri* - causes bacterial canker disease in *Citrus* plants.

Xanthomonas campestris pv. *campestris* - causes black rot of crucifers.

Division - Firmicutes: Gram-positive bacteria are included in this Division.

Class - Firmibacteria: Mostly single-celled, gram-positive bacteria are classified under this Class.

Endospore forming rods. The bacteria of this group are gram-positive. They are mostly single-celled and the cells are straight or slightly curved rods or spiral-shaped. All of them produce endospores.

Family - Bacillaceae. The bacteria belonging to this Family are rod-shaped. They are gram-positive and form endospores.

Genus - *Bacillus*. The bacteria of this Genus are straight rods and are non-motile. They are gram-positive and form endospores. They are mostly soil-inhabitants, but also found in water, food, manure etc.

(e-g) *Bacillus subtilis* - produces the antibiotic - **'Polymycin'**, which is used against some bacterial diseases in humans.

Bacillus brevis - produces the antibiotic - **'Thryothrycin'**, which is used for the treatment of some bacterial diseases in humans.

Bacillus cereus - causes food infections by the production of toxic substances. Provides biocontrol of damping off diseases of legumes.

Coryneform group of bacteria. The bacterial cells are either straight or slightly curved rods. All are gram-positive. Some species are motile.

Genus - *Clavibacter (Corynebacterium)*. The bacteria of this Genus are straight or slightly curved rods. Generally they are non-motile, but some species are motile by means of one or two polar flagella. All the bacteria of this Genus are gram-positive. Many species cause diseases in plants, while a few cause diseases in humans.

(e-g) *Clavibacter michiganense* f.sp. *sepedonicum* - causes ring rot of potato

Clavibacter michiganense f.sp. *michiganense* - causes bacterial canker and wilt of tomato.

Clavibacter diphtherae - causes Diphtheria in humans.

Class - Thallobacteria: Mostly filamentous and branching bacteria. All are gram-positive.

Order - Actinomycetales. The bacteria belonging to this Order are mostly filamentous and branching like fungi. The filaments may be very short or long and thread-like. Some of them are capable of forming endospores. A few of them possess flagella. The bacteria of this Order are mostly soil-inhabitants in nature. Some species are responsible for causing various diseases in humans, animals and plants, while some are highly beneficial, as they produce several antifungal and antibacterial antibiotics, which are used for the treatment of many diseases in humans, animals and plants. Some of them are highly capable of disintegrating complex organic matter present in the soil into simpler forms. There are three important Families under this Order.

Family 1 - Actinomycetaceae. The thalli of the bacteria of this Family are filamentous, short, delicate and aseptate and appear like fungal hyphae. The branches of the filaments are characteristically twisted at the end regions. On maturity, many transverse septa are formed and the hypha is divided into many short segments. Each segment is either rod-shaped or elliptical in shape. Subsequently these segments split and separate and each segment becomes an independent bacterium. In some other cases, septa are not formed instead spherical spores are formed at the tip of each branch one behind the other in chains.

Genus - *Actinomyces*. The bacteria of this Genus produce simple or branched filaments, which are non-motile, very narrow, delicate and septate at irregular intervals. The spores are abstricted in succession from the tips of the branches downwards. The spores look more like segments of the filaments than organized spores. Sometimes, the segments, which are cut off, resemble conidia or resting spores. They do not form endospores. The organisms are mostly aerobic. Many of them cause diseases in animals and humans **(Fig. 68)**

(e-g) *Actinomyces bovis* - causes **'Actinomycosis'**, a disease of the mouth in cattle.

Family 2 - Mycobacteriaceae. The bacteria of this Family are either delicate filaments or straight or slightly curved rods. In the filamentous forms, no branches are formed. They do not have flagella and do not form endospores. They are aerobic in nature.

Genus - *Mycobacterium*. Some species of this Genus cause very serious diseases in humans and animals **(Fig. 68)**

(e-g) *Mycobacterium tuberculosis* - causes tuberculosis in humans and a few animal species.

Mycobacterium leprae - causes leprosy in humans.

Family 3 - Streptomycetaceae. The bacteria belonging to this Family produce slender, aseptate, branched filaments. Conidia are abstricted from the characteristic, spirally twisted tips of the branches downwards in chains of three to many spores. The bacteria are aerobic in nature and are mostly saprophytes and soil-inhabitants. A few bacteria live parasitically on animals and plants, and cause diseases. Many species produce different antibiotics, which are active against bacteria, fungi, algae, viruses or protozoa.

Genus - *Streptomyces.* Many species of this Genus produce antifungal and antibacterial antibiotics. A few of them cause diseases in plants **(Fig. 68)**

(e-g) *Streptomyces scabies* - causes common scab of potato.

Streptomyces cinnamomens - produces the antibiotic - **'Aureofungin'**, which is used for the control of many fungal diseases, especially basal stem rot or 'Thanjavur wilt' of coconut.

Streptomyces griseus - produces the antibiotic - **'Streptomycin'**, which is used for the control of many diseases in humans and animals, particularly Tuberculosis.

Streptomyces erythreus - produces the antibiotic - **'Erythromycin'**, which is also widely used for controlling certain diseases in humans and animals.

Symptoms of Bacterial Diseases

Just as fungal plant pathogens, bacterial plant pathogens also produce specific and typical disease symptoms. However, based on the virulence of the pathogen, environmental conditions and the relationship and interaction between the host and the parasite, the symptoms may vary to a certain extent. Several pathogens can be known by the very symptoms they produce. Some of the major symptoms produced by bacterial pathogens are detailed below:

Local lesions. The most common disease symptoms produced by bacterial plant pathogens are leaf spots of various sizes and shapes. Some pathogens may cause spotting on the stems, unripe and ripe fruits etc., besides leaves. Generally, the spots are circular, irregularly circular or angular. When the weather is humid, the spots appear water-soaked and slimy and innumerable bacteria are found in the ooze exudate, from which secondary infection takes place. The spots enlarge in the course of time and the central portion of the spots become necrotic and turns brown or black in color. In some cases, the spots are surrounded by a yellowish halo. In dicotyledonous plants, the spots are limited by the veins, as a result the spots become angular. Almost all bacterial leaf spots and blights of leaves, stems and fruits are caused by the bacteria of the Genera - *Pseudomonas* and *Xanthomonas* **(Fig. 69)**

(e-g) Circular leaf spot of bean - caused by *Pseudomonas syringae* pv. *syringae*.

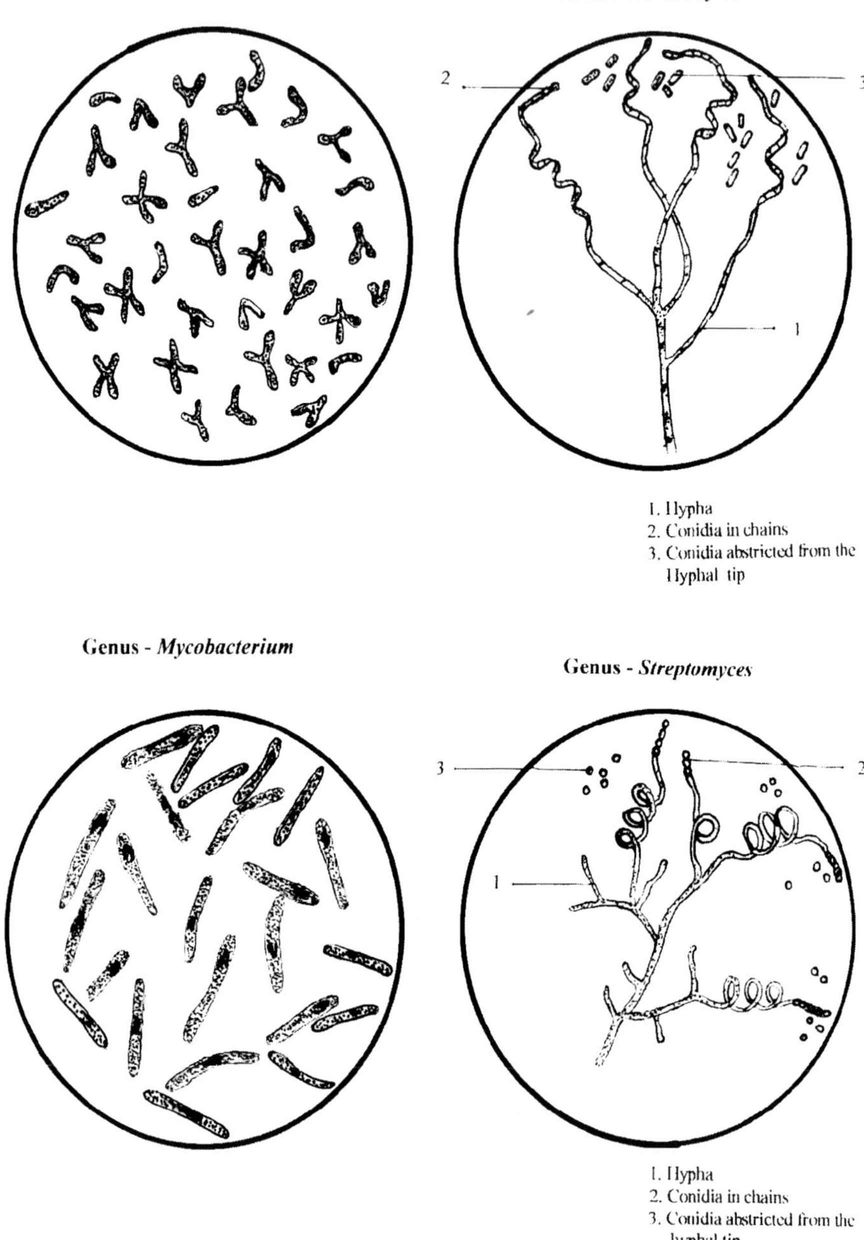

Fig. 68 : Morphology of some species of bacteria

Circular leaf spot of mango - caused by *Pseudomonas mangiferae* pv. *indicae*.

Irregular spots of tomato fruits - caused by *Xanthomonas campestris* pv. *vesicatoria*

Angular leaf spot of cotton - caused by *Xanthomonas campestris* pv. *malvacearum*.

Angular leaf spot of cucumbers - caused by *Pseudomonas syringae* pv. *lacrymans*.

2. **Blights.** In some bacterial diseases, the spots continue to enlarge rapidly and the diseases are then called **'blights'**. In severe infections, the spots may be so numerous that they destroy most of the plant surfaces and appear blighted or the spots may enlarge and coalesce, thus producing large areas of dead plant tissues and blighted plants. In monocotyledonous plants, the bacterial spots when restricted by the longitudinal parallel veins of the leaves appear as long stripes or streaks. Here also in humid or wet weather, the infected tissues often exude masses of bacteria that cause secondary infection. When the bacterial exudation dries on the surface of the leaves, it gives a shiny appearance to the leaves and the leaf surface feels rough and coarse. Severely affected plants dry and present a burnt appearance **(Fig. 69)**

   (e-g) Bacterial leaf blight of rice - caused by *Xanthomonas campestris* pv. *oryzae*.

   Bacterial leaf streak of rice - caused by *Xanthomonas campestris* pv. *translucens*.

3. **Bacterial vascular wilts.** Vascular wilts caused by bacteria, affect mostly herbaceous plants, such as several vegetables, field crops, ornamental plants etc. The pathogens enter the host plants only through natural openings in the plants or through wounds caused by insects, nematodes or by agricultural implements and rarely through non-cutinised areas, such as root hairs Once they enter into the host, they multiply rapidly in the xylem vessels and migrate through these vessels. In this process they interfere with the translocation of water and nutrients, as a result the aboveground parts of the plant wilt and ultimately die. The bacteria also produce some enzymes and toxic substances, which damage the cell walls of the xylem vessels and the adjacent parenchyma tissues, thereby cause the formation of cavities in the disintegrated tissues, which are filled with bacteria, gum-like substances and cellular tissues. Sometimes, the bacteria move from

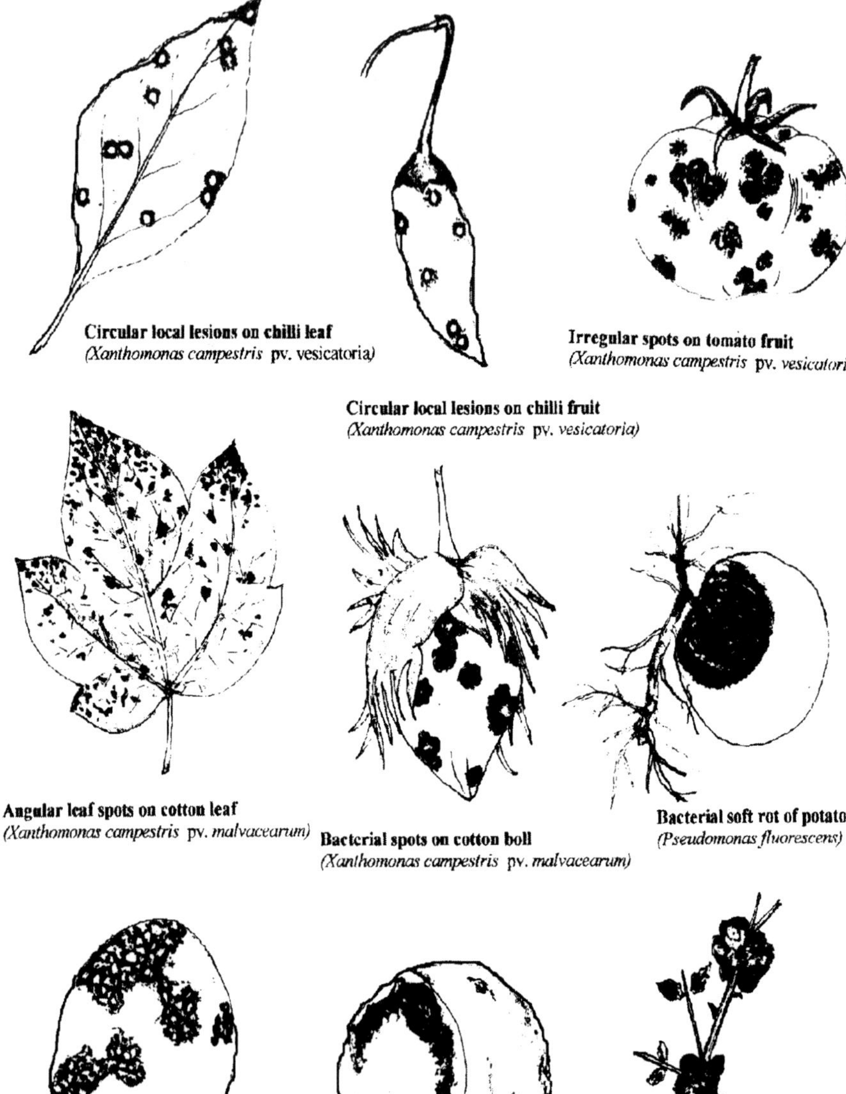

Fig. 69 : Symptoms of bacterial diseases

the vascular bundles, spread throughout the intercellular spaces and ooze out through the leaf stomata or through cracks in the stem regions **(Fig. 69)**

(e-g) Wilt of tomato - caused by *Clavibacter (Corynebacterium) michiganense* f. sp. *michiganense*.

Wilt of cucurbits - caused by *Erwinia tracheiphila.*

Bacterial wilt of solanaceous plants and 'Moko' disease (bacterial wilt) of banana - caused by *Pseudomonas solanacearum*

4. **Soft rots**. Soft rots occur mostly in succulent plant parts, such as fruits, tubers, bulbs, soft stems etc., either in the field or in storage. The rotting tissues emit a foul smell due to the release of some volatile substances during the disintegration of the affected tissues by the bacteria. Rotting tissues become soft and watery. Slimy masses of bacteria frequently ooze out from cracks in the tissues. The rotting areas enlarge rapidly in diameter and depth and the color of the affected areas also turns blackish and somewhat depressed. The bacteria involved are either saprophytic or weak secondary parasites. The parasites can enter the tissues only through wounds caused by insects or by agricultural implements during intercultivation operations or at the time of harvest or during transport **(Fig. 69)**

   (e-g) Soft rot of carrots and several fleshy fruits, vegetables etc, - caused by *Erwinia carotovora* pv. *carotovora.*

   Soft rot of potato, onion and fleshy fruits - caused by *Pseudomonas fluorescens.*

5. **Bacterial galls**. Galls are produced on the stems and roots of plants mostly by the bacteria of Genus - *Agrobacterium, Pseudomonas* etc. The galls may be formed, as a result of unorganized overgrowths of plant tissues or due to proliferation of tissues, which results in the formation of irregular galls **(Fig. 69)**

   (e-g) Crown gall tumors on rose stem - caused by *Agrobacterium tumefaciens.*

6. **Bacterial cankers.** Cankers on the stem, branches or twigs are invariably accompanied by such symptoms on leaves, fruits and other parts of the plants. The lesions in most of the cases appear corky.

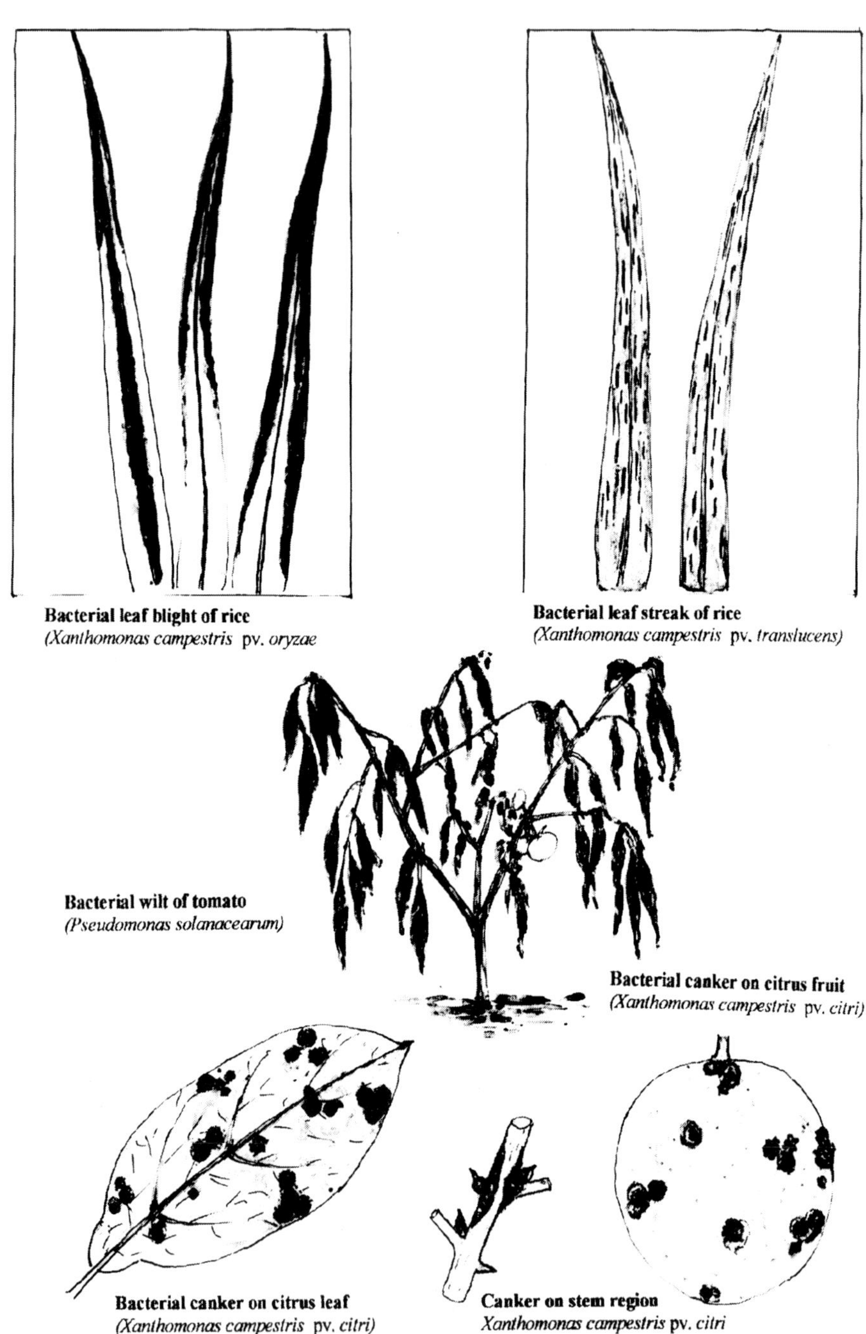

Fig. 69 : Symptoms of bacterial diseases

The center of the lesions is brown and sunken with ruptured, necrotic tissues. The margin of the lesions may often be surrounded by yellowish halo **(Fig. 69)**

(e-g) *Citrus* canker - caused by *Xanthomonas campestris* pv. *citri.*

Mode of Entry of Bacterial Pathogens

Bacteria are not capable of entry into the host plants by direct penetration. They enter into plants mostly through various types of wounds and less frequently through natural openings, such as stomata, hydathodes, nectarthodes etc. **(Fig. 70)**

The wounds utilized by bacteria for their entry into the host may be fresh or old and may consist of lacerated or killed tissues. The bacteria may multiply for a short time in such tissues before they advance into the healthy tissues. Wounds may be caused by biting and chewing insects or sucking insects, mammals, cultural operations, such as pruning, transplanting, harvesting etc., and self inflicted injuries, such as leaf scars, as well as lesions caused by other pathogens. Bacteria after getting access to such wounds multiply in the fresh wound sap or in a film of rain- or dew-water present on the wound and subsequently invade the adjacent plant cells. Most of the bacteria secrete enzymes, toxins or growth hormones that may disintegrate and kill the nearby cells.

Many bacteria enter plants through stomata, while some others enter through other natural openings, such as hydathodes, nectarthodes and lenticels. Generally, stomata are found in large numbers on the under surface of leaves and are open during the daytime, but more or less closed at night. Bacteria present in a film of water over a stoma can easily swim or wriggle through the stoma and reach the substomatal cavity, where they start multiplying in large numbers.

Generally, hydathodes are permanently open pores at the margins and tips of leaves, which are connected to the veins and secrete droplets of water containing various nutrients. Some bacteria use these pores as a means of entry into the leaves **(e-g)** *Xanthomonas campestris* - causes black rot of cabbage.

Some bacteria enter the blossoms through the nectarthodes or nectaries, which are also open pores as hydathodes.

Lenticels are natural openings on fruits, stems and tubers that are filled with loosely connected cells to allow passage of air. The lenticels are open during the growing season and a few bacteria enter through these pores. Most pathogens that enter through lenticels can also enter through wounds.

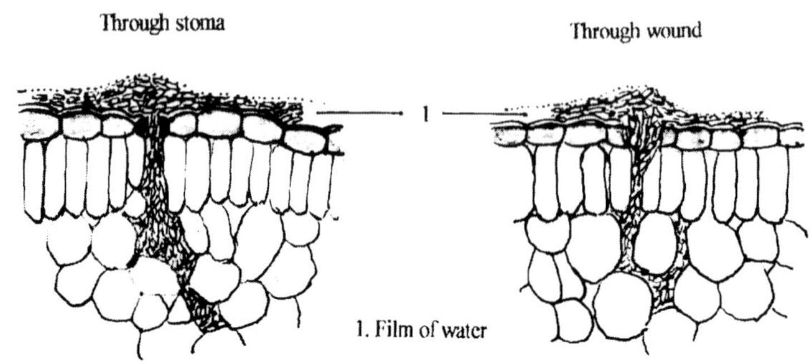

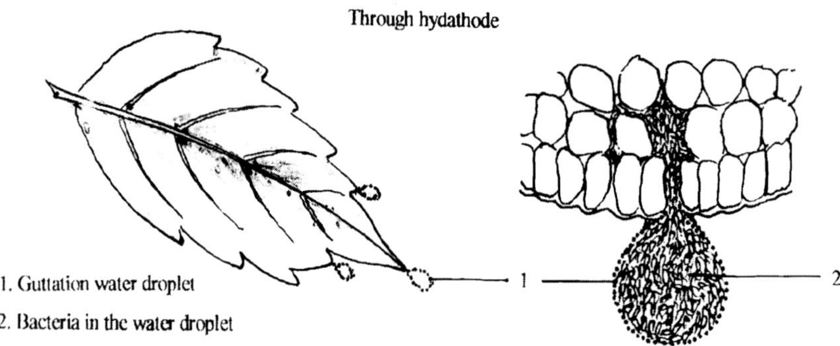

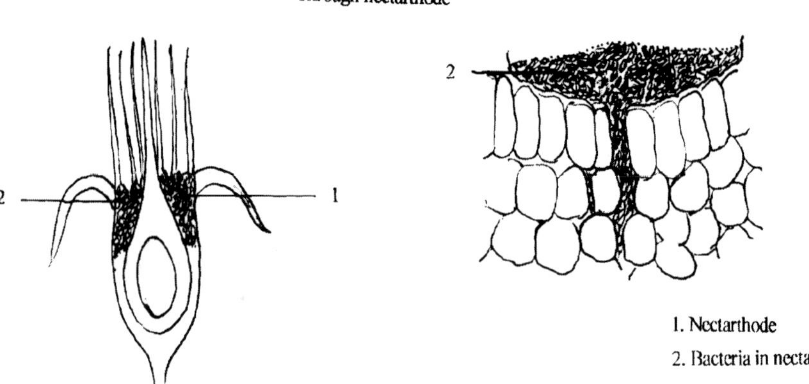

Fig. 70 : Mode of entry of bacterial pathogens into the host

After the entry of bacterial pathogens into the host, they establish themselves in the susceptible host tissues and procure nutrients from them, thereby causing infection. During this phase, the pathogens multiply in large numbers and colonize the plant. Successful infection results in the production of visible changes in the appearance of infected plants, usually at the sites of infection, which are known as **'disease symptoms'**. The symptoms may appear from a few days to a few weeks after infection or inoculation. The time interval between inoculation and the appearance of disease symptoms is called the **'incubation period'**. This period may vary depending upon the susceptibility of the host, host-parasite interaction, growth stage of the host and the prevailing environmental conditions, which may be favorable either to the host or the pathogen. During infection, the pathogens release several biologically active substances, such as enzymes, toxins and growth regulators. In this process the pathogens may obtain nutrients, either from the living cells without killing them for a long time or kill the cells and utilize their contents or kill the infected cells and disintegrate the surrounding tissues. When the host is susceptible and is in a vulnerable stage for the attack by the pathogen, the pathogen being virulent and the environmental conditions are favorable for the rapid multiplication of the pathogen, the disease develops rapidly. Bacteria invade tissues intercellularly. However, when the cell walls are dissolved they may grow intracellularly. Bacteria causing vascular wilts invade the xylem vessels and multiply within the vascular cells.

Some bacteria may invade a single cell or a few cells or a small area of the plant and remain localized. Such infections may remain localized throughout the growing season as spots or may enlarge gradually as blights. Some infections may enlarge quite rapidly, involving the entire plant organs, such as flowers, fruits, leaves etc. or a part of the plant or the plant as a whole.

Dissemination of Bacterial Pathogens

A few bacteria, by virtue of their motile nature can move very short distances and are able to move from one host to another, which is very close to it. Otherwise dissemination of almost all the bacterial pathogens is accomplished passively by several agents, such as air, water, insects, mammals or humans, seeds etc.

Dissemination through soil. Many facultative saprophytes live in the plant debris or in the organic matter in the soil. These bacteria are transmitted from one field to another through soil adhering to agricultural implements, hooves of cattle, transport vehicles and even by humans.

Dissemination through seeds and other vegetative propagants. Many bacterial pathogens are transmitted through seeds and other propagants internally or externally even to far off places.

Dissemination by air. Dissemination of bacterial pathogens by air occurs less frequently, but can occur under specific conditions. The bacteria causing fire blight of apple and pear produce fine strands of dried bacterial exudate containing bacteria and these strands may be broken off and disseminated by wind. Bacteria present in the soil may be blown away along with plant debris or soil particles in the dust. Wind also helps in disseminating bacteria by blowing away rainsplash droplets containing bacteria. Wind may carry away insects that may be smeared with bacteria. Wind also causes adjacent plants or plant parts to rub against one another and this helps the spread of bacteria by direct contact.

Dissemination by water. Water is an important agent of dissemination of bacterial pathogens. Bacteria present in the soil or plant refuse are disseminated by rain or irrigation water that flows on the surface or moves through the soil. Most of the bacteria are exuded in a sticky fluid and are disseminated by rain, which either washes them downward or splashes them in all directions. Further, raindrops or drops from overhead irrigation pick up any bacteria present in the air and bring them downward, which may land on susceptible host plants. Though water is less important than air in long distance transport of pathogens, during floods, storms etc., they may be carried over very long distances. However, water dissemination is more efficient in the sense that pathogens land on already wet host surfaces, which facilitate their movement and germination.

Dissemination by insects, nematodes, mammals etc. Many insects carry bacterial pathogens externally from plant to plant and deposit them on the plant surfaces or in the wounds they make while feeding. In many diseases, such as soft rot, blights etc., the insects are smeared with the bacterial pathogens along with the sticky exudates as they move among plants and subsequently the bacteria are transmitted.

Plant Viruses

The awareness about plant virus diseases came into prominence after the destructive outbreak of **'potato leaf roll'**, which was prevalent in Great Britain from 1770 onwards. During the nineteenth century, several other important crop diseases, now known to be due to virus agencies were reported. Mosaic and `Sereh` diseases of sugarcane, tomato mosaic etc. were detected during this period.

Mayer of Holland (1886) described a disease in tobacco and based on the marbled effect of light and dark green patches on the leaves termed that as **'mosaic disease'**. He showed that tobacco mosaic could be mechanically transmitted from a diseased to a healthy plant. The Russian Botanist **Iwanowsky** (1892), demonstrated that infection of a healthy plant by the juice from a diseased plant occurred even after the juice had been filtered through finest bacterial filters and that the infective matter multiplied abundantly in the new host. He was the first to give a clear-cut evidence of a virus and for the first time the existence of invisible disease producing agencies, smaller than anything yet known was postulated. This agent was later called a **'virus'**, a Latin word denoting **'poison potion'** or **'poison'**. The Dutch Bacteriologist **Beijerinck** (1898), after conducting detailed investigations, propounded his theory of the **'contagium vivum fluidum'**, an infectious living fluid or non-particulate agent capable of causing diseases. **Erwin F. Smith** (1888) showed that the destructive disease known as **'peach yellows'** could be transmitted by grafting or budding but not by any other methods. In 1901, the Japanese Pathologist **Takami** succeeded in transmitting rice yellow dwarf disease by the rice brown plant hopper. In 1916, **Quanjer** of Holland proved that potato leaf roll was an infectious communicable disease. Following these findings, similar diseases were reported in several other important crops. In 1917, **d`Herelle** reported that such agents could destroy bacteria also and he named them as **'bacteriophages'**. **Twort** (1922) suggested the existence of some ultramicroscopic forms of life, more primitive than the bacteria or protozoa. **Schelsinger** (1933) was the first to determine the composition of a virus. He showed that a bacteriophage consists of only protein and DNA. In 1935, the American Biologist, **Stanley** crystallized the virus causing tobacco mosaic disease and demonstrated that the crystals retained their infectivity when inoculated into healthy plants. He thus showed that viruses were not like typical cells. **Hershey** and **Chase** (1952) demonstrated that infection by virus is the result of penetration of viral DNA into the host cells. After 1940, with the invention of Electron microscope and other sophisticated equipments, it was possible to view and photograph virus particles and also to study their characteristics in much more detailed manner.

Definition of Viruses

Scientists the worldover have given various definitions for viruses. But, some of the chief characteristics generally accepted by every one are: that viruses are very minute, ultramicroscopic and smaller than bacteria; they contain only a single nucleic acid i.e. the genetic material; they are capable of multiplication from the genetic material; they are obligate parasites and are able to produce diseases; they are incapable of growth; they lack enzymes for energy metabolism and they do not contain ribosomes.

Green (1935) has defined viruses as 'ultramicroscopic parasitic organisms, which are capable of reproduction just as other microorganisms'.

According to **Bawden** (1964), 'viruses are very minute microorganisms, which cannot be seen through ordinary microscopes and are obligate parasites that can live and multiply only inside the host cells and are capable of causing diseases'.

Lwoff (1957) defined that 'viruses are infectious, potentially pathogenic nucleoproteins with only one type of nucleic acid, which reproduce from the genetic material, are unable to grow and divide and devoid of enzymes'.

Luria and **Darnell** (1968) defined that 'viruses are entities, the genome of which are element of nucleic acid that replicate inside the living cells using the cellular synthetic machinery and causing the synthesis of specialized elements that can transfer the viral genome to other cells'.

Nature of Viruses

In the simplest form, viruses consist of nucleic acid and protein, with the protein forming a protective coat around the nucleic acid. In other words, a virus is only a nucleoprotein, which is capable of causing diseases. Some viruses are infective even at very small concentrations. It multiplies only in living cells and it is too small and transparent to be seen individually with an ordinary microscope. All viruses are pathogenic and cause several diseases in all living organisms, from single-celled microorganisms to large plants and animals. Of the total of more than 2,000 viruses known, one-fourth of them cause plant diseases. Although viruses exhibit characteristics of other microorganisms, such as ability to cause diseases, possessing genetic functions and capacity to reproduce, they behave as chemical molecules also.

Although viruses exist in several forms, they are mostly either rod-shaped or polyhedral or variants of these two basic structures. There is always only RNA (ribose nucleic acid) or only DNA (deoxyribose nucleic acid) in each virus and in most plant viruses only one kind of protein is present. Some viruses, however may have two or more different proteins. RNA is the most common nucleic acid present in plant viruses and is involved in protein synthesis.

Viruses do not divide and do not produce any kind of specialized reproductive structures, such as spores. Instead they multiply by inducing the host cells to form more viruses. Viruses cause diseases not by destroying host cells with toxins but by utilizing host cellular substances during multiplication, thereby disrupting cellular processes and occupying space in the host cells. Because of these, cellular metabolism in the host plants is adversely affected, which results in disease development.

Viruses are much smaller than bacteria and their size range from 20 - 300 nm. In general, up to one million viruses can be accommodated within a bacterial cell. The *tobacco mosaic virus* measures 280.0 x 8.6 nm., while the *tomato bushy stunt virus* measures 27.4 nm in diameter and the *potato ring rot virus* measures 19 nm in diameter. The *smallpox virus*, which is supposed to be a very large one, measures 300 nm in diameter. The size of viruses is determined by electron microscopy, ultracentrifugation and by filtration through collodion membranes.

Many plant viruses have split genomes consisting of two or more distinct nucleic acid strands or segments encapsulated in different sized particles made of the same protein substances. Some viruses, such as *tobacco rattle virus* consist of two rods, a long one and a shorter one, while some like the *alfalfa mosaic virus* consist of four rods of different sizes. In such multicomponent viruses, all of the nucleic acid strand components must be present in the plant cells for the virus to multiply and cause infection.

Properties of Plant Viruses

Viruses are infective microorganisms that show several variations from the typical microbial cell. They are neither cells nor do they consist of cells. The simplest viruses are nucleoprotein particles consisting of genetic material, either DNA or RNA, surrounded by a protein coat or capsid. Plant viruses are of different shapes and sizes. They are elongate (rigid rods or flexuous threads) or spherical (isometric or polyhedral) or cylindrical or bacillus-like rods (helical) **(Fig. 71)**. The diameter of isometric spherical viruses range from 17 - 70 nm , while the bullet-shaped, bacilliform viruses measure up to 300 nm. in length and 95 nm. in width, the short, rigid, rod-shaped viruses measure 114 - 215 nm. in length and 23 nm. in width, and the long, flexuous particles measure up to 2,000 nm. in length and 10 nm. in width. The icosahedral virus particles have a geometrical shape, with 20 triangular faces and 12 corners and three protein subunits are arranged near the corners of each triangular face.

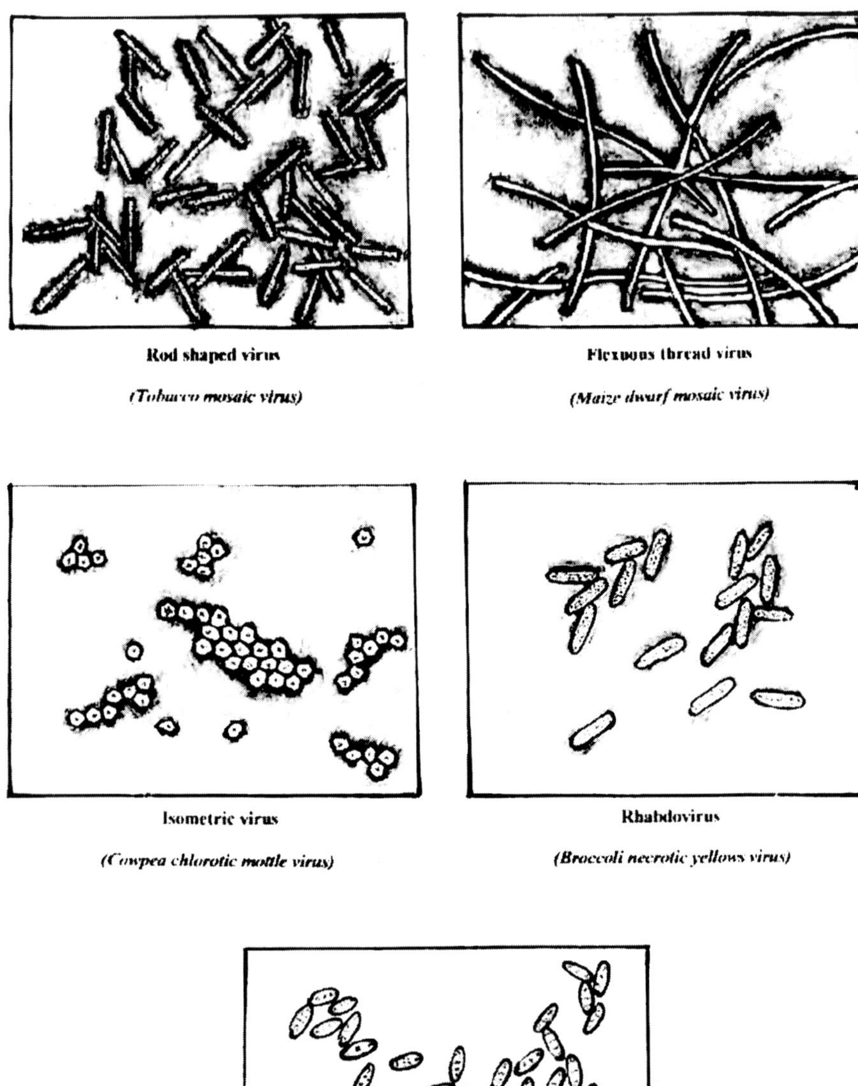

Rod shaped virus

(Tobacco mosaic virus)

Flexuous thread virus

(Maize dwarf mosaic virus)

Isometric virus

(Cowpea chlorotic mottle virus)

Rhabdovirus

(Broccoli necrotic yellows virus)

Bullet-shaped or Bacilliform virus

(Alfalfa mosaic virus)

Fig. 71 : Shapes of some plant viruses as seen through Electron Microscope

The **'virion'** (virus particle) consists of a protein coat or **'capsid'**, which encloses a genome of either RNA or DNA. The entire structure is called the **'nucleocapsid'**. The protein coat or capsid is composed of a number of sub-units called **'capsomeres'**. The capsid may be a polyhedron or a helix or a combination of both, as in some bacteriophages. There is always only RNA **(ribose nucleic acid)** or only DNA **(deoxyribose nucleic acid)**. However, some viruses consist of more than one size of nucleic acid and proteins, while some others contain enzymes or membrane lipids. The nucleic acid makes up 5 - 40 per cent of the virus and protein makes up the rest of the virus. The spherical viruses contain higher percentage of nucleic acid than the elongated ones. In this respect they differ from typical cells, which are made up of proteins, carbohydrates, lipids and nucleic acids. Some viruses are surrounded by a membranous envelope that is derived from the host cell. It also serves to protect the virus and helps in the transmission of the virus from one cell to another. The envelope consists of a lipid bilayer and two types of proteins viz., glycoproteins and matrix proteins, each with special functions. The spikes on the outer surface of the virions consist of glycoproteins. The envelope and capsid proteins are specified by the viral genes. The lipid and carbohydrate of the glycoprotein are derived from the host cells. Polyhedral and helical viruses may have naked capsids or the capsids may be enveloped. Polyhedral viruses occur with tetrahedral, octahedral or icosahedral symmetry, but mostly they have icosahedral symmetry.

Viruses do not have any cytoplasm and so, cytoplasmic organelles found in typical cells are also absent. They do not have any limiting cell membrane. They utilize the ribosomes of the host cells for protein synthesis during reproduction. Viruses cannot multiply outside a living cell. They do not have enzyme system and protein synthesis machinery and hence are metabolically inactive outside the host cell. Viral nucleic acid replicates by utilizing the protein synthesis machinery of the host. Viruses do not have the power of growth or division. A fully formed virus does not grow in size by the addition of new molecules. By itself the virus cannot divide. Only its genetic material is capable of reproduction and that too only inside a host cell. Viral nucleic acids, which contain instructions for the structure and function of the viruses, show considerable diversity. The DNA or RNA components of viruses may be single- or double- stranded, linear or circular. Some may have plus (positive) polarity, while others have minus (negative) polarity. Many of the smaller viruses can be crystallized and thus they behave like chemical molecules. Viruses are adversely affected by extremes of environmental conditions, such as very high or very low temperatures, radiation etc., but not affected by antibiotics. New strains of viruses may occur as a result of spontaneous mutation.

With respect to the number of strands, four types of nucleic acids are found in viruses: **(Fig. 76)**

Single-stranded DNA (ssDNA)
Double-stranded DNA (dsDNA)
Single-stranded RNA (ssRNA)
Double-stranded RNA (dsRNA)

Single-stranded DNA (ssDNA) may be linear as in *Parvoviruses* or circular as in many bacteriophages. During replication, the ssDNA becomes double stranded and is known as the replicative form.

Double-stranded DNA (dsDNA) occurs in a variety of forms. In many animals and bacteriophages dsDNA is linear. In the *Papovovirus* and in some phages, it is in the form of a closed circle.

Single-stranded RNA (ssRNA) is found in many animal viruses and polyhedral (Icosahedral) plant viruses. The strand may be plus (infectious) as in most plant viruses, RNA bacteriophages, *Picornaviruses* and *Togaviruses* or the strand may be negative (non-infectious) as in *Rhabdoviruses* and *Paramyxoviruses*.

Plus ssRNA directly acts as messenger RNA (mRNA) and translates protein on the ribosomes for virus multiplication and hence it is infectious. On the other hand, minus ssRNA first transcribes a complementary mRNA strand from the host, which has opposite polarity. Then the plus mRNA translates protein from the host. The minus strand alone is not infectious **(Fig. 75)**

Double-stranded RNA (dsRNA) is found in animal viruses, Cytoplasmic polyhedrosis viruses of insects and in some plant viruses, such as *Wound tumor viruses, Rice dwarf virus* etc. Usually plant viruses contain only RNA, either ssRNA or dsRNA.

Satellite viruses (SV). Some viruses, such as the satellite viruses are associated with and dependent upon another virus **(helper virus** or **activator virus)** that allows the SV to multiply and cause infection. The term **'satellite'** is used to describe such nucleic acid molecules, which are unable to multiply in the host cell without the aid of a specific helper virus. However, in all such cases, the helper virus alone can exist quite independently. The satellite virus particle may contain deficient nucleic acid or defective nucleic acid, which in the absence of the helper virus cannot form a capsid protein or cause infection. Thus, the *Satellite tobacco ring spot virus (STRSV)* can infect a plant and multiply only in the presence of *Tobacco ring spot virus (TRSV)*, which is the helper virus and the capsid protein of *STRSV* is coded by the helper *TRSV* and the two viruses are indistinguishable in size, shape and serological tests.

Multiplication of Viruses

Plant viruses can enter into the plant cells only through wounds caused mechanically or by vectors or into an ovule through an infected pollen grain. During multiplication of an RNA virus, the nucleic acid of the virus is first freed from the protein coat. It then induces the host cell to produce the **'viral RNA polymerase'**. This enzyme, polymerase utilizes the viral RNA **(+)** as a template and produces a complementary or a mirror copy RNA **(-)**. The primary RNA **(+)** and the complementary RNA **(-)** are temporarily connected to form a double stranded RNA **(+ -)**, which soon separates to produce the original virus RNA **(+)** and the mirror image RNA **(-)** strand. Subsequently, this mirror image RNA **(-)** serves as a template to synthesize more virus RNA **(+)** strands **(Fig. 72)**. Thus, each virus particle multiplies simultaneously, as a result the viral population increases at an alarming rate inside the host cells. Subsequently each RNA **(+)** strand is enveloped by the protein capsid, which is also synthesized by the host cell and new virus particles are formed. Thus, by synthesizing the complementary RNA **(-)** strand, the host itself initiates the replication process of virus particles within the cells.

The replication of some other viruses differs considerably. In the case of viruses with segmented RNA, enclosed in two or more virus particles, all the particles must be present in the host cell for the virus to multiply and cause infection. In the case of single stranded RNA *Rhabdovirus*, the RNA is negative and so, is not infectious. This RNA must be transcribed by a virus carried enzyme called **'transcriptase'** into a positive messenger RNA [mRNA **(+)**] strand in the host **(Fig. 72)**. In the double stranded RNA isometric viruses, the RNA is segmented within the same virus and is non-infectious. So, it depends for its replication in the host on a transcriptase enzyme, which is carried within the virus. Once the negative RNA strand is transcribed into a positive RNA strand, then this RNA **(+)** strand replicates as mentioned before. The sites of the host cell in which nucleic acid and protein are synthesized and the two components to assemble and produce the virions vary according to the virus group. For most RNA viruses, the virus RNA, after it is freed from the protein coat replicates itself in the cytoplasm, where it serves as a messenger RNA and in cooperation with the ribosomes produce the virus proteins. In other viruses, such as the ones with ssDNA, synthesis of viral nucleic acid and proteins and their assembly into virions take place in the host nucleus, from which the virus particles are released into the cytoplasm.

After inoculation, it takes about 10 hours for the first complete virions to appear in the plant cell. These virus particles may exist, either singly or in groups and may form amorphous or crystalline inclusion bodies within the cytoplasm or nucleus of the host cells as the case may be.

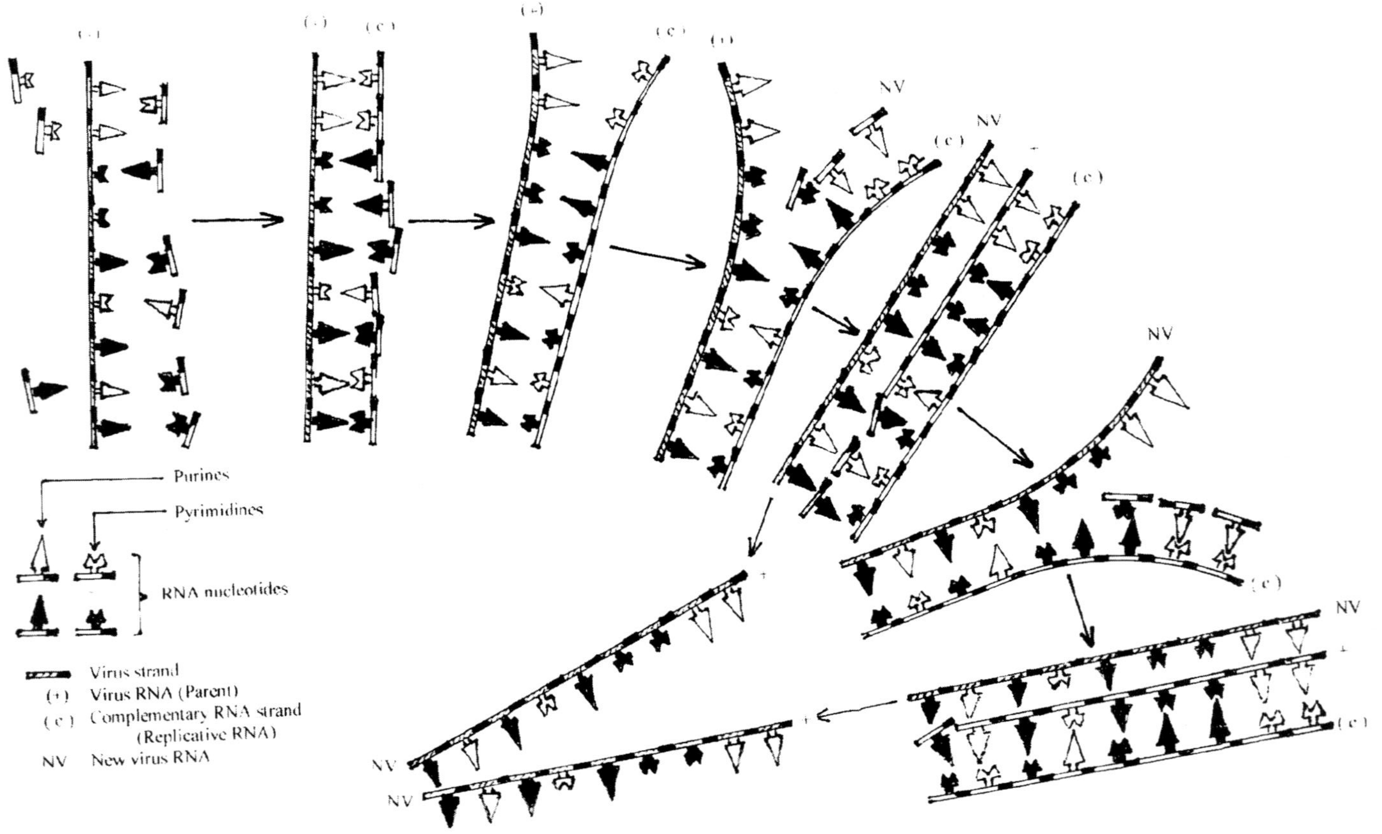

Fig. 72 : Diagrammatic representation of viral RNA replication

Translocation of Viruses in Plants

After a virus infects a plant, it moves from one cell to another and multiplies in most of the cells. Viruses move from cell to cell through minute passages in the cell wall called **'plasmodesmata'** connecting adjacent cells **(Fig. 73)**. They multiply in each parenchyma cell they infect. In leaf parenchyma cells, the virus moves about 1.0 mm or 8 - 10 cells per day.

After initial infection, viruses reach the phloem and through it are transported rapidly to all plant parts. Usually, it takes 2 - 5 days or more for the virus to move out of an infected leaf. After entering the phloem, it moves rapidly through it toward the growing apical meristem or other food utilizing parts, such as tubers, rhizomes etc. In the phloem the virus spreads systemically throughout the plant and again reenters the parenchyma cells adjacent to the phloem through plasmodesmata.

The development of local lesion symptoms is an indication of localized infection within the lesion area. In several diseases, the lesions continue to enlarge and the virus spreads systemically beyond the borders of the lesions.

Mosaic virus infected plant cells may contain 1,00,000 to 10,000,000 virus particles per cell. Systemic spread of some viruses may involve all living cells of a plant.

Transmission of Plant Virus Diseases

Plant viruses are transmitted from plant to plant in a number of ways. The chief modes of transmission include mechanical or sap transmission, as well as biological means through vegetative propagation, seed, pollen, phanerogamic parasites and vectors, such as specific insects, mites, nematodes, fungi etc.

Transmission by Vegetative Propagation

When plants are propagated vegetatively by budding or grafting, by cuttings or by using tubers, corms, bulbs, rhizomes etc., the viruses present in the mother plant from which these propagants are taken will be transmitted to the progeny in all certainty. Most of the fruit and several ornamental trees and shrubs, many field crops, such as potatoes, sugarcane etc. and many flowering plants are propagated vegetatively. As such, this means of transmission of viruses is considered to be most important.

Transmission of viruses may also occur through natural root grafts of adjacent plants, particularly trees. This mode of transmission is sometimes seen in some well-established orchards.

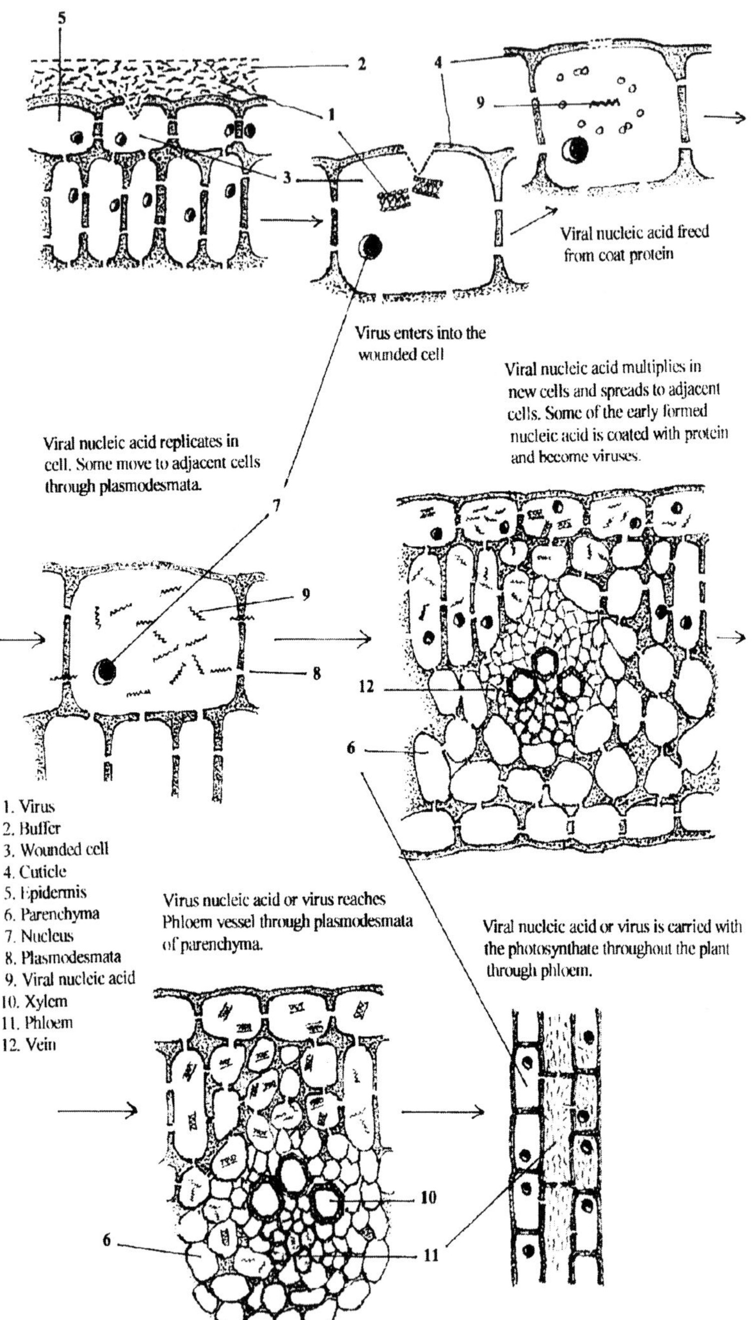

Fig. 73 : Systemic distribution of viruses in plants after mechanical inoculation

Mechanical or SAP Transmission

Plant viruses are unable to pass through the cuticle of the host plant and enter the cells below without any assistance. For infection to occur, the virus has to enter the tissues of the host through a sub-lethal wound. In nature strong wind, sandstorm etc. may cause injuries in both healthy and diseased plants and the virus from the diseased plants may enter into the adjacent healthy plants through these injuries. Further, when plants are wounded during cultural operations by tools and implements, hands or clothes or by animals feeding on the plants, the sap carrying the virus is transmitted to wounded healthy plants. *Tobacco mosaic virus, Potato virus X* and *Cucumber mosaic virus* are transmitted through sap in the field.

Mechanical transmission of plant viruses is of great importance in studying the viruses that cause plant diseases. This is usually accomplished by grinding the leaf or flower petals of a diseased plant and the sap extracted. The sap is then strained to remove any tissue fragments that may be present in the sap. Often a buffer solution is added to stabilize the virus. A solution of 1 % di-potassium hydrogen phosphate containing 0.1 % sodium sulfite is an ideal buffer for transmitting many viruses. The infected materials are usually ground in the buffer at a low temperature of 0°C. The infectious sap is then applied on the surface of leaves of young plants by means of a glass spatula or cheese cloth or fine brush. Prior to application of the sap, the leaf surface is wounded by gently rubbing with an abrasive, such as carborundum powder. The viruses present in the sap, enter the leaf cells through the wounds caused and initiate fresh infection. In local-lesion hosts, symptoms usually appear 3 - 7 days or more after inoculation. In systemically infected hosts, symptoms usually appear 10 - 14 days or more after inoculation.

Seed Transmission

More than 100 viruses are transmitted by seed to a smaller or greater extent. Transmission of mosaic disease through seed is known in bean, clover, cucumber, lettuce etc. Seed transmission cannot occur, if the virus is restricted to the phloem alone *(Beet curly top virus)*, as there is no vascular connection from the mother plant to the seed embryo. In all cases of seed transmission, the virus is also found in the pollen. Seed infection plays a major role in the transmission and survival of many important virus diseases. Seed transmission provides an ideal starting point for the establishment of a disease in a field crop, as it enables infection to occur even in the earliest seedling stage, leading to severe infection as the plant grows. Further, seed infection results in the uniform distribution of infected seedlings throughout a field crop, which may provide a reservoir of virus source for subsequent secondary infection by other means of transmission. Further, it provides an ideal means by which the virus can survive the winter or other unfavorable periods.

In the case of *Bean common mosaic virus*, infected seeds provide the only means of virus survival. *Cucumber mosaic virus* is known to be transmitted through seeds of the weed, *Stellaria media* and the weed seeds which remain dormant for prolonged periods in the soil provide a source for the long-time persistence of the virus during the temporary absence of the primary host.

In addition to the importance of seed infection in the local spread of certain viruses, seed transmission is extremely important in the International transmission of plant viruses. Viruses, such as *Bean common mosaic virus* and *Pea seed-borne virus* have been distributed to many countries through infected seeds.

Some viruses, such as *Tobacco mosaic virus* in tomato seeds are carried in or on the seed coat (testa) and are transmitted to the seedlings when they are transplanted. The virus enters the seedling tissue through cells that are injured during the transplanting process. Several other viruses, such as *Barley stripe mosaic virus* and *Bean yellow mosaic virus* in soybean are also transmitted by external contamination of the testa.

In contrast to such transmission by external contamination, true seed transmission or embryo transmission is by far the most important means of seed transmission and most of the viruses transmitted through seed are carried in the embryo.

Pollen Transmission

Besides embryos becoming infected as a result of virus infection of the mother plant, female gametes may also become infected through pollination of the healthy mother plant by infected pollen. The pollen-borne viruses may enter the ovule along with the male gamete by passing through the pollen tube into the embryo sac. Some viruses, such as the *Bean common mosaic virus* and *Barley stripe mosaic virus* are often transmitted through infected pollen. In some cases, pollen-carried viruses can spread through the fertilized flowers down into the mother plant, which also becomes infected with the disease, as in the case of cherry infected with *Prunus necrotic ringspot virus*.

Insect Transmission

Insects are by far the most important group of plant virus vectors, both in terms of the number of viruses they transmit and their prevalence. More than 70 % of insect vectors of plant viruses belong to the Order -Hemiptera and the aphids (Family - Aphididae) are the most important vectors of this group. Other insect vectors belonging to this Order, such as leafhoppers (Cicadellidae), plant hoppers (Delphacidae), tree hoppers (Membracidae), white flies (Aleurodidae), true bugs (Pentatomidae) and mealy bugs (Coccoidae) are also vectors. All the above mentioned insects have piercing and sucking mouthparts. Some other plant virus vectors, such as thrips (Order - Thysanoptera), which have rasping and sucking mouth parts, beetles (Order - Coleoptera) and grass hoppers (Order - Orthoptera), which have biting and chewing mouth parts also transmit a few plant viruses.

Insects with sucking mouth parts carry plant viruses on their stylets after feeding on the infected plant for a very short time, often only a few minutes or seconds. Such stylet-borne viruses, persist in the vector only for a few to several hours. Therefore they are called as **'non-persistent viruses'**. The vector can transmit the viruses immediately after acquisition by inserting its stylets into a healthy plant for feeding. Non-persistent viruses are carried on or near the mouthparts of the insect and do not multiply within the insect. Such viruses are usually readily sap transmissible by mechanical means also.

In the case of some virus diseases, such as beet yellows, the insect vectors must feed on the infected plant for fairly long time before they can acquire enough viruses for transmission. These insects can then transmit the viruses to a healthy plant after considerably long feeding periods of several minutes to several hours. Such viruses persist in the vector for a few days, usually 1 - 4 days and are called **'semi-persistent viruses'**. These viruses are non-persistent in the sense that they do not circulate within the vector. The semi-persistent viruses are usually associated with phloem cells.

With still other viruses, the insect vectors accumulate the viruses internally and after passage of the viruses through the insect tissues, they introduce the viruses into healthy plants again through their mouthparts. Such viruses are known as **'persistent'** or **'circulative viruses'**. In the case of circulative viruses, usually a long acquisition feeding time of 6 - 24 hours is necessary for the insect vector to acquire the viruses. Then a latent period (incubation period) of about 12 hours or more from the acquisition-feeding period is required for the insect to become viruliferous and capable of transmitting the viruses on to a healthy plant. Once the insect acquires the viruses, it retains the ability to transmit the viruses for at least a week, but frequently much longer and sometimes it is able to transmit the viruses for the rest of its life. Some circulative viruses may multiply in their respective vectors and are called **'propagative viruses'** and are sometimes transmitted through the eggs of the infected vector to its progeny. Such type of virus transmission is called **'transovarial transmission'**. Viruses transmitted by insects with chewing type of mouthparts, such as beetles may also be circulative or may be carried on the mouthparts.

Aphids are the most important insect vectors of plant viruses and transmit a vast majority of stylet-borne viruses. Though aphids are capable of transmitting viruses through all the above three modes, non-persistent mode of transmission is of considerable economic importance and far more numerous than semi-persistent or persistent means of transmission. Aphids generally acquire the stylet-borne viruses after feeding on a infected plant for only a few seconds (30 seconds or less) and can transmit the viruses by feeding on a healthy plant for a similarly short time of a few seconds. The length of time aphids remain viruliferous after acquisition of a stylet-borne virus varies from a few minutes to several hours, after which they can no longer transmit the viruses unless they acquire the viruses once again by feeding on a diseased plant. In aphids transmitting stylet - borne viruses, the viruses are carried on the tips of the stylets and are lost while probing and feeding on the host cells and they do not persist after moulting and are not carried in their eggs. Several aphid species can transmit the same stylet-borne viruses and the same aphid species can transmit several different viruses. However, in many cases the vector-virus relationship is quite specific. In the few cases of aphid transmission of circulative viruses, the aphids cannot transmit the virus immediately, but require several hours of incubation period after the acquisition feeding. But once they start transmitting the virus, they continue to do so for many days without any further acquisition feeding.

Leafhoppers, planthoppers and treehoppers transmit about 50 plant viruses. Viruses transmitted by leafhoppers mainly cause yellowing and leaf rolling symptoms in the infected host plant and only a few of these viruses are sap-transmissible. The viruses are mostly concentrated in the phloem cells and their vectors feed mainly in the phloem tissues and the viruses cause disturbances in plants that arise primarily in the region of the phloem. All such viruses, except a few are circulative and several are known to be propagative, while some persist through the moult and some others are transovarial. Most leafhopper and planthopper vectors require a feeding period of one to several days before they become viruliferous, but once they have acquired the virus, they may remain viruliferous for the rest of their lives. A latent period of 1 - 2 weeks from the acquisition feeding time is required for a leafhopper or planthopper to begin transmission.

In contrast to the persistent leafhopper transmitted viruses, *Rice tungro virus* (transmitted by *Nephotettix cincticeps*) and *Maize chlorotic dwarf virus* (transmitted by *Graminella nigrifrons*) behave like semi-persistent viruses. They persist only for a few days in their vectors, have no latent period and are not carried through the moult.

The virus diseases transmitted by white flies (Aleurodidae) are usually referred to as **'rugaceous'** and cause mosaic and leaf distortion symptoms in infected plants. *Bemisia tabaci* is the most important and widespread vector. The vectors feed mainly on phloem tissues. White flies require a minimal acquisition feeding time of 10 - 60 minutes and a latent period of 4 - 8 hours. But the time required to transmit the viruses to a healthy plant is much less than the acquisition time. The vector may retain the viruses for 2 - 25 days. The virus is not propagative and not transovarial. Both nymphs and adults can transmit the viruses. These viruses are not usually sap-transmissible.

Beetles (Coleoptera) transmit about 45 different plant viruses and the beetle transmitted viruses have no other vectors. Diseases caused by *Cowpea mosaic virus* are widely distributed and infect many economically important crops, such as bean, cow pea and soybean. The acquisition period is 24 hours or less. However, in some cases a single bite of an infected leaf makes a beetle viruliferous. No latent period is required prior to transmission. The virus may be retained in the beetle for a period of 1 - 21 days.

Mealy bugs (Pseudococcidae) transmit a few viruses. *Planococcoides njalensis* and *Planococcoides citri* transmit most viruses in the group including *Cacao swollen shoot virus*. The vectors acquire the viruses by feeding on the phloem cells of the infected host plant. The transmission is semi-persistent in nature. Acquisition feeding time ranges from 48 - 72 hours in most of the cases. Infection of the host plant occurs following transmission feeding periods as short as 15 minutes. The insect can retain the viruses for a maximum of 3 - 4 days and nymphs are more effective vectors than the adults, because they have got more mobility than the adults, which are not particularly mobile.

Thrips (Thysanoptera). Only *Tomato spotted wilt virus* is transmitted by thrips. Four species, including *Frankliniella fusca* (Thripidae) transmit the viruses in a persistent manner. The acquisition feeding time required ranges from 15 minutes to 4 days. There is a latent period of 4 - 16 days and the vector remains viruliferous for life. Only nymphs are capable of acquiring and transmitting the viruses and there is no transovarial passage of the viruses.

Mite Transmission

In the Class - Arachnida, only mites of the Family - Eriophyidae are known to transmit at least six plant viruses. Of these, *Wheat streak mosaic virus,* transmitted by *Aceria tulipae* is the most important one. The viruses are found in the body cavity around the intestine and in the salivary glands. The acquisition feeding time required is 15 minutes. After a latent period of 15 minutes, the vector can transmit the viruses. They persist in the vector for upto 9 days and are **transstadial** (viruses retained in the vector even after moulting). There is no transovarial transmission.

Nematode Transmission

About 20 plant viruses are transmitted by one or more species of five Genera of soil-inhabiting, ectoparasitic (free-living) nematodes. All the nematode vectors belong to the Order - Dorylaimida and include species of the Genera - *Trichodorus, Paratrichodorus, Longidorus, Paralongidorus* and *Xiphinema*. Nematodes of the Genera - *Longidorus, Paralongidorus* and *Xiphinema* transmit several polyhedral-shaped viruses known as *Nepoviruses,* such as *Grape fan leaf virus, Tobacco ring spot virus* and some other viruses. Nematodes of the Genera - *Trichodorus* and *Paratrichodorus* transmit at least two rod-shaped *Tobraviruses, Tobacco rattle virus* and *Pea early browning virus.* Virus transmission by nematodes is non-persistent in nature and the viruses do not multiply within the vector. The vectors have probing mouthparts consisting of a single stylet. The stylet penetrates the plant cells during feeding and the cell contents, along with the viruses from the absorption sites are sucked into the pharynx. Transmission of viruses occurs by the back-flow of materials including the viruses when the nematode feeds on the roots of a nearby healthy plant. This back-flow occurs during the initial phases of feeding, when saliva secreted from the oesophageal bulb of the nematode is injected into the host cells along with the viruses.

Juveniles, as well as the adult nematodes are capable of transmitting the viruses however, the viruses are not carried through the juvenile moults or through the eggs. After moulting, the juveniles or the resulting adults must acquire the viruses to become viruliferous once again.

Fungus Transmission

Soil-inhabiting fungi transmit at least 15 plant viruses. Fungal vectors belonging to two groups of obligate parasites transmit these viruses. Of this *Olpidium* sp. (Chytridiales), transmit *Tobacco necrosis virus*, *Satellite virus* and *Cucumber necrosis virus*, which has isometric particles. *Polymyxa* sp. and *Spongospora* sp. (Plasmodiophorales) transmit many viruses, such as *Wheat mosaic virus* and *Potato mop top virus*, which have rod-shaped particles. In the case of *Olpidium* sp., the virus particles present in the soil-water get attached to the outer walls of the zoospores and cilia and pass with the protoplasm of the zoospore into the host root cell through the zoospore infection canal. On the other hand, *Polymyxa* and *Spongospora* sp. transmit the viruses through the resting spores produced by the fungi. The viruses can remain viable in the resting spores for long periods and when the resting spore germinates and releases zoospores, the viruses are also transmitted to new roots by the zoospores.

Dodder Transmission

Phanerogamic parasites, such as dodder (*Cuscuta* sp.) also transmit plant viruses from one plant to another through the bridge formed between two plants. The viruses are usually transmitted passively through the phloem of the dodder plant from the infected plant to the healthy one.

Classification of Plant Viruses

In 1962, **A.Lwoff**, **R.Horne** and **P.Tournier** proposed a system of classification of viruses referred to as the 'LHT' system of classification and was adopted by 'The Provincial Committee of Nomenclature of Viruses', formed by the 'International Association of Microbiological Society'. The LHT system of classification is based on (i) the nature of nucleic acid (RNA or DNA), (ii) symmetry of viral particle (Helical, icosahedral, cubic, cubic-tailed), (iii) presence or absence of envelope, (iv) diameter of capsid, and (v) number of capsomeres forming the capsid.

'The Plant Virus subcommittee' established in 1966 by the 'Executive Committee of the International Committee of Nomenclature' recommended a classification of viruses into 16 sub groups including 4 monotypic groups (**Harrison** et. al., 1971). The groups are: 1.*Tobravirus* group 2.*Tobamovirus* group 3.*Potexvirus* group 4.*Carlavirus* group 5.*Potyvirus* group 6.*Cucumovirus* group 7.*Tymovirus* group 8.*Comovirus* group 9.*Nepovirus* group 10.*Bromovirus* group 11.*Tombusvirus* group 12.*Caulimovirus* group 13.*Alfalfa mosaic virus* 14.*Pea enation mosaic virus* 15.*Tobacco necrosis virus* 16.*Tomato spotted wilt virus*. In this classification, only the well-studied viruses have been included in the above groups.

Siegal and **Hariharasubramanian** (1974) propounded another system of classification, which is mainly that of **Harrison** et. al., with the addition of some other virus groups.

In the current system of nomenclature and classification, the plant viruses are divided into Groups, Families and Genera. The system of classification followed by **George N. Agrios** (1997) is followed in this text :

KINGDOM - VIRUSES

RNA VIRUSES

SINGLE-STRANDED POSITIVE RNA (+SSRNA)

ROD-SHAPED PARTICLES

One ssRNA

Genus - *Tobamovirus* **(e-g)** *Tobacco mosaic virus*

Two ssRNAs

Genus - *Tobravirus* **(e-g)** *Tobacco rattle virus*

Two - four ssRNAs - Fungus transmitted, rod-shaped viruses

Genus - *Furovirus* **(e-g)** *Soil-borne Wheat mosaic virus*

Three ssRNAs

Genus - *Hordeivirus* **(e-g)** *Barley stripe mosaic virus*

FILAMENTOUS PARTICLES

One ssRNA

Genus - *Capillovirus* **(e-g)** *Apple stem grooving virus*

Genus - *Carlavirus* **(e-g)** *Carnation latent virus*

Genus - *Trichovirus* **(e-g)** *Apple chlorotic leaf spot virus*

Genus - *Potexvirus* **(e-g)** *Potato virus X*

One or two ssRNAs

Family - Potyviridae

Genus - *Potyvirus* **(e-g)** *Potato virus Y*

Genus - *Rymovirus* **(e-g)** *Ryegrass mosaic virus*

Genus - *Bymovirus* **(e-g)** *Barley yellow mosaic virus*

LONG FLEXUOUS FILAMENTS

One ssRNA

Genus - *Closterovirus* **(e-g)** *Beet yellow virus*

ISOMETRIC PARTICLES

One ssRNA

Family - Sequiviridae

Genus - *Sequivirus* **(e-g)** *Parsnip yellow fleck virus*

Genus - *Waikavirus* **(e-g)** *Rice tungro spherical virus*

Family - Tombusviridae

Genus - *Tombusvirus* **(e-g)** *Tomato bushy stunt virus*

Genus - *Carmovirus* **(e-g)** *Carnation mottle virus*

Families (Not established)

Genus - *Machlomovirus* **(e-g)** *Maize chlorotic mottle virus*

Genus - *Necrovirus* **(e-g)** *Tobacco necrosis virus*

Genus - *Luteovirus* **(e-g)** *Barley yellow dwarf virus*

Genus - *Marafivirus* **(e-g)** *Maize rayado fino virus*

Genus - *Sobemovirus* **(e-g)** *Southern bean mosaic virus*

Genus - *Tymovirus* **(e-g)** *Turnip yellow mosaic virus*

Two ssRNAs

Family - Comoviridae

Genus - *Comovirus* **(e-g)** *Cowpea mosaic virus*

Genus - *Nepovirus* **(e-g)** *Tobacco ringspot virus*

Genus - *Fabavirus* **(e-g)** *Broad bean wilt virus*

Families (Not established)

Genus - *Dianthovirus* **(e-g)** *Carnation ringspot virus*

Genus - *Enamovirus* **(e-g)** *Pea enation mosaic virus*

Three ssRNAs

Family - Bromoviridae

Genus - *Bromovirus* **(e-g)** *Brome mosaic virus*

Genus - *Cucumovirus* **(e-g)** *Cucumber mosaic virus*

Genus - *Ilarvirus* **(e-g)** *Tobacco streak virus*

Genus - *Alfamovirus* **(e-g)** *Alfalfa mosaic virus*

SINGLE-STRANDED NEGATIVE RNA (- SSRNA)

One (-) ssRNA

Family - Rhabdoviridae

Genus - *Cytorhabdovirus* **(e-g)** *Lettuce necrotic yellows virus*

Genus - *Nucleorhabdovirus* **(e-g)** *Potato yellow dwarf virus*

Three (-) ssRNAs

Family - Bunyaviridae

Genus - *Tospovirus* **(e-g)** *Tomato spotted wilt virus*

Four (-) ssRNAs

Peculiar particle morphology (Folded or filamentous)

Genus - *Tenuivirus* **(e-g)** *Rice stripe virus* **(Fig. 76)**

DOUBLE-STRANDED RNA (dsRNA)

ISOMETRIC VIRUSES

Two dsRNAs

Family - Partitiviridae

Genus - *Alfacryptovirus* **(e-g)** *White clover cryptic virus* 1

Genus - *Betacryptovirus* **(e-g)** *White clover cryptic virus* 2

Ten - twelve dsRNAs

Family - Reoviridae

Genus - *Fijivirus* **(e-g)** *Rice fiji disease virus*

Genus - *Oryzavirus* **(e-g)** *Rice ragged stunt virus*

Genus - *Phytoreovirus* **(e-g)** *Wound tumor virus*

DNA VIRUSES

DOUBLE-STRANDED DNA (dsDNA)

ISOMETRIC CIRCULAR dsDNA VIRUSES

Genus - *Caulimovirus* **(e-g)** *Cauliflower mosaic virus*

NON-ENVELOPED BACILLIFORM PARTICLES

Genus - *Badnavirus* **(e-g)** *Rice tungro bacilliform virus*

SINGLE-STRANDED DNA (ssDNA)

GEMINATE (Twin particles)

Family - Geminiviridae

Genus - *Geminivirus* **(Fig. 76)**

Subgroup 1 - Monocotyledonous hosts, leafhopper vectors, monopartite genome

(e-g) *Maize streak virus*

Subgroup 2 - Dicotyledonous hosts, leafhopper vectors, monopartite genome

(e-g) *Beet curly top virus*

Subgroup 3 - Dicotyledonous hosts, mostly white fly vectors, bipartite genome

(e-g) *Bean golden mosaic virus*

SINGLE ISOMETRIC PARTICLES

(e-g) *Banana bunchy top virus*

General Characteristics of Important Genera of Plant Viruses

Genus - *Tobamovirus*. The single, rigid, rod-shaped helical viruses measure about 300 nm. in length and 15 -18 nm. in diameter and contain a single molecule of ssRNA coiled inside a protein capsid. The RNA strand is spirally wound and follows the pitch of the protein helix. The protein sub-units of the virus are arranged on the helical RNA molecule with a distinct central hole. Shorter rods, which may be present, are not infectious. Although the RNA is infective by itself, it is less infective than the intact virion, because unprotected RNA is subjected to the action of enzymes (nucleases) and is degraded. The protein capsid functions as a protective tube around the RNA **(Fig. 74, 76)**

(e-g) *Tobacco mosaic virus (TMV); Tomato mosaic virus; Cucumber green mottle virus.*

Genus - *Tobravirus*. The viruses of this Genus have rigid, rod-shaped particles of two lengths, measuring 180 - 215 nm. and 46 - 114 nm. x 22 nm. in width. The long and short particles each contain a separate ssRNA molecule. Both particles are necessary for infection. The long RNA particle can replicate itself, but lack information to code for coat protein. However, it may code for replicase, which is necessary for the replication of both long and short RNA particles. The short particle contains the code for coat protein and replication, but cannot replicate by itself. Because of the division of functions between the long and short particles, it is necessary that both the particles should be present for the replication of the complete virus, which only can cause stable infection. These viruses have a wide host range and are mainly transmitted by nematodes **(Fig. 76)**

(e-g) *Tobacco rattle virus (TRV); Pea early browning virus; Pepper ring spot virus.*

Genus - *Furovirus*. The Genus consists of viruses that are rigid, rod-shaped particles of 2 - 4 ssRNA genomes of different lengths. The particles of different lengths contain different RNA molecules. They have a narrow host range and are transmitted by soil-inhabiting fungi **(Fig. 76)**

(e-g) *Soil-borne Wheat mosaic virus; Peanut clump virus.*

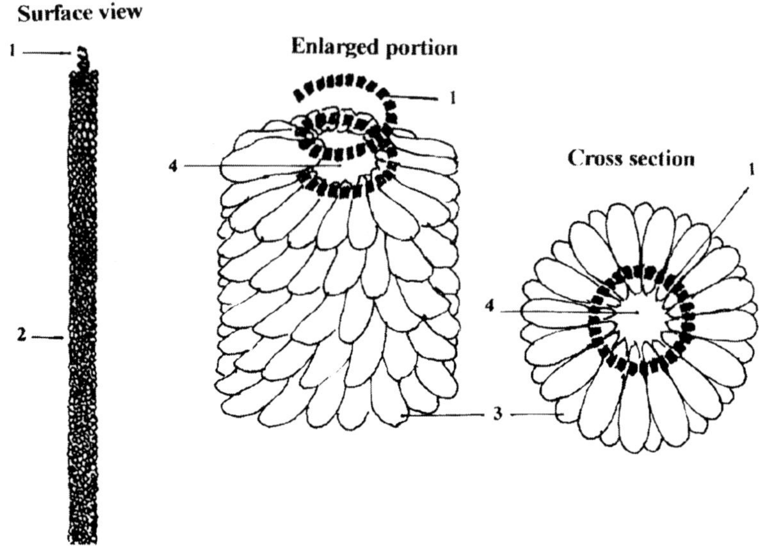

Fig. 74 : Structure of *Tobacco mosaic virus*

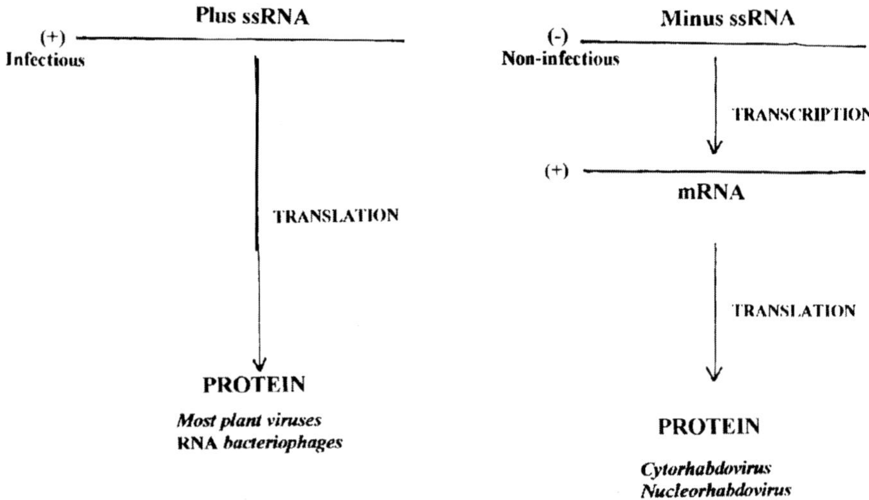

Fig. 75 : Infection by plus and minus strands of ssRNA viruses

Genus - *Hordeivirus*. This Genus consists of viruses, which are rigid, rod-shaped particles measuring 100 - 150 nm. in length x 20 nm. in width. They have a ssRNA genome within particles of three different lengths. Particles of different lengths contain different RNA molecules. Two or all the three particle components are required for infectivity. They have a narrow host range and are transmitted mechanically by sap or through seeds **(Fig. 76)**

(e-g) *Barley stripe mosaic virus.*

Genus - *Capillovirus*. The Genus consists of viruses that are flexuous, rod-shaped particles with a single ssRNA **(Fig. 76)**

(e-g) *Apple stem grooving virus; Potato virus T.*

Genus - *Carlavirus*. The Genus consists of flexuous rod-shaped particles of helical symmetry, measuring 620 - 700 nm. in length x 13 nm. in width, with a single ssRNA. Members of this Genus have a narrow host range and frequently show mild or no symptoms of infection. Except in some legumes, no economically important diseases are caused. They are transmitted by sap inoculation and by aphids in a non-persistent manner. In the vector, the viruses persist for less than two hours **(Fig. 76)**

(e-g) *Carnation latent virus (CLV); Alfalfa latent virus; Cowpea mild mottle virus; Eggplant mild mottle virus.*

Genus - *Potexvirus*. The virus particles are flexuous rods with helical symmetry and 470 - 580 nm. in length and 13 nm. in width. The genome is a single ssRNA. They are readily transmitted by sap inoculation and no vectors are known **(Fig. 76)**

(e-g) *Potato virus X (PVX)* - it causes mild mosaic of potato, the symptoms of which are not conspicuous. It is also called *Potato latent virus, Potato mottle virus* or *Solanum virus* 1; *Cassava common mosaic virus*; *Papaya mosaic virus; Clover yellow mosaic virus.*

Genus - *Potyvirus*. The flexuous, rod-shaped particles measure 680 - 900 nm. in length x 11 nm. in width and contain a single ssRNA molecule. Members of this Genus are of considerable economic importance, because they infect a wide range of plants. They are transmitted by aphids in a non-persistent manner and by sap inoculation. Some are transmitted through seeds **(Fig. 76)**

(e-g) *Potato virus Y (PVY)* - this virus in combination with *PVX*, causes Rugose mosaic virus disease of potato causing severe damage to the crop.

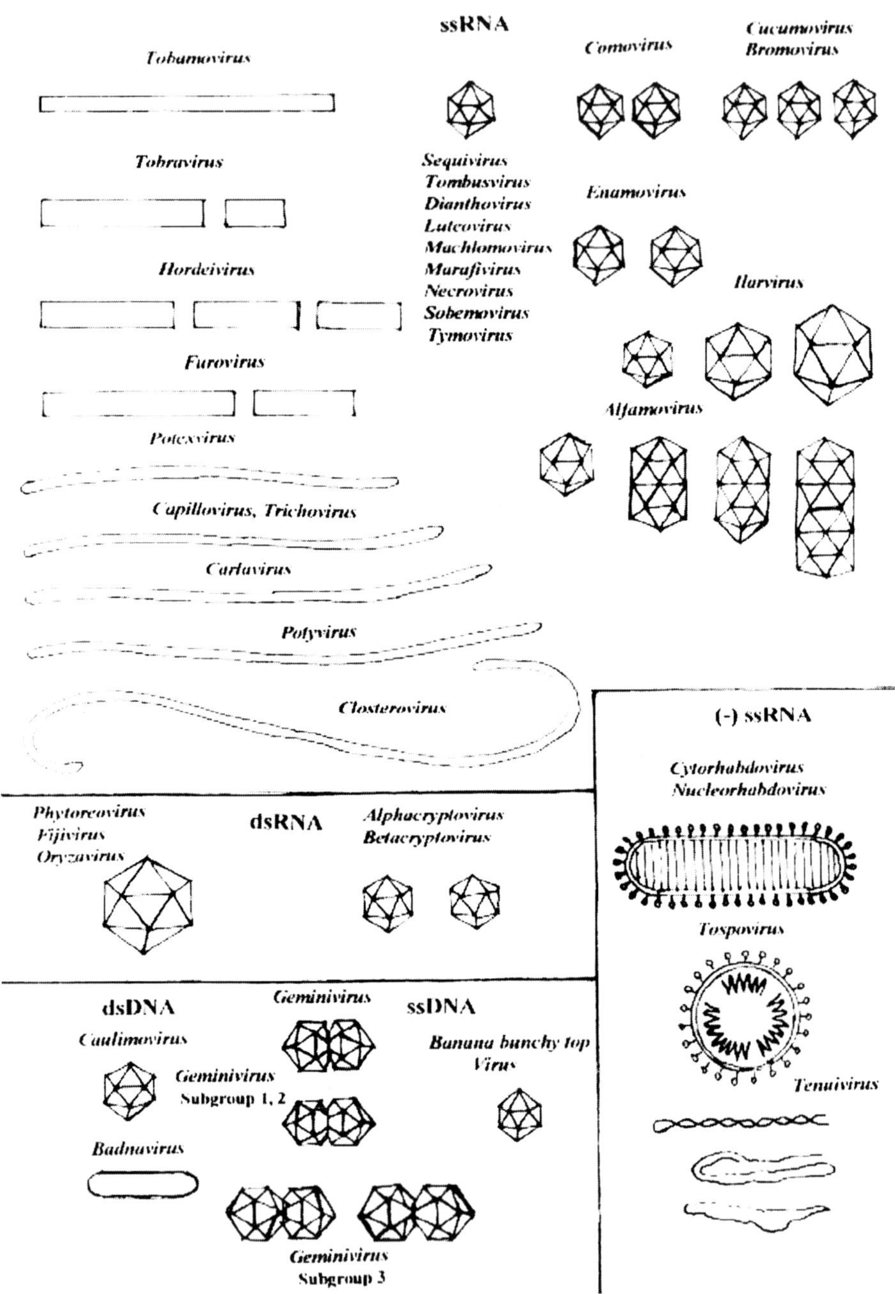

Fig. 76 : Schematic diagram of genera of plant viruses

The foliage is mottled, wrinkled, puckered and markedly reduced in size. The leaf margins roll downwards and the entire plant is dwarfed. In case of severe attack, the leaves are blighted. *PVY* is also called *Solanum virus* 2; *Potato virus A (PVA)*, another member of this Genus, in combination with *PVX* causes crinkle disease of potato, which is characterized by larger yellow patches on the foliage that becomes brittle. *PVA* is also called *Solanum virus* 3; *Bean yellow mosaic virus; Beet mosaic virus; Carnation vein mottle virus; Celery mosaic virus; Lettuce mosaic virus; Onion yellow dwarf virus; Papaya ringspot virus; Soybean mosaic virus; Sugarcane mosaic virus; Watermelon mosaic virus; Tulip flower breaking virus.*

Genus - *Rymovirus*. The Genus consists of flexuous, rod-shaped particles with one or two ssRNA molecules

(e-g) *Ryegrass mosaic virus; Oat necrotic mottle virus; Wheat streak mosaic virus.*

Genus - *Bymovirus*. The viruses of this Genus are flexuous, rod-shaped particles with one or two ssRNA molecules

(e-g) *Barley yellow mosaic virus; Wheat spindle streak mosaic virus; Oat mosaic virus.*

Genus - *Closterovirus*. Members of this Genus have very long, flexuous, rod-shaped particles, measuring 600 - 2,000 nm. in length x 10 nm. in width. The genome consists of a single ssRNA molecule. The viruses have a wide host range and some of them cause important economic diseases. Aphids transmit some members in a semi-persistent manner **(Fig. 76)**

(e-g) *Beet yellows virus; Beet yellow stunt virus; Carnation necrotic fleck virus; Carrot yellow leaf virus; Citrus tristeza virus* - causes Citrus decline and stem pitting in *Citrus* trees; *Wheat yellow leaf virus.*

Genus - *Waikavirus*. Members of this Genus consist of isometric particles of size 30-34 nm. in diameter, with a single ssRNA molecule. The virus has a narrow host range within the Graminae and is transmitted by leafhoppers in a semi-persistent manner. It is not transmitted by sap inoculation. It causes 'tungro' virus disease in rice. This *Rice tungro spherical virus (RTSV)* in combination with *Rice tungro bacilliform virus (RTBV)* causes rice 'tungro' virus disease. Only the rice green leafhopper vector transmits the virus in a non-persistent manner. The virus particles persist in the vector for a short period of 5-6 days

(e-g) *Rice tungro virus (RTSV); Maize chlorotic dwarf virus (MCDV).*

Genus - *Tombusvirus*. Members of this Genus have a monopartite ssRNA genome with isometric particles of about 30 nm. in diameter. The host range is wide and they can be transmitted by sap inoculation or through virus particles in the soil **(Fig. 76)**

(e-g) *Tomato bushy stunt virus; Turnip crinkle virus; Carnation Italian ring spot virus; Petunia asteroid mosaic virus.*

Genus - *Necrovirus*. The isometric virus particles contain monopartite ssRNA and are 26 - 28 nm. in diameter. The viruses have a wide host range and are distributed worldwide. It is readily transmitted by sap inoculation and also by the soil fungus *Olpidium brassicae* **(Fig. 76)**

(e-g) *Tobacco necrosis virus; Cucumber mosaic virus.*

Genus - *Luteovirus*. The genome consists of monopartite ssRNA within isometric particles of about 25 nm. diameter. The host range of individual members of this Genus is fairly extensive. The viruses are not transmitted by sap inoculation, but are transmitted by aphids in a persistent manner **(Fig. 76)**

(e-g) *Barley yellow dwarf virus; Beet mild yellowing virus; Pea leaf roll virus; Bean leaf roll virus; Potato leaf roll virus.*

Genus - *Sobemovirus*. Members of this Genus have monopartite ssRNA genome with isometric particles about 28 nm. in diameter. They have a restricted host range and may be seed transmitted. They are also transmitted by sap inoculation and by beetles **(Fig. 76)**

(e-g) *Southern bean mosaic virus; Turnip rosette virus.*

Genus - *Tymovirus*. The monopartite ssRNA genome is contained in isometric particles, 28 - 30 nm. in diameter. The host range of *TYMV* is restricted to the Cruciferae and other members of this Genus have a narrow host range. They are readily transmitted by sap inoculation and by flea beetles **(Fig. 76)**

(e-g) *Turnip yellow mosaic virus (TYMV); Cacao yellow mosaic virus; Eggplant mosaic virus; Wild cucumber mosaic virus; Okra mosaic virus.*

Genus - *Comovirus*. The genome consists of an isometric bipartite ssRNA. The two types of RNA are contained in separate isometric particles, each measuring about 24 nm. in diameter. Both types of particles are required for infection. Other particles containing no RNA may also be present in the infected host. The host range of each member of this Genus is relatively restricted and they induce mainly mottling and stunting symptoms in the host. All members are readily transmitted by sap inoculation and by beetles in a persistent way. Some members are seed transmitted **(Fig. 76)**

(e-g) *Cowpea mosaic virus (CPMV); Radish mosaic virus; Broad bean mosaic virus; Bean rugose mosaic virus; Cowpea severe mosaic virus.*

Genus - *Nepovirus*. The angular, isometric particles are approximately 28 nm. in diameter and the genome is bipartite, consisting of two ssRNA molecules in distinct particles. Both types of particles are required for infection. A third particle containing no RNA is frequently found in infected plants. The viruses of this Genus have a wide host range causing ringspot and mottle symptoms. Latent infection is common in some hosts. These viruses are transmitted by sap inoculation and by soil-inhabiting nematodes. Seed transmission is also common in many hosts. Pollen transmission is also found to occur in many members

(e-g) *Tobacco ringspot virus (TRSV); Tomato ringspot virus; Grapevine chrome mosaic virus; Cacao necrosis virus; Hibiscus latent ringspot virus; Mulberry ringspot virus; Grapevine fan leaf virus; Arachis mosaic virus.*

Genus - *Fabavirus*. The genome consists of bipartite ssRNA. Both the types of RNA are contained in separate isometric particles, each measuring about 24 nm. in diameter. Both the types of particles are required for infection. Other particles containing no RNA may also be present in the infected plants. The host range is very much restricted and they cause mottling, stunting and wilt in their hosts. They are mostly transmitted by sap inoculation and by beetles in a persistent manner

(e-g) *Broad bean wilt virus.*

Genus - *Dianthovirus*. Members of this Genus have a bipartite ssRNA genome. The isometric particles are approximately 31- 34 nm. in diameter. They are readily transmitted by sap inoculation and through soil microbes **(Fig. 76)**

(e-g) *Carnation ringspot virus (CRSV); Red clover necrotic mosaic virus; Sweet clover necrotic mosaic virus.*

Genus - *Enamovirus*. The virus has a bipartite ssRNA genome, encapsidated separately in isometric particles about 30 nm. in diameter. Both types of particles are required for infection. The host range of this virus is narrow. The virus is transmitted by sap inoculation and by aphid vectors in a persistent manner. Infected plants develop mosaic and enation symptoms **(Fig. 76)**

(e-g) *Pea enation mosaic virus (PEMV).*

Genus - *Bromovirus*. The Genus contains viruses with a tripartite genome of ssRNA, separately encapsidated in isometric particles, about 26 nm. in diameter. All three components of RNA are required for infection. The members of this Genus have a narrow host range and are transmitted by sap inoculation and rarely by beetles **(Fig. 76)**

(e-g) *Brome mosaic virus (BMV); Broad bean mottle virus; Cowpea chlorotic mottle virus.*

Genus - *Cucumovirus*. The viruses of this Genus have tripartite genomes, encapsidated in three types of isometric particles of about 28 nm. in diameter. All three particles are necessary for infection. The members are easily transmitted by sap inoculation and by aphids in a non-persistent manner. Seed transmission also occurs in some hosts. CMV has a wide host range and is of considerable economic importance in crops throughout the World **(Fig. 76)**

(e-g) *Cucumber mosaic virus (CMV); Peanut stunt virus; Tomato aspermy virus.*

Genus - *Ilarvirus*. The members of this Genus have a tripartite ssRNA genome, each contained in separate isometric particles of size varying from 26 - 35 nm. in diameter. The three RNA molecules are not infectious on its own, but infection occurs only when a fourth and smaller RNA molecule or the coat protein is present. The members have a wide host range and are easily sap transmitted. Some are transmitted through seed and pollen, while one member is transmitted by thrips **(Fig. 76)**

(e-g) *Tobacco streak virus (TSV); Apple mosaic virus; Citrus leaf rugose mosaic virus; Cherry rugose mosaic virus; Rose mosaic virus; Black raspberry latent virus; Citrus variegation virus; Prunus necrotic ringspot virus.*

Genus - *Alfamovirus*. The tripartite genome is composed of three larger molecular particles of ssRNA, while a fourth small particle of RNA, the coat protein messenger RNA is also present and each is separately encapsidated. The three larger RNA particles, together with the fourth small RNA or the coat protein are required for infectivity. The three larger particles are bacilliform in shape, measuring 58 x 18, 48 x 18 and 36 x 18 nm. in size and the fourth is isometric measuring 18 nm. in diameter. The viruses have a wide host range and cause economically important diseases in legumes and other crops. They are readily transmitted by sap inoculation and by aphids in a non-persistent manner. In some hosts they are seed transmitted **(Fig. 76)**

(e-g) There is only one member in this Genus - *Alfalfa mosaic virus*

Genus - *Cytorhabdovirus*. The type member of this Genus is *Lettuce necrotic yellow virus.* The virus particles consist of an outer membrane containing lipid and protein, and an inner helical core containing protein and a single molecule of -ssRNA. They are bacilliform or bullet-shaped particles measuring 160 - 380 nm. in length x 50 - 95 nm. in width and are the largest among plant viruses. Generally they are not transmitted by sap inoculation, but are transmitted by aphids and leafhoppers in a persistent manner. The viruses multiply within the vector. Some plant viruses, such as *LNYV*, are of economic importance **(Fig. 76)**

(e-g) *Lettuce necrotic yellow virus (LNYV); Barley yellow striate mosaic virus; Beet leaf curl virus; Coffee ringspot virus.*

Genus - *Nucleorhabdovirus*. The virus particles consist of a helical core containing protein and a single molecule of -ssRNA. They are bacilliform or bullet-shaped particles, measuring 160 - 380 nm. x 50 - 95 nm. in size. Aphids and leafhoppers transmit them in a persistent manner **(Fig. 76)**

(e-g) *Potato yellow dwarf virus; Wheat chlorotic streak virus.*

Genus - *Tospovirus*. The type member of this Genus is *Tomato spotted wilt virus.* The isometric particles are 85 nm. in diameter and are enclosed within a lipoprotein envelope. The -ssRNA genome is composed of three molecules of RNA. *TSWV* is transmitted by sap inoculation and by thrips in a persistent manner. It has a wide host range of at least 160 plant species and causes diseases of economic importance in tomato, tobacco, potato and various other crops **(Fig. 76)**

(e-g) *Tomato spotted wilt virus (TSWV)*

Genus - *Tenuivirus*. The viruses have peculiar particle morphology and are sometimes folded. Usually they are filamentous and 3.0 - 12.0 nm. in width with four -ssRNA particles **(Fig. 76)**

(e-g) *Rice stripe virus (RSV); Rice hoja blanka (white leaf) virus; Rice grassy stunt virus; Maize stripe virus*

Genus - *Fijivirus*. The type member of this Genus is *Fiji disease virus.* The icosahedral, double- stranded RNA (dsRNA) genome has 10 segments encapsidated within an isometric particle, measuring approximately 70 nm. in diameter. Members of this Genus infect only plants of the Family - Graminae and plant hoppers (Fulgorids) are the vectors **(Fig. 76)**

(e-g) *Sugarcane fiji disease virus; Cereal tillering disease virus; Maize rough dwarf virus.*

Genus - *Oryzavirus*. The isometric particles are similar to those of the *Fijivirus*. The host range is very narrow and only some plants of the Family - Graminae are infected. The brown plant hopper is the vector **(Fig. 76)**

(e-g) *Rice ragged stunt virus* is the only member of this Genus.

Genus - *Phytoreovirus*. The double-stranded RNA (dsRNA) genome has twelve segments encapsidated within an isometric particle of size about 70 nm. in diameter. The virus is transmitted only by leafhoppers, in which they are propagative **(Fig. 76)**

(e-g) *Wound tumor virus (WTV)* - this virus has been found to infect about 43 plant species. It causes tumors on roots and stems by infection through accidental wounds. Infection also takes place through leafhopper vectors; *Rice dwarf virus.*

Genus - *Geminivirus*. The members of this Genus have unique particle morphology in that their isometric particles of size 18 - 20 nm. diameter occur mainly in pairs. The ssDNA genome is unique among plant viruses. Members of this Genus infect a wide range of plant species, but individual members have a narrow host range. Some members of this Genus are transmitted by leafhoppers and others by white flies. They are of considerable economic importance in food and fibre crops.

Subgroup 1. These viruses have a monopartite genome. They infect monocotyledonous hosts and are transmitted by leafhopper vectors **(Fig. 76)**

(e-g) *Maize streak virus.*

Subgroup 2. These viruses with monopartite genome infect dicotyledonous hosts and are transmitted by leafhopper vectors **(Fig. 76)**

(e-g) *Beet curly top virus; Cassava latent mosaic virus; Tomato golden yellow mosaic virus; Tomato yellow dwarf virus; Tobacco leaf curl virus; Tobacco yellow dwarf virus.*

Subgroup 3. These viruses with bipartite genome infect dicotyledonous plants and are transmitted mostly by white fly vectors **(Fig. 76)**

(e-g) *Bean golden mosaic virus; Bean summer death virus; Mungbean yellow mosaic virus.*

Geminate single isometric particle virus **(Fig. 76)**

(e-g) *Banana bunchy top virus*

Genus - *Caulimovirus*. Members of this Genus are large, icosahedral, isometric, circular dsDNA particles having a genome consisting of one molecule of dsDNA with three discontinuities, two in one strand and one in the other. The genome is encapsidated in isometric particles measuring about 50 nm. in diameter. They have a restricted host range and are aphid transmitted in a non-persistent manner. Some members are transmitted by sap inoculation **(Fig. 76)**

(e-g) *Cauliflower mosaic virus (CaMV); Dahlia mosaic virus; Strawberry vein banding virus; Cassava vein mosaic virus.*

Genus - *Badnavirus*. This consists of non-enveloped, bacilliform particles, 150 - 300 nm. in length and 30 - 35 nm. in width. This virus along with *Rice tungro spherical virus (RTSV)* causes rice tungro virus disease **(Fig. 76)**

(e-g) *Rice tungro bacilliform virus (RTBV); Banana streak virus; Cacao swollen shoot virus*

Common Symptoms of Virus Diseases

'Symptoms' are the manifestation of the effects that a virus causes on the growth, development and metabolism of an infected host plant, which are mostly visible to the naked eye. Symptoms are of major importance, because they are the main means by which a virus disease is diagnosed and named. The nature of severity of the disease symptoms determines the economic importance of a particular virus in terms of yield loss and reduced quality. Virus infection alters the cell metabolism of the host plant resulting in biochemical and physiological changes. Such abnormal metabolism leads to some specific changes in the plant parts, which are exhibited as symptoms.

The occurrence of more than one type of symptoms in a diseased host due to infection by a specific virus or pathogen is called a **'syndrome'**. If the host is infected by more than one type of virus or pathogen simultaneously, the cumulative effect on symptom expression may be much more severe and is called **'synergistic effect'**. Such severe symptom expression does not occur as a result of infection by any one of the virus or pathogen individually.

Viruses can enter a host only through some kind of wound or injury caused as a result of rubbing together of infected and healthy leaves or most often through injuries caused due to vector feeding. The primary symptoms that develop at the site of virus entry in the infected leaves leading to distinct areas of diseased cells is called **'local lesions'**. Local lesions may vary in size from small pinpoints to large chlorotic patches or necrotic patches.

In some cases, the virus is unable to spread in the plant beyond the site of primary infection and the cells in the local lesions die. This type of restricted response is known as **'hypersensitivity'** reaction, which is observed in resistant hosts. In some other cases, no local lesions are visible as a result of primary infection and this is known as **'latent infection'** and the hosts are called **'symptomless carriers'**. In a susceptible host, the viruses are not confined to the site of primary infection, but spread from cell to cell within the leaf mesophyll through the plasmodesmata connections between the cells. Ultimately, the viruses reach the vascular system and spread rapidly through the entire plant. This leads to the secondary or **'systemic infection'**. Usually, the viruses move along the phloem vessels, but a few viruses move in the xylem vessels also. The viruses move from cell to cell in the leaf mesophyll and produce secondary or **'systemic symptoms'**.

Almost all virus diseases cause some degree of dwarfing or stunting of the entire plant and reduction in total yield, and may shorten the life span of the infected plants. However, viral diseases rarely kill the infected plants. The most conspicuous symptoms of virus infections are usually those appearing on the leaves, but some viruses may cause striking symptoms on the stem, fruits and roots with or without any symptoms on the leaves. The most important visual symptoms caused by virus infection, called macroscopic symptoms are described below:

1. **Stunting and dwarfing.** Reduced plant size is a common symptom of most virus infections and is most likely to be found in association with other symptoms. Growth may be reduced evenly throughout the plant or the stunting may be confined to certain parts or organs of the plant. Along with the aerial parts of the infected plant, root growth may also be stunted

   (e-g) Pea stunt disease caused by *Red clover vein mosaic virus*; Rice dwarf disease; Peanut stunt disease; Barley yellow dwarf disease **(Fig. 78)**

2. **Mosaic.** Mosaic is one of the most important symptoms caused by virus infection in many plant species. The term **'mosaic'** refers to leaf symptoms in which cells in some areas of the leaf or some specific organ are infected and discolored, while cells in other areas are not discolored. The infected areas are usually pale green or chlorotic due to loss or reduced production of chlorophyll. The shape and pattern of this type of symptoms vary considerably. If the discolored portions of the leaf are round, the symptoms are termed as **'mottle'**. Chlorotic flecking, spotting and blotching may also occur

(e-g) Tobacco mosaic **(Fig. 77)**; Tomato mosaic; Cucumber mosaic; Beet mosaic; Sugarcane mosaic; Soybean mosaic; Cassava mosaic; Papaya mosaic; Cucumber green mottle mosaic; Alfalfa mosaic etc.

In the case of some other mosaic diseases, such as Cauliflower mosaic of Brassicae, regular light and dark green banding symptoms occur. In the case of some virus diseases, which infect monocotyledonous plants, such as Barley stripe mosaic, Maize streak, Wheat spindle streak etc., the mosaic symptoms appear as light and dark green stripes or streaks. Mosaic symptoms may also occur on the stems or fruits of infected plants as in the case of Marrow (vegetable of the gourd family) infected with *Cucumber mosaic virus.*

3. **Yellow mosaic.** The leaves of infected plants develop more of irregular yellow patches than the green portion and the infected plants are stunted

   (e-g) Turnip yellow mosaic; Clover yellow mosaic; Bean yellow mosaic **(Fig. 77)**; Barley yellow mosaic.

4. **Chlorosis.** In the infected plants, the whole leaf may become chlorotic due to decreased chlorophyll production and breakdown of chloroplasts. Most chlorotic symptoms are linked with internal histological disorders, such as abnormal changes in the palisade cells and intracellular vacuoles. Symptoms of chlorosis usually start in the interveinal areas and spread through the entire leaf lamina. In some cases, the chlorosis may be confined to the area of the vein and is referred to as **'veinal chlorosis'** or **'vein yellowing'**.

   (e-g) Wheat yellow leaf; Potato yellow dwarf; Barley yellow dwarf; Cowpea chlorotic mottle; Beet yellows.

5. **Vein clearing.** Here the cells adjacent to the vein becomes translucent, while the interveinal areas remain green

   (e-g) Okra vein clearing **(Fig. 78);** Lettuce vein clearing.

6. **Vein banding.** In the infected plant, the area adjacent to the veins of the leaf remain green, in contrast to the remaining areas of the leaf, which may be chlorotic

   (e-g) Chilli vein banding caused by *Tobacco etch virus* **(Fig. 78)**

7. **Ring-spotting.** The ring-spotting symptom is a common feature of some viruses. The disease is restricted to a ring or broken ring of infected cells. The infected cells may be chlorotic or necrotic and

sometimes the rings may occur in concentric circles. These symptoms are most common on the leaves of infected plants, but may also occur on the stems, fruits and pods

(e-g) Cabbage necrotic ring spot; Chlorotic ring spot and broken ring spot of celery; Tobacco chlorotic ring spot; Tobacco concentric ring spot; Carnation Italian ring spot; *Hibiscus* latent ring spot; Mulberry ring spot; Tomato ring spot **(Fig. 77)**; Strawberry ring spot; Coffee ring spot; Papaya ring spot.

8. **Necrosis.** There are many types of virus induced necrosis, which occur in systemically infected tissues that are caused by the death of cells. This type of necrosis may occur in the form of small or large lesions on the leaves. In some cases, the veins may become necrotic and the necrosis may spread to the stem and root apices and ultimately the plant may be killed. Necrosis may occur on fruits or seeds also

   (e-g) Cabbage leaf necrosis caused by *Turnip mosaic virus*; Oat necrotic mottle; Carnation necrotic fleck; Tobacco necrosis; Cucumber necrosis; Cacao necrosis; Clover necrotic mosaic.

9. **Enations or tumors.** Due to infection by some viruses, characteristic tumor-like outgrowths are formed on the leaves and roots. The outgrowths on the leaves are referred to as **'enations'** and these appear like warts on the upper- or lower surface of the leaf. Such growth is caused by abnormal cell proliferation, which may be due to virus induced hormonal imbalance

   (e-g) Pea enation mosaic; Sugarcane 'Fiji' disease; Maize rough dwarf - wart-like outgrowths are produced on the lower surface of leaves due to abnormal proliferation of the underlying phloem tissues; Tumors on clover roots caused by *Wound tumor virus*.

10. **Leaf distortion.** In some virus infections, the leaf lamina is affected and it may become irregularly distorted or become strap-like. Such abnormal growth is the result of hormonal imbalance within the leaf and the symptoms are similar to the leaf disorder that may occur as a result of hormonal spray damage

    (e-g) Bean common mosaic in *Phaseolus vulgaris*; Strawberry latent ring spot in celery; *Tobacco mosaic virus* in tomatoes; *Bean leaf roll virus* in *Vicia faba*; Beet leaf curl; Tobacco leaf curl **(Fig. 77)**

11. **Leaf roll.** The leaf lamina roll and curl upwards

    (e-g) Potato leaf roll **(Fig. 77)**; Bean leaf roll, Pea leaf roll.

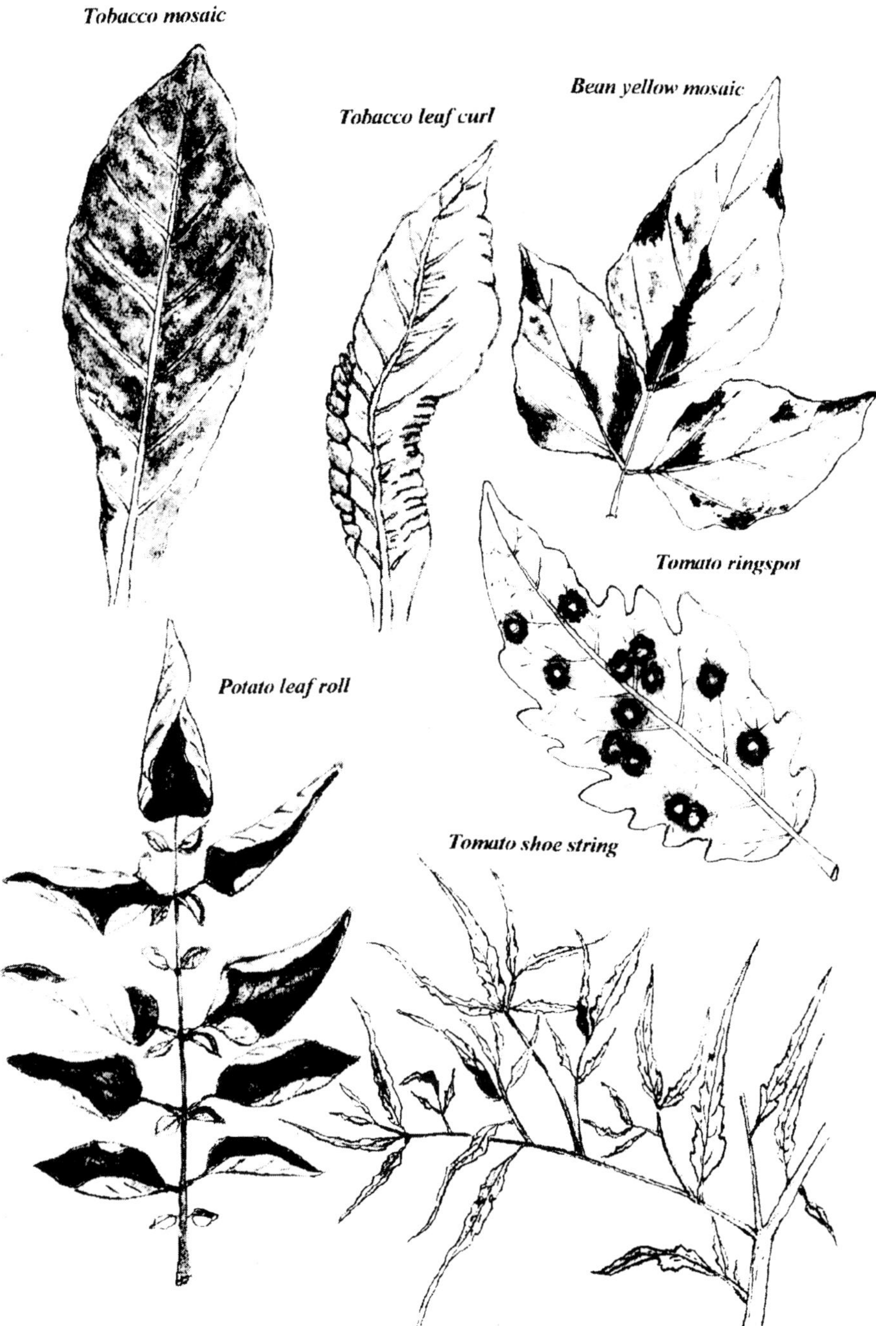

Fig. 77 : Symptoms of virus diseases

12. **Stem and root distortion.** In this type of malformation, the number of cells in the infected area may increase in number (hyperplasia) or the cells may enlarge in size abnormally (hypertrophy). In contrast, in some cases the number of cells may decrease abnormally (hypoplasia) or the cells or a particular organ may be reduced in size (atrophy)

    (e-g) Cacao swollen shoot - the virus causes abnormal cell proliferation in the stem of cacao plants leading to abnormal swelling of the stem, stem pitting of apple and stem pitting of citrus - the virus causes hypoplasia of cells in the stem. The pitting is visible as elongated pits or furrows on the surface of the stem, when the bark is removed **(Fig. 78)**. This may be caused by *Citrus tristeza virus*; Apple stem grooving; *Cucumber mosaic virus* infected plants produce fruits of cucumber, which are stunted and distorted.

13. **Rat tail or shoe string.** The leaves of tomato plants infected with *Cucumber mosaic virus* are distorted and often become filiform, shoe-string-like or rat tail-like.

    (e-g) Leaves of celery plants infected with *Strawberry latent ring spot virus* become stunted and appear strap-like; Tomato shoe string caused by *Cucumber mosaic virus* **(Fig. 77)**

14. **Petal or flower break.** Virus induced color-break symptoms of flowers and petals are caused by some viruses. The color-break symptoms may take the form of streaking, flecking or sectoring of the petal tissues with a color different from the normal. The color-break may be due to the loss of anthocyanins, which cause an underlying pigment to be exposed. In addition to color-break symptoms, the virus-infected plants may produce fewer, deformed and stunted flowers.

    (e-g) Color-break of tulip flower petals caused by *Tulip mosaic virus* **(Fig. 78)**; Color-break of gladiolus flowers infected with *Bean mosaic virus* or *Cucumber mosaic virus*.

15. **Fruit, seed and pollen abnormalities.** Plants infected with viruses causing mosaic, ring spotting and necrosis may affect the fruits and seeds also. Fruits from infected plants may be fewer, smaller, malformed or of changed texture

    (e-g) Plums infected with *Sugarbeet curly top virus* produce fewer fruits than normal plants; fruits of *Cucumis sativa* infected with *Cucumber mosaic virus* are malformed; fruits of apple tree infected with *Apple rough skin virus* have abnormal skin texture **(Fig. 78)**

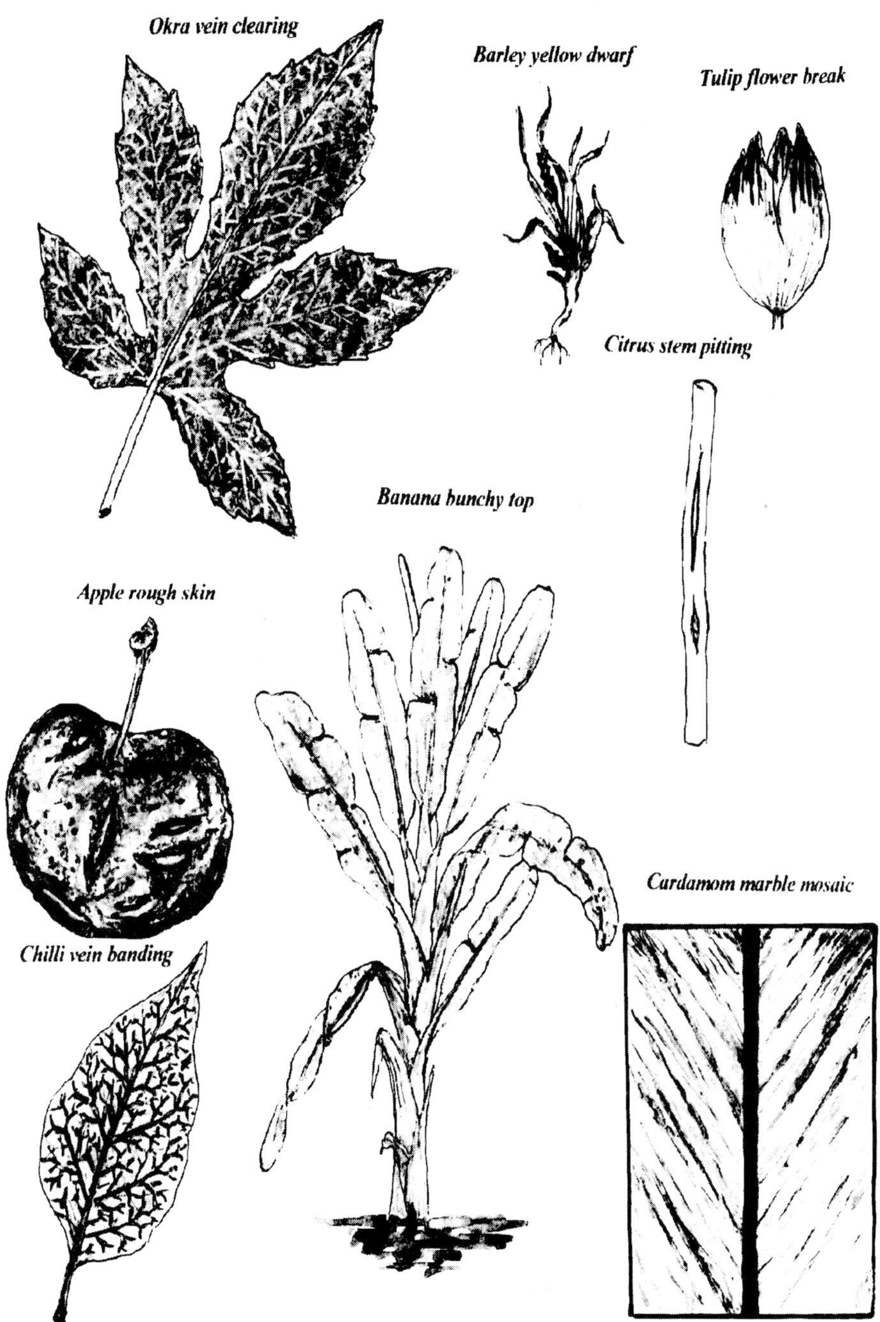

Fig. 78 : Symptoms of virus diseases

Mother plant infected with some viruses may show drastic effect upon seed production **(e-g)** Seed production of lettuce is greatly reduced when infected with *Lettuce mosaic virus* and sometimes the virus can cause complete abortion of the lettuce seeds. In other cases, when the virus is carried in the seed, the germination is badly affected.

Pollen from an infected plant is frequently sterile or its vitality is poor **(e-g)** The rate of pollen germination of *Nicotiana rustica* infected with *Cherry leaf roll virus* is much slower; the flowers of red gram infected with *Red gram sterility mosaic virus* become sterile and the flowers drop off without producing any seeds.

Viruses Producing Specific Symptoms

16. **Bunchy top or cabbage top**. The virus-infected plants are severely stunted. The leaves are reduced in size and the crown becomes composed of several leaves unfurling at the same time and appears as a rosette. The leaves show dark green streaks along the secondary veins

    (e-g) Bunchy top of banana **(Fig. 78)**

17. **Marble mosaic or 'Katte'**. Leaves of infected plants show a general chlorosis, with slender, interrupted, parallel pale green streaks running along the veins from the mid-rib to the margin. In advanced stages of infection, the streaks appear as a mosaic pattern on a marble stone and hence it is called **'marble mosaic'**

    (e-g) Marble mosaic or 'Katte' caused by *Cardamom mosaic virus* **(Fig. 78)**

18. **Rosette**. The plants infected by the virus become very much stunted. The internodes become shortened and the leaflets very much reduced in size, upright, chlorotic, curled and distorted. The plants present a bushy appearance

    (e-g) Groundnut rosette caused by *Groundnut rosette virus.*

19. **Tungro**. The virus infected plants are stunted and the color of the lower leaves may turn yellow or orange-yellow commencing from the leaf tip and margins and progress downwards

    (e-g) Rice 'tungro' virus disease caused by *Rice tungro spherical virus* and *Rice tungro bacilliform virus*

Common Symptoms of Viroid Diseases

'Viroids' are known to cause atleast 20 plant diseases till now. Among them 'coconut cadang-cadang disease', 'potato spindle tuber', '*Chrysanthemum stunt*' and 'apple scar skin' are more important economically. So far, no animal or human diseases caused by viroids have been identified.

1. **Potato spindle tuber.** Infected potato plants appear erect, spindly and dwarfed. The leaves are much reduced in size and erect, and the leaflets are darker green and sometimes show rolling and twisting. The tubers are elongated with tapering ends and they are smaller, with more number of conspicuous tuber eyes. Yield is also considerably reduced.

2. ***Citrus* exocortis.** *Citrus exocortis viroid* attacks many types of oranges, limes, lemons and citrons. It is commercially important when infected budwood of orange, lemon, grapefruit and other *Citrus* trees are grafted on exocortis-susceptible rootstocks. Infected susceptible plants develop narrow, vertical, thin strips of partially loosened outer bark that gives the bark cracked and scaly appearance when they are about 4 - 8 years old. Infected susceptible plant may also show yellow blotches on young infected stems. The infected plants appear stunted and give poor yield. *CEV* is readily transmitted from diseased to healthy trees by budding knives, pruning shears or any other cutting tools, by hand and possibly by scratching and gnawing of animals. It is also transmitted by dodder and by sap to herbaceous plants. *CEV* retains its infectivity on contaminated knife blades for at least 8 days. The viroid is highly resistant to heat inactivation and to almost all chemical sterilants.

3. **Coconut cadang-cadang. 'Cadang-cadang'** (dying disease) of coconut and other palms is very common in the Philippines. Palm trees usually become infected with the viroid disease after they have begun to flower. The first symptoms appear on the coconuts, which become rounded and develop scarification on their surface, while the leaves begin to show bright yellow spots. After 3 or 4 years, the inflorescence are killed and coconut production ceases. Gradually the leaves develop more and larger yellow spots, making the whole fronds look chlorotic from a distance. Within 5 - 7 years, the color of the crown becomes yellowish or bronze and the number and size of fronds continue to be reduced. Ultimately, the bud dies, falls off and leaves the palm trunk barren like a pole.

The *Cadang-cadang viroid* survives in infected coconut and in other palm trees, mostly in all the tissues including the husks and embryo of the nuts. It may be transmitted through nuts and pollen, and several other methods, such as through mouthparts of insects, through cutting tools, tapping knives etc. Cadang-cadang viroid disease cannot be controlled by any available means and there are no resistant coconut cultivars.

Symptoms of Phytoplasma and Spiroplasma Diseases

Important plant diseases caused by **'Phytoplasmas'** are 'aster yellows', 'apple proliferation', 'apricot chlorotic leaf roll', 'coconut lethal yellowing', 'grapevine yellows', 'elm yellows', 'pear decline' and 'stolbar' or 'big bud disease' of solanaceous plants. '*Citrus* stubborn' and 'corn stunt' are important diseases caused by **'Spiroplasmas'**.

1. **Aster yellows. 'Aster yellows'** causes chlorosis and dwarfing of the infected plant, abnormal production of shoots, sterility of flowers, malformation of organs and reduction in yield. The pathogen is found to infect carrot, onion, tomato, lettuce and several other ornamental, vegetable and weed plants.

   The symptoms of 'aster yellows' may vary according to the host. On carrot, initial symptoms appear as vein clearing and yellowing of the young leaves. Infected plants then produce many shoots and the tops look like 'witches' broom'. The internodes and petioles become short. The young leaves become smaller and often dry, while the petioles of older leaves become twisted and break off. The entire plant looks bronzed and reddened. The floral parts are deformed. The carrots are small, tapered, abnormally shaped and have wooly secondary roots. Infected carrots emit an unpleasant odor. The roots are predisposed to soft rot in the field and in storage. The phytoplasma survives in perennial ornamentals, vegetables and weeds. Aster yellows is transmitted by budding or grafting and by several leafhoppers. The *Phytoplasma* is mainly present in the phloem of infected plants. The leafhoppers acquire the *Phytoplasma* while feeding from the phloem of infected plants and after incubation period transmits the *Phytoplasma* to a healthy plant while feeding from the phloem. Controlling the disease is very difficult and not practicable.

2. **Lethal yellowing of coconut palms. 'Lethal yellowing'** appears as a blight that kills palm trees within 3 - 6 months after the appearance of initial symptoms. Besides coconut palm, the disease affects several other kinds of palms and produces identical symptoms. The first

symptoms are the premature dropping of coconuts of all sizes. The successive inflorescence that appears has blackened tips, the male flowers are black and dead, and there is no fruit setting. Soon, the lower leaves turn yellow and the yellowing progresses upward from the older to the younger leaves. The older leaves then wither prematurely, turn brown and cling to the tree, and all the younger leaves also become yellow. Within a short period all the leaves including the bud die. Finally, the entire top of the palm topples down leaving the trunk barren like a pole.

The phytoplasmas occur mainly in young phloem cells. No vectors are known with certainty. The plant hopper, *Myndus crudus* is a suspected vector. The *Phytoplsma* is sensitive to the antibiotic - Tetracycline and injecting Tetracycline solution is found to be effective in controlling the disease. Malayan dwarf and some other varieties possess resistance to this disease.

3. ***Citrus* stubborn disease. '*Citrus* stubborn disease'** affects mostly sweet oranges and grapefruit. The disease affects leaves, fruits and stems of all commercial varieties regardless of rootstock. In general, affected trees show a bunchy, upright growth of twigs and branches, with short internodes and large number of shoots. The trees show slight to severe stunting. The leaves become progressively small, often mottled or chlorotic. Affected trees bloom at all seasons, but produce fewer fruits, some of which are small and malformed. Affected fruits drop off prematurely. Fruits are often sour or bitter and have an unpleasant odor. The seeds in the affected fruits are poorly developed and aborted.

   The pathogen, *Spiroplasma citri* is found in the phloem. The pathogen affects many dicots and monocots, such as crucifers and several stone fruits, such as peach and cherry. Infected pea and bean plants wilt and die. *Spiroplasma citri* is highly sensitive to Tetracycline. Budding and grafting transmit *Citrus* stubborn disease. Several leafhoppers, such as *Circulifer tenellus, Scaphytopius nitridus* and *Neoaliturus haemoceps* serve as vectors.

Symptoms of Diseases Caused by Fastidious Vascular Bacteria

1. **Ratoon stunting of sugarcane. 'Ratoon stunting disease'** occurs in almost all sugarcane growing areas of the world. Losses due to this disease are much greater in the ratoon crop than in the planted crop.

Sometimes, the losses may go upto 30 per cent or more. Infected plants show slow initial growth, appear stunted and may have fewer and thinner stalks. No other external diagnostic symptoms are exhibited. If sufficient irrigation facilities are available, the plants may not show stunting to a large extent. However, infected plants often show internal symptoms. In very young shoots, pinkish discoloration is seen just below the meristem of the shoot above its attachment to the sett. In matured canes, discoloration is seen in the vascular bundles at the nodes, but not extending into the internodes. Affected vascular bundles are plugged with bacteria contained in a colored gummy substance.

The pathogen, *Corynebacterium (Clavibacter) xyli* subsp. *xyli* is a gram-positive, fastidious xylem-inhabiting coryneform bacterium, measuring 0.3 - 0.5 x 1.0 - 4.0 µm. in size. The pathogen overseasons in infected sugarcane plants and propagative materials, such as seed canes. New plants arising from infected seed canes develop the disease. The bacteria may also be transmitted through cutting knives and implements used for cultivation and harvesting. The disease has spread to different countries through infected sugarcane germplasm.

The disease can be controlled by use of disease-free seed material, heat treatment of planting material in hot water at 50°C for 2 hours or hot air at 54°C for 8 hours or in a mixture of steam and hot air (Aerated steam) at 52°C for 1 hour, sanitation of cutting equipments and by the use of disease resistant cultivars.

Symptoms of Diseases Caused by Algal Parasites

Red rust of tea *(Cephaleuros parasiticus)*. 'Red rust' attacks all kinds of tea, both young and old. However, it causes severe damage to young tea plants by attacking and killing stem tissues in patches. This often results in die-back of the stems. Older leaves and tea capsules are also attacked. Red rust lesions on the stems can be recognized by their circular- to oval-shaped, purplish-black color and longitudinal cracks on the surface. The patches are mostly found on the upper side of horizontal branches, but on all sides of the vertical ones. The plants infected by red rust on the stems invariably present a characteristic appearance in that the leaves produced above the affected region become variegated with yellow patches. The disease is more common on the older tea leaves. The spots produced are more or less circular, about 10 - 15 mm. in diameter, with a slightly raised, crusty appearance on the upper surface and a purplish margin. Older spots become dry and the central portion turns light dirty brown or grayish.

Fructifications appear at the center of the spots, either on the upper surface or on both the surfaces of the leaves. Affected patches on the stem become noticeable when the alga produces fructifications, which appear on one to three year old stems. The fruiting patches are brick red or orange in color. From these patches a dense growth of erect, tiny, orange colored hairs, the sporangiophores arise. At the tips of each sporangiophore, 4 or 5 minute knobs (sporangia) are formed. Inside each sporangium there are about 8 -10 motile endospores. On maturity of the sporangium, the outer wall bursts open and orange-colored, spherical or oval endospores (zoospores) with two, long, slender flagella emerge. When the weather is wet, the zoospores swim about on the water film on the surface of the leaves and are distributed by wind, rain splashing etc. The spores germinate and penetrate into the tissues of the host and multiply at the expense of the host cells and gradually penetrate deeper. The host plant attempts to ward off the infection by growing a layer of corky tissues below the affected area, giving the spots a raised, crusty appearance. When weather conditions are not favorable, the alga remains dormant inside the tissues of leaves and stems. When favorable conditions return the alga starts producing fresh crops of spores, which cause new infection.

Severely affected branches should be cut and destroyed to avoid recurrence of the disease. Spraying with Bordeaux mixture-0.75 % or copper oxychloride fungicide at 2.5 gm./ lit. of water is effective in controlling the disease. Maintaining the plants healthy and strong by applying recommended dose of fertilizers and adopting proper irrigation methods will help to minimize the incidence of the disease.

Phanerogamic Parasites

Genus - *Cuscuta*. Genus - *Cuscuta* belongs to the Family - Cuscutaceae. It is a holo stem parasite. It is generally known as the **'dodder'**. Dodder is also referred to as **'strangleweed'**, **'pull-down'** and **'hellbind'**. Dodder affects the growth and yield of infected plants and sometimes the infected plants are completely destroyed. Dodder may also transmit viruses from virus-infected plants to healthy plants. Several species of dodder exist. Some species prefer legumes, while others attack many other broad-leaved plants and legumes. Crops that suffer losses from dodder include alfalfa, clover, berseem, flax, onion, sugarbeet, potato, several ornamentals and hedge plants.

Orange or yellow vine-strands of dodder grow and entwine the stem and other aboveground parts of the host. The growing tips reach out to adjacent plants and a single dodder plant may infect the hosts in an area of up to 10 feet diameter. Such dodder-infected patches continue to enlarge and in perennial hosts like alfalfa they become larger every year. The twisting, slender stem of dodder is tough, curling, thread-like and devoid of leaves, but in place of leaves have minute scales. The stem is usually yellowish or orange, sometimes tinged with red or purple or sometimes almost white. From the stem clusters of tiny white, pink or yellowish flowers are produced. Gray to brown seeds are produced in abundance by the flowers in capsules and the seeds mature within a few weeks after bloom.

Dodder seeds overwinter in the soil in infected fields or get mixed with the seeds of the host plants. During the growing season, the seed germinates and produces a slender, yellowish shoot, but no roots. The leafless shoot tries to establish contact with a susceptible host. If no contact is made with a susceptible host plant, the stem falls to the ground, remains dormant for a few weeks and then dies. If contact is made, it entwines the host plant, sends haustoria into it and climbs up the plant. The haustoria penetrate the stem or leaf, reach the vascular tissues and absorb nutrients and water from the host. Soon after contact with a host is established, the base of the dodder shrivels and dries. Then the dodder loses all connections with the ground and becomes a complete parasite. The infected host plants become weakened by the parasite, their vigor declines and growth suppressed, and may finally die.

Dodder seeds, which fall to the ground remain viable for a long period and may be spread to nearby areas by animals, water, implements etc. and over long distances by contaminated seeds.

Dodder is effectively controlled by preventing its introduction into a field by use of dodder-free seeds, by cleaning implements thoroughly before moving them from infested fields to new areas free from dodder infection and by limiting the movement of cattle from infested to other fields. Application of cattle manure contaminated with seeds of dodder should be avoided. In dodder infested fields, growing of dodder-susceptible crops should be avoided for a few years **(Fig. 79)**

(e-g) *Cuscuta gronovi*.

Genus - *Dendrophthoe*. Genus - *Dendrophthoe* belongs to Family - Viscaceae. It is a semi stem parasite. It is generally referred to as **'loranthus'** or **'mistletoes'**. Because it attacks mango trees to a large extent it is also called **'mango parasite'**. Besides mango, it attacks guava, sapota, cashew, jack, lime, pomegranate and several other fruit trees, moringa, plantation trees, waste-land trees, road side shade trees etc. Infected twigs and branches develop swellings and cankers of various sizes on the infected areas and produce **'witches' brooms'**. The parasite produces wedge-shaped haustoria, which grow into the bark, cambium and xylem of the branches and at these locations swellings and cankers are formed. Infected trees may survive for many years, but they show reduced growth and vigor, and portions of the branch beyond loranthus infection often become deformed and may even die in course of time. The yield is also very much reduced

These mistletoes are parasitic evergreens that have well-developed leaves and stems upto 1 - 2 cm. in diameter. Sometimes the stems grow along the host branches and establish contact at several points on the branches and send in haustoria from each point of contact. The height of mistletoe plants varies from a few centimeters to a meter or more. They produce typical green leaves that can carry on photosynthesis and manufacture their own requirements of carbohydrates. However, they produce haustorial sinkers, rather than roots, which grow inside the branches and stems of hosts and absorb water and mineral nutrients.

The parasite produces small, elongated, tubular, grayish-white or reddish, dioecious flowers in bunches. The fruits arising from the flowers are berry-like, fleshy, slightly elongated, reddish in color, sweet and sticky, and with a single seed in each fruit. The seeds are spread by birds that eat the berries containing the seed and excrete the sticky seed on the surface of branches of trees on which they perch. Squirrels and other animals also spread the seeds. The sticky seeds that stick on to the beak of birds or the fur coating of animals also tend to disseminate the seeds. The seeds are disseminated over long distance mainly by birds. The seed that adheres to the branch of a susceptible host germinates, sends haustoria into the host and establishes itself in the host.

Adopting physical methods, such as pruning of infected branches or severely affected trees may control mistletoes. Spraying a mixture of water, diesel oil and liquid soap is found to be effective in controlling the parasite. In one liter of water, 300 - 400 ml. of diesel oil and 5.0 ml. of liquid soap are mixed thoroughly and then sprayed on the parasite **(Fig. 80)**

(e-g) *Dendrophthoe falcata.*

Genus - *Orobanche*. Genus - *Orobanche* belongs to the Family - Orobanchaceae. It is a holo root parasite. It attacks tobacco plants to a large extent. It is generally called as **'broomrapes'** of tobacco. Broomrapes occur worldwide and attack several hundred species of herbaceous dicotyledonous plants including eggplant, tomato, cabbage, cauliflower, radish, sunflower etc. Plants affected by broomrapes usually occur in small patches and the host plants may be stunted. The stunting depends upon the stage at which the host is attacked and the number of broomrapes attacking the host. The losses due to broomrape attack may range from 10 - 70 per cent of the crop. The pathogen is a whitish to yellowish annual plant and 15 - 50 cm. tall. It has a thick, fleshy stem, with dense, brown, thin scale-like leaves. It produces numerous, pretty, white, yellowish-white or slightly purple tubular, snapdragon-like flowers, which are longer than the scaly leaves. From the flowers, seedpods are formed, which are about 5.0 mm. long, each containing several hundreds of minute, black seeds.

Broomrapes overwinter as seeds, which may remain viable in the soil for more than 10 years. Seeds germinate only when roots of certain plants, not necessarily those of susceptible hosts grow near them. The root exudates initiate germination of the seeds by breaking their dormancy. On germination, the seed produces a radicle, which grows toward the root of the susceptible host plant, becomes attached to it and forms a shallow, cup-like appressorium that surrounds the root. From the appressorium, a mass of undifferentiated cells penetrate the host root, extend to and occasionally into the xylem, and absorb nutrients and water from it. Some of these cells differentiate into parasite xylem vessels and connect the host xylem with the vascular system of the parasite. Other undifferentiated cells become attached to the phloem cells and obtain nutrients from them. Soon the parasite develops a stem, which appears above ground. Meanwhile, the original root produces secondary roots that grow outward and get themselves attached to other host roots and infect them. From these points of contact new roots and stems of the parasites are produced. Thus a cluster of broomrape plants arise around the infected host plant. Several such broomrapes may grow simultaneously on the roots of the same host plant.

The best method to control broomrapes is to prevent introduction of its seeds into new areas. The parasites can be physically removed and destroyed before they produce seeds and contaminate the soil. Leaving the infected field fallow during summer months and repeated summer ploughings may help to expose the seeds to the vagaries of nature and destroy them. Crops not susceptible to infection by this parasite may be raised for a few years. Crops, such as chillies, black gram, green gram, gingelly, sorghum etc. are not attacked by this parasite. Root exudates of some crops, such as flax and chillies induce germination of seeds of broomrape, but they do not produce flowers. As soon as the broomrapes germinate, they may be physically removed and destroyed **(Fig. 81)**

(e-g) *Orobanche cernua; O. ramosa*

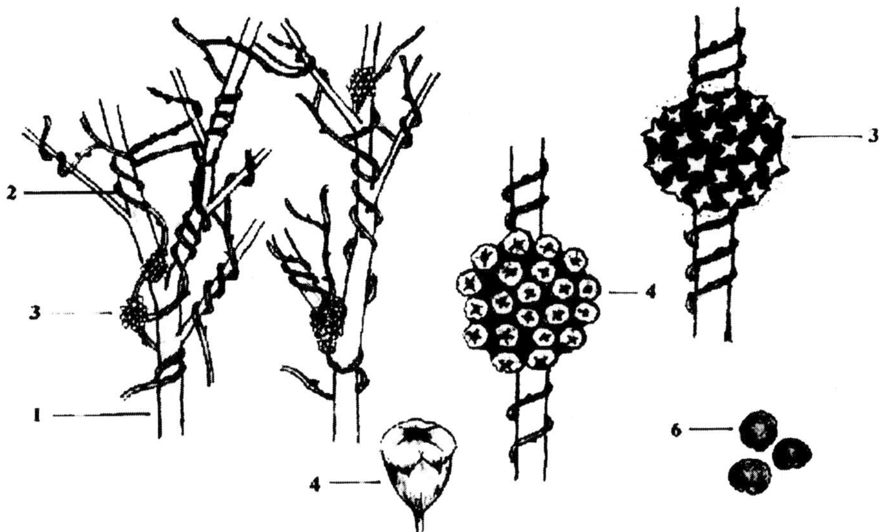

1. Host 2. Parasite 3. Flowers 4. Seed capsule 5. Berry 6. Seed

Fig. 79 : Stem Parasite
Genus - *Cuscuta*

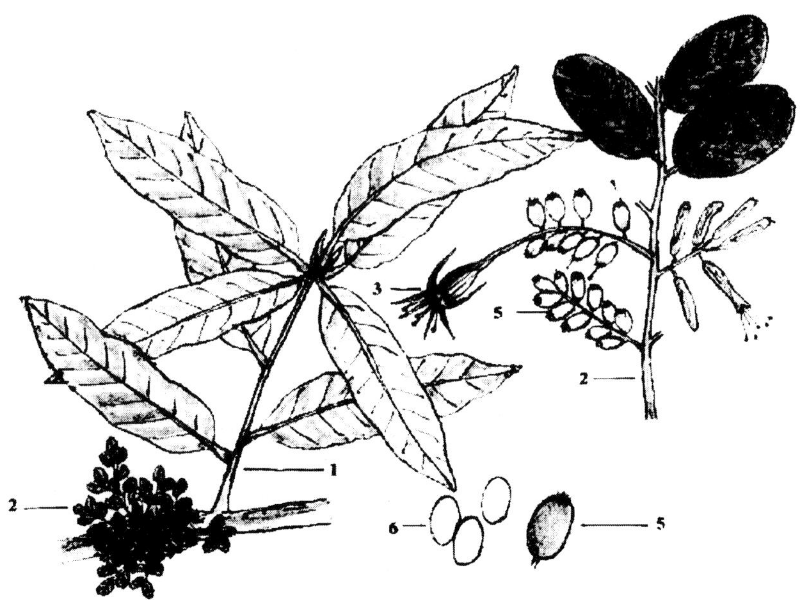

1. Host 2. Parasite 3. Flowers 4. Seed capsule 5. Berry 6. Seed

Fig. 80 : Stem Parasite
Genus - *Dendophthoe*

Genus - *Striga*. Genus - *Striga* belongs to the Family - Scrophulariaceae. It is a semi root parasite. It is commonly known as **'witchweed'** and it parasitizes important economic crops, such as corn, sugarcane, rice, cowpea, tobacco and some lesser millets. Losses due to this parasite may go up to 100 per cent.

Affected plants remain stunted, wilt, and turn yellow and may die, if they are heavily parasitized. Infected root bears many witchweed haustoria, which are attached to the root and feed on it. One to several witchweed plants may be found around the infected plants. Witchweed is a small, pretty plant. It has a bright green, slightly hairy stem and grows to a height of 15 - 30 cm. It produces many branches. The leaves are rather long and narrow in opposite pairs. The flowers are small, usually red, yellowish or white and always have yellow centers. Flowers appear just above the leaf attachment to the stem and are produced throughout the season. From the flowers capsules are formed, each containing more than a thousand tiny, brown seeds. The root of witchweed is white and has no root hairs. It obtains most of the nutrients, such as minerals, water and organic substances from the host plant through haustoria produced from the root. The witchweed plant is green and can manufacture some of its own requirements of food.

The parasite overwinters as seeds, which require a resting period of 15 - 18 months before germination. Seeds close to host roots germinate and grow towards the host roots, attracted by the exudates of their roots. Once the rootlet of the witchweed comes in contact with the host root, its tip swells into a bulb-shaped haustorium. The haustorium dissolves and penetrates the host root and reaches the vascular vessels of the host root, and absorbs water and nutrients. The parasite produces several roots, which send more haustoria into the host roots. The seeds of the parasite are spread by wind, by water, by contaminated seeds, by contaminated soil carried through agricultural implements and farm machinery, and cattle.

Witchweed is very difficult to control. Introduction of witchweed to areas, which are not infected earlier, should be avoided. Witchweed plants may be physically removed and destroyed before they produce seeds. This practice is to be followed 4 - 6 times at 2 months interval. Non-host plants, such as groundnut, cotton, soybean and other legume crops may be raised in infected fields. The root exudates from the roots of these plants induce the germination of witchweed seeds and in the absence of suitable hosts, the parasites may die of starvation. The weedicide, 2-4-D (Fernoxon) is capable of killing the parasite. Spraying a solution containing 450 gm. of the weedicide in 500 lit. of water per acre on the soil surface destroys the

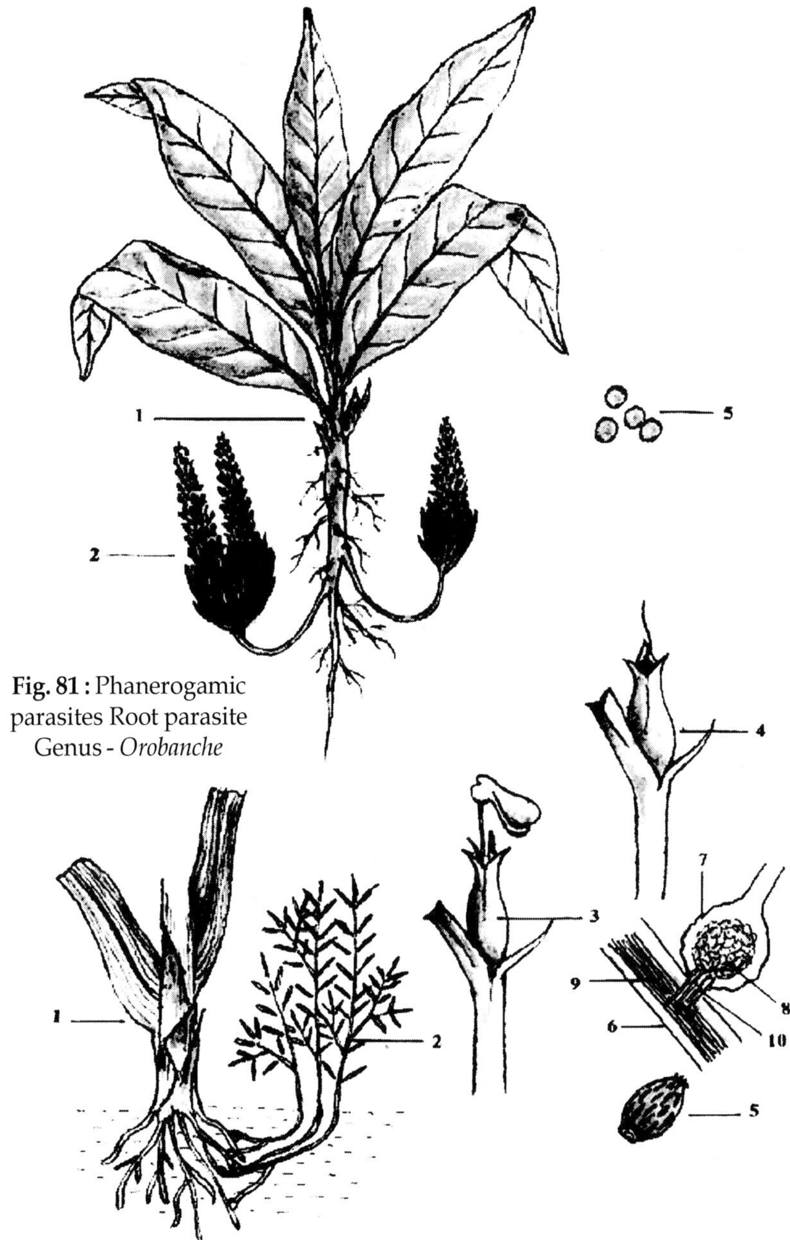

Fig. 81 : Phanerogamic parasites Root parasite
Genus - *Orobanche*

1.Host 2.Parasite 3.Flower 4.Seed capsule 5.Seed 6.Host root 7.Haustorium of the parasite 8.Nucleus 9.Host root vessels 10.Tracheids from haustorium

Fig. 82 : Root parasite
Genus - *Striga*

germinating parasite. However, this weedicide is very harmful to cotton, tobacco and a few other crops. Ethylene at the rate of 400 ml. per acre may be injected into the soil. This organic chemical fumigant stimulates the germination of the seeds of the parasite. The germinated plants may be physically removed and destroyed. Use of resistant varieties and seed treatment with selected herbicides are other methods of controlling the parasite **(Fig. 82)**

(e-g) *Striga densiflora, S. lutea, S. euphrasioides.*

□□□

Chapter 2

Principles of Plant Disease Management

'Plant disease management' involves the integration of various measures adopted to protect the plants by preventing them from contact with the pathogenic organisms, preventing the growth of the pathogens causing diseases, preventing sporulation of the pathogens and spread of diseases, destroying insect vectors responsible for transmitting diseases and such other prophylactic and curative measures. The ultimate aim of these measures is to avoid large-scale occurrence of diseases, reduce the severity of diseases and minimizing the losses caused by diseases. In disease management, importance should be given to control all the diseases attacking a particular crop at a particular time, especially those which occur in severe form in all the seasons and cause extensive damage to the crop and yield loss. Each and every crop is subjected to attack by one or more than one diseases. At the same time the control measures for controlling each of the disease may differ. In this respect a knowledge of the pathogen causing the disease, the symptoms of the disease, mode of infection and spread of the disease, the growth stage at which the crop is attacked, interaction between the host (crop) and the pathogen and weather conditions favorable for the occurrence and spread of the disease is absolutely necessary, so as to adopt suitable control measures to combat the disease and to minimize the damage to the crop, and reduce the yield loss.

The occurrence and spread of diseases are very much influenced by the environment, host and the pathogen. When the environmental conditions are suitable for the pathogen and not quite suitable for the host, the incidence of the disease is liable to be more. Further, if the host plants are not maintained properly and are weak, then they are more vulnerable to attack by the pathogen and the disease incidence may be severe. When the inoculum of the pathogen is present abundantly, the chances of attack by the pathogen are also more. Taking these facts into consideration, plant disease management is depicted as **'manipulation of the disease triangle' (Fig. 60)**

All control measures should be aimed at protecting the plants from becoming diseased and not at curing the plants after they have become diseased. The principal methods adopted for controlling plant diseases are broadly categorized under five different heads viz. :

1. Evasion
2. Exclusion
3. Eradication
4. Protection
5. Immunization

These methods involve the employment of cultural, physical, chemical, biological and genetical agents.

1. Evasion

'Evasion' or **'avoidance'** of a disease is possible by adopting certain measures, which enable the host to avoid contact with the pathogen or measures, which ensure that the susceptible stage of the host and favorable conditions for the pathogen do not synchronize. This can be achieved by measures, such as selection of area, selection of field or land, adjusting the time of sowing, selection of varieties, selection of seed and planting materials and modifying the cultural practices.

(i) **Selection of area.** Many fungal and bacterial diseases are more severe in wet areas than in dry areas. Crops, which are susceptible to these diseases, if grown in such areas are liable to attack by the pathogens. Smut disease of pearl millet caused by *Tolyposporium penicillariae* and ergot caused by *Claviceps microcephala* are more severe in wet areas and in areas where there is continuous rainfall for several days, especially during the flowering stage of the crop. Similarly in areas, where there is rainfall for long duration and plenty of dewfall in the early mornings during the active growth stages, cultivation of rice varieties, such as IR.50, TKM.9, J.13 etc., which are highly susceptible to rice blast caused by *Pyricularia oryzae* should be avoided.

(ii) **Selection of land or field.** Selection of land or field is also very important in avoiding certain diseases. Many soil-borne diseases can be effectively checked by this method. In the case of several soil-borne pathogens, if a contaminated field is selected for cultivating a susceptible crop, there is every chance of the disease occurring in a severe form, as there is already a build-up of inoculum in the soil. So, in a field where a soil-borne disease had occurred in a particular crop in the previous season or year, growing the same crop in that field in the ensuing season should be avoided. The sugarcane red rot pathogen, *Colletotrichum falcatum* can survive in the soil for a period of 2 months or even more in the absence of the crop. So, if sugarcane is cultivated in the same field in the succeeding season, the crop may be subjected to attack by the pathogen. Similarly many other soil-borne diseases, such as ergot, downy mildew and smut of pearl millet, *Fusarium* wilt of redgram, sheath blight of rice, *Fusarium* wilt of cotton, bacterial wilt and late blight of potato, *Fusarium* wilt and bacterial wilt of banana can be prevented, if the same crop or other susceptible crops are not cultivated in the same field year after year. If the susceptible host is not cultivated in the field continuously for a few seasons or years, the pathogen may die of starvation due to non-availability of a suitable host.

Drainage conditions of a field also play an important role in evasion of diseases. Low-lying and water-logged fields predispose the crops to attack by many pathogens. Sugarcane red rot, downy mildew of pearl millet etc. occur more severely in such fields. So, whenever such crops are grown in these conditions, proper drainage facilities should be provided. Selection of land is very important in establishing orchards. Apple trees should not be planted in fields previously occupied by oak trees. If apple trees are planted in such fields, the trees may be affected by collar rot caused by *Rosellinia necatrix*, which attacks oak commonly. Land with good soil texture and enough soil depth should be selected for setting up orchards.

Otherwise, the root growth and efficiency of the roots will be affected, as a result the plants may develop deficiency diseases within a few years.

(iii) **Selection of sowing time.** Susceptible varieties or plants are usually attacked by pathogens only when environmental conditions are favorable for the pathogens and when the plants are at the susceptible growth stage. If the coincidence of the susceptible growth stage of the crop and the environmental conditions favorable for the pathogen is avoided by manipulating the time of sowing slightly, the incidence of the disease can be reduced to a considerable extent. Pulse crops, such as grams, peas etc., when sown immediately after the rains are severely attacked by the root rot pathogen, *Rhizoctonia* species. This is because of the high soil moisture and high soil temperature immediately after the rains, which is conducive for the development of the disease. A slight delay in the sowing time is helpful in evading the disease. Continuous rainfall during the flowering stage of pearl millet predisposes the plants to attack by ergot and smut diseases. So, the sowing time may be slightly modified, so that the flowering time does not coincide with the monsoon rains. Late sown wheat crop is attacked by the black rust fungus, *Puccinia graminis tritici* to a larger extent, because of the presence of plenty of uredospores in the air.

(iv) **Selection of disease-escaping varieties.** Some varieties of crops, although susceptible to a particular disease, may escape the disease attack due to certain inherent growth characters. These varieties may not have genetic resistance, but may have certain genetically controlled growth characters, such as fast growth, early maturity etc., by which they may escaped from severe attack. Early maturing varieties of pea may escape from powdery mildew and rust diseases. Pod formation is almost complete before the onslaught of the disease and so, the yield loss is considerably reduced. The damage due to black rust of wheat is also very much reduced in early maturing wheat varieties.

(v) **Selection of seeds and planting materials.** Several diseases are spread through infected seeds and planting materials, such as seedlings, suckers, grafts, cuttings, tubers, rhizomes, bulbs, corms etc. As such, selection of disease-free seeds and planting materials play an important part in evading a number of diseases. Brown leaf spot and bacterial leaf blight of rice, bacterial blight of cotton, red rot of sugarcane, *Fusarium* wilt and bacterial wilt of banana, brown rot of potato, rhizome rot of ginger, foot rot and leaf rot of pan (betelvine) and many virus diseases of various crops are mainly carried through seeds and planting materials. Use of seeds and planting materials from disease-free crops obtained from disease-free areas or from other countries where the disease is not prevalent, use of certified seeds, raising disease-free nurseries under controlled conditions for getting planting materials etc. may help to evade several diseases.

(vi) **Modification of cultural practices.** Spacing between plants, amount of irrigation, fertilizer and manure application, time and method of sowing or planting, cropping pattern, ratooning, depth of sowing, mixed cropping and such other cultural practices have much influence on the occurrence, spread and severity of several diseases. Making appropriate changes in these practices help to reduce disease incidence and yield loss.

Thick sowing may lead to damping off of seedlings of tomato, eggplant, tobacco and chillies. Close planting of rice results in severe incidence of blast and bacterial leaf blight

Irrigation with too much of water results in luxuriant growth of the foliage, which makes the plants more susceptible to foliar diseases, especially leaf spots and leaf blights. *Pythium* and *Phytophthora* species thrive well at high soil moisture levels. Cereal rusts occur more severely in heavily irrigated crops and in very wet soils. On the contrary, inadequate irrigation results in poor growth and weaker plants, which also predispose the plants to attack by many pathogens. Sprinkler irrigation spreads bacteria and fungal spores in the same way as raindropsplashes. Wherever there is heavy sporulation of fungal pathogens or exudation of bacterial masses on the leaf surface, sprinkler irrigation should be restricted.

Application of adequate quantities of fertilizers is another important factor that influences the occurrence and spread of diseases. Application of excessive amount of nitrogenous fertilizer encourages the growth of the foliage and predisposes the plants to foliar diseases. If nutrients are not supplied in balanced proportions, that may lead to deficiency or toxicity diseases and other diseases caused by pathogens.

Depth of sowing also has a bearing on the incidence of diseases. Deeper sowing results in delayed germination and weaker seedlings. This may predispose the seedlings to attack by root rot and collar rot pathogens. Deep sowing predisposes groundnut seedlings to attack by *Aspergillus niger* and *A. pulverulentus* and cause seed rot or collar rot. The incidence of bunt and smut diseases of wheat is increased, when the seeds are sown deeper.

Deep summer ploughing helps to bury the diseased plant debris deeper into the soil and thus reduces the inoculum potential in the soil. Ploughing before sowing helps to control root rot disease of groundnut caused by *Sclerotium rolfsii*.

Flooding and impounding water in the field for a few days, when there is no crop helps to kill *Fusarium* spores present in the soil. *Fusarium* wilt of banana can be controlled by this method. Mixed cropping is found to control root rot of cotton caused by *Rhizoctonia bataticola* to a large extent. Ratooning is found to increase the severity of many diseases in a number of crops. Ratooning of sugarcane increases the severity of smut and red rot. Similarly, ratooning of cotton increases the incidence of bacterial blight

Growing a susceptible crop in places or regions where a particular pathogen or disease is not already present, growing a crop in places or regions where environmental conditions are unfavorable for a particular pathogen or development of a disease are also measures that can be adopted under evasion. Diseases caused by several species of *Fusarium, Rhizoctonia, Sclerotium* etc. can be avoided by these methods.

2. Exclusion

The methods under **'exclusion'** aim at exclusion of the pathogen from the host, so that contact between the pathogen and the host is prevented. Seed treatment, seed certification and quarantine measures are some of the methods followed under exclusion.

(i) **Seed treatment**. Heat treatment or fumigation with certain poisonous gases or chemical treatment of seeds and seed materials prior to sowing or planting helps to eliminate externally or internally seed-borne pathogens that may be present in such materials. By this method, a pathogen or disease may be excluded from being introduced from one place or region to other places.

(ii) **Crop inspection and certification**. Crops grown exclusively for seed purposes are regularly inspected for the presence of diseases that are transmitted through seeds or planting materials, such as cuttings, grafts etc. and diseased plants are removed and destroyed as soon as they are detected. The seeds or planting materials obtained from such disease-free crops may be distributed as certified disease-free seeds. Advocating this practice helps to prevent the spread of diseases from one region to another or from one state to another or from one country to another.

(iii) **Notification of diseases.** In case of certain diseases having a limited distribution within the country, their occurrence should be notified to the concerned authorities, so that suitable control measures are taken to prevent their further spread. In this respect, disease surveillance of crops plays an important role. Such notification enables other farmers to take up precautionary measures to control the disease. The incidence of rice blast, rice 'tungro' disease and such epidemic diseases should be notified.

(iv) **Quarantine regulations.** Quarantine regulations are defined as legal restrictions on the movement of pest infested and disease infected agricultural commodities from one place to other places, where the pests or diseases are not prevalent. Diseases that can spread through the agency of seeds and other planting materials alone can be restricted by quarantine regulations, while diseases that are spread by natural forces, such as air, water or insects cannot be restricted by quarantine regulations. Quarantine laws have been enacted in almost all the countries for preventing the introduction of new pests and diseases. In some countries, internal quarantine laws have been enacted to prevent the spread of certain specific pests and diseases by destroying the affected plants or eradicating the alternate hosts. Such laws were enacted in France in 1660 and in Denmark in 1903 for the rapid eradication of barberry plants, which is the alternate host of wheat black rust. Similar laws were enacted in some of the African countries for the destruction of cacao plants affected by 'cacao swollen shoot disease'.

Seeds and other propagating materials are imported from other countries for various research works or for introduction of new crops or high yielding crop varieties for increasing the production. Sometimes grains and other commodities are imported for food or for seed. Through these materials some plant pathogens may also be imported inadvertently. Such newly introduced pathogens may sometimes get acclimatized to the new environments and may become serious pathogens in due course. Even when quarantine laws are in force, several new diseases have been introduced into our country. In India, **'The destructive insect and pest act'** was enacted in 1914 and since then several amendments have been made to this act from time to time. A list of diseases introduced into India from other countries is given in **Table 3**.

Table 3. Diseases introduced into India from other countries

Diseases	Year of introduction	Introduced from
Coffee rust *(Hemileia vastatrix)*	1879	Sri Lanka
Potato late blight *(Phytophthora infestans)*	1883	Europe
Wheat flag smut *(Urocystis tritici)*	1906	Australia
Grapes downy mildew *(Plasmopara viticola)*	1910	Europe
Cucurbit powdery mildew *(Erysiphe cichoracearum)*	1910	Sri Lanka
Maize downy mildew *(Sclerospora phillipinensis)*	1912	Java
Rice blast *(Pyricularia oryzae)*	1918	Asia
Black rot of crucifers (Xanthomonas campestris)	1929	Java
Rubber powdery mildew *(Oidium heveae)*	1938	Malaya
Tobacco black shank *(Phytophthora nicotianae)*	1938	Holland
Banana bunchy top virus disease	1940	Sri Lanka
Crown gall of apple & pear *(Agrobacterium tumefaciens)*	1943	England
Potato wilt *(Synchytrium endobioticum)*	1953	Netherlands
Onion smut *(Urocystis cepulae)*	1958	Europe
Bacterial blight of rice (*Xanthomonas campestris* pv. *oryzae*)	1959	Philippines
Potato golden nematode *(Heterodera rostochiensis)*	1961	Europe

In India, there are 16 quarantine stations operating under the **'Directorate of Plant Protection and Quarantine'**, eight at seaports, six at airports and two on land frontiers. Seeds, plants, propagative materials etc. brought into India from other countries are thoroughly examined at these stations for the presence of any pest or disease. If these materials are found to be infected or suspected to be infected, they are destroyed immediately. If they are not found to be infected, they are kept under observation at these stations for a few days and then treated with appropriate plant protection chemicals. Only after these formalities are over **'Phytosanitary certificate'** is issued and after that they are allowed to be taken out to their destination. All seeds or propagating materials imported from other countries should be accompanied with a Phytosanitary certificate from the concerned authorities of the country from which they are sent. Without Phytosanitary certificate, no seed or seed materials are allowed into the country. As per this act, import of cacao from Africa, rubber plants or seeds from America and sugarcane from Australia is totally banned.

Besides such intercontinental quarantine regulations, there are domestic quarantine laws, which prohibit movement of certain agricultural commodities within the country. In India, **'The Domestic Quarantine Act'** was enacted in 1919. As per this act, movement of some plants or their produce from one part of the country to other parts of the country is prohibited, so as to prevent the introduction and spread of certain diseases. Movement of banana suckers from the states of Kerala and Orissa to other states is prohibited, so as to prevent the spread of 'banana bunchy top virus disease'. Transport of potato tubers from West Bengal to other states is prohibited to prevent the spread of 'Wart disease of potato'. Movement of coconut seedlings from Tamil Nadu and Kerala is prohibited to prevent the spread of 'Coconut root wilt'.

Besides seeds and seed materials, diseases may spread from one country to another through soil, logs of wood, packing materials, such as wooden boxes, crates, bags etc.

(v) **Exclusion of insects that spread diseases.** Several virus and mycoplasma diseases are transmitted by various insect and non-insect vectors, such as leafhoppers, aphids, white flies, thrips, mites, nematodes etc. Similarly many bacterial and fungal diseases are spread to some extent by some insects mechanically. Controlling such disease-spreading insect and non-insect pests by adopting suitable control measures helps in excluding the inoculum and thereby prevents spread of the diseases.

3. Eradication

The methods adopted under **'eradication'** aim at attacking and killing the pathogens directly or creating conditions unfavorable for their existence. Very often, the methods adopted under evasion and exclusion may not be quite sufficient to ward off certain pathogens or diseases. Under such circumstances, eradication methods are helpful in controlling the diseases or preventing further spread of the diseases. For successful implementation of these measures, knowledge about the life cycle of the pathogens, their mode of survival & perpetuation, their host range, growth habitats etc. is essential. The methods involve cultural, chemical and other methods, such as rouging, sanitation, crop rotation, eradication of alternate and collateral hosts, heat and chemical treatment of seeds, propagating materials and plants, soil treatment, biological control etc.

(i) **Rouging.** This practice involves the removal and destruction of diseased plants or plant parts from the fields at the early stages of infection. By removing such foci of infection, the pathogen present in the plant or plant parts is destroyed and at the same time by preventing inoculum production, the spread of the disease can also be checked. In orchards, it may not be possible to destroy the diseased trees completely and in such cases, disease affected parts may be removed by plant surgery and destroyed. Rouging is followed to control several diseases, such as loose smut of wheat, loose smut and covered smut of barley, sorghum and maize, smut of pearl millet and sugarcane, red rot of sugarcane, green ear and ergot of pearl millet, *Fusarium* wilt of cotton, red gram and banana, many virus diseases, such as yellow vein mosaic of lady's finger, marble mosaic disease of cardamom and phanerogamic parasites, such as loranthus attacking mango, guava, sapota, cashew etc.

(ii) **Eradication of alternate and collateral hosts.** Some plant pathogens require two widely different host species belonging to different botanical Families to complete their life cycle. By destroying such an alternate host, which is usually a perennial wild plant, the pathogen is prevented from completing its life cycle and thus the spread of the disease can be checked. Eradication of barberry plants, the alternate host of black rust of wheat from the hilly terrains helps to control the disease.

Many plant pathogens have several alternate and collateral hosts on which they can survive and continue their life in the absence of the primary host. From these hosts, which are usually wild or weed hosts, spores are produced all through the year and can serve as primary inoculum to initiate infection in the primary host. By destroying such secondary hosts, the production of spores and spread of the diseases can be checked. The rice blast fungus, *Pyricularia oryzae* has many wild rice hosts and weed hosts, such as *Panicum repens, Digitaria marginata, Setaria intermedia, Brachiaria mutica* etc. The yellow vein mosaic of lady's finger survives on the wild host *Hibiscus tetraphyllus.* The pearl millet ergot fungus, *Claviceps ciliaris* has *Panicum antedotale* and *Cenchrus ciliaris* as alternate hosts. *Bean yellow mosaic virus* also has many alternate hosts.

(iii) **Crop rotation.** When a particular crop or related crops or crops that may be affected by a certain pathogen is cultivated continuously in a field for several years, that may result in the build-up of inoculum of the pathogen in the soil, especially in the case of soil-inhabiting pathogens. The gradual increase in the inoculum potential over the years makes the soil unfit for cultivation of such crops and the soil is said to suffer from **'soil sickness'**. In such sick soils, if disease resistant varieties or crops, which are not affected by the pathogen are grown for 2 - 3 years, the pathogen being unable to attack the newly introduced crops may die of starvation in the absence of proper nourishment. In this respect, crop rotation plays an important part in reducing the population of the pathogen in the soil. Crop rotation helps to control a number of soil-borne diseases, such as wilt of red gram, linseed, pea, gram, sugarcane and cotton, root rot of vegetables, red rot of sugarcane, ergot and smut of pearl millet, bunt of rice and wheat etc. 'Tikka' disease of groundnut is reduced to a large extent by crop rotation with maize or soybean. Rotation of wheat with leguminous crops reduces the incidence of 'Take all' disease of wheat.

Further, each crop releases certain root exudates that contain different biochemical substances. These biochemical substances are capable of destroying some pathogens, either by direct action or by encouraging the growth and multiplication of some antagonistic soil microorganisms that may kill the pathogen. So, crop rotation with diversified crops is considered to be a very important method for the control of many soil-borne diseases. However, crop rotation is more effective in eliminating soil-borne pathogens, such as *Fusarium* and *Verticillium*, and a few bacterial pathogens, which can survive in the soil only for a short period in the absence of a regular host, while pathogens, which can survive in the soil for a long time in the absence of a regular host cannot be effectively controlled by this method.

Crop rotation is important in some other respects also. When a particular crop is grown in the same area continuously for several years, depletion of certain essential nutrients may occur. This may result in the development of deficiency diseases and weaker plants, which may become vulnerable to attack by other pathogens. Monocropping may also result in the accumulation of certain toxic substances and organic acids, which may spoil the soil quality. So, it is advisable that botanically different crops with different nutritional requirements having resistance to the pathogen should be included in the cropping sequence.

(iv) **Sanitation**. Like plant sanitation, field sanitation is also quite important for the control of many diseases caused by soil-borne microorganisms. Several plant pathogens survive in the diseased plant parts for a long time, while many others exist as dormant structures in the soil. Collection and destruction of diseased leaves, flowers, fruits and twigs fallen on to the ground by burning is very effective in controlling many diseases. Similarly the stubble left over in the field after the harvest of the crop should also be removed and destroyed by burning, as they are known to harbor many plant pathogens. Deep ploughing and burying the crop debris deep into the soil also helps in destroying many plant pathogens. Summer ploughing and bringing the spores to the soil surface and exposing them to direct sunlight is also effective in destroying the inoculum. Many of the infectious diseases, such as red rot of sugarcane, downy mildew of pearl millet and sorghum, *Verticillium* wilt of cotton, powdery mildew of wheat, barley and peas, foot rot and leaf rot of pan (betelvine), late bight of potato, *Alternaria* blight of potato and eggplant, bacterial blight of rice and cotton, sheath blight of rice, rice 'tungro' etc. can be controlled to a large extent by this method. Disinfecting the knives used for cutting setts of sugarcane, tapioca, banana suckers etc. by dipping them frequently in antiseptic solution helps in preventing the spread of some diseases, especially virus and bacterial diseases. Keeping the land fallow for a few seasons also helps in eradicating some pathogens by starving them.

(v) **Heat or chemical treatment.** The pathogens present in the plants or in different plant parts can be either killed or inactivated by heat or chemical treatment. Hot water treatment of seeds is used successfully for the control of some fungal and bacterial diseases. Treating the seeds in hot water at 54°C for 10 minutes controls loose smut of wheat *(Ustilago tritici)*; at 52°C for 15 minutes controls loose smut of barley *(Ustilago nuda)*, covered smut of barley *(Ustilago hordei)* and leaf stripe of barley *(Helminthosporium gramineum)*; at 52° - 54°C for 15 - 30 minutes controls halo blight of runner beans *(Pseudomonas phaseolicola)*; at 52°C for 15 - 30 minutes controls *Alternaria* blight *(Alternaria brassicae)* and black rot of cabbage *(Xanthomonas campestris)*. Hot water treatment is also effective in controlling ratoon stunting disease of sugarcane caused by *Clavibacter xyli*.

Hot air treatment, moist hot air treatment and aerated steam treatment are also advocated for the control of many diseases. Heat treatment or **'thermotherapy'** of sugarcane seed cuttings (setts) is extensively practiced for the control of many fungal, bacterial and virus diseases of sugarcane, such as smut, red rot, ratoon stunting, mosaic and other diseases that are carried through setts. Hot air treatment of sugarcane setts is administered at 54°C for 8 hours, hot water treatment at 50°C for 2 hours, moist hot air treatment at 54°C for 4 hours and aerated steam treatment at 52°C for 1 hour.

Treating the seeds with organomercurials, such as agrosan and ceresan, besides other fungicides, such as captan, thiram, sulfur dust and dithiocarbamates is a common practice adopted for eliminating externally seed-borne pathogens of several crops. Because of high levels of toxicity, the use of organomercurials has been banned. Now systemic fungicides are available, which can eliminate the inoculum present on the surface of the seeds, as well as deep-seated fungal or bacterial infection. Some antibiotics are also used successfully for seed treatment to control certain diseases. Treating rice seeds with carbendazim or tricyclazole or pyroquilon controls rice blast *(Pyricularia oryzae)*. Seed dressing with carboxin and oxycarboxin controls loose smut *(Ustilago tritici)* and black rust *(Puccinia graminis tritici)* of wheat respectively. Thiram seed soak controls foot rot and blight of pea *(Ascochyta pinodes* and *A. pisi)*.

(vi) **Soil treatment**. Soil treatment is found to destroy or inactivate several soil-borne pathogens. Soil treatment involves the use of chemicals, heat energy, flooding, fallowing, soil amendments etc. Heat and chemical treatment of soil have been widely used successfully for the eradication of plant pathogens in green-houses and nursery beds. Some of the chemicals used for soil treatment can eliminate many soil-borne pathogens, while some others can eliminate only certain pathogens. Pentachloronitrobenzene (PCNB) is found to be effective in eliminating several sclerotia-producing soil-pathogens. Soil treatment with carbon disulfide can eliminate *Armillaria mellea* causing root rot of *Citrus* and many other fruit trees. Chemicals, such as formalin can control most of the soil pathogens, but they are mostly phytotoxic. So, these chemicals should be applied to the soil a few weeks prior to sowing and only after the residual toxicity is removed from the soil completely by frequently stirring the soil, seeds should be sown.

Fumigants, such as dichloropropane-dichloropropene (DD mixture), methyl bromide, chloropicrin, vapam, nemagon etc. are mostly used to control soil-inhabiting nematodes and some soil-borne pathogens. Some of the granular chemicals, such as carbofuran, thimet etc. are also used to destroy soil-inhabiting nematodes. Bordeaux mixture or copper oxychloride is widely used in the nurseries for the control of damping off of seedlings. Captan, thiram, wet ceresan and pentachloronitrobenzene are also used for soil drenching to control many soil-borne diseases. Soil drenching is very costly because large quantity of fungicidal fluid is required to completely wet the soil around the root zone of plants. So, soil drenching is mostly done in the nursery beds and in green-houses. In the field, drenching is done only around the infected plant and the plants immediately surrounding the infected plant to reduce the cost of the operation. Burning of trash in the fields is also helpful in eradicating most of the soil pathogens.

Flooding and impounding water in the field continuously for a few days, when there is no standing crop help to eradicate many plant pathogens in the field. Under flooded conditions, the anaerobic bacteria present in the soil produce certain toxic metabolites and carbon dioxide, which can kill many soil pathogens including nematodes. This method is advocated for the control of some of the *Fusarium* species.

(vii) **Biological control**. Biological control of plant pathogens has received much attention in the recent past. Several fungal and bacterial mycoparasites capable of eliminating or suppressing certain pathogens have been identified and some of them are being used in field scale. *Trichoderma viride, T. harzianum, T. lignorum, Gliocladium virens, Pseudomonas fluorescens, Bacillus subtilis, Agrobacterium radiobacter* etc. have been tried with some amount of success for the control of a few diseases.

4. Protection

Many diseases, which can spread very rapidly, are carried to distant places through the agencies of wind, water, insects etc. Under such circumstances, the measures adopted under evasion, exclusion and eradication may not be quite sufficient in preventing the occurrence and spread of such diseases. So, the plants have to be protected from infection by pathogens. The measures taken under protection are aimed at destroying or inhibiting the pathogen, creating a toxic barrier between the host and the pathogen by using plant protection chemicals **(chemotherapy),** making slight changes in the environmental conditions, which are unfavorable for the development of the pathogen or any other such protective means.

(i) **By use of Fungicides.** Chemical formulations that kill or inhibit the growth and development of pathogenic fungi are called **'fungicides'**. Fungicides used for the control of plant diseases are classified as **'protectants'**, **'eradicants'** and **'therapeutants'**. Protectant fungicides are mostly prophylactic in their action. These fungicides when applied as sprays, dusts or seed dressers on the surface of plants, seeds or soils, do not penetrate the plant tissues essentially, but remain on the surface as a protective toxic covering, thus preventing infection by the pathogens. Eradicant fungicides can enter into the host tissues to a little extent and can eradicate the dormant structures of the pathogen, as well as the active pathogen from the host. They can remain effective on and inside the host for sometime and function as protectants also. Therapeutant fungicides are capable of entering into the plant tissues when applied through seeds, foliage or roots and can inhibit the development of the disease. They can be applied even after the occurrence of the disease. Chemotherapeutants are usually systemic in nature and can affect deep-seated infection. Systemic fungicides are mostly specific and control only specific diseases. Antibiotics are also systemic in action. **'Therapy'** includes solar heat energy, hot air and hot water treatments, besides chemical treatment.

(ii) **Eradication of vectors.** Many diseases caused by viruses, mollicutes and fastidious bacteria are transmitted by several insect and non-insect vectors, such as leafhoppers, aphids, white flies, thrips, mites, nematodes etc. actively in a non-persistent, semi-persistent or persistent manner. Spores of many fungal pathogens and bacteria are dispersed by insects, such as ants, flies, honeybees, beetles, wasps etc. mechanically. Controlling the vector population helps to reduce the disease incidence to a large extent. However, when the vectors are present in large numbers or when the vectors have already spread the diseases, adopting control measures against the vectors may not be effective in controlling the diseases. Even if a large number of the vectors are killed, they may transmit the pathogen before their death. Further, if a few vectors escape chemical treatment, they may transmit the pathogen to many plants. Insecticides having quick, knockdown effect should be used to control the vectors and control measures should be taken up in the early stages of pest infestation, so as to avoid build-up of vector population.

(iii) **Treating diseased plants and plant produce**. Although it is very difficult and sometimes rather impossible to cure diseased plants, some diseases can be cured by the application of certain systemic fungicides and antibiotics. These chemicals are absorbed into the plant tissues and are systemically translocated to all the parts of the plants. Several fungicides, such as benomyl, carbendazim, thiabendazole, carboxin, tridemorph, fenarimol, metalaxyl etc. are systemic, eradicant fungicides. These fungicides, when absorbed by the foliage or roots get mixed up with the protoplasm and eradicate deep-seated infection, at the same time induce temporary resistance to the plants and protect the plants from further infection. Some of these fungicides are capable of converting toxic substances produced by the pathogens into substances non-toxic to the plants. Besides curing the infection, these chemicals prevent further infection of healthy tissues. Most of these fungicides after absorption by the roots are carried through the xylem vessels acropetally to the top and remain in the leaf margins till they are disintegrated. When applied to the foliage, these fungicides are not carried through the phloem vessels to the root region, but remain in the leaf edges. Only a very few systemic fungicides are translocated basipetally to the root region. On the contrary antibiotics are translocated both upwards and downwards.

In case of fruit trees and other useful trees, cutting off and destruction of severely diseased branches (tree surgery), scraping off of diseased portions on the stem and branches, cleaning them thoroughly with antiseptic solutions, such as mercuric chloride - 0.1 % or sodium hypochlorite solution - 0.5 - 1.0 % or ethyl alcohol - 70 % and applying fungicidal paste on the injured portions protect the trees from further spread of the disease. In commercial orchards, vineyards, tea gardens etc., wounds are created during routine pruning operations and such wounds cannot be treated individually. Under such circumstances the wounds are protected from fresh infection by spraying with suitable fungicides.

(iv) **Thermotherapy**. Heat energy is also used for protecting plant products, such as grains, fruits and vegetables from infection by fungi and bacteria. All grains, legumes, nuts etc. carry with them many fungi and bacteria that can cause decay and spoilage in the presence of sufficient moisture. Drying the grains to bring down the moisture content to about 12 % affords protection from infection by fungi and bacteria. Many fruits can also be stored dry for a long time and kept free from infection by fungi and bacteria. Grapes, plums, dates and figs can be dried in the sun or by hot air and then stored. Even slices of fleshy fruits, such as apples, peaches, apricots etc. can be protected from infection and decay by microorganisms, if they are sufficiently dried.

Refrigeration also helps in protecting many agricultural commodities, such as succulent fruits and vegetables from infection and spoilage by microorganisms.

(v) **Supplying nutrients**. Supplying nutrients required by the plants in adequate quantities encourages the growth vigor of the plants. Such vigorously growing plants develop some amount of resistance to several pathogens. Diseased plants may also recover from the diseases to some extent, when adequate macro- and micronutrients are supplied in sufficient quantities. Application of excess of nitrogen over and above the requirement of plants, encourages the vegetative growth to a larger extent and the cells of the leaf tissues become larger in size and thin-walled. As a result of these, the plants become more prone to attack by several foliar diseases. When nitrogen is not supplied in adequate quantities, the growth of the plants is adversely affected and they become weaker. In such cases, even if the occurrence of foliar diseases is lesser, the plants become more susceptible to many other diseases. Deficiency of potash leads to accumulation of more water in the cells of the tissues, which predisposes the plants to many diseases. Supply of adequate quantities of lime results in hardening of cell walls by pectin, which resists infection by the wilt pathogens. Similarly, supply of sufficient quantities of micronutrients, such as boron, zinc, manganese, iron, magnesium etc. also helps in reducing the severity of several diseases and at the same time protects the plants from infection by pathogens.

5. Immunization

'Immunity' is an absolute quality in which the pathogen is exempted from infecting the host plant and establish parasitic relationship with the host. Of all the methods adopted for the control of plant diseases, immunization is the easiest, economical and most practical method. Resistant varieties not only resist the occurrence of diseases, but growing such varieties results in the saving of time, labor and money. Control of virus diseases, wilt diseases, rust diseases etc. is usually very difficult and often impracticable through evasion, exclusion, eradication and protection. Under such circumstances, development of disease-resistant varieties is considered to be the most effective method of disease control. However, it is almost impossible to develop varieties resistant to all the diseases. A variety, which may be resistant to a particular disease, may be highly susceptible to one or more other diseases. So, more importance is given to develop varieties resistant to the most important and destructive diseases. Development of varieties resistant to diseases is possible by selection, hybridization and mutation. These types of resistance is governed by the genes and hence the resistance may last for a long time, till the resistance is broken by the development of new races of the pathogen in due course of time. Resistance governed by genes is called **'true resistance'**. True resistance may be inherited by the presence of a single resistant gene, **'monogenic resistance'** or a few resistant genes, **'oligogenic resistance'** or many resistant genes, **'polygenic resistance'.** In true resistance, the host and the pathogen are in a state of incompatibility with one another, either due to production of some chemical substances toxic to the pathogen or due to development of various defense mechanisms by the host.

Resistance induced by genes develops certain physiological, morphological or functional changes in the host, which resist infection and development of diseases. In the case of resistance induced by physiological changes in the host, the host produces certain substances, such as **'phytoalexins'**, **'simple phenolic compounds'** etc., which inhibit infection by the pathogen or else the host is made to react in such a way that the food required is not made available to the pathogen. Resistance induced by morphological or structural changes in the host, such as formation of cork layers and abscission layers, formation of tyloses, deposition of gums or waxy coating on the host surface, development of hairiness, thickening of cuticle and epidermis etc., which function as barriers, prevent infection by the pathogens. In the case of resistance induced by functional changes in the host, the host may become sensitive to infection and may prevent entry of the germ tubes produced by the spores of the pathogen by closing down of the stomata.

Certain biochemical agents induce another type of resistance, which is temporary in nature. Genes do not control this type of resistance and so, the resistance is short-lived and is called **'acquired resistance'**. Some chemical substances, nutrients, systemic fungicides and antibiotics may cause such type of resistance. These substances when applied to the host are absorbed by the host tissues and exert resistance to invasion by the pathogens.

There are two kinds of true resistance, **'horizontal resistance'** and **'vertical resistance'**. When a crop variety is uniformly resistant to all the races of a particular pathogen, that type of resistance is called **'generalized'**, **'non-specific'**, **'field'**, **'lateral'**, **'quantitative'** or **'durable resistance'**, more commonly referred to as **'horizontal resistance'**. Horizontal resistance is controlled by many genes and hence is polygenic or multigenic in nature. Each of these genes individually may not be effective against the pathogen, but may play only a minor role in inducing resistance. But the many genes involved together exert their influence on the physiological functioning of the host plant in triggering the various defense mechanisms. Horizontal resistance towards a particular pathogen may vary from one variety to another variety of the same host plant. While one variety may show horizontal resistance to all the races of a particular pathogen, the level of horizontal resistance may not be the same with other varieties. Further, horizontal resistance may vary to a larger extent under different environmental conditions. Generally horizontal resistance provides incomplete protection, in other words, it does not protect the plants completely from getting infected, but the protection induced is more durable and more or less permanent. Because multiple genes control horizontal resistance, the pathogen has to undergo many mutations to completely break down the resistance of the host. Horizontal resistance tends to prevent epidemics in the field. Although horizontal resistance is quantitatively inherited by multiple genes, some of the genes contributing to horizontal resistance may be identical qualitatively to the genes responsible for vertical resistance also.

Many plant varieties are quite resistant to some races of a pathogen, but are susceptible to other races of the same pathogen. So, in such resistance, there is a definite differentiation between races of the pathogen. The resistance is effective only against specific races of the pathogen and ineffective against the other races. This type of resistance is called **'strong'**, **'major'**, **'specific'**, **'qualitative'** or **'differential resistance'**, but is more commonly referred to as **'vertical resistance'**. Vertical resistance is always controlled by one or a few resistant genes (monogenic or oligogenic resistance). These genes referred to as resistant genes **(R genes)** are responsible for the recognition of the pathogen by the host and thus play a major role in the expression of the disease.

Even a single resistant gene may induce complete resistance to a particular pathogen. However combination of more than one resistant gene in the same plant (R1 R2), (R1 R3), (R2 R3), (R1 R2 R3) etc., is desirable. This may result in inducing resistance to all the pathogen races, as each of the gene may provide resistance to different races of the pathogen. A plant species may have many resistant genes against a particular pathogen. But a variety of the plant species may have one or a few of these resistant genes. For instance, cotton has 20 - 40 genes for resistance against the bacterial blight pathogen *(Xanthomonas campestris* pv. *malvacearum)*. But each variety of cotton may not have that many resistant genes, but may have only one or a few of such genes. Each gene for resistance, for example the resistant gene 'R1' in the host plant induces resistance to all the races of the pathogen that contain the corresponding gene for **'avirulence'**, 'A1' and not the races of the pathogen, which do not contain the corresponding gene for avirulence, 'A1'. In the absence of the resistant gene 'R1', the variety becomes susceptible to all the races of the pathogen.

True resistance, whether horizontal or vertical is usually controlled by genes located in the chromosomes in the cell nucleus of the plant. However, in several plant diseases, resistance is controlled by genetic material contained in the cell cytoplasm and such resistance is referred to as **'cytoplasmic resistance'** (e-g) Corn leaf blight *(Helminthosporium maydis)* and corn yellow leaf blight *(Phyllosticta maydis)*.

Varieties with horizontal resistance, usually have complete resistance to all the races of a particular pathogen and provides complete protection from the disease, but a single or a few mutations in the pathogen may produce a new race that may infect the previously resistant variety and hence the protection may not be permanent. Generally vertical resistance is not affected much under different environmental conditions. Further, under certain environmental conditions or circumstances, some susceptible varieties of a crop may remain free from infection or symptoms and appear to be resistant. This is referred to as **'apparent resistance'**. Such apparent resistance may be due to **'disease escape'** and **'disease tolerance'**.

Disease escape. 'Disease escape' may occur whenever a genetically susceptible variety is not exposed to the factors responsible for causation of a disease, such as virulent pathogen and favorable environmental conditions simultaneously at the proper time and for sufficient duration. This results in the failure of interaction between the host and the pathogen at the appropriate time and thus the plants may escape the disease. A susceptible crop variety may escape from soil-borne pathogens when the seeds germinate and emerge faster or when the seedlings harden earlier and when the soil temperature and moisture conditions are not favorable for the pathogen to attack the plants. Sometimes, a susceptible plant variety may escape a disease because at the most susceptible growth stage of the crop, the pathogen may be absent, inactive or the environmental conditions may not be favorable for the pathogen. For example, young plants and tissues are affected much more severely by pathogens causing damping off, seedling blight, powdery mildews and most of the pathogenic bacteria and viruses. However, species of *Alternaria, Botrytis* and certain other pathogens attack mostly parts of grown-up plants. Location of the fields, spacing between plants, mixed cropping etc. may also play an important role in disease escape. Wound pathogens may not get easy access to the host tissues in the absence of wounds caused by strong winds, lashing rains, dust storms, insects etc. and may escape diseases.

Environmental factors also play a major role in plant disease escape. Many plants escape diseases caused by species of *Pythium* and *Phytophthora,* if the temperature is high and soil moisture is low, while low temperatures and high soil moisture make the plants escape from diseases caused by species of *Fusarium* and *Rhizoctonia.* Lack of moisture in the soil is the most common cause of disease escape in plants. Plants grown in most of the dry areas or during years of low rainfall usually remain free from late blight, downy mildews, scab and anthracnose, because these diseases require a film of water on the plant surface or high relative humidity during every stage of their life cycle. Similarly, in dry soils, club root of crucifers caused by *Plasmodiophora brassicae* and damping off caused by species of *Pythium* and *Phytophthora* are very rare. Disease escape is also attributed to some other environmental factors, such as strong wind, soil pH etc. Wind may favor disease escape by carrying the spores away from the crop plants. Under conditions of high soil pH, cruciferous crops may escape from infection by *Plasmodiophora brassicae* and under low soil pH, potato crops escape from *Streptomyces scabies.* Early maturity may also cause disease escape in many crops. Earliness in potato and wheat help the crops escape from late blight and rusts respectively. Similarly, lateness, rapid growth, resistance to bruises and injuries, tolerance to low and high temperatures, unattractiveness to vectors etc. are also often bred into crop varieties to help them escape specific diseases.

Disease escape from a number of diseases may be manipulated by adopting or modifying some of the cultural practices, such as use of disease-free and vigorous seeds, choosing proper soil, adjusting the sowing time, depth of sowing and spacing, fallowing, crop rotation, field and plant sanitation, intercropping, correcting soil pH etc.

Disease tolerance. Sometimes a susceptible crop variety, even though it is infected by a pathogen, can produce a good crop. This ability of the variety to withstand the disease infection and still produce a good crop is referred to as **'tolerance'**. Specific genes usually control tolerance. In tolerant varieties, the pathogen is allowed to develop and multiply, but the host may lack receptor sites for infection by the pathogen or may inactivate or manoeuvere the toxic substances produced by the pathogen and thus is able to produce a good crop. Tolerance to disease is commonly observed in many plant virus diseases in potato and apple. However, tolerant plants or varieties can produce an even better crop in the absence of the disease.

Mechanism of Disease Resistance in Plants

Each and every plant species is subjected to attack by several fungi, bacteria, mollicutes, viruses and nematodes. Although the affected plants may suffer damage to a lesser or greater extent, many plants may survive the attack and manage to grow and produce good yields. The ability of plants to survive attack by various pathogens is due to the presence of some type of defense mechanism, such as **(1)** Morphological characteristics that act as physical barriers and prevent entry and establishment of the pathogen inside the host **(2)** Induction of biochemical reactions in the cells and tissues of the plant and production of certain substances, which are toxic to the pathogen or inhibit the growth of the pathogen in the plant. However, the defense mechanisms, whether morphological or biochemical or a combination of both may vary between species and even plants within the same species. The defense mechanisms also vary depending upon the age of the host plant, nutritional conditions of the plant and the prevailing environmental conditions.

(1) Pre-existing Morphological and Biochemical Defense Systems

(a) **Morphological defense structures**. The surface of the host plant, which the pathogen has to penetrate to cause infection, is the first line of defense. Presence of waxy coating on the surface, thickness of the cuticle covering the epidermal cells, the thickness and structure of the epidermal cell walls, number and size of stomata and lenticels are the limiting defense structures that may hinder the entry of the pathogen into the host. Wax coating on the surface of leaves and fruits forms a water-repellent surface and prevents formation of a film of water, which is necessary for the germination of spores and entry of many fungi into the host, as well as for the multiplication of bacteria. Hairiness on the plant surface may also have a similar water-repellent effect. A thick cuticle may resist penetration and subsequent infection in case of many pathogens that enter the host by direct penetration of the cuticle. The thickness and hardness of the outer wall of the epidermal cells may also inhibit penetration of the cells by the pathogen and makes it difficult or impossible for the pathogen to advance deeper into the tissues. Many fungi and bacteria can enter the plants only through the stomata. The time of opening of the stomata, structure of the stomata and the guard cells also play an important role in conferring resistance in the case of some varieties. Wheat varieties in which the stomata open late in the day are found to have resistance to stem rust, because of desiccation of the germ tubes in the sun's heat. Narrow stomatal opening, broad and elevated guard cells also confer resistance by preventing entry of some bacterial pathogens. Presence of extended layers of sclerenchyma cells in cereals, such as wheat inhibits infection by the stem rust fungus. In the case of several leaf blight and angular leaf spot pathogens, their advance is limited by the veins, because of the restraint effected by the xylem bundle sheath and sclerenchyma cells of the veins.

(b) **Pre-existing chemical defenses.** Several substances produced by the host cells before or after infection play even a much more important role in defending the plants against infection by the pathogen than morphological defense structures in a number of cases.

Some of the inhibitory substances released by the aboveground and underground parts of the plants may prevent infection by certain pathogens. Resistant varieties of red-scaled onions produce toxic phenolic compounds, such as **'protocatechuic acid'** and **'catechol'**, which inhibit the germination of the conidia and prevent infection by the onion smudge fungus, *Colletotrichum circinans*. Similarly, fungitoxic substances exuded on the leaves of tomato plants inhibit the germination of the spores of *Botrytis*. Some plants resistant to certain pathogens have some inhibitory substances in their cells. Several **'phenolic compounds'**, **'tannins'** and some **'fatty acid-like compounds'** present in young leaves, fruits and seeds induce resistance to young tissues against pathogens. Many such substances inhibit the action of hydrolytic enzymes produced by the pathogens, thus preventing maceration of the host tissues. As the leaves grow older, they become more and more susceptible to attack by the pathogens because of reduction in the production of such inhibitory substances. The presence of **'saponins'**, such as **'tomatine'** in tomato and **'avinacin'** in oats help the plants to exclude fungal pathogens, which lack the enzyme **'saponinase'** to break down the saponins. Further, several plant proteins produced by resistant plants have also been found to have inhibitory effect on pathogens. These plant proteins inhibit the action of **'proteinases'** or **'hydrolytic enzymes'** produced by the pathogens that cause host cell wall degradation. Plant cells also contain certain hydrolytic enzymes, such as **'glucanases'** and **'chitinases'**, which may damage the pathogen cell wall and thus prevent infection.

(c) **Defense through lack of essential substances required for the pathogen.** For successful infection of a host, the pathogen should be able to recognize the host. If the pathogen does not recognize the plant as its host, it may not be able to infect the plant. Various substances, such as **'oligosaccharides'**, **'proteins'**, **'glycoproteins'** and **'polysaccharides'** are involved in the host - pathogen compatibility. In the host - pathogen interaction, the pathogen produces a host-specific toxin, which is responsible for the development of typical symptoms. If the pathogen-toxin is not able to react with specific receptors or sensitive sites in the host cells, the pathogen may not be able to infect the plant.

Further, resistant varieties or plants may not produce substances essential for infection and development of the pathogen. In case such substances are produced in smaller amounts than the actual requirement of the pathogen, then the severity and expression of the disease is much less. Bacterial soft rot of potato, caused by *Erwinia carotovora* is less severe on potatoes having low reducing sugar content.

(2) Induced Morphological and Biochemical Defenses

Recognition of the pathogen by the host. For successful development of morphological or biochemical defenses to combat the pathogen, it is essential that the host should be able to recognize the pathogen at an early stage. As soon as a particular plant molecule recognizes and reacts with a elicitor molecule derived from the pathogen, a series of morphological and biochemical changes begin in the plant cell, so as to ward off the pathogen and to counteract the toxic metabolites, such as enzymes and toxins produced by the pathogen. Various fungal and bacterial pathogens release substances, such as **'glycoproteins'**, **'carbohydrates'**, **'fatty acids'** and **'peptides'** in their immediate surroundings. Some of these substances make the plants recognize the presence of the pathogen and to counteract the action of these substances, the host triggers its defense mechanism. However, early recognition of the pathogen by the host is more effective in activating the defense mechanisms and preventing infection and development of many diseases, such as leaf spots and blights, as well as stem, fruit and root diseases.

(a) Induced morphological or structural defenses

Inspite of the presence of several pre-existing or internal defense systems in the host plants, most of the pathogens manage to penetrate and gain entry into their hosts and cause varying degrees of infection. However, even after penetration of the pre-existing defense structures, plants usually respond by forming one or more induced defense structures to prevent further infection by the pathogen. The defense structures involved may be of **'cytoplasmic defense reaction'**, **'cell wall defense structures'**, **'histological defense structures'** or **'necrotic'** or **'hypersensitive defense reaction'**

(i) Cytoplasmic defense reaction

In this type of defense mechanism, the cytoplasm of the host cells under attack by the pathogen is induced to react in such a way, that the mycelium of the pathogen is destroyed. The plant cell cytoplasm may surround the hyphae and prevent their further growth or the cell protoplast may disappear as the fungal growth increases or the cell cytoplasm may become granular and dense and causes disintegration of the mycelium. This type of host reaction may happen in case of attack by slow growing and weak fungal pathogens.

(ii) Cell wall defense structures

In this type of defense, the cell walls of the cells invaded by the pathogen are induced to undergo certain morphological changes. The cell walls may thicken in response to the attack by the pathogen by producing certain cellulose compounds impregnated with phenolic substances that increase resistance to penetration by the pathogen or the cell walls of the parenchyma cells coming in contact with the pathogen swell and produce a fibroid substance that surrounds and traps the pathogen and prevents its further development or the cells invaded by the pathogen may produce cellulosic papillae on the inner side of the cell walls and prevent further penetration of the cell by forming a sheath around the fungal hypha.

(iii) Histological defense structures

1. **Formation of cork layers**. Infection by disease causing microbes, such as fungi, bacteria, viruses and nematodes often induce plants to form several layers of cork cells due to stimulation of the host cells, which form a barrier between the affected and healthy tissues. The cork layers not only inhibit further invasion by the pathogen beyond the point of initial infection, but also block the spread of toxic metabolites that may be secreted by the pathogen and stop the flow of nutrients and water from the healthy to the infected area, thus depriving the pathogen of nourishment. The tissues within the areas delimited by such cork layers, including the pathogen die and form necrotic lesions. These necrotic lesions or spots are quite uniform in size and shape for a particular host-pathogen combination. In some host-pathogen combinations, the necrotic tissues are pushed outward by the healthy tissues underneath to form scabs, thus barricading the pathogen from the host completely.

2. **Formation of abscission layers**. Due to infection by pathogenic organisms, the host plants are induced to form small gaps between two layers of cells known as **'abscission layers'** on young, active leaves that bifurcate or barricade the foci of infection from those of the healthy tissues. As the infection tends to increase, the lamella between these two layers of cells is dissolved and the central infected area is cut off from the rest of the leaf. Gradually the affected area shrivels, dies and falls off carrying with it the pathogen, thereby the rest of the leaf is protected from further invasion.

3. **Formation of tyloses**. During invasion by pathogenic organisms, the host plants are induced to produce **'tyloses'**, which are overgrowths of the protoplast of adjacent parenchymatous cells. These tyloses protrude into the xylem vessels through several pits. The tyloses with cellulosic walls increase in size and numbers, and they clog the xylem vessel completely, thereby block further advance of the pathogen.

4. **Deposition of gum**. Many plants produce different types of gummy substances immediately after infection by pathogens or injury. Gum secretion is more common in fruit trees, such as *Citrus*, stone fruit trees etc. The gums, which are deposited in the intra- and intercellular spaces around the locus of infection, form a barrier and enclose the pathogen completely. The pathogen is thus isolated and is starved to death.

5. **Hypersensitive response.** As soon as a host cell is infected, the nucleus of the host cell is induced to move toward the invading pathogen and soon disintegrates. Brown resinous granules are formed in the cytoplasm of the cell and the cell dies. The invading hypha also dies along with the cell and further spread is stopped. The **'hypersensitive'** or **'necrotic type'** of defense is very common in several diseases caused by fungi, bacteria, viruses and nematodes.

(b) Induced biochemical defenses.

(i) **Hypersensitive response (HR).** Defense through hypersensitive response is not only morphological, but also biochemical in nature. Hypersensitive response results in the death of the host cell at the site of infection. This response is responsible for limiting the growth of the pathogen, thereby provides resistance to the host plant against infection by the pathogen. Hypersensitive response induced resistance is noticed in the case of many obligate and facultative fungal, bacterial, virus, mollicute and nematode parasites.

'Hypersensitive response' is the ultimate defense response initiated by the host, as soon as the pathogen-produced signal molecules known as **'elicitors'** are recognized. Recognition of the elicitors leads to several biochemical reactions in the infected and surrounding host cells. The most common biochemical reactions involved are increase in the oxidative reactions and ion movement, disruption of cell membranes, production of **'phenolics'**, synthesis of antimicrobial substances, such as **'phytoalexins'** and proteins, such as **'chitinases'**. The elicitor produced due to the presence of a pathogen gene, triggers the defense mechanism and develops resistance in the host and makes the pathogen avirulent. The gene, which induces resistance in the host, is called a **'resistant gene'**. Some soybean varieties carry a specific resistant gene that shows hypersensitive response to attack by the bacterium, *Psuedomonas syringae* pv.*glycinea*. The presence of this resistant gene makes the pathogen gene avirulent, thereby making the host resistant. The 'R' gene of these soybean varieties codes for an enzyme involved in the synthesis of a substance known as **'syringolides'**, which elicits the hypersensitive response and makes the pathogen-gene avirulent. Several such genes for resistance occur in plants, such as corn, tomato, tobacco, flax etc. The corn 'R' gene codes for an enzyme that inactivates the toxin of the pathogen, *Cochliobolus carbonum* causing leaf spot of corn.

(ii) **Active oxygen radicals, lipoxygenases and disruption of cell membranes.** The plant cell membrane consists of a phospholipid bilayer, where many kinds of protein and glycoprotein molecules are embedded. The protein molecules allow ions and metabolites to enter and exit the cell. Certain biochemical reactions also take place inside the cell. Further, the cell membrane acts as an induction site for defense mechanisms by triggering the hypersensitive response. Cell membrane defense response may result in the release of **'reactive oxygen radicals'** and certain **'lipoxygenase enzymes'**. These activated oxygen radicals and enzymes trigger the defense mechanism and induce hypersensitive response of the host cell.

(iii) **Reinforcement of host cell walls.** In many instances, infection by pathogens leads to production of certain substances that reinforce the cell walls, which resist further invasion. The most important defense substances produced by the plant cell walls are **'callose'**, certain **'glycoproteins'**, phenolic compounds, such as **'lignin'** and **'suberin'** and mineral elements, such as **'silicon'** and **'calcium'**.

(3) Production of Antimicrobial Substances

(a) **Pathogenesis-related proteins** (PR). These are certain groups of plant proteins that are toxic to the pathogens. Different plant organs, such as leaves, seeds and roots may produce different kinds of such proteins. Attack by pathogens or wounds induce production of such pathogenesis-related proteins in many cases, which are antimicrobial in nature.

(b) **Phytoalexins. 'Phytoalexins'** are toxic antimicrobial substances produced by healthy cells adjacent to cells damaged, as a result of infection by pathogens. These phytoalexins, when produced in sufficient quantities, induce resistance and restrict further development of the pathogen. Some of the common phytoalexins identified and studied include **'phaseolin'** in bean, **'pisatin'** in pea, **'glyceollin'** in soybean, **'rishitin'** in potato and **'gossypol'** in cotton.

(c) **Phenolic compounds.** Certain toxic phenolic compounds, such as **'chlorogenic acid'**, **'caffeic acid'** and **'ferulic acid'** are produced by the cells of resistant varieties, which inhibit infection by the pathogens.

(d) **Phenol-oxidizing enzymes.** Some of the phenol-oxidizing enzymes, such as **'polyphenol oxidases'**, **'peroxidases'** etc., produced by the host plants, oxidize phenolic compounds into **'quinones'**, which are more toxic to pathogens than the phenols themselves. These phenol-oxidizing enzymes are produced in larger quantities in the infected tissues of resistant varieties.

(e) **Detoxification of phenolic toxins.** The host plants, especially in the case of resistant varieties apparently produce some of the **'toxins'** produced by the pathogens and these toxins detoxify the toxins produced by the pathogens.

(f) **Acquired resistance by artificial inoculation with microbes** or **by chemical treatments.** Several chemical compounds, such as **'salicylic acid'**, **'arachidonic acid'**, **'dichloroisonicotinic acid'**, **'benzothiazoles'** etc. may induce localized and systemic resistance in plants. These chemicals, when applied in very small doses, through roots, foliage or by injection may develop acquired resistance. Several fungicides, such as fosetyl Al, metalaxyl, probenazole and triazoles also possess some resistance-inducing activity, when applied at doses required for the control of plant diseases.

(4) Defense through Genetic Engineering.

Defense through genetic engineering with plant or **pathogen-derived genes.** Some of the plant-derived genes, as well as pathogen-derived genes induce resistance by triggering the hypersensitive response in the host plant. Detection, identification, isolation and transfer of genes for disease resistance from a resistant plant to a susceptible plant are the basis involved in genetic engineering. Such resistant genes have been identified in corn and tomato, and have been found to induce resistance in susceptible plants.

Methods of Selection of Resistant Genotypes

(i) **Selection from existing crops.** When a susceptible variety of a crop, with high yielding potential and other agronomically desirable qualities is cultivated in a field under disease-stress conditions, a few plants may show resistance to a particular disease, at the same time these plants may have all other good qualities of the parent. Such plants are selected and seeds collected from them separately. These seeds are grown again under different disease-stress conditions and seeds from disease-free plants are collected. Thus the seeds from the selected plants are tested repeatedly. By this process new disease-resistant varieties with all desirable agronomic qualities can be obtained. Many varieties of cotton, flax and cabbage, resistant to *Fusarium* wilt have been evolved by this method. Sugarbeet varieties resistant to curly top, a virus disease have also been obtained by such selection.

(ii) **Pure line selection.** Here selection is made in pure lines of agronomically desirable crop varieties as mentioned above. The stock obtained from such repeated selections may be genetically pure. Blast resistant cultivars of IR.50 rice variety have been obtained by this method.

(iii) **Plant introduction.** In many instances, the genes responsible for imparting resistance to a particular disease or race of a particular pathogen may not be found in local or domestic varieties. Under such circumstances, plant materials with resistant genes are introduced from other countries, especially from places, where the disease occurs commonly. Such materials may provide rich resources of resistance genes.

(iv) **Hybridization.** Among the existing varieties of a crop, one variety may be resistant to a particular disease, but may not possess other desirable agronomic qualities. On the other hand, there may be good varieties with all desirable agronomic qualities, but they may be susceptible to the disease. By crossing the resistant variety with the other susceptible varieties, it is possible to combine the good qualities of both the parents and evolve varieties having disease resistance along with other desirable agronomic qualities. Several varieties having high yield potential and resistance to particular diseases have been evolved in rice, wheat, oats, sugarcane, cotton and many other crops by this method.

(v) **Mutation.** Evolution of new varieties by spontaneous gene mutation due to radiation may occur under certain conditions and as a result of such gene mutation, sometimes disease-resistant varieties may be evolved. The mutants being genetically evolved, breed true and all the characters of the parent are carried over to the subsequent generations. Such mutants may be treated as new varieties.

Mutation can be induced artificially by exposing the seeds or plants to abnormal radiation, such as X-rays, ultra violet rays, gamma rays etc. and also to mutagenic chemicals, such as **'colchicine'**, **'ethyl methane sulphonate'** etc. The mutants with desirable agronomic qualities and disease resistance may be selected and bred as detailed above. Thus, new genetically stable, disease-resistant varieties can be obtained.

(vi) **Tissue culture techniques.** Tissue culture of disease resistant plants is very useful in the case of clonally propagated plants, such as sugarcane, potato, banana, cassava, apple, tea etc. Large number of plantlet production from meristem and other tissues helps in evolving resistant genotypes

(vii) **Genetic engineering techniques.** Genetic material (DNA) from disease resistant genotypes can be introduced into susceptible plant cells or protoplasts to evolve new resistant varieties.

Table 4. Examples of some disease resistant crop varieties

Crop	Disease	Resistant varieties
Rice	Blast disease	Co.4, Co.25, Co.37, Co.43, TKM.1, ADT.36, ADT.39, ADT.46, IR.20, Tetep, Tadukan, Peta, Zenith, Sigadis, B.589 A4-18, Dawn, H.105 and T.172
Rice	Bacterial leaf blight	TKM.6, IR.20, IR.36, IR.54, BJ.1, Mashuri, Prasad, Saket 4, Sasyasree, Tadukan, Zenith and Sigadis
Rice	Brown leaf spot	BAM.10, Co.20 and SR 26 B
Rice	*'Tungro' virus disease*	Co.45, IR.36, IR.50, IR.64, Peta, Sigadis, HR.21, TKM.6, Pankhari 203
Rice	Sheath rot	GEB.24 and Jagannath
Sugarcane	Red rot	CoC.771, CoC.774, CoC.775, CoC.777, CoC.779, CoC.8201, CoC.90063, Co.6304, Co.6907, Co.62101, Co.62198, Co.Si.86074, Bo.91 and Bo.99
Sugarcane	Smut	Co.449, Co.527, Co.6806, Co.7704, Co.7807, Co.62175, Co.62198, CoC.671, CoC.771, CoC.772, Co.C.90063, CoC.91061 and Co.Si.86071
Potato	Late blight	Kufri Naveen, Kufri Jeevan, Kufri Alenkar, Kufri Khasi Garo and Kufri Moti
Lady's finger	Vein clearing	Pusa Savani
Black gram	Yellow mosaic	K.66-110, Mash-1, NP.16, NP.21, Pant U.19, Pant U.26, Pant U.30 Pant U.30, PLU.27, PLU.476, UG.135, VBN.1 and VBN.2
Green gram	Yellow mosaic	ML.1, ML.5, ML.267, Pant mung 2, Pant mung 3, PDM.54 and Pusa 9072
Cotton	Bacterial blight	BJA.592, HG.9, Reba B.50, AR.4, B56-181, B.61-2037, K.4005, PK.1572 and EL.156 E
Wheat	Black rust or Stem rust	Sonora 64, Lerma Rojo 64A, K.65, Sharbati Sonora, Kalyan Sona, Sonalika, Hira, Lal Bahadur, HD.1918 (Pratap), HD.2009 (Arjun), HD.2122, HD.4530, HD.1102 and UP.262

Other General Methods of Protection

The methods under evasion, exclusion, eradication and protection aim at preventing the pathogens from gaining access to the host or eradicating the pathogens, either before they gain access to the host or after they infect the host. Besides the above mentioned methods, adoption of certain improved agricultural practices, including provision of proper nourishment to the host help to impart some amount of resistance to the host plants by enhancing their growth vigor. Proper crop management, water management and soil management are quite important in enhancing the growth vigor of the host plants.

Crop Management

Making certain alterations in the regular crop husbandry and thereby modifying the environment, help the host plants to resist or to escape attack by many pathogens. Controlling plant diseases by crop management involves the basic principles of exclusion, eradication and protection and such methods are being followed from olden days. Crop management practices are more important in the case of low-income crops, such as pulses, oilseeds and lesser millets and in the case of susceptible varieties of crops.

Crop rotation, use of good quality and disease-free seeds, application of organic manure, deep ploughing, summer ploughing, field sanitation, destruction of disease infected plants, eradication of alternate and collateral hosts, raising the crop at the appropriate time, so as to prevent the coincidence of vulnerable growth stage of the crop and weather conditions favorable for disease occurrence are methods followed under crop management, which may help to control many diseases to a large extent. Use of biological agents to control diseases also comes under crop management.

Water Management

Just as crop management, water management also produces certain changes in the environment, which may indirectly help to prevent infection and spread of certain diseases. Excessive irrigation may directly affect the host plants by making conditions unfavorable for their growth, which may render them more vulnerable to attack by pathogens. Besides that, excess of water adversely affects the activities of soil microorganisms and indirectly render the host plants more susceptible to attack by pathogens. On the contrary, when there is lack of sufficient water, normal growth of the host plants is affected and the plants may become weak. Such weak and unthrifty plants are more vulnerable to attack by pathogens. The time interval between successive irrigations and number of irrigations are also quite important as the amount of irrigation water. During the active growth stages of the plants, timely irrigation should be given, so as to keep the soil sufficiently moist to promote healthy growth of the plants. Under such conditions, the plants are not subjected to any adversities or hardships or stress. When the plants are healthy and strong, the incidence of diseases and their severity may be considerably less.

Soil Management

Just as crop and water management, soil management also produces certain changes in the environment, which may help to control certain diseases. Soil amendment by the addition of organic matter to the soil is an ideal method to improve the soil environment. When the organic matter decomposes in the soil, it may induce several physical, chemical and biological changes in the soil. Such changes in the soil environment may affect many soil - borne pathogens in various ways. Besides affecting the pathogens, the changed environment may benefit the root growth of the host plants also. The biochemical substances, antibiotics and phenolic substances released when the organic matter decomposes are absorbed by the host roots and may induce resistance to the host plants.

Further, when the organic matter decomposes in the soil, microorganisms belonging to the Genera - *Trichoderma, Penicillium, Streptomyces, Bacillus* and such other beneficial organisms grow and multiply in the soil in abundance and exude several antibiotics, which are capable of killing disease causing organisms. Fully decomposed organic matter also serves as nutrients for the plants and promotes their growth.

Several soil-inhabiting fungi are also capable of parasitizing and killing other soil-borne pathogenic fungi. When the organic matter in the soil decomposes, such fungal parasites multiply in the soil enormously and destroy the pathogenic fungi and nematodes in the soil. A few fungal organisms are capable of attacking some parasitic nematodes directly and kill them. They grow around the nematodes and absorb the body contents of the nematodes through specialized organs. The organic matter in the soil also encourages the growth and multiplication of such beneficial fungi.

Cross Protection

'Cross protection' technique has been found to be effective mostly in the case of virus diseases. Cross protection involves protection of a plant from a more virulent strain of a particular virus by inoculating the plant with a mild strain of the same virus. Inoculation can be made either by rubbing, infiltrating or injecting. The chain of reactions caused in the host plant as a result of such inoculation, induces resistance against the virulent strain and protects the plant from infection. This technique has been found to be successful only in the case of a few virus diseases. Tomatoes can be protected with mild strains of *Tobacco mosaic virus*, *Citrus* plants with mild strains of *Citrus tristeza virus* and papaya with mild strains of *Papaya ring spot virus*. However, mild strains are not effective against all the severe strains found in different localities and mild strains of viruses have not been detected for many of the virus diseases. Further, in perennial crops, such as *Citrus* plants, the mild strains are capable of providing cross protection only for a limited period of a few years.

Direct Control of Plant Pathogens by Chemical Formulations

In the integrated system of disease management, use of chemical formulations for the control of plant diseases is quite inevitable, even though other methods of disease control are followed. In the case of disease-resistant varieties also, application of plant protection chemicals has been found to enhance the resistance capability.

Depending on the kind of pathogens affecting the plants, plant protection chemicals used to control them are called fungicides, bactericides, nematicides, viricides or herbicides. Fungicides are commonly used for the control of fungal diseases in the field, green-houses, kitchen-gardens, store-houses etc. Bactericides may inhibit the multiplication of bacterial pathogens or kill them.

All the so-called fungicides are not capable of killing the fungal pathogens. Most fungicides and bactericides are only **'protectants'**. When they are applied on to the plant surface prior to infection, their presence on the parts of the plant form a poison barrier between the plant and the pathogen and prevent the fungal spores to germinate and cause infection or arrest their growth, **'fungistatic'**. Some fungicides may prevent the fungal pathogens from producing spores, **'antisporulants'**, while some others may kill the pathogens, **'eradicants'**. Some fungicides are absorbed by the plant tissues and are translocated inside the plant tissues. They kill the pathogen present inside the plant tissues and cure the plant from the disease, **'therapeutants'**. Many of the older fungicides that are being commonly used are effective in preventing disease infection only in the plant parts on which they have been applied, **'local action'**. They are not absorbed or translocated inside the plant tissues, **'non-systemic'**. However, many of the newly introduced chemical formulations and antibiotics are absorbed by the plant parts and translocated inside the tissues, **'systemic'** and function as eradicants or therapeutants. Fungicides, such as benomyl, carbendazim, thiabendazole, carboxin and metalaxyl, and antibiotics, such as Streptomycin, Tetracycline, Terramycin and Aureofungin are systemic in action. Systemic chemicals are gradually replacing many of the contact, protectant fungicides because of their effectiveness and long lasting efficacy, whereas the non-systemic ones, which are completely exposed on the host surface, get disintegrated soon and loose their efficacy within a much shorter period of time.

Some fungicides are effective in controlling many pathogens and are called **'broad-spectrum fungicides'**. Copper fungicides, dithiocarbamates, captan and some such non-systemic fungicides can control many diseases. Some other fungicides are quite specific and can control one or a few diseases only. They are called **'narrow spectrum fungicides'** or **'selective fungicides'**. Most of the systemic fungicides and antibiotics are narrow spectrum chemical formulations.

Formulation of Fungicides

1. **Wettable powders** (WP). Most of the fungicides are formulated as wettable powders and applied as sprays. They may contain more of active ingredient, usually 50 - 95 %. Along with the active ingredient, some synergists or activators, stickers and spreaders are added. When the wettable powder is mixed in water, the powdered material is dispersed as minute particles and held in suspension in water, but they do not dissolve in water completely. When the fungicidal fluid is sprayed, the active material sticks on to the surface of the sprayed parts after the water evaporates, but does not enter into the plant tissues. Because of the possibility of more uniform distribution of the active material on the sprayed surface, wettable powders are more effective than dust formulations. Wettable powders should have long shelf-life, must disperse in water uniformly, should be able to stick and spread on the sprayed surface, especially on leaves with smooth surface without rolling down and should not form hard cakes in storage. The particles in suspension in the spray fluid may settle down gradually and hence a certain amount of agitation is necessary at the time of spraying to prevent the particles from settling down.

2. **Dusts** (D). Dust formulations usually contain 4 - 10 % of active material. However, sulfur dust is used in the form of 95 % powder. In the dust formulations, the finely powdered active ingredient is mixed with well-

powdered, inert carriers, such as chalk, talcum, kaolinite clay, gypsum, volcanic ashes or saw dust. The particles of dust formulations are of size 100ì or less. Dusts can be dusted directly on the plant parts by means of either manually operated dusters or power dusters. They should be used in the morning hours, when there is dew deposit on the surface of the foliage and when there is less of wind, so as to avoid drift. Dusting is easier, more economical than spraying and does not require water for application.

3. **Emulsifiable concentrates** (EC). In these formulations, the active material is dissolved in solvents and emulsifiers and are available in liquid form. When these formulations are mixed with water, the active materials do not dissolve in water, but are dispersed and held in suspension in water by the emulsifiers. Because the particles are in a suspended state, there is a possibility of the suspended particles to settle down. So, a certain amount of agitation is necessary while spraying. When these formulations are mixed in water and sprayed, first of all the solvents evaporate, followed by water and the active material alone sticks on to the sprayed surface

4. **Water-soluble concentrates** (WSC). In these formulations, the active ingredient is dissolved in solvents only and no emulsifiers are added. When mixed with water, both the active ingredient and the solvents get completely dissolved in water. So, while spraying no agitation is necessary. When these are sprayed, the solvents and the water evaporate leaving behind only the active ingredient on the sprayed surface.

5. **Granules** (G). In these formulations, the active ingredient is mixed with inert substances or carriers, such as lime, gypsum etc. and produced in the form of granules. As the carriers dissolve in water after application in the field, the active ingredient is released gradually, dissolves in the water and gets absorbed by the plant parts and being systemic, are translocated throughout the plant tissues. Granules are of varying sizes ranging from 250 - 1,250ì in diameter and may contain 4 - 10 % of active ingredient. Granules may be broadcasted in the field or applied in the planting furrows or in particular parts of the field where there is infection (spot application). Water is necessary for the granules to dissolve and get absorbed by the plants. So, there should be enough moisture in the field at the time of application of granular fungicides.

   Application of granular fungicides is very easy. There is no chance of the chemical being blown away by wind. They may not stick on to the surface of plant parts and hence fungicidal toxicity on the surface of the plant parts is very minimal. The active ingredient is released slowly and so their residual toxicity persists for a much longer period. For applying granules, there is no need for water, as in the case of spraying.

Granular fungicides have no significant adverse effects on the soil microflora.

6. **Slurry**. In the case of slurry also, the active ingredient is dissolved in solvents and emulsifiers and prepared as thick, glue-like formulation. Just as wettable powders, slurry also contains higher percentage of active material. When the formulation is mixed in water, the active material does not dissolve completely in water, but is dispersed and held in suspension in water by the emulsifiers. Here also the suspended particles may have a tendency to settle down and so, a certain amount of agitation is necessary while spraying. Cuman L is available in the form of slurry.
7. **Pastes**. Some fungicides are mixed with small quantities of water and prepared as pastes. These pastes are mostly used for application on wounds to prevent further spread of diseases. Bordeaux paste is commonly used to treat wounds after tree surgery.
8. **Solutions**. Some inorganic chemical compounds, such as mercuric chloride and nickel chloride are used to control certain diseases. These chemicals are completely soluble in water and hence no agitation is necessary while applying.
9. **Fumigants**. Fumigants are mostly used for the control of nematodes and a few specific pathogenic microorganisms. They are effective against certain fungi besides nematodes, insects and weeds. Chloropicrin, methyl bromide, mylone, vapam and vorlex are the commonly used soil fumigants.

Mode of Action of Fungicides

Most of the older chemicals, especially the non-systemic ones can either prevent sporulation of the pathogens or eradicate the pathogens before they actually infect the host plants and cause diseases. These chemical formulations are generally known as **'protectants'**. Chemical formulations used for the control of plant diseases act in three different ways:

i) Reduce the production of spores or other reproductive structures formed by the pathogens to the maximum possible extent or to kill such structures, so as to prevent large-scale spread and survival of the pathogens

ii) Inactivate or kill spores or other reproductive structures, which happen to reach the surface of the host plants before they germinate and infect the host plants.

iii) Eradicate the pathogens already present inside the host after infection. Some of the recently introduced fungicides and antibiotics can enter into the host plants, get translocated in the plant tissues and are capable of eradicating the pathogens present inside the plant tissues as dormant structures. Such chemical formulations are known as **'eradicants'**. Some chemicals can enter into the plants, systemically move inside the plant tissues and cure the diseased plants by killing the pathogens, which have already caused the diseases. These chemicals are called **'chemotherapeutants'**. Eradicants and therapeutants act as protectants also.

Classification of Fungicides

i) **Classification based on the chemical composition.** Based on the metallic component in the chemical composition, fungicides are generally classified as copper, sulfur, mercury, zinc, tin or benzene fungicides. Some of the metal-based fungicides, such as sulfur fungicides, mercury fungicides etc. include both inorganic and organic compounds.

ii) **Mode of action.** Based on the mode of action of the fungicides, they are classified as protectants, eradicants and therapeutants. Most of the seed-treating chemicals and fungicides used for the control of foliar diseases, such as leaf spots, leaf blights, powdery mildews etc. are **'protectants'**.

'Eradicants' remain both on the surface of the sprayed parts, as well as inside the plant tissues for a few days in an active form and destroy the dormant spores and the pathogens in the growing stages. Such chemicals function as protectants besides eradicants.

'Therapeutants' are capable of destroying the pathogen in the plant tissues even after disease development and cure the plants from the disease. Eradicants and therapeutants are systemic in nature. They are absorbed by the plant parts and are translocated inside the plant tissues. Many systemic fungicides and antibiotics function as therapeutants.

iii) **Mode of application.** Based on the mode of application of fungicides, they are classified as seed dressing fungicides, soil treating fungicides, foliar fungicides and fumigants.

Methods of Application of Plant Protection Chemicals

i) Spraying

Most of the chemicals used for spraying on the foliage and floral parts are aimed at controlling many fungal and bacterial diseases. These fungicides and bactericides, which are commonly used are protectants and must be present on the plant surface in advance of the pathogen, so as to prevent infection. The presence of such chemicals forms a poison barrier between the host plants and the inoculum and does not allow the fungus spores to germinate and infect the host or they may kill the spores on germination. Bactericides may inhibit the multiplication of bacteria or kill them. The eradicants, which have a direct effect on the pathogen that had already invaded the leaves, stems, fruits, or other plant parts, kill the pathogen inside the host or suppress the sporulation of the pathogen without killing it. Some fungicides, such as edifenphos, dodine etc. have a partial systemic action, because they can be absorbed by parts of the leaf tissue and translocated internally to a small extent inside the leaf area. Several fungicides, such as benomyl, thiabendazole, carboxin and metalaxyl are clearly systemic and are translocated internally throughout the host plant. Bactericides, such as Streptomycin, Tetracycline etc are also systemics. Newer systemic fungicides, such as metalaxyl, triadimefon, fenarimol etc. are therapeutants.

Compounds with low surface tension called **'surfactants'** are often added to the fungicides, so as to make them spread better on the sprayed surface. Some chemicals with good sticking quality called **'stickers'** are also added to the fungicides to increase the adherence of the fungicide to the plant surface.

Fungicides and bactericides applied as sprays are more efficient in providing a uniform, protective residue layer on the sprayed surface than dust formulations. Usually, spraying is considered to be four times more effective than dusting. Further, most fungal spores require a film of water on the leaf surface or high atmospheric humidity near saturation for their germination. Sprays provide enough moisture for the spores to germinate and then kill them. Many fungicides and bactericides, especially the protectants are effective only on contact with the pathogen. Hence, it is quite important that the entire surface of the plant is thoroughly covered with the chemical to ensure maximum protection. The interval between sprays in the case of matured tissues may vary from 10 - 14 days. However, in the case of fast growing plants or leaves, the spray interval should be shorter.

ii) Dusting

Formulations, which are insoluble in water or which cannot be dispersed in water are dusted on the crop surface directly as finely powdered dust. Dusting is more effective, when it is done under conditions of high humidity near saturation. The dusts adhere to the film of water present on the plant surface and act as wettable powders. Spraying or dusting of protective chemicals after the pathogen has infected and entered the host plant may not control the disease effectively.

iii) Seed treatment

Seeds and other plant propagative materials, such as cuttings, setts, suckers, tubers, bulbs, rhizomes, corms etc., used for vegetative propagation are often treated with chemical formulations to prevent infection by pathogens or decay after planting. It is considered to be the easiest and most economical method of treatment for controlling many seed-borne diseases, especially externally seed-borne diseases. Seed treatment may also protect the emerging seedlings from fresh infection from soil-borne pathogens for a short period by producing a zone of inhibition around the germinating seedlings.

Some pathogens are found adhering to the seeds or seed materials or admixed with seeds as contaminants. They are known as 'seed-contaminants' or more generally as 'externally seed-borne pathogens'. These pathogens can be controlled by treating the seeds or seed materials with certain chemicals called **'seed disinfestants'**. Formalin, mercuric chloride etc. are used for this purpose. Treating with such chemicals may not give long lasting protection to the seeds after sowing, because they get disintegrated soon and loose their toxicity. These chemicals may sometimes adversely affect the viability of seeds also.

Seeds or seed materials can be protected more efficiently from externally seed-borne pathogens by treating them with some chemicals known as **'seed protectants'**. These chemicals, besides eradicating the pathogens found on the surface of seeds, remain without loosing their toxicity for a long time, even after sowing. When such treated seeds or seed materials are planted, the chemicals diffuse into the soil and form a zone of inhibition around the seeds and prevent further infection of the germinating seedlings. These chemicals may also give protection to the seedlings from invasion by soil-borne pathogens. Non-systemic, organic protectant compounds, such as ceresan, agrosan, captan, thiram, chloroneb, maneb, mancozeb and pentachloronitrobenzene are very commonly used as seed-protectants. To give maximum protection to the seeds, these chemicals have to be incorporated with the seeds thoroughly, so that the seeds are completely covered. Systemic chemicals, such as carboxin, benomyl, thiabendazole, metalaxyl and triadimenol are also used for seed treatment. Seed treating chemicals can be applied on the seeds as dusts or as slurry or the seeds can be soaked in the chemical solution and air-dried before sowing.

Some pathogens are found inside the seeds or seed materials. Treating the seeds with certain specific chemicals known as **'disinfectants'** or **'eradicants'** can eliminate such internally seed-borne pathogens. Such chemicals are capable of killing the pathogens found on the surface of seeds, as well as inside the seeds without causing any harm to the seeds. Seed treatment with carboxin controls loose smut of wheat. Seed treatment with metalaxyl controls downy mildew of sorghum and oats.

While some chemicals used for seed treatment are capable of controlling specific diseases of some plants, others are more general in their efficacy and can control many diseases of a number of plants.

iv) Soil treatment

Drenching the soil with fungicidal solutions around the root zone of plants to wet the soil up to a depth of 10 - 15 cm. controls some of the soil-borne diseases. Drenching can be done before or after planting of the crop. For drenching an acre of field, about 1,000 liters of chemical fluid may be required and the cost involved may be very high. This practice is mostly done in nurseries and seedbeds prior to planting to prevent soil-borne diseases, such as damping off, seedling blights, crown and root rots and wilts. Sometimes, such drenching may be given after planting to control diseases. Protective chemicals, such as Bordeaux mixture, copper oxychloride, captan, thiram, wet ceresan, pentachloronitrobenzene etc. are commonly used for this purpose. Systemic fungicides, such as metalaxyl, triadimefon, ethazol and propamocarb are also used to control some of the soil-borne diseases. These systemic chemicals may provide long lasting control from a single pre-plant application. Downy mildews and rusts can be controlled to a large extent by soil-drenching with metalaxyl and triadimenol respectively. The residual toxicity of systemic chemicals may remain in the plant parts for a long time after application. Hence, use of such chemicals to vegetables, fodder crops and fruit crops prior to harvest should be avoided.

v) Wound treatment

When grown-up plants are cut or pruned, especially in orchards, open wounds are caused. Through these wounds, many pathogens may gain entry into the plants very easily and cause several diseases. Such cut-ends and wounds are treated with fungicidal pastes immediately after the wounds are caused. The fungicide is mixed in small quantity of water to make a thick paste. The paste is applied on the cut-ends or wounds to protect the plants from invasion by other pathogens. In the case of bushy plants, such as tea, coffee etc., where application of fungicidal paste on individual cut-ends is not possible, fungicidal solution is sprayed on the plants, so as to cover the cut-ends. Bordeaux paste and copper oxychloride paste or sprays are commonly used for this purpose.

vi) Fumigation

Some volatile chemicals are used as fumigants to eradicate soil-borne pathogens including nematodes, especially in nurseries and seedbeds. Formalin is one of the oldest soil fumigant used as formaldehyde for the control of damping off and seedling blights. Formalin solution containing 37 - 40 % of formaldehyde is used as a soil drench. It is effective against damping off diseases in the nurseries and seedbeds. Carbondisulfide is used as a soil fumigant mostly for the control of nematodes and a few soil-borne fungal pathogens. It is used in *Citrus* orchards for the control of *Armillaria mellea.* It is highly volatile and acts as an insecticide also. Chloropicrin is another soil-fungicide and is effective against some fungal diseases. Vapam (sodium N - methyl dithiocarbamate) is another commonly used soil fungicide. It is not volatile, but it disintegrates in wet soil to produce **'methyl isothiocyanide'**, **'carbon disulfide'**, **'hydrogen sulfide'**, **'methylamines'** and other products. It is effective against cotton wilt, damping off of papaya seedlings and root rot of beet. Methyl bromide is used as a soil-fumigant in nursery beds, vegetable gardens, orchards etc. for the control of soil-borne pathogens, including nematodes. DD mixture (dichloro propene - dichloro propane) is used extensively for nematode control, besides a few soil-borne pathogenic microorganisms. Dazomet (tetrahydro dimethyl thiazene thione) is another chemical used as soil-fumigant. It releases **'methyl isothiocyanate'**, when applied to soil and is effective against nematodes and some soil-borne fungi, such as *Pythium, Rhizoctonia, Fusarium, Verticillium* and *Colletotrichum.* Most of the soil-fumigants, when applied to the soil volatilize and the toxic fumes released kill the pathogens and nematodes present in the soil. They are usually applied to the soil by means of soil injectors at a depth of 15 - 20cm. and at distances 30 - 40cm. apart, 2 - 3 weeks before sowing. Such chemicals are usually harmful to the seeds and the germinating seedlings. So, a few days before sowing the soil is stirred thoroughly, so that the residual fumes may escape from the soil.

vii) Control of post-harvest diseases

Elemental sulfur is often used as dust for treating post-harvest produce. Sulfur undergoes sublimation during storage of the produce and protects fruits and vegetables from invasion by pathogens. In some cases sulfur dioxide gas is used to destroy the pathogens and protect the produce. Borax, biphenyl, sodium-o-phenyl phenate and fungicides, such as benomyl, thiabendazole and imazalil are also used commercially for controlling post-harvest diseases of fruits, especially *Citrus* fruits. Chlorinated water is also used to wash and treat tomatoes and certain other vegetables in commercial packing-houses. Elemental sulfur, sulfur dioxide, dichloran, captan and benzoic acid are used for the control of storage rots of stone fruits, pome fruits, bananas, grapes, strawberries, melons and potatoes.

Qualities of An Ideal Fungicide

1. It should have high level of toxicity to the pathogens, even in small concentrations.
2. It should not be toxic to the host plants, cattle and humans.
3. It should be able to remain on the surface of the sprayed parts or inside the plant tissue for many days without loosing its toxicity.
4. It should not loose its toxicity, when mixed with water for preparing spray fluid.
5. It should have good sticking and spreading qualities.
6. It should have long shelf life.
7. It should be able to control more than one disease simultaneously.
8. The cost should be less.

Chemicals Used as Fungicides

1. Sulfur Fungicides

Use of sulfur as a fungicide dates back to several centuries and elemental sulfur dust had been used to control many diseases and pests. Even now, many sulfur fungicides are being used widely for the control of several crop diseases. Both inorganic and organic sulfur compounds are being used as fungicides.

(i) Inorganic sulfur fungicides

(a) **Sulfur dust.** Elemental sulfur is powdered in a fine dust form and used for dusting on plant surfaces to control certain diseases. The particle size of sulfur dust ranges from 47 - 75ì. Commercial sulfur dust used for dusting contains 95 % of elemental sulfur. For dusting one acre of crop 10 - 15 kg. of sulfur dust is required. Rust diseases, powdery mildews, groundnut 'tikka' disease etc. can be effectively controlled by this chemical. Application of sulfur dust, when the temperature is very high is injurious to the crops. Cucumbers, apple trees and apple fruits are adversely affected by sulfur dust. Sulfur dust is also used as a carrier in many other dust formulations. Sulfur dust is not commercially available nowadays for use as fungicide.

(b) **Wettable sulfur**. Wettable sulfur is available in the form of water-dispersible powder. It contains 80 % of sulfur. It is available under the trade names Thiovit, Cosan, Sulfex, Elosan, etc. It is capable of controlling diseases, such as powdery mildews, rusts, groundnut 'tikka' disease etc. To prepare spray fluid the fungicide is mixed with water at the rate of 4.0 gm./ liter of water. These formulations may also have injurious effects on cucumbers and apple trees. Use of these formulations should be avoided when the temperature is very high, as they may cause scorching of leaves.

(c) **Lime-sulfur**. The following materials are required for the preparation of lime-sulfur:

Quick lime - 1.0 kg.

Sulfur dust - 2.0 kg.

Water - 10.0 liters

A small quantity of water is added to the sulfur dust and made into a thin paste. Quick lime is slaked in about 3.0 liters of water. When slaking of lime begins, the sulfur paste is added to it and the mixture is stirred well till the lime slakes completely. Enough water is added to the mixture to make the volume to 10 liters. The whole thing is boiled for half an hour and again the volume made to 10 liters, filtered and stored. For spraying, this stock solution is diluted with water in the ratio 1:30. Lime-sulfur contains 22 % of sulfur as **'polysulphide'**, which is the active material. Diseases of plantation crops, such as powdery mildews, anthracnose disease, brown spot etc. can be controlled effectively by this fungicidal preparation. Because of the difficulties in the preparation of this fungicide, it is no longer being used.

(ii) Organic sulfur fungicides (Dithiocarbamates)

Most of the organic sulfur fungicides are broad-spectrum fungicides and can control a number of diseases of various crops. They are widely used throughout the world. They do not cause any kind of damage to the sprayed plants. In many crops, these fungicides control more than one disease simultaneously. However, these fungicides have short shelf life. When stored for a long time they gradually disintegrate due to heat, light and moisture and the toxicity decreases. All these fungicides are derivatives of **'dithiocarbamic acid'** and hence are called **'dithiocarbamates'**

(a) **Thiram.** The technical name of thiram is 'tetramethyl thiuram disulfide'. It is available under the trade names Thiram, Thiride, Arasan, Hexathir, Panoram, Tersan etc. It is mostly used for the treatment of seeds and bulbs of vegetables, flowering plants, cereals and grasses. Besides seed treatment, it is used for the control of some foliage diseases, such as rusts. Externally seed-borne diseases of rice, pearl millet, cotton, groundnut, lady's finger, sorghum, tobacco, tomato, chillies and several other crops can be controlled by treating the seeds at 4.0 gm. of chemical / kg. of seeds. The fungicide is mixed with water at the rate of 2.0 gm./ liter and is used for spraying for the control of anthracnose of beans, grapes etc. The fungicidal solution at the same concentration can also be used for soil drenching for the control of damping off disease of tobacco, tomato, eggplant, chillies etc. Thiram is also effective in controlling some bacterial pathogens also.

(b) **Ferbam.** The technical name of ferbam is 'ferric dimethyl dithiocarbamate'. It is available under the trade names Ferbam, Fermate, Hexaferb, Fermocide, Coromet, etc. as 70 % and 76 %WP for spraying. It can control banana 'Sigatoka' disease, anthracnose of *Citrus* plants, downy mildew of chillies, leaf blight of tomato, black leaf spot and rust diseases of rose, besides many other diseases of vegetables and ornamental plants. For spraying, the chemical is mixed with water at the rate of 2.0 gm./ liter.

(c) **Ziram.** Its technical name is 'zinc dimethyl dithiocarbamate'. It is available under the trade names Ziram, Ziride, Zerlate, Hexazir, Milbam, Cuman L etc. as 50 %WP or in a slurry form for spraying. The chemical is mixed with water at the rate of 2.0 gm./ liter and is used for the control of anthracnose of cucumbers, fruit rot and die-back of chillies, leaf blight of tomato, early leaf blight of potato, sugary disease of pearl millet etc. Cuman L is available as a slurry.

(d) **Zineb.** Its technical name is 'zinc ethylene bisdithiocarbamate'. It is available under the trade names Zineb, Dithane Z 78, Lonocol, Hexathane, Blicin, Unizeb, Polyram, Parzate etc. as 65 %WP for spraying and 5 % or 15 %D for dusting. It is an excellent, broad-spectrum fungicide and used very widely for the control of several foliar and fruit diseases of many vegetables, such as tomato, potato, chillies etc, as well as grapevine, flowering plants, trees etc. It is very effective in controlling early and late blight of potato, leaf diseases of tomato, rice blast, fruit rot of chillies, a few downy mildews and rusts, anthracnose and leaf diseases of many other crops. **But zineb is not effective in controlling powdery mildews**. For preparing spray fluid, the chemical is mixed with water at the rate of 2.0 gm./ liter. It is compatible with copper fungicides. Zineb in combination with copper oxychloride is available commercially under the trade names **Miltox** and **Blitane**.

(e) **Maneb**. Its technical name is 'manganese ethylene bisdithiocarbamate'. It is available under the trade names Maneb, Dithane M 22, Tersan, Manzate etc. It is more effective than zineb and largely used for the control of various diseases of vegetables, bean anthracnose, downy mildew of cucumbers, leaf spots of chillies and *Citrus* plants, downy mildews, leaf blights, blister blight of tea, *Alternaria* blights of potato and eggplant, late blight of potato, downy mildews of cabbage, carrot, beet, cauliflower, rusts etc. The spray fluid is prepared by mixing the fungicide with water at the rate of 2.0 gm./ liter.

(f) **Mancozeb**. Its technical name is 'zinc-manganese ethylene bisdithiocarbamate'. It is available under the trade names Mancozeb, Dithane M 45, Manzate 200, Pencozeb etc. The addition of zinc in this fungicidal formulation reduces the phytotoxicity of maneb and improves its fungicidal properties. It also supplies manganese and zinc to the plants. It is an excellent, broad-spectrum fungicide and affords good control of several diseases of vegetables, fruit trees and cereals, anthracnose of beans, cucumbers, fibre crops, tomato and tobacco, downy mildew of bean, cucumbers and sorghum, rust diseases, *Alternaria* blights, groundnut 'tikka' disease etc. The fungicidal fluid is prepared by mixing the chemical with water at the rate of 2.0 gm./ liter. **However, it is not effective in controlling powdery mildews**. It is phytotoxic to tobacco seedlings, apple and some varieties of cucumbers. It is not compatible with copper oxychloride fungicides.

(g) **Nabam**. Its technical name is 'disodium ehtylene bisdithiocarbamate'. It is available under the trade names Dithane D 14, Parzate etc. It is very effective in controlling early and late blights of potato, *Alternaria* blight of tomato, as well as seedling rots and root rots caused by *Pythium, Fusarium* and *Rhizoctonia*. The spray fluid is prepared by mixing the chemical with water at the rate of 2.0 gm./ liter.

(h) **Vapam**. Its technical name is 'sodium methyl dithiocarbamate'. It is available as Vapam, Karbation, Vitafume etc. in a liquid form. It is a non-fumigant and is used for soil treatment. When applied to the soil, it releases **'methyl isothiocyanate'**, which is toxic to fungal pathogens, soil-inhabiting nematodes and insects. It can control *Fusarium* wilt of cotton, seedling rots caused by *Sclerotium, Rhizoctonia* etc. and some soil nematodes affecting the roots of potato, *Citrus* etc. The chemical is mixed with water at the rate of 1.0 ml./ liter and used for drenching the soil.

2. Copper Fungicides

After the invention of Bordeaux mixture by the French Scientist Millardet in 1885, copper fungicides had assumed great importance and had paved the way for the formulation and introduction of several other fungicides. Copper fungicides had been used for the past several centuries and even now they are widely used for the control of many diseases of several crops. All copper fungicides are inorganic compounds. They are all derivatives of copper sulfate, copper carbonate, copper oxide and copper oxychlorides

Bordeaux mixture was the first fungicide to be developed and used for the control of downy mildew of grapes. After that, it was used for the control many other diseases in various other crops. Still it is considered to be one of the best fungicides. Subsequently many other copper-based fungicides and several other fungicides had been developed. Bordeaux mixture is very effective in controlling many fungal and bacterial leaf spots, leaf blights, anthracnose diseases, downy mildews, cankers etc.

(i) Bordeaux mixture

Bordeaux mixture is prepared by mixing copper sulfate and lime in correct proportion. It is mostly used as 1.0 % or 0.75 % mixture. To prepare 1.0 % Bordeaux mixture, 1.0 kg. of powdered copper sulfate is dissolved in 50 liters of water in a vessel. Similarly, 1.0 kg. of slaked lime is mixed with 50 liters of water in another vessel. Then, the copper sulfate solution is poured slowly into the lime solution and the mixture is stirred constantly while mixing. **Lime solution should never be poured into the copper sulfate solution.** Instead of pouring the copper sulfate solution into the lime solution, both the solutions may be poured into a third vessel simultaneously, while the mixture is stirred continuously while mixing. For preparing Bordeaux mixture, metal vessels should not be used, as this may induce the copper in the copper sulfate solution to dissociate. Plastic, glass or mud vessels may be used for this purpose. When copper sulfate solution is added to lime solution, chemical reaction takes place and cupric hydroxide and calcium sulfate are formed.

$$CuSO_4 + Ca(OH)_2 \rightarrow Cu(OH)_2 + CaSO_4$$

'Cupric hydroxide' in Bordeaux mixture is the active principle, which is soluble and is toxic to the fungal spores and the germ tubes. Well-prepared Bordeaux mixture is deep blue in color and clear. In the prepared Bordeaux mixture, copper should not be in excess. If there is excess of copper, it may cause toxicity known as **'copper injury'** to the plants, when sprayed. To test whether there is excess of copper in the prepared fungicidal fluid, a well-polished iron knife is dipped into the mixture and kept for sometime. If there is excess of copper, it will get deposited on the surface of the knife as a reddish deposit. In such cases, more of lime solution is added till the copper deposition is stopped. Bordeaux mixture is not stable and hence, it should be used immediately after preparation.

However, when Bordeaux mixture is used in cool, wet weather, it is toxic to the plants and may cause scorching or burning of leaves.

Bordeaux mixture is used as soil drench for the control many soil-borne pathogens. Damping off of seedlings of tobacco, tomato, chillies, eggplant etc. is effectively controlled by soil drenching with 1.0 % Bordeaux mixture. Bordeaux paste is used for treating wounds in many horticultural crops and trees.

(ii) Burgundy mixture

It is another copper fungicide that can be prepared and used. It is a modification of Bordeaux mixture and instead of lime, sodium carbonate is used. To prepare Burgundy mixture, 1.0 kg. of powdered copper sulfate is dissolved in 50 liters of water and 1.0 kg. of sodium carbonate is dissolved in another 50 liters of water separately. These two solutions are then mixed together and stirred thoroughly. It is less phytotoxic than Bordeaux mixture and can be sprayed even on tender leaves. However, it is less effective than Bordeaux mixture and can control diseases, such as leaf spots, leaf blights, anthracnose etc.

(iii) Cheshunt compound

This fungicide can be prepared by mixing copper sulfate and ammonium carbonate. It is prepared by mixing 3 parts of powdered copper sulfate and 11 parts of powdered ammonium carbonate. The dry, powdered mixture is stored in an airtight container for 24 hours prior to mixing with water for spraying on plants. From this mixture, 30 gm. is taken, mixed in a small quantity of hot water, till the chemicals are dissolved in water completely. Then, water is added to this stock solution to make the volume to 10 liters and used for spraying. It is a very effective fungicide and can control a number of diseases that are controlled by Bordeaux mixture. It can be used for foliar spraying, as well as for soil drenching.

(iv) Copper oxychloride fungicides

Although Bordeaux mixture, Burgundy mixture and Cheshunt compound are very effective fungicides and can control many diseases of several crops, it is rather difficult to prepare them. Further, the prepared fungicidal solutions are not stable and should be used immediately after preparation. So, copper fungicides, such as copper oxychloride and cuprous oxide fungicides, which can be readily mixed with water and used, have been introduced. Copper oxychloride fungicides are widely used for the control of many diseases. It is available as Fytolon, Cupramar, Blitox, Perenox, Cuprocide, Micop, Parrycop, Blue copper etc. as 50 %WP for spraying or drenching. The diseases, which can be controlled by Bordeaux mixture, especially seedling rots, leaf spots, leaf blights, anthracnose, rusts etc. are controlled by foliar spraying with copper oxychloride fungicide. To prepare spray fluid, the chemical is mixed with water at the rate of 2.5 gm./ liter. It is also used as a soil drench for the control of damping off disease of tobacco, tomato, chillies, eggplant etc.

3. Mercury Fungicides

Mercury fungicides are capable of attacking and killing pathogens. They are mostly used for seed treatment for the control of externally seed-borne diseases. Mercury fungicides can destroy bacterial pathogens also. However, these chemicals remain in the soil, as well as on the surface of treated parts as toxic residues for a long period of time and may cause harm to cattle and humans. So, use of these chemicals has been considerably restricted and some of them have been totally banned. These chemicals are toxic to plants and hence are not used for spraying on the plant surface.

(i) Ceresan dry

Its technical name is 'phenyl mercury acetate'. It is marketed as Ceresan dust, Hexasan etc. For seed treatment, 2.0 gm. of the chemical is mixed with 1.0 kg. of seeds, so as to cover the seed surface thoroughly, at least 24 hours prior to sowing. Externally seed-borne diseases of cotton, pearl millet, sorghum, rice, groundnut, tobacco, tomato, wheat etc. can be effectively controlled.

(ii) Agrosan GN

Its technical name is 'phenyl mercury acetate + ethyl mercury chloride'. Externally seed-borne diseases of eggplant, chillies, cotton, bean, rice, pearl millet, groundnut etc. can be effectively controlled.by treating the seeds at the rate of 2.0 gm./ kg. of seeds. The seeds should be treated, at least 24 hours prior to sowing.

Tubers and bulbs can also be treated with ceresan dust or agrosan dust to eliminate the pathogens adhering to the surface of such plant parts. Potato scab and potato black scurf can be effectively controlled by this treatment.

(iii) Ceresan wet

Its technical name is 'methoxy ethyl mercury chloride'. It is a wettable powder and is available under the trade names Wet Ceresan, Agallol, Aretan etc. Aretan is commonly used for the treatment of rhizomes of turmeric, ginger etc. Aretan fungicidal solution is prepared by mixing 2.5 gm. of the chemical in 1.0 liter of water and the rhizomes are kept immersed in the fungicidal solution for 30 minutes and then planted. Sugarcane setts can be treated with Aretan solution at the rate of 2.5 gm./ liter of water or with Agallol solution at the rate of 5.0 gm./ liter of water. The setts are kept immersed in the fungicidal solution for 5 minutes prior to planting to control sett-borne diseases. Drenching the soil with wet ceresan can control soil-borne diseases, such as stem rot, root rot and wilt. The fungicidal solution is prepared by mixing the chemical at the rate of 1.0 gm./ liter of water. The soil around the stem and root region is drenched with the fungicidal solution.

4. Quinone Fungicides

(i) **Chloranil**. Its technical name is '2,3,5,6 - tetrachloro - 1,4 - benzoquinone'. It is available under the trade name Spergon as 25 %WP for spraying and 96 %D for dry seed treatment. The seed-treating chemical is used for seed and bulb treatment. Seed rot and seedling rot of beans, cabbage, cotton, peas, as well as smut disease of barley and cholam can be controlled by seed treatment. It is used at the rate of 2.0 gm./ kg. of seeds, at least 24 hours prior to sowing. For spraying, Spergon wettable powder is used at the rate of 1.25 - 1.5 gm./ liter of water.

(ii) **Dichlone**. Its technical name is '2,3 - dichloro - 1,4 naphthoquinone'. It is available as Dichlone, Phygon etc. as 50 %WP. It is used for seed treatment and for spraying on the foliage. Treating the seeds with the fungicide at the rate of 2.0 gm./ kg. of seeds can control externally seed-borne diseases of peas, sugarbeet and greens. It can control smut disease of sorghum also. As foliar spray, it is used at the rate of 1.25 - 1.5 gm./ liter of water for the control of many diseases, such as apple scab, peach leaf curl, anthracnose of beans, tomato leaf spots etc.

5. Benzene Fungicides

(i) **Pentachloro nitro benzene.** It is available under the trade names PCNB, Brassicol, Terrachlor, Quintozene etc. It is a long-lasting, soil-fungicide and very effective in controlling soil-borne diseases of vegetables, turf and ornamentals caused by *Rhizoctonia, Sclerotium, Plasmodiophora* and *Sclerotinia,* but not capable of controlling wilt disease caused by *Fusarium* and seedling rots caused by *Phytophthora* and *Pythium.* It has nematicidal properties also. Fungicidal fluid is prepared by mixing the chemical at the rate of 1.0 gm./ liter of water and used for soil drenching. Because of its high residual toxicity, its use has been restricted to a considerable extent.

(ii) **Chloroneb.** Its technical name is '1,4 - dichloro - 2,5 dimethoxy benzene'. It is available under the trade name Demosan as 65 %WP. It is very effective against diseases caused by *Rhizoctonia solani, R. rolfsii* and *Phytophthora cinnamomi.* It has got systemic effect also. It is mostly used for the control of seedling diseases of cotton, bean, sugarbeet and soybean. For spraying it is mixed with water at the rate of 1.25 - 1.5 gm./ liter.

(iii) **Chlorothalonil.** Its technical name is 'tetrachloro isophthalo nitrite'. It is available under the trade names Daconil, Bravo, Termil, Kavach, Speldrum etc. as 25 %WP for spraying. It is a broad-spectrum, contact fungicide and used for the control of early and late blights of potato and tomato, leaf spots, leaf blights, powdery mildews, downy mildews, apple scab, coffee berry diseases, yellow and black rusts of wheat, 'Sigatoka' disease of banana and groundnut leaf diseases. For spraying, the chemical is mixed with water at the rate of 1.25 - 1.5 gm./ liter.

(iv) **Dinocap.** Its technical name is 'methylheptyle dinitrophenyl crotonate'. It is available under the trade names Karathane, Crotothane, Arathane, Capryl, Mildex etc. as 48 %EC or 25 %WP for spraying. It is an excellent substitute for sulfur for the control of powdery mildews of many crops and is highly specific against them. It is also effective against mites. The fungicidal solution is prepared by mixing it at the rate of 1.25 - 1.5 ml./ liter of water.

(v) **Dexon.** Its technical name is 'p - (dimethylamino) benzene diazosodium sulfonate'. It is used against soil-borne phycomycetous fungi causing root rots and damping off diseases of plants.

(vi) **Dichloran.** It is sold as Botran as 50 %WP and 75 %WP. It is used as foliar and fruit fungicide, as well as post-harvest spray. It is used at the rate of 1.25 - 1.5 gm./ liter of water against diseases of vegetables and flowers caused by *Botrytis, Sclerotinia* and *Rhizopus.* It inhibits the protein synthesis metabolism in the pathogens and arrests cell multiplication and growth.

(vii) **Biphenyl.** It is sold as Biphenyl and is widely used for the control of post-harvest diseases of *Citrus* caused by *Penicillium, Diplodia, Botrytis* and *Phomopsis.* It is volatile and is applied for impregnating packing materials of fruits.

6. Heterocyclic Nitrogenous Compounds

(i) **Captan.** Its technical name is 'trichloromethyl thiocyclohexene dicarboximide'. It is available as Captan 50, Esso fungicide 406, Orthocide, Merpan, Hexacap, Vancide etc. as 50 %WP for spraying and 75 %WP for seed treatment. It has got some systemic properties also. Although it is a fairly good broad-spectrum fungicide, **it is not effective in controlling rusts and powdery mildews.** It is used mainly for treating seeds, but can also be used for foliar spraying and soil drenching. Use of this chemical at 4.0 gm / kg. of seeds controls seed-borne diseases of many crops, such as cotton, groundnut, rice, eggplant, sorghum etc. Spraying this fungicide at the rate of 1.25 - 1.5 gm./ liter of water is very effective in controlling leaf spots, leaf blights and fruit rots of many fruit crops, vegetables and ornamentals, banana 'Sigatoka' disease, grape downy mildew, blight diseases of potato and tomato, downy mildew of cucumbers etc. Soil drenching with this fungicidal solution at 1.0 gm./ liter of water controls many soil-borne diseases, especially damping off disease of tobacco, tomato, chillies, eggplant etc. It is not compatible with Bordeaux mixture and Lime-sulfur.

(ii) **Captafol.** Its technical name is 'tetrachloro ethylthiocyclohexene dicarboximide'. It is available under the trade names Captafol, Difolatan, Foltaf, Sanspor, Difosan etc. It has properties similar to captan and acts as a protectant and eradicant. It is used for spraying at the rate of 1.25 - 1.5 gm./ liter of water. It gives effective control of banana 'Sigatoka' disease, groundnut 'tikka' diseases, early and late blights of potato and tomato, anthracnose etc. It is highly resistant to weathering and persists on the host surface for a longer period than many other fungicides.

(iii) **Iprodione.** It is marketed as Rovral, Chipco 26019, Glycophene etc. It is a broad-spectrum, contact fungicide and inhibits spore germination and mycelial growth. It is effective against *Botrytis, Monilinia, Sclerotinia, Alternaria* and *Rhizoctonia.* As a foliar spray, it is applied at the rate of 1.25 - 1.5 gm./ liter of water. It is also used for seed treatment and as a post-harvest dip of fruits.

(iv) **Flutolanil.** It is available as Moncut, Prostar etc. It is used as a protective chemical against *Rhizoctonia, Sclerotium rolfsii, Corticium* and *Typhula.* It is applied as a spray.

(v) **Vinclozolin.** It is available as Ornalin, Ronilan, Vorlan etc. It is a contact and protective fungicide. It is effective against *Botrytis, Monilinia, Sclerotinia* and many other fungi. It is used mostly as a spray on strawberries, lettuce, turf and ornamentals, and on fruits.

(vi) **Folpet.** Its technical name is 'n-trichloromethyl thiophthalamide'. It is available under the trade name Phaltan as 50 %WP for spraying and 10D for dusting. For spraying, it is used at the rate of 1.25 - 1.5 gm./ liter of water. It is a substitute for captan and is effective against all the diseases that can be controlled by captan.

7. Organo-tin Fungicides

(i) **Du-ter.** Its technical name is 'triphenyl tin hydroxide'. It is available under the trade names Du-ter, Super Tin, Farmatin, Tubotin etc. Diseases caused by *Cercospora, Helminthosporium, Septoria, Alternaria, Pythium, Phytophthora* and *Rhizoctonia,* groundnut 'tikka' diseases, rusts, early and late blight of potato and tomato, besides many other diseases of vegetables, fruit trees, ornamentals etc. can be effectively controlled by this fungicide. It is mixed with water at the rate of 1.25 - 1.5 gm./ liter to prepare the spray fluid. It has got long residual toxicity.

(ii) **Brestan.** Its technical name is 'triphenyl tin acetate'. It is available as 40 %WP and 60 %WP under the trade name Brestan for spraying. Diseases caused by *Alternaria, Septoria* and many other foliar diseases, such as blast disease of rice and ragi, groundnut 'tikka' diseases, potato blights etc. can be controlled by this fungicide. The spray fluid is prepared by mixing the chemical at the rate of 1.25 - 1.5 gm./ liter of water. It has got some bactericidal effect also. It is much more effective than copper fungicides. It has got long residual toxicity and provides long-lasting protection. It inhibits spore germination and kills the germ tubes.

(iii) **Brestanol.** Its technical name is 'triphenyl tin chloride'. It is available as 45 %WP under the trade names Brestanol, Tinmate etc. It is equally effective as du-ter and brestan and controls several diseases of various crops, vegetables, fruits and ornamentals.

8. Nickel Fungicides

Nickel chloride. It is an inorganic compound used as a fungicide, especially for the control of blister blight of tea. The spray fluid is prepared by mixing the chemical at the rate of 1.25 - 1.5 gm./ liter of water.

9. Organic Systemic Fungicides

The introduction of systemic fungicides led to a revolution in the control of plant diseases. Carboxin and oxycarboxin were the first systemic fungicides to be introduced. Subsequently other systemic fungicides, such as pyrimidines and benzimidazoles were introduced. Nowadays, numerous systemic fungicides belonging to various groups are being used for the control of diseases. Systemic fungicides are very specific than non-systemic fungicides and can control only one or rarely a few diseases. They are absorbed by the plants through the foliage or roots and are translocated within the plant through the xylem. Most of them move upward (acropetal) in the transpiration stream and may accumulate at the leaf margins. A few of them, such as fosetyl-Al and thiabendazole can move downward (basipetal) also. Many of these fungicides are only locally systemic within the sprayed leaves and are not translocated throughout the plant. Systemics are effective when applied as seed treatments, root dips, soil drenches and trunk injections, besides foliar sprays. Almost all systemics are site-specific in their action. Systemic fungicides may act and destroy the pathogens in the following ways:

(i) They may inactivate the enzymes and toxins produced by the pathogens

(ii) They may inhibit production of enzymes or destroy the enzymes produced by the pathogens

(iii) Because of their greater permeability, they may easily enter through the cell walls into the cells of fungal hyphae of the pathogen, get accumulated in the cells and destroy the pathogen

(iv) They may cause damage to the membranes of the fungal hyphae and inhibit formation of fungal structures, such as appressoria, haustoria and cushion or may inhibit emergence of germ tubes.

Many pathogens, through simple mutation may develop resistance to some of the commonly used systemic fungicides within a few years of introduction. To avoid development of such resistant strains of the pathogens, it is always better to add one broad-spectrum contact fungicide along with the systemic fungicide at the time of application.

(i) Benzimidazoles

Benzimidazoles include some of the important systemic fungicides, such as benomyl, carbendazim, thiabendazole and thiophanate.

Benomyl. Benomyl is available under the trade names Benlate, Tersan 1991 etc. as 50 %WP for spraying, drenching etc. It is a safe, broad-spectrum fungicide, effective against a large number of important fungal pathogens. It controls several leaf spots, leaf blights, rots, scabs, as well as many seed-borne and soil-borne diseases. It is particularly effective for the control of powdery mildews of all crops, scab of apple, peaches etc., brown rot of stone fruits, root rots, *Cercospora* leaf spots, black spot of roses, blast disease of rice and ragi besides diseases caused by *Sclerotinia* and *Botrytis.* It suppresses infection by *Rhizoctonia, Thielaviopsis, Ceratocystis, Fusarium* and *Verticillium.* However, it is ineffective against fungi belonging to Oomycetes, some Basidiomycetes and some Imperfect fungi, which produce dark-colored spores, such as *Helminthosporium, Bipolaris, Drechslera, Alternaria* and *Curvularia.* It is ineffective against bacterial pathogens also. It is used for foliar spraying, seed treatment, root dipping, soil drenching and for fruit dipping. The spray fluid is prepared by mixing the chemical at the rate of 1.0 gm./ liter of water. Fungicidal solution at this concentration is used for soil drenching also. For seed treatment, the chemical is mixed with the seed at the rate of 2.0 gm./ kg. of seeds. The seedlings from such treated seeds grow more vigorously and show a greater degree of resistance to diseases. During disintegration process inside the plant tissue, benomyl is first transformed into carbendazim, which is also an effective fungicide. So, the residual toxicity of the fungicide remains in the plant tissue for many more days and imparts resistance to the plants. A large number of strains of fungal pathogens have developed resistance to this fungicide.

Carbendazim. Carbendazim is available under the trade names Bavistin, Derosal, Bengard, Zoom, MBC, Tagstin, Agrozim etc. as 50 %WP and 60 %WP. It is used as a spray fungicide and as a seed treating fungicide. It can control most of the diseases that can be controlled by benomyl. It is highly effective in controlling sheath blight of rice. Spray fluid is prepared by mixing the fungicide at the rate of 1.0 gm./ liter of water. For seed treatment, the chemical is mixed with the seed at the rate of 2.0 gm./ kg. of seeds. Seedlings from carbendazim treated seeds grow more vigorously and have more resistance to diseases. Rice seedlings from carbendazim treated seeds show resistance to blast disease for many days in the nursery.

Thiabendazole. Thiabendazole is available under the trade names Thiobendazole, Mertect, Tecto, Storite etc. as 60 %WP for spraying and for post-harvest treatment. It is also a broad-spectrum fungicide and is effective against many diseases caused by imperfect fungi, such as leaf spots of several crops, as well as diseases of bulbs and corms. Diseases caused by *Botrytis, Ceratocystis, Colletotrichum, Fusarium, Rhizoctonia, Sclerotinia, Septoria, Verticillium* etc. can be controlled effectively by this fungicide. It is commonly used for post-harvest treatment for the control of storage rots of *Citrus*, apples, pears, bananas and potato. However, it is not effective against some species of *Mucor, Phytophthora, Pythium, Oospora, Rhizopus, Phoma* etc.

Thiophanate and Thiophanate-methyl. Thiophanate is available under the trade names Topsin, Cercobin, Fungo, Enovit etc. as 50 %WP, while thiophanate-methyl is available as Topsin M, Cercobin M, Enovit methyl etc. as 70 %WP for spraying. Both are effective against many fungal diseases affecting the foliage and roots, especially vegetable crops. They are broad-spectrum, preventive and curative fungicides used as foliar spray to control powdery mildews, downy mildews, *Botrytis* diseases and several leaf and fruit spots, scabs and rots. They are also used as soil drench to control soil-borne diseases.

(ii) Oxanthiins or Carboximides

Oxanthiins were the first systemic fungicides developed in 1966. They include primarily carboxin and oxycarboxin. They destroy the fungal pathogens by interfering with their enzyme metabolism.

Carboxin. It is available under the trade name Vitavax. It is available as 75 %WP, 34 %EC or 10 %D. It is primarily a seed treating fungicide, but is also used for spraying. For seed treatment, Vitavax 10 %D is used at the rate of 2.0 gm./ kg. of seeds. By seed treatment, smut disease of wheat, barley, pearl millet and such other crops can be controlled. Using a mixture of thiram and Vitavax 10 %D for seed treatment is effective in controlling other seed-borne diseases, besides smut. The spray fluid is prepared by mixing Vitavax 75 %WP at the rate of 1.0 gm./ liter of water or Vitavax 34 %EC at 1.0 ml./ lit. of water. Spraying is effective in controlling groundnut rust, blister blight of tea, seedling blight of cotton, bean, black gram etc., as well as powdery mildews. It is also effective against damping off disease caused by *Rhizoctonia.*

Oxycarboxin. It is available under the trade name Plantvax as 75 %WP for spraying, 10 %D for seed treatment and 5 %G for soil application. It affords effective control of rust disease of various crops, powdery mildews, diseases caused by *Aspergillus, Helminthosporium, Curvularia, Cladosporium* etc. It is more effective than carboxin and controls more number of diseases. Seedling rots caused by *Pythium* are also controlled to some extent by this fungicide. Rusts of wheat, barley, bean and groundnut, blister blight of tea, root rot caused by *Rhizoctonia,* smuts of wheat, barley, pearl millet etc. and powdery mildews of many crops can be controlled by spraying Plantvax. For spraying, Plantvax 75 %WP is mixed at the rate of 1.0 gm./ liter of water. Sometimes it is used for seed treatment also.

Tridemorph. Its chemical name is 'n - tridecyl - 2,6 - dimethyl morpholin' and is sold as Tridemorph, Calixin, Beacon, Bardew etc. as 75 %EC for spraying. It is a recently developed systemic fungicide, which has excellent prophylactic and curative action against powdery mildew of cereals *(Erysiphe graminis),* 'Sigatoka' disease of banana *(Mycosphaerella musicola)* and other ascomycetous pathogens. It controls blister blight of tea *(Exobasidium vexans)* and pink disease of rubber *(Corticium salmonicolor)* effectively.

(iii) Organo-phosphorus fungicides

These fungicides include edifenphos, fosetyl-Al, kitazin and pyrazophos.

Edifenphos. Its technical name is 'ethyl diphenyl dithiophosphate' and is available under the trade names Edifenphos, Hinosan etc. as 50 %EC for spraying. It is a specific fungicide for the control of blast disease of rice and ragi. It is effective in controlling other leaf spot diseases also. The fungicidal spray fluid is prepared by mixing the chemical at the rate of 1.0 ml./ liter of water. It inhibits chitin synthesis in the fungal cells. It has got local systemic action.

Fosetyl-Al. It is available as 80 %WP under the trade name Aliette for spraying. It is highly effective against many foliar, root and stem diseases caused by Oomycetes, such as *Phytophthora, Pythium* and the downy mildews in several crops. It is applied as a foliar spray, soil drench, root dip and post-harvest dip. Treatment with this fungicide may be effective for 2 - 6 months. The fungicidal spray is prepared by mixing the chemical at the rate of 1.0 gm./ liter of water. A combination of this fungicide and mancozeb can control many diseases, besides the diseases caused by oomycetous pathogens.

Iprobenphos. Its technical name is 'ethyl benzyl phosphorothiolate'. It is commonly known as EBP, Kitazin etc. and is available as 48 %EC for spraying, 2 %D for dusting and 17 %G for soil application. It is highly effective in controlling blast disease of rice, finger millet and other cereals. Sheath blight and sheath rot of rice and grain discoloration diseases are also controlled by this fungicide. The spray fluid is prepared by mixing the chemical at the rate of 1.0 ml./ liter of water.

Pyrazophos. It is available as Afugan, Curamil etc. It has systemic, protective and curative properties and is effective against powdery mildews, as well as *Bipolaris* and *Drechslera* diseases of various crops.

(iv) Acylalanines

Acylalanines include mainly metalaxyl.

Metalaxyl. It is available as Ridomil, Acylon, Apron etc. Ridomil and Acylon are available as 25 %WP for spraying and 2G or 5G for soil application, while Apron is available as 25 %WP for spraying and 35 %SD for seed treatment. Metalaxyl is one of the best systemic fungicides and is highly effective in controlling many pathogens belonging to Oomycetes. It is a curative fungicide and can be applied even after the infection has started. Damping off of seedlings and seedling rots caused by *Pythium* and *Phytophthora*, as well as downy mildews of pearl millet, sorghum, maize, peas etc. can be controlled by treating the seeds with the chemical at the rate of 2.0 gm./ kg. of seeds. Downy mildews of several crops, such as pearl millet, sorghum, maize, cucumbers, grapevine, onion, crucifers etc., late blight of potato and tomato, blight of pepper etc. can be effectively controlled by spraying the chemical at the rate of 1.0 ml./ liter of water. Metalaxyl in combination with a broad-spectrum contact fungicide, such as mancozeb or zineb is effective in controlling several diseases simultaneously, besides downy mildew. Ridomil MZ, available as 72 %WP contains 8 % metalaxyl and 64 % mancozeb. Similarly, seed treatment with metalaxyl in combination with a broad spectrum, non-systemic, contact seed treating fungicide, such as thiram or captan is effective against many seed-borne diseases, including damping off of seedlings and downy mildews. It is also used for soil drenching for the control of the diseases mentioned above. Many strains of some fungal pathogens have developed resistance to metalaxyl and as such, it is always better to add one broad-spectrum, contact fungicide along with it.

Metalaxyl sold as Apron 35 %SD is used for seed treatment for the control of many seed-borne and soil-borne pathogens belonging to Oomycetes.

(v) Pyrimidines

Pyrimidines include diamethirimol, ethirimol, bupirimate, fenarimol and nuarimol.

Diamethirimol (Milcurb), **ethirimol** (Milstem) and **bupirimate** (Nimrod) are effective against powdery mildews of various crop plants.

Fenarimol (Rubigan) and **nuarimol** (Trimidal) are effective against powdery mildews and also several other leaf spots, rusts and smut fungi.

(vi) Triazoles

Triazoles include several excellent systemic fungicides, such as **triadimefon** (Bayleton), **triadimenol** (Bayton), **bitetanol** (Baycor), **butrizol** (Indar), **propiconazole** (Tilt), **etaconazole** (Vangard), **myclobutanil** (Rally) and **difenoconazole** (Score). All these fungicides possess long, protective and curative effect against a wide range of root and seedling diseases, and foliar diseases, such as leaf spots, blights, powdery mildews, rusts, smuts and other diseases caused by many fungi belonging to Ascomycetes, Basidiomycetes and Fungi Imperfecti. They are used as foliar sprays and for seed and soil treatment. The spray fluid is prepared by mixing 1.0 gm. of the fungicide in 1.0 liter of water.

10. Miscellaneous Systemic Fungicides

Ethazole. Ethazole sold as Truban, Terrazole, Koban etc. is a seed, soil and turf fungicide, effective against damping off, as well as root and stem rots caused by *Pythium* and *Phytophthora.* In combination with PCNB or thiophanate methyl, it is sold as Banrot, which is a broad-spectrum fungicide capable of controlling *Fusarium* and *Rhizoctonia* diseases also.

Imazalil. Imazalil is sold as Fungaflor. It is effective against many fungi belonging to Ascomycetes and Fungi Imperfecti, causing powdery mildews, leaf spots, fruit rots and *Fusarium* wilts. It is applied mostly as a foliar spray, but is also used for seed treatment and for post-harvest treatment of fruits. It has good preventive and curative properties.

Prochloraz. It is available as Prochloraz and is effective against diseases caused by Ascomycetes and Fungi Imperfecti, such as powdery mildews, leaf spots, blights and fruit rots. It is used for foliar spraying and also for seed treatment.

Propamocarb. It is sold as Banol, Previcur etc. It is effective against diseases caused by *Pythium* and *Phytophthora,* as well as downy mildews, some rusts and a few other diseases. It is applied as a seedling dip, soil drench, foliar spray and also as a seed dressing fungicide.

Triforine. It is sold as Cela, Funginex, Saprol etc. It is effective against many diseases caused by Ascomycetes and Fungi Imperfecti, causing powdery mildews, foliar and fruit spots, fruit rots, anthracnose and some rusts. It is used as a foliar spray.

Dodine. Its technical name is 'n-dodecyl guanidine'. It is sold as Cyprex, Guanidol, Melprex, Syllit etc. as 65 %WP for spraying. It is an excellent fungicide against apple scab, besides certain foliage diseases of cherry, strawberry and roses. It is also effective against several other diseases caused by ascomycetous and deuteromycetous fungi. It has got local systemic properties. It is a good eradicant and has long-lasting protective effect. Some strains of apple scab fungus, resistant to this fungicide have appeared in recent times. It is not compatible with Bordeaux mixture, lime, oils and Streptomycin.

Tricyclazole. It is available as Beam, Bim etc. as 20 %WP for spraying, 75 %WP for seed treatment, 1 %D for dusting and 4 %G for soil application. It is mostly used as a seed treatment chemical at the rate of 2.0 gm./ kg. of seeds. It is an excellent fungicide for the control of rice blast and seedlings from treated seeds are protected for about 30 days after sowing. For spraying, the wettable powder formulation is mixed with water at the rate of 1.0 gm./ liter.

Pyroquilon. It is sold as Fongorene and is available as 50 %WP for spraying and 5 %G for soil application. It is also an excellent seed-treating chemical. Seed treatment at the rate of 2.0 gm./ kg. of seeds protects rice seedlings from blast disease for about 30 days after sowing. For spraying, the fungicide is mixed with water at the rate of 1.0 gm./ liter of water.

Isoprothiolane. It is available as Fugi-one. It is a specific fungicide against rice blast and can control stem rot of rice also.

Qualities of An Ideal Systemic Fungicide

1. It should be capable of attacking the pathogen directly and destroy it. The secondary products produced, as the fungicide disintegrates inside the plant tissue should also have fungicidal properties to destroy the pathogen.

2. It should be able to induce some changes in the physiological functioning of the host plant and induce some type of biological, chemical or physical resistance by triggering the defense mechanism in the host plant, so as to resist the attack by the pathogens.

3. It should be stable and able to remain in the plant tissue for a long period of time and prevent infection by the pathogen and eradicate the pathogen, if the pathogen has already established inside the host.
4. It should not have any harmful effect on the host plant.
5. It should have less residual toxicity in the edible parts of the plant and in the seeds.

To develop a systemic fungicide with all these characteristics is rather very difficult. However, most of the systemic fungicides that are in use are supposed to be harmless to the host plants, cattle and humans

11. Antibiotics

'Antibiotics' are substances produced by some microorganisms that are toxic to some other microorganisms. Most antibiotics are produced by Actinomycetes, while certain fungi and branching bacteria produce others. When antibiotics are applied for disease control, they are absorbed and translocated systemically in the tissues. They move upwards, downwards and laterally in the plant tissues. They may control plant diseases by acting on the pathogen directly or on the host. So far, very few antibiotics are available for the control of plant diseases. Penicillin, Streptomycin, Cycloheximide and Tetracycline are effective against bacterial pathogens, while Griseofulvin, Actidione, Phymaricin, Blasticidin, Nystatin, Bulbiformin and Aureofungin are effective against fungal pathogens. Antibiotics possess protective, eradicant and curative properties. When they are used at higher concentrations, they may be harmful to the host plants. They are capable of controlling specific pathogens. Some may control a few bacterial pathogens, while some others may control only a few fungal pathogens. Actinomycetes, which can control bacterial pathogens, are not effective against fungal pathogens.

(i) **Agrimycin.** It is available as Agrimycin 100, Phytomycin and Orthostreptomycin. Agrimycin contains 15 % Streptomycin sulfate and 1 - 5 % Terramycin. It is effective against fire blight of apple, *Citrus* canker, bacterial blight of cotton and black leg of potato. It is mostly used as a spray fungicide, but is also used for seed treatment for the control of seed-borne diseases of potato, cotton, cereals, cauliflower etc. When this antibiotic is sprayed in combination with copper oxychloride fungicide, it is more effective in controlling the diseases. To control rice bacterial leaf blight and bacterial leaf streak, cotton bacterial blight, *Citrus* canker and several other bacterial diseases attacking fruit crops, it is sprayed at the rate of 200 - 400ppm (0.2 -

0.4 gm./ liter of water). To control rice bacterial blight and rice leaf streak diseases, 120 gm. of Agrimycin is mixed in 200 liters of water and sprayed in one acre of crop. Addition of 500 gm. of copper oxychloride along with 120 gm. of Agrimycin is highly effective in controlling the above diseases. Bacterial leaf blight (black arm) of cotton can be controlled by mixing 40 gm. of Agrimycin and 800 gm. of copper oxychloride in 300 liters of water and sprayed over an area of one acre.

(ii) **Streptomycin**. It is produced by the Actinomycete - *Streptomyces griseus.* It prevents protein synthesis in bacteria, thereby kills them. Streptocycline, which is a combination of Streptomycin or Streptomycin sulfate and Terramycin, is a broad-spectrum bactericide capable of controlling a wide range of bacterial pathogens causing spots, blights and rots. *Citrus* canker, bacterial leaf spot of tomato and pepper, halow blight of French bean, fire blight of apple, *Fusarium* wilt of cotton etc. are controlled effectively by this antibiotic. It can be used for spraying, for seed treatment, root dip, for treating tubers, cuttings etc. and for protecting fruits, tubers etc. For spraying Streptomycin is used at the rate of 100 - 500ppm (0.01 - 0.05 gm./ liter of water). Bacterial leaf blight and leaf streak of rice can be controlled by spraying 6.0 gm. of Streptomycin mixed in 200 liters of water. For seed treatment to control leaf blight and leaf streak of rice, 2.0 gm. of the chemical is mixed in 120 liters of water and the seeds are soaked in the antibiotic solution for 8 hours and sown. It is used as a seed dip of potato seed pieces for the control of bacterial rots and cuttings of betelvine plants for the control of foot rot. It is also effective against some fungal diseases, such as late blight of potato and tomato.

(iii) **Cycloheximide**. It is commonly referred to as 'Actidione' and is available as Actidione, Actispray, Actidione PM and Actidione RZ. It is produced by *Streptomyces griseus* and *S. nouresi.* It inhibits protein synthesis in the fungal pathogens. It is used as a spray and as a protectent of harvested fruits and vegetables. It is not effective against bacterial pathogens. For spraying it is used at 20 - 100 ppm (0.02 - 0.1 gm./ liter of water). It is used for the control of cherry leaf spot, powdery mildew of rose and beans, rust of wheat, covered smut of oats, bunt of wheat, brown rot of peach besides post-harvest rots caused by *Rhizopus, Botrytis* etc. Because of high residual toxicity, its use for post-harvest treatment is very much limited.

(iv) **Griseofulvin**. It is produced by *Penicillium griseofulvum, G. patulum, P. nigricans* and *P. jancyewski* It is toxic to several plant pathogenic fungi, such as *Botrytis fabae, Alternaria solani,* some powdery mildews and rusts. Foliar spraying with this chemical at 100 - 1,000 ppm (0.1 - 1.0 gm./ liter of water) controls powdery mildew of rose and bean, downy mildew and powdery mildew of cucumbers, early blight of tomato, blight of pea, apple rot etc.

(v) **Aureofungin**. It is available in a water-soluble form as Aureofungin Sol. It is a broad-spectrum antifungal antibiotic produced by *Streptoverticillium cinnamomeum* var. *terricola (Streptomyces cinnamomens)*. It is absorbed by the aerial plant parts or roots and translocated inside the plants systemically. For spraying, it is used at the rate of 20 gm. /120 liters of water. It is capable of controlling several diseases, such as *Citrus* gummosis, downy mildew, powdery mildew and anthracnose of grapes, sugary disease of pearl millet, groundnut 'tikka' diseases, leaf rusts of wheat, blast disease of rice and ragi, brown spot of rice, fruit rot of mango, tomato etc. It can be used for seed treatment to control some diseases. To control downy mildew of pearl millet, 2.0 gm. of the chemical is dissolved in 60 liters of water and the seeds are soaked in the solution for 2 hours, shade dried and then sown. Rice blast and rice brown leaf spot can be controlled by soaking the seeds for one night in water, followed by soaking the seeds in a solution containing 1.2 gm. of Aureofungin Sol and 1.2 gm. of copper sulfate in 60 liters of water for one hour, allowed to sprout and then sown. Potato seed pieces are soaked in a solution containing 2.0 gm. of Aureofungin Sol in 60 liters of water for 30 minutes for the control of rot diseases. To prevent fruit rots of tomato, mango etc., the fruits are dipped in a solution containing 5.0 gm. of Aureofungin Sol in 10 liters of water. It can also be applied through the roots for the control of *Ganoderma* root wilt of coconut. For this 2.0 gm. of Aureofungin Sol and 1.0gm. of copper sulfate are dissolved in 100 ml. of water and the solution injected or fed through the roots. Copper sulfate added to Aureofungin acts as a synergist.

(vi) **Nystatin**. It is known as Mycostatin, Fungicidin etc. It is an antifungal antibiotic produced by *Streptomyces noursei*. Anthracnose of beans and downy mildew of cucumbers are controlled by foliar spraying with the antibiotic. It is used for seed treatment against stripe disease of barley. Post-harvest dip with this antibiotic controls peach brown rot and banana anthracnose.

(vii) **Bulbiformin**. It is another antifungal, antibiotic produced by *Bacillus subtilis*. Soils amended with *Bacillus subtilis* controls wilt disease of pigeon pea caused by *Fusarium udum*.

(viii) **Blasticidin**. It is produced by *Streptomyces griseochromogenes*. It is effective in controlling rice blast even at very low concentrations of 50 - 100 ppm.

(ix) **Tetracyclines**. Tetracyclines are antibiotics produced by a number of species of *Streptomyces*. Among the Tetracyclines, Terramycin (Oxytetracycline) is produced by *Streptomyces rimosus* and Aureomycin (Chlorotetracycline) by *Streptomyces aureofaciens*. Terramycin is used as a soil drench or as root dip for the control of crown gall of apple. In combination with Streptomycin, it controls fire blight of apple and pear. Tetracyclines can control diseases caused by *Mycoplasmas* and *Fastidious vascular bacteria*.

(x) **Kasugamycin**. It is produced by *Streptomyces kasugaensis*. It is commercially available as Kasumin and is specific for the control of blast disease of rice, finger millet and other millets.

Table 5. List of antibiotics produced by microorganisms for disease control

Antibiotics	Produced by	Target pathogens
1.Griseofulvin	*Penicillium griseofulvum* *Penicillium patulum* *Penicillium nigricans* *Penicillium jancyewski*	Fungi
2.Ustilagic acid	*Ustilago zeae*	Fungi
3.Aspergillic acid	*Aspergillus flavus*	Bacteria
4.Penicillin	*Penicillium notatum* *Penicillium chrysogenum*	Bacteria
5.Cycloheximide	*Streptomyces griseus* *Streptomyces nouresi*	Fungi
6.Aureofungin	*Streptomyces cinnamomens*	Fungi
7.Chloromycetin	*Streptomyces venezuelae*	Bacteria
8.Nystatin	*Streptomyces nouresi*	Fungi
9.Neomycin	*Streptomyces fradiae*	Bacteria
10.Streptomycin	*Streptomyces griseus*	Bacteria
11.Terramycin (Oxytetracycline)	*Streptomyces rimosus*	Bacteria
12.Aureomycin (Chlorotetracycline)	*Streptomyces aureofaciens*	Bacteria
13.Blasticidin	*Streptomyces griseochromogenes*	Fungi
14.Erythromycin	*Streptomyces erythreus*	Bacteria
15.Polymycin	*Bacillus subtilis*	Bacteria
16.Bulbiformin	*Bacillus subtilis*	Fungi
17.Thryothrycin	*Bacillus brevis*	Bacteria
18.Kasugamycin	*Streptomyces kasugaensis*	Fungi

12. Carbonate Compounds

Sodium bicarbonate, ammonium bicarbonate and potassium bicarbonate plus 1 % neem oil or sunflower oil or fine mineral oil spray is inhibitory and fungicidal to the powdery mildew pathogen of rose *(Sphaerotheca pannosa)*, black spot pathogen of rose *(Diplocarpon rosae)* and to several fungal pathogens of cucumber, gray mould pathogen *(Botrytis cinerea)* etc.

13. Phosphorus Compounds

Spraying with solutions of monopotassium phosphate or dipotassium phosphate gives satisfactory control of powdery mildew diseases of cucumber and grapes.

14. Film Forming Compounds

Film forming compounds, such as anti-transpirant polymers, mineral oils, surfactants, vegetable oils etc. have been found to give satisfactory control of some plant diseases. Most of these film forming materials are permeable to gases, non-phytotoxic, resist weathering and are biodegradable. They reduce infections by altering the characteristics of the leaf surface and interfere with the adhesion of the pathogen to the host and with the recognition of infection sites on the host. Mineral oils of petroleum origin have been commercially used for the control of banana 'Sigatoka' leaf spot disease *(Mycosphaerella musicola)*. Powdery mildew of apple *(Podosphaera leucotricha)* can be effectively controlled by spraying with oils obtained from seeds of sunflower, corn, soybean etc.

Quantity of Fungicidal Fluid Required for Spraying

The quantity of fungicidal fluid required for spraying, depends to a large extent on the growth stage and age of the crop, variety of the crop, extent of foliage etc. A young crop may require lesser quantity of spray fluid than grown-up crop, which may require a higher quantity. Crops with wider canopy and more foliage may require more spray fluid. In groundnut, the spreading varieties require more spray fluid than the bunch varieties. As compared to rice crop, sugarcane crop may require about 2½ times more of spray fluid. In the case of rice, lesser quantity of spray fluid is required for short varieties, with less tillering capacity and narrow leaves than tall varieties, with high tillering capacity and broader leaves. Well-fertilized, irrigated and well-maintained crops, with luxuriant growth and dense foliage require more spray fluid. Further, irrigated crops may require more spray fluid than rainfed crops. Taking all the above facts into consideration, the quantity of fungicidal fluid required for spraying on specific crops is ascertained. The quantity of spray fluid required for spraying on different crops in general is furnished in **table 7**. For dusting, 10-15 kg. of dust formulation is required for 1.0 acre and for soil drenching a quantity of 1,000 liters per acre is required.

Table 6. Quantity of spray fluid required for spraying on different crops

Crop	Quantity of spray fluid required (liter / acre)
Rice	200 - 250
Ground nut	250 - 300
Cotton	300 - 500
Millets	200 - 350
Potato	300 - 350
Pulse crops	250 - 350
Sugarcane	350 - 500
Wheat	250 - 350
Chillies, tomato, eggplant, tobacco	200 - 350
Grapevine	500 - 600
Banana	250 - 600
Coconut	5.0 lit. / tree
Arecanut	3.0 lit. / tree
Mango, guava, sapota	5.0 - 10.0 lit. / tree
citrus plants	5.0 - 10.0 lit. / tree

Preparation of Fungicidal Spray Fluid

The percentage of active ingredient present in the different fungicidal formulations varies from one another. To prepare spray fluids, the formulations have to be mixed with appropriate quantities of water prior to spraying. In the case of fungicides, usually the actual percentage of active ingredient present in each of the formulation is not taken into consideration while preparing the spray fluid. The active ingredient is always taken as 100 % and based on that spray fluid is prepared. Thus, in the formulation of edifenphos, which actually contains 50 % of active ingredient, for the purpose of preparing spray fluid the active ingredient is taken as 100 %. So, to prepare 0.1 % of edifenphos spray fluid, 1.0 ml. of the formulation is mixed with 1.0 liter of water. The actual quantity of water to be mixed with the fungicidal formulations to get spray fluids containing the desired percentage of active ingredient is furnished in **table 7**.

Table 7. Quantity of fungicidal formulation to be mixed with water to get spray fluid of desired concentration

Fungicide	Percentage of Active ingredient in the spray fluid (a.i. in the formulation taken as 100 %)	Quantitv of fungicidal formulation to be mixed with water (gm./ml. Per lit. of water
Wettable sulfur	0.4 %	4.0 gm. / lit. of water
Copper oxychloride	0.25 %	2.5 gm. / lit. of water
Dithiocarbamates (Ziram, Thiram, Ferbam, Zineb, Maneb, Mancozeb, Nabam)	0.2 %	2.0 gm. / lit. of water
Quinone fungicides (Chloranil, Dichlone)	0.125 - 0.15 %	1.25 - 1.50 gm. / lit. of water
Benzene fungicides (Chloroneb, Chlorothalonil, Dichloran, Dinocap)	0.125 - 0.15 %	1.25 - 1.5 gm. / lit. of water
Heterocyclic nitrogen compunds (Captan, Captafol, Folpet, Iprodione)	0.125 - 0.15 %	1.25 - 1.5 gm. / lit. of water
Organo tin fungicides (Du-ter, Brestan, Brestanol)	0.125 - 0.15 %	1.25 - 1.5 gm. / lit. of water
Systemic fungicides (benomyl, carbendazim, thiabendazole, thiophanate, thiophanate methyl, oxanthiins, edifenphos, fosetyl-Al, iprobenphos, acetylalanines)	0.1 %	1.0 gm. / lit. of water

Instead of using a high volume sprayer for spraying, if a semi-low volume or low volume sprayer is used, the same quantity of the fungicidal formulation used in high volume spraying should be used in semi-low or low volume spraying also. Because, lesser quantity of water is required in semi-low or low volume spraying, proportionately lesser quantity of the fungicidal formulation should not be used. If 200 gm. of fungicidal formulation in 200 liters of water is used to cover a unit area using a high volume sprayer, the same quantity of 200 gm. of the formulation should be used to cover the same area by using a semi-low or low volume sprayer, even though water required for preparing the spray fluid may be much less.

Plant Protection Appliances

For effective control of plant diseases, besides selection of appropriate chemical formulations, selection of proper plant protection appliances to carry out the operations is also quite important. Fungicides are mostly used as sprays and dusts, using sprayers and dusters respectively. Different types of sprayers and dusters are being used for such operations and in both, there are manually operated, as well as power operated equipments. In sprayers there are both hydraulic and pneumatic types.

Sprayers. The main function of a sprayer is to atomize or break the fungicidal solution into fine droplets and to ensure uniform distribution of the fungicide in a specific quantity of spray fluid on the surface of parts to be sprayed. Based on the quantity of spray fluid used on a specific area, the spraying is termed as **'high volume spraying'**, **'semi-low volume spraying'**, **'low volume spraying'** or **'ultra low volume spraying'**.

For high volume spraying, a quantity of 200 - 300 liters of spray fluid is required to spray one acre of small crops, such as rice, groundnut, finger millet, wheat etc. While spraying, the spray fluid is broken into droplets of size 250 - 400µ. When several such droplets fall on the sprayed surface, the droplets may aggregate to form bigger droplets and may role down to the ground leading to wastage. In semi-low volume spraying, which is done with the aid of ordinary power sprayers, one acre can be covered with 50 - 100 liters of spray fluid. The spray fluid is mixed with air and forced out at high speed, as a result the fluid is broken into very small droplets of size 150 - 200ì. In low volume spraying, which is also called 'concentrate spraying', one acre can be covered with 5 - 50 liters of spray fluid. The spray fluid is sprayed under high pressure and is broken into minute droplets of size 70 - 150ì. Here the chances of droplets aggregating and rolling down from the sprayed surface is very much reduced, at the same time more area can be covered in lesser time. In ultra low volume spraying, no water is added to the formulation and it is sprayed as such. To cover one acre 0.2 - 2.4 liters of the formulation is sufficient. The chemical is mixed with air and sprayed at high pressure and it is broken into very minute, mist-like droplets of size 30 - 50ì, which settles down on the crop surface as mist.

Parts of a Typical Sprayer

(i) **Spray fluid tank.** The spray fluid tank is provided inside the sprayer itself in the pneumatic hand sprayers, knapsack sprayers etc., whereas in foot sprayers, rocker sprayers, bucket sprayers etc., the fluid tank is not provided inside the sprayer. The spray fluid is kept in separate vessels outside, into which the suction hose is put. In the sprayers provided with a tank inside the sprayer unit, the tank is made of non-corrosive metal, such as brass, bronze or stainless steel, which can withstand the corrosive action of chemicals or of high-density polyethylene material. The tank capacity may vary from 0.5 - 1.0 liter in small hand sprayers, 10 - 25 liters in hand operated knapsack sprayers, semi-low volume sprayers or power mist blowers and up to 2,700 liters in big trolley-mounted or tractor-mounted power-operated sprayers. The tank is provided with a mouth with screw cap to fill up the spray fluid, a strainer to filter the spray fluid, a tap at the side of the bottom to remove the unused spray fluid and an agitator to constantly agitate the spray fluid at the time of spraying.

(ii) **Agitator.** Emulsifiable concentrates and wettable powders, when mixed with water to form a spray fluid, do not go into a solution completely, but are dispersed in water as finely suspended particles. When the spray fluid is not agitated, these suspended particles may tend to settle down and hence the active ingredient may not be distributed uniformly. To prevent settling down of these suspended particles, the spray fluid is constantly agitated while spraying and for this purpose, a paddle-like agitator is provided inside the fluid tank in such a way that when the sprayer is operated, the agitator also moves inside the tank and agitates the spray fluid constantly. In sprayers, which do not have built-in fluid tank, such as rocker sprayers, foot sprayers, bucket sprayers etc., the spray fluid is frequently agitated manually with a rod.

(iii) **Filters.** Dust, refuse, caked materials etc. found in the spray fluid may clog the nozzle and obstruct spraying operations. To prevent this, a strainer is provided in the mouth, which filters off all such materials while filling up the tank with the spray fluid. Similarly, fine filters are also provided at the points where the spray fluid passes from the fluid tank into the pump and from the pump into the delivery tube.

(iv) **Pump**. The pump helps to force the spray fluid from the fluid tank through the delivery tube under pressure as small droplets on to the area to be sprayed. The efficiency of a sprayer depends to a large extent on the action of the pump.

In **'pneumatic sprayers'**, which work on the principle of air pressure, an air pump or pneumatic pump is used. Air is pumped into the tank containing the spray fluid, as a result pressure is applied on top of the spray fluid and the compressed air forces the spray fluid to come out of the tank through the delivery tube under high pressure. In such type of sprayers, no pressure is exerted on the spray fluid directly. On the other hand in **'hydraulic sprayers'**, the pressure from the pump is exerted directly on the spray fluid, as a result the spray fluid is forced out from the pump through the delivery tube as small droplets.

(v) **Power source**. For the sprayers operated by hands or legs, only manpower is used. For power sprayers mechanical power from petrol or kerosene powered machines are used.

(vi) **Pressure gauge**. In pneumatic sprayers, pressure gauges are provided to measure the air pressure inside the fluid tank or inside the air chamber. Manually operated sprayers, such as 'Marut' sprayers, rocker sprayers, foot sprayers etc., as well as power operated pneumatic sprayers work efficiently at pressures ranging from 15 - 20 pounds.

(vii) **Valves**. To regulate the air or spray fluid entering the pump and to regulate the flow of spray fluid moving on to the delivery tube from the tank, ball-type valves with valve seats are provided. A cut-off valve is also provided in the lance to cut off the spray fluid from being sprayed.

(viii) **Hose**. In sprayers where the spray fluid tank is not provided within the sprayer, such as rocker sprayers, foot sprayers, bucket sprayers etc., a suction hose is provided to suck in spray fluid from the vessel into the pump assembly. In all sprayers, a delivery hose is provided to take the spray fluid from the pump to the spray lance. The hoses are made of pliable nylon or rubberized tubes and are made to withstand high pressure, to last longer and to resist the corrosive action of materials, such as petroleum oils, grease and other chemical compounds.

(ix) **Spray lance**. It is a long, thin, jointless tube, either straight or smoothly bent slightly at the terminal end and mostly about 90 cm. long in manually operated sprayers and made of non-corrosive metals, such as brass. It is attached to the delivery tube with a screw clip and at its terminal end the nozzle is screwed. A handle and a cut-off valve are provided near the place where it is attached to the delivery tube.

(x) **Nozzle**. The nozzle is another important part of a sprayer and is responsible to break the spray fluid into small droplets. Several types of nozzles are used to meet different requirements. Depending upon the type of the nozzles, the size of the spray droplets and the quantity of spray fluid vary.

In general, the nozzle consists of a basal bracket, a body, a circular disc or orifice plate, washer, a whorl plate and a screw cap with a central hole. The nozzles are of two types functionally viz., the **'hollow cone nozzle'** and the **'solid cone nozzle'**. In the hollow cone nozzle, the spray fluid, which comes out of the pump under pressure, strikes against the whorl plate, swirls and comes out as a hollow boom or a hollow cone with its center hollow, whereas in the solid cone nozzle, the spray fluid comes out through the orifice plate and the whorl plate in such a way that the hollow of the cone is also filled with the spray fluid forming a solid cone.

(xi) **Air-chamber**. In some of the pneumatic, manually operated sprayers, such as the rocker sprayers, foot sprayers etc. and pneumatic power sprayers, a separate air chamber is provided. When the pump is operated, air fills up in the air chamber and pressure is developed inside. The compressed air forces the spray fluid through the delivery tube. In such sprayers, as long as there is sufficient pressure inside the air- chamber, the spray fluid comes out continuously.

Types of Sprayers

Manually Operated Hydraulic Sprayers

1. **Hand syringe or garden syringe**. It is the oldest and simplest type of sprayer consisting of a long cylinder or barrel made of tin or polythene and a plunger with a wooden handle. At the anterior end of the plunger, a leather valve is attached in-between a metal washer and a screw nut. The leather valve fits closely within the wall of the barrel. At the anterior end of the barrel, a nozzle with a small hole is provided. The spray fluid is kept in a separate vessel. It works as an ordinary injection syringe. The nozzle is placed inside the spray fluid and the plunger is pulled up, as a result a vacuum is created inside the barrel and the spray fluid enters the barrel to fill up the vacuum. Then the nozzle is pointed towards the area to be sprayed and the plunger is pushed in. The pressure exerted directly on the spray fluid forces it to come out

through the nozzle as droplets. Thus, every time the spray fluid is first drawn into the barrel and then sprayed. The spray droplets are fairly large in size and the spraying is not continuous. It is used to spray potted plants and small kitchen gardens.

2. **Bucket sprayer**. It consists of a long barrel made of brass, a plunger, valves and a foot pedal to hold the sprayer pressed to the ground in an upright position. There is a suction hole with a valve at the side near the bottom of the barrel and a delivery tube at the side near the top end of the barrel, which is closed at the top end. A valve is provided at the delivery tube to which the delivery hose is attached. The spray fluid is kept in a suitable vessel. The barrel is placed inside the vessel and the sprayer is held in an upright position with the foot pedal. During the upward stroke, the lower valve opens and the spray fluid enters the barrel to fill the vacuum created. During the downward stroke, the lower valve closes and due to the pressure exerted on the spray fluid inside the barrel, the upper valve opens and the spray fluid is forced out through the delivery tube and sprayed through the nozzle. Spraying is done only during the downward stroke and as such the spraying is not continuous. Two workers are required to undertake spraying operation, one to operate the sprayer and the other for spraying. It is suitable to spray small crop areas, kitchen gardens, small bushes etc. **(Fig. 83)**

3. **Hydraulic knapsack sprayer**. This type of sprayers can be conveniently carried on the back and for this, the tank is shaped like a bean-seed with a slight curve. It works on the same principle as the bucket sprayer but the pump is fitted inside the tank itself. The capacity of the tank may vary from 10 - 20 liters and is made of brass or high-density polyethylene material. With the help of a lever and handle attached to the plunger it can be operated with one hand. An agitator is also provided inside the tank. When the plunger is moved up and down by operating the lever, the agitator also moves up and down and agitates the spray fluid. There is a delivery tube at the top of the barrel to which a short delivery hose is attached and the spray lance is attached at the terminal end of the delivery hose. A cut-off valve is provided in the lance. One worker is enough to operate the sprayer by working the plunger with one hand and do the spraying with the other hand simultaneously. Because the plunger is continuously operated up and down, the spraying is also continuous. It is ideal for spraying on rice, vegetable crops and other small crops and 1.0 - 2.0 acres can be covered in a day **(Fig. 84)**

Manually Operated Pneumatic Sprayers

1. **Pneumatic hand sprayer**. In this type of sprayers the fluid tank serves as the pressure chamber as well. The spray fluid is forced out through the delivery tube, only because of the air pressure built in above the spray fluid and hence some space should be left empty in the fluid tank. A pressure gauge is fitted to the tank to measure the built-in pressure. When the plunger is drawn up, air from outside enters the pump and when the plunger is pushed down, the air inside the pump is forced into the fluid tank and fills the empty space above the spray fluid. When the pump is worked in this way several times, sufficient pressure is built up in the empty space above the spray fluid and the spray fluid is forced out through the delivery tube, when the cut-off valve provided in the spray lance is opened. The cut-off valve with a lever arrangement helps to open and close the delivery tube, one end of which opens at the bottom of the fluid tank for the spray fluid to enter the delivery tube and the other end opens at the nozzle for the spray fluid to be sprayed. As long as there is enough air pressure inside the tank, spraying can be done continuously. When the pressure becomes less the pump is worked again to increase the pressure. These sprayers can be used to spray garden plants, flower plants, kitchen gardens and in the house to destroy household insects. The tank capacity ranges from 0.5 - 1.0 liter.

2. **'Marut' backpack knapsack sprayer**. In this type of sprayer also the fluid tank serves as the pressure chamber. To withstand high pressure, the tank is made of strong brass sheet. In the center of the tank, a pump is fitted with a plunger and wooden handle in a perpendicular position. A mouth, with a screw cap to fill up the tank with the spray fluid and a pressure gauge are provided at the top of the cylindrical tank. A delivery tube is provided at the side near the bottom of the tank. One end of the delivery hose is attached to the delivery tube, while the other end is attached to the spray lance with a nozzle and cut-off valve. It works on the same principle as the pneumatic hand sprayer. The tank capacity varies from 3.5 - 17.0 liters and pressure ranging from 2 - 4 kg. /cm ^{3} is developed inside the tank. The sprayer is placed on the ground and the pump is worked, so that sufficient pressure is built inside the tank. Then the sprayer is carried on the back and spraying is done. One worker is enough to carry out spraying operation and an area of 1.0 - 2.0 acres may be covered in a day. It is used to spray small crop areas and is largely used by the Public Health Department **(fig. 87)**

3. **Rocker sprayer**. This type of sprayers can be operated by standing and moving the long rocking lever forward and backward. It consists of a pump, pressure chamber or air chamber, suction hose, delivery hose, lance, nozzle and valves, and the whole sprayer assembly is mounted on a wooden board. There is no tank attached to the sprayer and so, the spray fluid is kept in a separate vessel in which the suction tube is put. When the rocking movement of the handle operates the pump, pressure is developed inside the pressure chamber and the spray fluid, which is sucked into the pump through the suction hose is forced through the delivery tube due to the pressure developed in the pressure chamber. However, to get a continuous, uniform spray the pump has to be operated continuously. To operate the sprayer, two workers are required, one for operating the pump and the other for spraying. Long delivery hose may be provided for the sprayer so that larger areas can be covered without shifting the sprayer often. Tall crops and fairly high trees can be sprayed with this sprayer by using a straight, high-tree lance. In the pressure chamber an air pressure upto 14 - 18 kg. /cm.^{3} may develop and an area of 3.0 - 4.0 acres can be covered in one day **(Fig. 85)**
4. **Foot sprayer**. This type of sprayer is operated from a standing position by operating the foot pedal up and down. The sprayer, which is fitted to a metal stand in an upright position, consists of a vertical pump, a plunger attached to a lever and pedal, pressure chamber at the top of the pump barrel, suction tube, delivery tube, spray lance, nozzle, valves etc. The pedal attachment is provided with springs, so that when the pedal is pushed down, it automatically comes to its original position. There is no tank attached to the sprayer and the spray fluid is kept in a separate vessel into which the suction tube is put. It works on the same principle as the rocker sprayer. In the pressure chamber, pressure develops up to 17 - 20 kg. /cm.^{3}. Long delivery hose may be attached to the sprayer. Tall crops, high trees etc. can be sprayed with this sprayer by using a straight, high-tree spray lance. Two workers are required to undertake spraying operations, one for working the pump and the other for spraying and an area of 3.0 - 5.0 acres can be covered in one day **(Fig.86)**

Manually Operated Mist Blowers

Hand atomizer or flit pump. The working of this type of sprayer is almost similar to that of pneumatic sprayers. It is a very handy sprayer used for the control of mosquitoes and other household pests, besides control of diseases of garden plants and potted plants. The spraying is not continuous.

Fig. 83 : Bucket sprayer

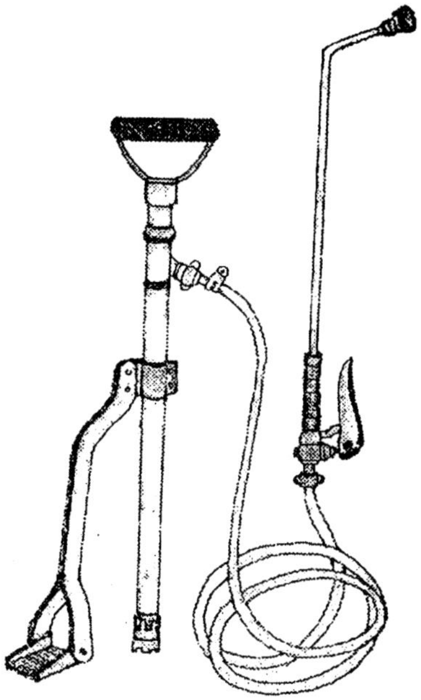

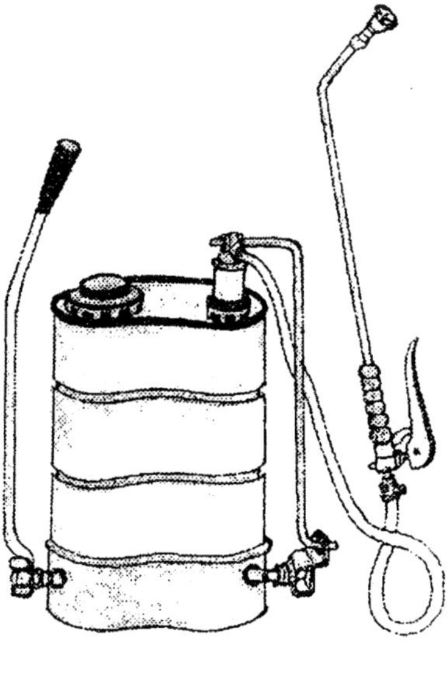

Fig. 85 : Rocker sprayer

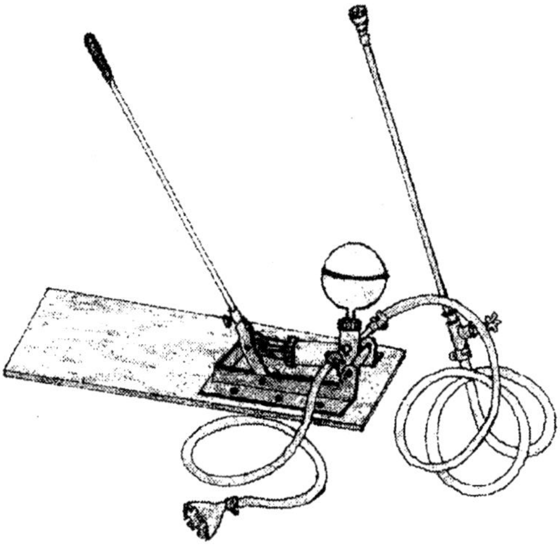

Fig. 86 : Foot sprayer

Iydraulic sprayers

Fig. 84 : Knapsack sprayer

Fig. 86. Foot sprayer

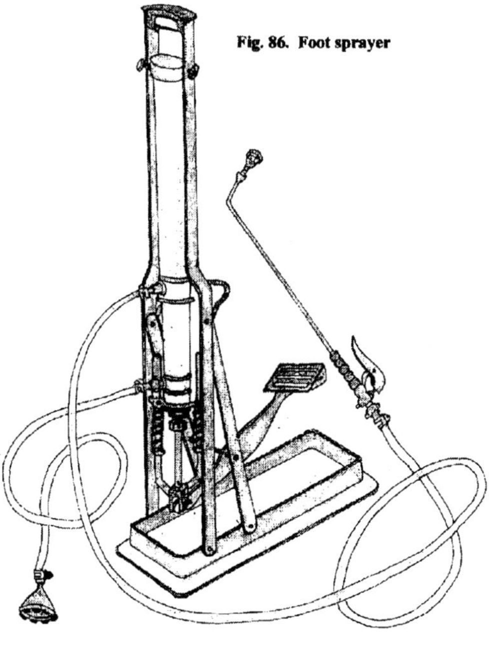

The sprayer consists of a cylindrical tube or barrel 25 cm. in length and 3.75 cm. in diameter and made of tin sheet or high-density polyethylene material. The posterior end of the barrel is covered with a screw cap, which is provided with a hole in the center through which passes the plunger. The closed anterior end of the barrel is provided with a small hole in the center, which serves as the nozzle. The plunger is made of a metal rod with a wooden handle at the posterior end. A metal washer and a screw nut with a leather valve in-between them, which fits closely within the wall of the barrel serves as the piston. The spray fluid tank is also made of tin sheet or polyethylene material having a capacity of 227 ml. and is provided with a screw cap known as the filler cap. The barrel is attached to the filler cap of the tank on the lower side of the anterior end. A small tube, 3.12 mm. in diameter passes through the center of the filler cap and is at right angle to the barrel. This is the delivery tube, one end of which opens near just half in front of the nozzle, while the other end opens near the inner bottom of the tank. The spray fluid is filled in the tank after removing the filler cap and then screwed back tightly.

When the plunger is drawn back, a good quantity of air enters the barrel through the nozzle. When the plunger is pushed down, pressure is exerted on the air inside the barrel. As the air cannot escape through the piston, it is forced out through the nozzle at great force. When the air goes out with such great force, some of the air inside the delivery tube is also taken out causing a vacuum in the delivery tube, which is filled up by the spray fluid coming up from the tank. This fluid thus entering the delivery tube is also blown out along with the air in the form of a fine mist **(Fig. 88)**

Power Operated Sprayers

Power mist blower. In such sprayers, which are also known as **'semi-low volume sprayers'**, the spray fluid comes out of the sprayer as very minute droplets. The size of the droplets varies from 150 - 200ì and the droplets aggregating together to form bigger droplets and rolling down from the sprayed parts and getting wasted is avoided. The water required to prepare the spray fluid in such sprayers is 20 - 80 % less than in the case of conventional high volume sprayers and so, the concentration of the chemical in the spray fluid is proportionately very high. However, because the spray fluid is micronized into very fine droplets and sprayed uniformly on the crop surface, no spray injury is caused to the sprayed parts. Though the quantity of water to be mixed with the fungicide formulation is much less (50 - 100 liters), the quantity of chemical to be used should be calculated only on the basis of the water required for spraying with a high volume sprayer. In some such semi-low volume sprayers, an additional fan-like atomizer is also attached near the exit point of the hose. The air coming out of the delivery hose at great force makes the fan attachment to rotate very fast, as a result the spray droplets coming out through the delivery hose are further micronized into still finer droplets. In such type of sprayers, a quantity of 6.8 - 42.5 cu. ft. of air is forced out through the delivery hose every minute at a speed of 200 - 400 km./ hour.

In these sprayers, two separate units are mounted on a single metal stand. One is the motor and the other is the sprayer unit. The motor of the power mist blower is mostly run on petrol and is powered by an engine with 1.2 - 3.0 HP and rotates at 5,500 revolutions per minute. When the motor runs at such high speed, the fan attached to the crank also rotates at high speed, drawing air from outside and forces it out through the delivery hose at very high force. The delivery hose is connected to the fan chamber.

The sprayer unit consists of a cylindrical fluid tank of 10 - 12 liters capacity, which is placed horizontally. A small petrol tank is also situated by the side of the fluid tank. The tanks are made of high-density polypropylene material. The fluid tank is provided with a mouth and screw cap for filling with spray fluid and a delivery tube at the bottom with a tap to regulate the flow of the spray fluid. A thin plastic tube attached to this delivery tube opens inside the posterior end of the delivery hose. The petrol tank is also provided with a mouth and screw cap and a delivery tube with a tap. A thin plastic tube is connected to the petrol tap and the other end is connected to the carburettor of the engine.

When the motor is run, air from the fan chamber comes out through the delivery hose at a great force. The spray fluid, which comes out through the delivery tube by gravitational force into the delivery hose in small quantities is micronized into small droplets by the air current and forced out as mist. An area of 7.5 - 10.0 acres can be covered in one day.

These sprayers can easily be converted into power dusters by changing a few parts. A small plastic tube with several holes is attached to the blower unit, which allows air into the dust filled tank and forces the dust to come out through the delivery hose. By adjusting the attachment provided near the handle of the delivery hose, the quantity of dust to be dusted can be regulated **(Fig. 89)**

Low volume and ultra-low volume sprayers. In low volume spraying, the fungicidal formulation is mixed with very small quantity of water, while in the case of ultra-low volume spraying, the chemical formulation is sprayed as such, without the addition of water. While using a low volume sprayer, the same quantity of chemical formulation required to spray a unit crop area using a high volume sprayer is used. To spray one acre of crop area, the chemical formulation is mixed with 5 - 50 liters of water. While spraying, the spray fluid is broken into very minute particles of size 70 - 150μ. With an ultra low volume sprayer, one acre can be covered with a quantity of 0.2 - 2.4 liters of the chemical formulation without addition of any water and here the fungicidal formulation is broken into mist-like particles of size 30 - 50μ. This kind of spraying is possible with very high-pressure pneumatic sprayers provided with special types of nozzles. For this purpose, trolley-mounted sprayers or tractor-mounted sprayers are used. In trolley-mounted sprayers, power is obtained from diesel or kerosene powered engines, while in the case of tractor mounted sprayers, power is obtained from the tractor engine itself. An area of 20 - 25 acres can be covered in one day with such sprayers. These types of sprayers are mostly used in big commercial orchards, plantations, estates etc.

Dusters. Dust formulations are applied by using hand operated or power operated dusters. The most commonly used manually operated dusters are the hand rotary dusters. These dusters can carry 5.0 - 6.0 kg. of dust formulations in the hopper or container of the dusters. The flow of dust can be regulated by adjusting the feed regulator. One operator is enough to operate the duster and carry out dusting simultaneously and an area of 1.0 - 2.0 acres can be covered in one day **(Fig. 90)**

Aerial spraying and dusting. Special types of aircrafts are used to carry out plant protection operations. Vast areas ranging from 500 - 2,000 acres can be sprayed or dusted in one day. Fixed wing aircrafts and helicopters are used for these purposes. The **'Beaver'** type of fixed wing aircraft is used commonly and this can cover an area of 1,000 acres per 6 flying hours, when fitted with low volume spray nozzles and up to 3,000 acres, when fitted with ultra-low volume nozzles. It can carry up to 630 liters of spray fluid and are powered by 220 - 450 HP engines. These aircrafts can carry 100 - 1,000 kg. of dust formulations and discharge the dust at rates varying from 0.5 - 25.0 kg, /acre. Application is done at speeds varying from 105 - 180 km per hour and they can fly at heights ranging from 3.0 - 11.0 meters above ground level when spraying or dusting. The swathe covered in each sortie is 15 - 60 meters wide.

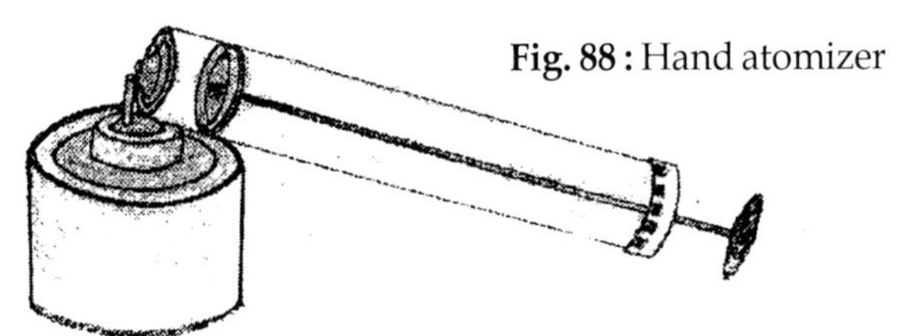

Fig. 88 : Hand atomizer

Pneumatic sprayer

Fig. 87 : 'Marut' sprayer

Fig. 89 : Power sprayer

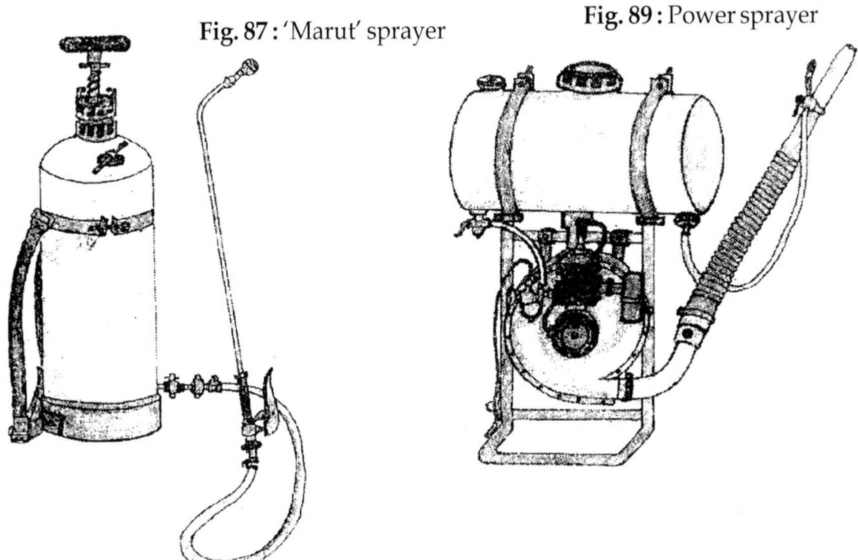

Fig. 90 : Hand rotary duster

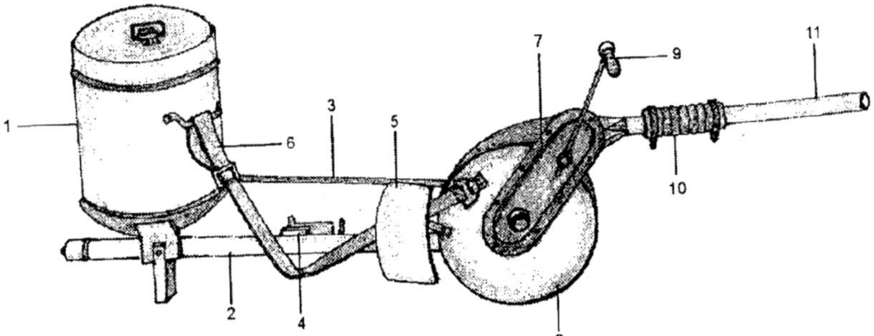

1. Hopper 2. Suction tube 3. Agitator shaft 4. Dust regulator 5. Breast plate 6. Shoulder belt 7. Gear box 8. Fan chamber 9. Crank with handle 10. Delivery hose 11. Delivery tube

The helicopters on the other hand can cover an area of 500 - 850 acres in 6 flying hours, when fitted with ultra-low volume spray nozzles. They can carry 180 - 270 liters of spray fluid or 195 kg. of dust formulations. The swathe covered in each sortie is about 13 meters wide and they can fly at heights of about 1.25 meters above ground level and at a speed of about 37.5 km per hour. Vast stretches of short crops, tall crops, fruit trees, high trees etc. can be sprayed or dusted easily and effectively by aircrafts.

The expenditure incurred for aerial spraying or dusting is comparatively less than that for ground spraying per unit area and vast areas can be covered within a short span of time.

Compatibility of Fungicides with other Agrochemicals

When a crop is affected by more than one disease or by a disease and a pest, it may be necessary to use different types of fungicides or a fungicide in combination with an insecticide to control them both simultaneously. However, instead of applying the different plant protection chemicals separately, if they can be mixed together and applied in a single operation, labor, time and expenditure can be saved to a considerable extent. When some fungicides are mixed together or when a particular fungicide is mixed with an insecticide, reaction may set in and the chemicals may loose their efficacy to control either the disease or the pest or both and sometimes may even cause injury to the crop, such as blighting of the foliage, shedding of flowers, fruits etc. On the other hand, when some fungicides or fungicides and insecticides are mixed together, no chemical reaction may take place and the efficacy of the individual chemical to control the disease and the pest may be exercised simultaneously without causing any injurious effects to the crop. Thus, when two or more fungicides or a fungicide and an insecticide, even after mixing together are effective in controlling both the disease and the pest without causing any injurious effect to the crop, then the chemicals are said to be **'compatible'**. So, it is of great advantage to know the compatibility of plant protection chemicals and other agrochemicals, so that whenever necessary they can be mixed together and applied in a single operation to control the disease and the pest simultaneously.

Bordeaux mixture is not compatible with a number of insecticides and fungicides, such as endosulfan, methyl demetan, malathion, methyl parathion, diazinon, phosphamidon, carbaryl, pyrethroids, monocrotophos, captan, difolatan etc. Bordeaux mixture mixed with lindane, dithiocarbamate etc should be used immediately after preparation. If fungicides, such as captan and difolatan are mixed with insecticides, such as malathion, methyl demeton etc., the mixture should be used immediately. If it is kept for a longer period of time after mixing, chemical reaction may set in and their effectiveness will be lost. Mixing of wettable sulfur and antibiotics with insecticides should be avoided. Though dithiocarbamates are compatible with most of the plant protection chemicals, they are not compatible with chemicals containing heavy metals, such as copper, iron, mercury, calcium etc.

The compatibility of some of the important insecticides, fungicides and other agrochemicals are furnished in **table 8**.

Phytotoxicity of Fungicides

'Phytotoxicity' may be defined as 'any kind of adverse side effects or injury to the host plant that may be caused due to the application of fungicides or any other agrochemicals'. Normally application of fungicides at the correct concentration may not have any injurious or toxic effects on the host plant. However, if some fungicides are applied at higher concentrations, they may cause phytotoxicity in some form or other. Phytotoxicity may be caused as a result of spraying some fungicides on certain crops, seed treatment with some chemicals, soil treatment with some fumigants, weedicide spraying and also due to some other causes, such as smoke, frost, extreme heat etc.

Bordeaux mixture will cause toxicity, if it is not well prepared. In the prepared solution, if copper is in excess, it may produce **'copper-burn'**. Further, if Bordeaux mixture is sprayed in cool, wet weather it may be toxic to the plants and cause scorching of leaves. Mercury fungicides, such as wet ceresan produce toxicity and scorching of the foliage and hence should not be used for spraying on the foliage. Application of sulfur dust when the temperature is very high, leads to toxicity and scorching of leaves. Further, sulfur dust is toxic to cucumbers and apple trees. Apple fruits may also develop necrotic spottings, when sulfur is applied on their surface. Similarly spraying of wettable sulfur also produces injurious effects on the foliage of crops, when the temperature is very high. Wettable sulfur is also toxic to cucumbers and apple trees. Mercuric chloride is used as a seed-treating chemical for the control of externally seed-borne diseases of a few crops at a concentration of 1 : 1,000. If the chemical is used at higher concentrations, it may be harmful to the seeds and the germinating seedlings. Many fungicides, especially the antibiotics and a few sytemic fungicides, if applied at concentrations higher than the recommended dosage may cause toxicity to the host plants.

Table 8 : Compatibility of Fungicides with other Agrochemicals

	Plant protection chemicals	1	2	3	4	5	6	7	8	9	10	11	12	13	14	15	16	17	18	19	20	21
1	Lindane	-	+	+	+	+	+	+	+	+	+	±	+	+	±	+	+	+	x	+	+	+
2	Dieldrin,Heptachlor	+	-	+	+	+	+	+	+	+	+	+	+	+	+	±	+	+	+	+	+	+
3	Endosulfan	+	+	-	+	+	+	+	+	?	+	x	+	?	+	+	?	?	?	+	?	+
4	Dicofol	+	+	+	-	+	+	+	+	+	+	+	+	+	+	+	x	+	x	+	+	+
5	Methyl parathion	+	+	+	+	-	+	+	+	+	+	x	+	+	+	±	x	+	x	+	+	+
6	Methyl demeton	+	+	+	+	+	-	+	+	+	+	x	+	+	±	±	±	+	+	+	x	+
7	Malathion,Diazinon	+	+	+	+	+	+	-	+	+	+	x	+	+	+	±	x	+	x	+	+	+
8	Phosphamidon	+	+	+	+	+	+	+	-	+	+	x	+	+	+	+	x	+	x	+	+	+
9	Dimethoate	+	+	x	+	+	x	+	+	-	x	x	x	+	+	+	+	x	x	+	+	+
10	Carbaryl	+	+	+	+	+	+	+	+	+	-	x	x	+	+	+	x	x	x	+	+	?
11	Bordeaux mixture	±	+	x	+	x	x	x	x	+	x	-	+	+	+	x	+	+	+	x	+	x
12	Wettable sulphur	+	+	+	+	+	+	+	+	+	+	+	-	+	+	±	+	+	+	+	+	?
13	Copper oxychloride	+	+	?	+	+	+	+	+	+	+	+	+	-	±	±	+	+	+	+	x	+
14	Dithiocarbamates	+	+	x	+	x	±	+	+	+	+	+	+	±	-	+	x	+	x	+	+	+
15	Captan,difolatan	+	±	+	+	±	+	±	+	+	+	x	±	±	+	-	x	+	x	+	+	+
16	Manganese sulphate	+	+	?	x	x	?	x	x	x	x	+	+	+	x	x	-	x	+	x	+	+
17	Urea	+	+	x	+	+	+	+	+	?	+	+	+	+	+	+	x	-	+	+	+	±
18	Zinc sulphate	x	+	x	x	x	+	x	x	x	x	+	+	+	x	x	+	+	-	x	+	?
19	Pyrethroids	+	+	+	+	+	+	+	+	+	+	x	+	+	+	+	x	+	x	-	+	x
20	2-4.D	+	+	+	+	+	x	+	+	+	+	+	+	x	+	+	+	+	+	x	-	x
21	Monocrotophos	+	+	+	+	+	+	+	+	+	?	x	?	?	+	+	x	+	?	x	x	-

\+ compatible x Not compatible ± Should be used immediately after mixing ? To be used with caution

Chemicals, such as formalin, vapam etc. used as fumigants for soil treatment should be applied to the soil 2 - 3 weeks before sowing. A few days before sowing, the soil is stirred well to let out the residual fumes completely from the soil. If there is any residue in the soil, it may be toxic to the germinating seedlings.

Weedicides are highly toxic to some of the plants. 2,4-D is very toxic to cotton plants, even at very low concentrations. The leaves become dwarfed and malformed and the plants become stunted. Spraying of hormones to prevent flower fall may also develop toxicity, if the concentration becomes high.

Antibiotics are mostly used in very small concentrations for disease control. If they are used at higher concentrations, they may produce toxicity.

Some of the physical factors, such as smoke, frost etc. may also produce toxic symptoms on fruits. The fruits of mango trees near brick kilns develop black, necrotic spots on the basal portion of the fruits. Frost also causes burning symptoms on fruits in the hilly regions.

Resistance of Plant Pathogens to Plant Protection Chemicals

Several plant pathogens have developed strains that are resistant to certain fungicides, because of continuous or repeated and widespread use of such chemicals. Protectant chemicals, such as thiram, maneb, captan etc., even though used for many years, have not developed any resistant strains, because these fungicides affect several vital processes of the pathogens, which are controlled by many genes and too many gene changes are necessary to produce a resistant strain. However, in recent times, even some of the strains of a few serious pathogens have developed resistance to dithiocarbamates. Resistance to some fungicides belonging to benzene were noticed as early as 1960, when *Penicillium* strains, *Tilletia* strains, *Rhizoctonia* strains etc., resistant to benzene fungicides were found. Subsequently, a strain of *Venturia inaequalis* causing apple scab resistant to dodine appeared and hence the chemical became ineffective against the fungus over very large apple growing areas. Strains of *Erwinia amylovora* causing fire blight were found to have resistance to the systemic antibiotic - Streptomycin. After the introduction and widespread use of many systemic fungicides, especially benomyl, strains of numerous fungal pathogens resistant to one or more of these fungicides appeared. Strains of oomycetous pathogens resistant to the systemic fungicide - metalaxyl have also been observed, as a result of continuous and widespread use of this single-site fungicide.

Nowadays, many important fungal pathogens, such as *Alternaria, Botrytis, Cercospora, Colletotrichum, Fusarium, Verticillium, Sphaerotheca, Mycosphaerella, Aspergillus, Penicillium, Phytophthora, Pythium* and *Ustilago* are known to have produced strains resistant to one or more of the systemic fungicides. Whenever single-site chemicals, such as systemic fungicides are used repeatedly, resistant strains of most of the fungi may appear by a single mutation. Such mutations may occur either by chance or in some cases by the action of the fungicide itself. This can be overcome to a large extent by using mixtures of specific systemic fungicides and wide-spectrum protectant fungicides. For the control of *Botrytis* or *Sclerotinia*, a mixture of benomyl and either captan, dichloran or iprodione can be used and for the control of downy mildews, a mixture of metalaxyl and mancozeb can be used. Alternating sprays with systemic and non-systemic, protectant fungicides is also an effective method in controlling such diseases, besides preventing the development of resistant strains by the pathogens.

Precautions and Safety Measures in Handling Fungicides

Unlike most of the insecticides, acaricides, nematicides and rodenticides, fungicides are comparatively less toxic to human beings and livestock. Fungicides may enter the body system of humans and livestock, either by ingestion through mouth, inhalation by nose or absorption by bare skin and exert their poisonous effect. If such poisonous chemicals are not handled properly or if sufficient care is not taken during application, dangerous consequences may result. In large-scale use of fungicides, hazards may arise due to accidental or intentional poisoning. Operational hazards may occur at the time of application, while post-operational hazards occur due to toxic residues. To avoid such hazards, precautionary measures should be taken at every stage, such as storing, handling and applying.

(i) In storage

Fungicides, fumigants etc. should be stored securely under lock and key, with the labels on the container intact and kept away from all types of food and livestock feeds. They should be kept out of reach of children and livestock.

Liquid concentrates should not be transferred from their original containers to other containers having some other label or with no label.

The fungicides should be stored away from direct sunlight and moisture. Wettable powders should be stored in moisture-proof containers to prevent caking.

Table 9. List of few pathogenic strains resistant to specific plant protection chemicals.

Fungicides	Disease and host	Strains of pathogen
Benomyl	Powdery mildew of grapevine	*Uncinula necator*
	Gray mould rot of apple, pear, grapes, bean etc.	*Botrytis cinerea*
	Powdery mildew of eggplant	*Erysiphe cichoracearum*
	Soft rot of apple	*Penicillium expansion*
	Scab of apple	*Venturia inaequalis*
	Soft rot *of Citrus*	*Penicillium digitatum*
	Powdery mildew of cucumbers	*Sphaerotheca pannosa*
Metalaxyl	Late blight of potato	*Phytophthora infestans*
	Downy mildew of grapevine	*Plasmopara viticola*
	Downy mildew of cucumbers	*Pseudoperonospora cubensis*
Iprodione	Gray mould rot of apple, pear, grapes etc.	*Botrytis cinerea*
Triadimenol	Powdery mildew of barley	*Erysiphe graminis* f. sp. *hordei*
	Smut of barley	*Ustilago avenue*
Thiophanate methyl	Gray mould rot of apple, pear, grapes, beans etc.	*Botrytis cinerea*
	Powdery mildew of eggplant	*Erysiphe cichoracearum*
	Softrot of *Citrus*	*Penicillium digitatum* *Penicillium italicum*
	Scab of apple	*Venturia inaequalis*
	Brown leaf spot of rice	*Helminthosporium oryzae*
	Anthracnose of banana	*Colletotrichum musae*
Edifenphos	Blast of rice Dodine	*Pyricularia oryzae*
Dodine	Scab of apple	*Venturia inaequalis*
Mancozeb	Brown leaf spot of rice	*Helminthosporium oryzae*
	Root and collar rot of Several crops	*Rhizoctonia bataticola*
Dinocap	Powdery mildew of cucumbers	*Sphaerotheca fuliginea*
Sreptomycin	Fire blight of apple and pear	*Erwinia amylovora*

Empty fungicide containers should be disposed by burning or deep burying. They should not be used for storing food products, edible oils, food grains etc.

Fungicides should not be handled with bare hands. Rubber gloves, protective clothing etc. should be used while handling them.

Smoking or use of open fires in the godowns should be prohibited.

(ii) In the field

To open fungicide containers and for preparing fungicide solutions, separate knives, spoons, funnels, vessels etc. should be used.

Only experienced, physically and mentally sound persons should be engaged to undertake fungicide application.

While handling or applying fungicides, eating, smoking, chewing etc. should be avoided

Protective clothing should be used, while spraying or dusting

At the time of spraying or dusting, care should be taken, so that the discharge nozzle should always face the direction of the wind and not against the wind.

Fungicides should not be used when the crop is in open bloom, as they may be dangerous to honey bees or other pollinators.

Fungicides with long residual toxicity should not be applied to vegetables, fruit trees and fodder crops, when they are in the fruit bearing or harvesting stages.

Only recommended and correct dosage of fungicides should be applied.

Bare hands should never be used for mixing the fungicidal solution

Care should be taken to avoid fungicidal solutions from mixing with water in water sources, while preparing fungicidal solutions or while washing the plant protection equipment.

Clogged nozzles should not be blown with the mouth to remove dust or dirt.

Toxic materials coming in contact with the bare skin or eyes should be avoided.

Grains treated with seed-treating fungicides should not be used for human consumption or fed to livestock even after washing them.

When there is strong wind current, spraying or dusting should be avoided.

During fungicidal application, if headache, giddiness, tiredness, vomiting sensation or other sick symptoms occur, the operation should be discontinued immediately and a doctor should be consulted.

(iii) After completion of work

After completion of the days work, hands, feet, face and other uncovered parts of the body should be thoroughly washed with soap and water.

The plant protection equipment should be washed and cleaned thoroughly and stored securely.

Management of Seed-borne Diseases

Many plant diseases are carried through seeds. The pathogens may be present in the form of spores, resting spores, resting bodies or mycelium as seed contaminants or they may be carried externally on the seed coat or inside the tissues of the seed. The oospores of the downy mildew pathogens of pearl millet *(Sclerospora graminicola)*, sorghum *(Sclerospora sorghi)*; sclerotia of ergot pathogen of pearl millet *(Claviceps fusiformis)*, wheat and rye *(Claviceps purpurea)*; spore balls of bunt of wheat *(Tilletia caries)* etc. are carried along with the seeds as **'contaminants'**. They get admixed with the seeds at the time of harvest, threshing or drying of grains. Spores of rice blast pathogen *(Pyricularia oryzae)*, wheat stinking smut pathogen *(Tilletia caries)*, barley covered smut pathogen *(Ustilago hordei)*; sclerotia of groundnut dry root rot pathogen *(Rhizoctonia bataticola)*; cotton *Fusarium* wilt pathogen *(Fusarium oxysporum* f.sp. *vasinfectum*, *Verticillium* wilt pathogen *(Verticillium dahliae)*, seedling blight pathogen *(Alternaria macrospora)*, bacterium causing black arm of cotton *(Xanthomonas campestris* pv. *malvacearum)*, potato black scurf *(Rhizoctonia solani* and *R. bataticola)* etc. may be present adhering to the seed coat. Such seeds are said to be **'infested'** with the pathogen and such diseases are said to be **'externally seed-borne diseases'**. The infestation may occur, either in the field or at the time of harvesting, threshing or drying of the seeds. In other cases, the pathogens enter the seeds during their formation and remain dormant inside the endosperm or embryo of the seeds as mycelium. Rice brown leaf spot pathogen *(Helminthosporium oryzae)*, bacterial leaf blight pathogen *(Xanthomonas campestris* pv. *oryzae)* and bacterial leaf streak pathogen *(Xanthomonas translucens* f.sp. *oryzae)*; wheat loose smut pathogen *(Ustilago segetum)*; barley loose smut pathogen *(Ustilago nuda)*; chillies die-back and fruit rot pathogen *(Colletotrichum capsici)*; cabbage black rot pathogen *(Xanthomonas campestris)*; blight pathogens of crucifers *(Ascochyta tenuis, A. brassicae and A. raphani)*; gram blight pathogen *(Ascochyta rabiei)*; cotton black arm bacterium *(Xanthomonas campestris* pv. *malvacearum)*; *Bean mosaic virus* etc. are carried inside the seed tissues. Such seeds are said to be **'infected'** with the pathogens and the diseases caused by them are said to be **'internally seed-borne diseases'**. Some pathogens, such as *Ascochyta brassicola* and *A. raphani* are capable of penetrating the seeds up to the embryo of the seeds at the time of seed formation. Loose smut of wheat and loose smut of barley are exclusively internally seed-borne.

Many pathogens are carried in the propagating materials, such as transplants, setts, stem cuttings, tubers, bulbs, rhizomes etc. Sugarcane red rot *(Colletotrichum falcatum)*, whip smut *(Ustilago scitaminea)*, wilt *(Cephalosporium sacchari)*, pineapple disease *(Ceratocystis paradoxa)*, ratoon stunting *(Clavibacter xyli* sub sp. *xyli)* and grassy shoot *(Mycoplasma-like organism)* are carried in the setts. Potato late blight *(Phytophthora infestans)*, bacterial brown rot *(Pseudomonas solanacearum)*, ring rot *(Clavibacter michiganense* sub sp. *sepedonicum)* and virus diseases, such as leaf roll, rugose mosaic, mild mosaic etc. are carried in seed potatoes. Turmeric rhizome rot *(Pythium aphanidermatum)* and ginger rhizome rot *(Pythium aphanidermatum, P. myriotylum* and *P. monospermum)* are carried in the seed rhizomes. *Banana bunchy top virus* is carried in the suckers used for planting. Betelvine foot rot *(Phytophthora parasitica* var. *piperina)* is carried in cuttings used for planting. The organisms causing these diseases are carried inside the tissues of the living plant organs in a dormant stage or in an active stage.

Seed-borne diseases may be controlled, either by seed treatment with certain seed-treating chemical formulations, by the use of physical agents, such as hot water, hot air, solar heat etc. or by certain biological agents. Organo-mercurials, such as ceresan dry, agrosan, aretan, agallol, wet ceresan etc. were extensively used for the control of many seed-borne fungal and bacterial pathogens of several crops. Seed treatment with organo-mercurials is found to be very effective because of their broad-spectrum activity against a number of pathogens. They remain unchanged on the seed surface and in the soil and afford long lasting protection to the seeds and the germinating seedlings. They have vapour action and the fumes can penetrate deeper into the seeds and hence are more effective. However, because of their long residual toxicity and adverse effects on humans and livestock, they are no more recommended for disease control.

Nowadays, several other non-mercurial, broad-spectrum, less toxic, contact fungicides, such as thiram, captan, zineb, maneb, brassicol, difolaton etc., systemic fungicides, such as carboxin, oxycarboxin, benomyl, carbendazim, pyroquilon, tricyclazole, thiophanate methyl, melalaxyl etc. and antibiotics, such as Streptomycin, Aureofungin etc. are being used for seed treatment.

Many externally seed-borne pathogens of rice, pearl millet, cotton, groundnut, sorghum, tomato, chillies, lady's finger, eggplant etc. can be controlled by seed treatment with thiram, captan, zineb, maneb, brassicol etc. at 4.0 gm./ kg. of seeds, at least 24 hours prior to sowing. Seed pelleting with thiram, captan and maneb in a small quantity of water is found to be more effective than dry seed treatment. However, for some particular diseases and for pathogens, which are internally seed-borne, specific chemicals have to be used to get better results. Treating the seeds with carboxin at 2.0 gm./ kg. of seeds is very effective in controlling smuts of cholam, pearl millet, wheat black rust etc. Even internally seed-borne loose smut of wheat and barley can be controlled by seed treatment with carboxin. Seed treatment with metalaxyl at 2.0 gm./ kg. of seeds is effective in controlling seed-borne pathogens belonging to Oomycetes, such as downy mildews of pearl millet, sorghum etc. Seed treatment with ethirimol at 2.0 gm./ kg. of seeds controls powdery mildews of wheat and barley. Seed rots and seedling rots of beans, cabbage, cauliflower etc. can be controlled by seed treatment with chloranil and dichlone at 2.0 gm./ kg. of seeds. Seed treatment with chloroneb at 2.0 gm./ kg. of seeds controls many seed-borne diseases caused by *Rhizoctonia*. Ethazole seed treatment at 2.0 gm./ kg. of seeds controls many damping off and stem rot diseases caused by *Pythium* and *Phytophthora*. Seed treatment with tridemorph or thiabendazole at 2.0 gm./ kg. of seeds controls *Ascochyta* blight of chickpea, peas etc. Seed treatment with benomyl or carbendazim at 2.0 gm./ kg. of seeds is effective in controlling blast disease of rice, ragi and tenai. In some cases, wet seed treatment is found to be more effective in controlling certain diseases. Soaking rice seeds in carbendazim, pyroquilon or tricyclazole solution at 2.0 gm./ liter of water for 24 hours, then allowing the seeds to sprout and then sowing controls blast disease of rice. The seedlings from such treated seeds, are more vigorous, healthy and have resistance for up to 30 days after sowing. Steeping sugarcane setts in a solution of carbendazim or benomyl - 50 gm. + urea - 1.0 kg. in 100 liters of water for 15 minutes is effective in controlling red rot, wilt, sett rot, pineapple disease, ratoon stunting and grassy shoot. Dipping the setts in mercuric chloride - 0.1 % or formalin - 1.0 % solution for 5 minutes prior to planting controls sugarcane smut.

Some antibiotics are also capable of controlling certain diseases. Treatment with Aureofungin - 1.2 gm. + copper sulfate - 1.2 gm. in 60 liters of water for one hour, after soaking the seeds for 8 hours in cold water gives good control of rice blast and brown spot of rice. Soaking the seeds in a solution of Streptomycin - 2.0 gm. in 120 liters of water for 8 hours controls bacterial blight and bacterial leaf streak of rice. Treating cotton seeds in Streptomycin solution at 1.0 gm./ liter of water for 4 - 8 hours controls black arm of cotton. Seed treatment with Streptomycin controls bacterial blight of pea also.

In the case of cotton, acid delinting is found to control many fungal and bacterial diseases. Cotton seeds are delinted with commercial concentrated sulfuric acid at 1.0 kg. of acid / 10 kg. of seeds. The seeds are then treated with a fungicide, such as thiram, captan or zineb at 4.0 gm./ kg. of seeds. This treatment is effective in controlling *Fusarium* wilt, *Verticillium* wilt. *Alternaria* blight and black arm disease.

Thermotherapy is also effective in controlling many diseases. Hot water treatment at 54°C for 10 minutes controls loose smut of wheat and barley. Soaking the seeds in cold water for 4 hours, followed by drying under direct sunlight for 4 hours can control grain smut, loose smut and head smut of sorghum. Rice brown spot is controlled by seed treatment in hot water at 55°C for 10 minutes. Heat treatment is widely used for the control of sugarcane diseases. Treating the setts in hot water at 50°C for 2 hours or hot air at 54°C for 8 hours or aerated steam at 52°C for 1 hour is effective in controlling all sett-borne diseases. Treating cotton seeds in hot water at 56°C for 10 minutes after delinting with commercial concentrated sulfuric acid controls black arm disease.

Seed treatment with some biological agents is also effective in controlling certain diseases. *Rhizoctonia* seedling blight can be controlled by seed treatment with *Trichoderma viride* at 4.0 gm./ kg. of seeds. Dipping betelvine cuttings in a spore suspension of *Trichoderma viride* controls foot rot. Treating rice seeds with *Pseudomonas fluorescens* at 10 gm./ kg. of seeds controls rice blast.

Management of Soil-borne Diseases

Many fungal pathogens are either facultative parasites or facultative saprophytes and continue to live in the crop residues left over in the soil in the field after harvest and produce spores. Similarly, some pathogenic bacteria are also able to live in the soil and in the crop debris. Under favorable environmental conditions and when suitable hosts are available, these pathogens become active and infect the host. Many other fungi produce special reproductive structures, such as oospores, cleistothecia, perithecia, pycnidia, sclerotia, microsclerotia, chlamydospores, sporocarps etc., which remain in the soil in a dormant stage for a long time, sometimes for several years, till environmental conditions are favorable for their germination and suitable hosts are available. The most important soil-borne fungal pathogens are *Pythium, Phytophthora, Fusarium, Sclerotium* and *Rhizoctonia,* which cause seed rot, damping off, seedling blight, root rot and collar rot of several crop plants. They affect the soft, immature parenchymatous tissues and cause disintegration of the tissues of the plant parts, as a result the seedlings may collapse and die ultimately. These pathogens produce pectolytic, cellulolytic and other cell-wall degrading enzymes. They are capable of producing certain toxins that are toxic to the seedlings and plants. They also possess high degree of competitive ability against other soil microbes.

The organisms causing *Fusarium* wilt disease of red gram *(Fusarium udum),* chickpea *(Fusarium oxysporum* f.sp. *ciceri),* French bean *(Fusarium solani* f.sp. *phaseoli),* cotton *(Fusarium oxysporum* f.sp. *vasinfectum),* banana *(Fusarium oxysporum* f.sp. *cubense),* tomato *(Fusarium oxysporum* f.sp. *lycopersici)* etc. are typical soil-borne organisms that inhabit the soil naturally. They live saprophytically in the stubble of diseased plants, crop debris and dead organic matter in the soil for indefinite periods. The macroconidia and chlamydospores produced by them also persist in the soil in a viable state for a long time, sometimes for many years.

The organisms causing *Verticillium* wilt of cotton and eggplant *(Verticillium dahliae),* tomato *(Verticillium dahliae* and *V. albo-atrum)* are also natural inhabitants of soil and can lead a saprophytic life in plant debris and organic matter present in the soil. The microsclerotia produced by these organisms persist in the soil for several months. The mycelium of organisms causing sheath blight of rice, cotton seedling blight and black scurf of potato *(Rhizoctonia solani),* charcoal rot of corn, soybean etc. and root rot of several other crops, such as groundnut, cotton, tomato, bean, maize, sweet potato, pigeon pea, black gram, green gram, cowpea, sesame, sunflower etc. *(Rhizoctonia bataticola),* stem rot of rice *(Sclerotium oryzae),* betelvine wilt *(Sclerotium rolfsii),* onion white rot *(Sclerotium cepivorum)* etc. can survive in the diseased plant debris as saprophytes and may cause fresh infection. The sclerotia produced by these organisms can remain viable in the soil for several months, germinate and cause fresh infection, when conditions are favorable. Some organisms, such as *Aspergillus niger* and *A. pulverulentus* causing crown rot of groundnut, *Rhizopus artocarpi* causing inflorescence rot and *Rhizopus nigricans* causing soft rot of sweet potato are present in the dead organic matter in the soil. The organisms causing damping off of seedlings, seedling rot, root rot and foot rot of many crops, such as eggplant, tobacco, tomato, cabbage, chillies, cotton, sugarbeet etc., soft rot of vegetables and fruits and rhizome rot of ginger, turmeric etc. *(Pythium aphanidermatum, P. de baryanum, P. myriotylum, P. butleri, P. vexans, P. graminicola* and *P. monospermum)*, late blight of potato and tomato *(phytophthora infestans)*, seedling blight of castor *(Phytophthora parasitica)*, black shank of tobacco *(Phytophthora parasitica* var. *nicotianae)*, foot rot of betelvine *(Phytophthora parasitica* var. *piperina)*, gummosis and fruit rot of citrus *(Phytophthora citrophthora, P. palmivora* and *P. parasitica)*, blight of yam *(Phytophthora colocasiae)*, foot rot of pepper *(Phytophthora capsici)* etc. are natural inhabitants of soil and live saprophytically on dead plant and animal wastes, crop debris and other organic matter and infect the crop when conditions are favorable. The chlamydospores and oospores produced by some of these fungi can remain in the soil in a viable state for a long time and may serve as primary inoculum to initiate the disease even after a much longer period of time.

Some other fungal pathogens are capable of living on the stubble and other plant debris and continue to produce conidia, which may attack the aerial parts of the plants. Blast of rice *(Pyricularia oryzae)*, blast of finger millet and tenai *(Pyricularia setariae)*, brown spot of rice *(Helminthosporium oryzae)*, leaf blight of sorghum *(Helminthosporium turcicum)*, leaf blight of wheat *(Alternaria triticina)*, leaf blight of sesame, tomato and chillies *(Alternaria solani)*, blight of rapeseed *(Alternaria brassisicola, A. brassicae)*, blight of cotton *(Alternaria macrospora)*, leaf spot and fruit rot of eggplant *(Alternaria melangenae* and *A. solani)*, blight of chick pea *(Ascochyta rabiei)*, blight of pea *(Ascochyta pinodes* and *A. pinodella)*, *Macrophomina phaseolina* causing charcoal rot of sweet potato, potato, maize, soybean, redgram, chickpea, cowpea, black gram, greengram, bean etc., die-back and fruit rot of chillies *(Colletotrichum capsici)*, leaf spot of ginger *(Colletotrichum zingiberis)*, *Phyllosticta* leaf spot of ginger *(Phyllosticta zingiberi)*, leaf spot of turmeric *(Taphrina maculans)*, 'tikka' leaf spot of groundnut *(Cercospora personata* and *C. arachidicola)*, sett rot of sugarcane and basal rot and fruit rot of pineapple *(Ceratocystis paradoxa)* etc. can survive as mycelium in the crop debris for a long time and continue to produce conidia, which become air-borne and infect the aerial parts of the host

Several fungal pathogens produce thick-walled, hard, fruiting bodies that can withstand adverse environmental conditions and remain viable in the soil for months together. Under favorable conditions, these fruiting bodies germinate and cause primary infection or they may produce secondary spores on germination, which may cause primary infection. Most of these organisms infect the aerial parts of the host plants. Some fungi belonging to Myxomycetes produce resting spores that can remain in the soil for even a few years. *Physoderma zea maydis* causing brown spot of maize, *Plasmodiophora brassicae* causing club root of cabbage, *Synchytrium endobioticum* causing wart disease of potato etc. produce such resting spores.

Many other fungi belonging to Oomycetes produce resting spores, as a result of sexual reproduction. Most of these organisms are obligate parasites and cannot live in the soil independently. However, the resting spores produced by them can remain in the soil in a viable state for a considerably long time, sometimes for several years and serve as initial inoculum. Downy mildew of sorghum *(Sclerospora sorghi)*, downy mildew and green ear disease of pearl millet *(Sclerospora graminicola)*, downy mildew of sugarcane *(Sclerospora sacchari)*, downy mildew of pea *(Peronospora pisi)*, downy mildew of rapeseed, cabbage etc. *(Peronospora parasitica)*, downy mildew of cucurbits *(Pseudoperonospora cubensis)*, white rust of rapeseed, cabbage etc. *(Albugo candida)* etc. can produce thick-walled oospores capable of withstanding extremes of climatic conditions.

Many fungi belonging to Ascomycetes also produce resting spores as a result of sexual reproduction. Powdery mildew of wheat *(Erysiphe graminis* f.sp. *tritici)*, powdery mildew of pigeon pea *(Oidiopsis taurica)*, powdery mildew of pea, cabbage etc. *(Erysiphe polygoni)*, powdery mildew of tobacco and cucurbits *(Erysiphe cichoracearum)*, powdery mildew of rose *(Sphaerotheca pannosa)* etc. produce cleistothecia; 'Sigatoka' disease of banana *(Mycosphaerella musicola)* produces perithecia; sugary disease or 'ergot'of pearl millet produces 'ergot'. These spore bodies can persist in the soil for a long time.

Similarly, the Basidiomycetes produce smut spores, which can remain in the soil for prolonged periods. Loose smut of sorghum *(Sphacelotheca cruenta)*, head smut of sorghum *(Sphacelotheca reiliana)*, long smut of sorghum *(Tolyposporium ehrenbergii)*, smut of pearl millet *(Tolyposporium penicillariae)*, bunt of wheat *(Tilletia caries)*, 'Karnal' bunt of wheat *(Tilletia indica)*, smut of onion *(Urocystis cepulae)* etc. produce smut spores, which can persist in the soil for long periods. The sporocarp of pathogens causing basal stem rot of coconut, arecanut etc. *(Ganoderma lucidum)*, brown root disease of rubber *(Fomes noxius)* etc. also persist in the soil and produce basidiospores and cause fresh infection.

Besides fungal pathogens, many bacterial pathogens are also capable of surviving in dead plant and animal wastes, organic matter etc. in the soil and may cause diseases under favorable conditions. The pathogens causing bacterial blight of rice *(Xanthomonas oryzae* pv. *oryzae),* bacterial blight of pea *(Pseudomonas syringae* pv. *pisi),* bacterial wilt of banana, brown rot of potato, wilt of ginger etc. *(Pseudomonas solanacearum),* bacterial leaf spot of tomato, chillies etc. *(Xanthomonas campestris* pv. *vesicatoria),* bacterial blight of cabbage *(Xanthomonas campestris* pv. *campestris),* bacterial soft rot of carrot, onion etc. *(Erwinia carotovora* pv. *carotovora),* common scab of potato *(Streptomyces scabies)* etc. live in soil as saprophytes.

The seeds of phanerogamic parasites that attack sugarcane (*Striga* sp.), tobacco *(Orobanche cernua* and *O. ramosa)* etc. also persist in the soil for several years and attack the crops.

Soil conditions, such as soil moisture, temperature, pH, physical and chemical properties, organic matter content and available nutrients, presence of antagonistic microorganisms etc. play an important role in the development of soil-borne diseases. In general, the pathogens, which produce zoospores, such as *Plasmodiophora, Spongospora, Synchytrium, Aphanomyces, Phytophthora, Pythium* etc., thrive in wet soils. Similarly, wet soils are favorable for the wilt pathogens *(Fusarium* and *Verticillium).* The bacterial pathogen *(Pseudomonas solanacearum)* causing bacterial wilt of banana, brown rot of potato, wilt of ginger etc. thrives better in wet soils. Soil-borne nematodes require soil moisture for their locomotion, so as to reach the roots of their hosts. On the contrary, many cereal smuts, *Streptomyces scabies,* causing common scab of potato etc. prefer drier soils.

Temperature is another factor, which controls the occurrence of soil-borne diseases. The pathogen causing wilt of cabbage and tomato *(Fusarium oxysporum)* is favored by high temperature of 28°C, while some other pathogens causing *Verticillium* wilt of cotton *(Verticillium dahliae),* stem canker of potato *(Corticium solani),* root rot of tobacco *(Thielaviopsis basicola)* etc. are favored by low temperature of 20°C. Some bacterial pathogens causing brown rot and soft rot of potato grow best when the temperature is as high as 37°C. Temperature also controls the incubation period of many pathogens. The pathogen causing *Fusarium* wilt of cotton, completes its life cycle in 12 days at a temperature of 27°C and in 58 days at 16°C. The root knot nematode of vegetables *(Meloidogyne javanica)* completes its life cycle in 80 days at 14°C and in 16 days at 27°C. Under conditions, when the life cycle is completed in a shorter period, the pathogen multiplies much faster in the soil.

Soil alkalinity or acidity may affect the pathogenic organisms, either directly or indirectly. For many fungal pathogens, soil acidity is found to be more suitable. The pathogens causing *Fusarium* wilt of cotton, tomato etc., club root of cabbage, *Sclerotium* root rot pathogen etc. survive much better in acidic soils, while some other pathogens causing *Verticillium* wilt of cotton and tomato, apple scab etc. survive better in alkaline soils. For several nematodes also acidic soils are found to be more favorable. Under such conditions they grow and multiply very rapidly in the soil.

Presence of more oxygen in the soil encourages the mycelial growth of pathogens, such as *Phytophthora, Pythium* etc. and the reproduction of soil nematodes. Presence of excess amount of carbon dioxide in the soil adversely affects the growth and survival of *Rhizoctonia* and *Sclerotinia*. Soil texture also plays an important role in the survival of pathogenic organisms in the soil. Light soils are more favorable for *Rhizoctonia,* while heavy clay soils are more favorable for *Fusarium.*

Control of soil-borne diseases is rather difficult and sometimes quite impossible in the case of some diseases. However, an integrated approach comprising of certain cultural practices, soil treatment and biological control measures help in reducing the population of the pathogen in the soil and controlling the diseases to a considerable extent.

Cultural practices include, field sanitation, deep ploughing to turn over the soil, summer ploughing, flooding, crop rotation, removal of alternate and collateral hosts, removal of severely affected plants, application of organic manure, application of plant nutrients etc. The mycelium, resting spores, such as oospores, cleistothecia, perithecia, pycnidia, acervuli and sclerotia of several pathogenic fungi, as well as many bacterial pathogens remain active or in a dormant state in diseased leaves, flowers, fruits, twigs etc. fallen on the ground. Periodical removal and destruction of such plant parts helps in eradicating the inoculum from the soil. Removal of stubble after the harvest of crops helps to reduce the inoculum in the soil to a large extent. This method is effective in the case of rice blast, rice brown leaf spot, rice sheath blight, rice bacterial blight, rice 'tungro' virus disease, cotton and red gram *Fusarium* wilts, black arm of cotton etc. Deep ploughing turns over the soil and exposes the inoculum to the vagaries of weather and destroys them. Summer ploughing also helps to expose the inoculum to direct sunlight, which may kill them. Flooding and inundating water for a few days in the field, when there is no crop helps to eradicate many pathogens, especially species of *Fusarium.* Anaerobic conditions produced, as a result of flooding results in the production of carbon dioxide and other toxic substances, such as acetic acid by the soil microbes, which may adversely affect the soil pathogens.

Crop rotation is another method by which the build-up of inoculum in the soil can be reduced. Very often, continued cultivation of the same crop or related crops may result in the build-up of inoculum in the soil to very high levels, that makes it impossible to cultivate the same crop in the soil. Such soils are called **'sick soils'**. *Fusarium* wilts of red gram, linseed, banana, cotton etc., red rot, wilt and whip smut of sugarcane, 'ergot', smut and downy mildew of pearl millet, 'tikka' disease of groundnut, sclerotial diseases of several crops and some soil-borne bacterial diseases are examples of diseases that may cause soil sickness. Crop rotation with botanically unrelated crops, which are not susceptible to the pathogens for a few years may be effective in reducing the inoculum potential in the soil.

Addition of organic matter in the form of farmyard manure, compost, green manure, oil cakes etc. helps to control many soil-borne diseases. When such organic matter decomposes in the soil, the quality and texture of the soil is improved, as a result the microbial population in the soil is increased. Some of these microorganisms, especially the Actinomycetes are antagonistic towards the pathogens and destroy them. *Fusarium* wilt of banana, cotton etc. and several diseases caused by *Rhizoctonia,* basal stem rot of coconut and some nematodes can be controlled to some extent by this method. Providing nutrients in sufficient quantities promotes good growth of the plants, while deficiency of nutrients adversely affects plant growth. Deficiency of nitrogen makes the plants weak and predisposes them to attack by many pathogens. Such weak plants are more susceptible to diseases, such as *Fusarium* wilt of tomato, bacterial wilt of potato, tomato, eggplant etc., *Pythium* damping off and seedling blight of tobacco, chillies, eggplant etc. Application of lime helps to control *Rhizoctonia* root rots, *Sclerotium* stem rots, *Fusarium* wilts and *Erwinia* soft rots.

Soil-borne diseases may be controlled to some extent by the application of certain chemical formulations. Seed treatment with some of the broad-spectrum, contact fungicides, besides controlling the seed-borne inoculum, dissipate into the soil around the seeds to a small extent and produce a zone of inhibition around the seeds, thus protecting the germinating seedlings from soil-borne pathogens for a short period. Seed treatment with metalaxyl protects germinating pearl millet seedlings from downy mildew infection. Drenching the soil with some fungicides to wet the soil to a depth of 4.0 - 6.0 inches may be effective in controlling many soil-borne diseases. Drenching the whole field is impracticable and the cost of this operation is also very high, because 1,000 liters of fungicidal solution is required to drench one acre of field. Further, the chemicals get disintegrated due to physical, chemical and biological activities in the soil and lose their toxic effect on soil-borne pathogens within a very short time. However, drenching is followed in small areas, especially in the nurseries, seedbeds, green-houses etc. Bordeaux mixture - 1 % is often used to control pre-emergence and post-emergence damping off of seedlings of eggplant, tobacco, tomato, chillies etc. Copper oxychloride, thiram, brassicol, captan etc. are also used for the control of many soil-borne diseases caused by *Pythium* and *Phytophthora* in the nurseries. A few systemic fungicides, such as metalaxyl, thiophanate, dexon, fosetyl-Al etc. can also be used for drenching seed beds and nurseries to control soil-borne diseases caused by Oomycetes. Drenching with carboxin controls *Rhizoctonia* diseases. Soil drenching should be done when there is sufficient moisture in the soil and not when the soil is dry. If drenching is done in dry soils, the fungicidal solution may not penetrate deeper into the soil to reach the root zone of the host plant. Some inorganic chemicals, such as mercuric chloride and fumigants, such as formalin, vapam etc. are also used for soil drenching. These chemicals should be applied 2 - 3 weeks before sowing and a few days before sowing, the soil should be stirred thoroughly to let out the residual fumes from the soil.

Seed treatment and soil treatment with some of the bio-control agents, such as *Trichoderma* species, *Verticillium lacanii, Bacillus subtilis, Penicillium oxalicum, Pseudomonas fluorescens* etc. have been found to be effective in the control of certain diseases

Management of Foliar Diseases

A vast majority of fungal, bacterial and virus diseases affect the aerial parts of plants, especially the foliage, while a few pathogens affect the flowers, fruits and seeds. Diseases, such as leaf spots, leaf blights, blast, anthracnose, powdery mildews, downy mildews, white rust, rusts and a few smuts affect the foliage. Some of the seed-borne and soil-borne pathogens also cause foliar diseases. Foliar diseases usually do not cause death of the plants, but affect the green areas of the leaves by producing spots, blights, necrosis, leaf fall, shot holes etc., thereby affect the synthesis of carbohydrates through photosynthesis. This adversely affects the growth of the plants and causes reduction in yield.

The organisms causing foliar diseases produce specific symptoms and from such symptoms, most of the disease causing organisms can be identified up to the Genera. Detailed microscopic, cultural and inoculation studies have to be made to identify the species. Some foliar diseases, such as blast of rice, sheath blight of rice, bacterial blight of rice, 'Tungro' virus disease of rice, bacterial blight of cotton, black rust of wheat etc. may cause death of the severely infected plants. The organisms causing foliar diseases produce large number of spores and the spores are easily carried by wind, rainsplash, insects etc. and hence are very difficult to control.

Numerous fungicides have been introduced for the control of plant diseases. Many fungicides are broad-spectrum, contact fungicides, which can afford protection to the host plants from attack by many fungal pathogens. But these chemicals may not have any curative properties. However, there are many systemic fungicides, which act as eradicants and therapeutants, as well as protectants. But most of the systemic fungicides are very specific and can afford effective control of one or a few diseases. The antibiotics used for the control of foliar diseases, mostly bacterial diseases are also very specific in their action and can control only particular diseases. Antibiotics also act as eradicants and therapeutants besides protectants.

The systemic fungicides enter the plants through leaves, roots and other plant parts and are translocated inside the tissues along with the protoplasm. Most of the systemic fungicides move upward, while only a few have downward or lateral movements. The antibiotics are also translocated inside the plant tissues.

The contact, protectant fungicides can exert their action only if they come into contact with the spores. Spraying or dusting of such fungicides should be done, so as to provide a thorough coating on the surface of the foliage. In this respect, spraying is better than dusting, as it provides a better coverage than dusting. Many fungicides, especially the contact fungicides are broad-spectrum ones and are able to control many diseases, while others have narrow-spectrum activity and can control a few, specific diseases. The systemic fungicides and antibiotics are very specific and can control only particular diseases. Copper fungicides, such as Bordeaux mixture and copper oxychloride, inorganic sulfur fungicides, such as sulfur dust and wettable sulfur, dithiocarbamates, such as zineb, maneb, mancozeb, ferbam, ziram etc., benzene compounds, such as dichloran and chlorothalonil, heterocyclic compounds, such as captan and iprodione, organo tin compounds, such as Du-ter and Brestan are broad-spectrum fungicides and can control many foliar diseases of several crops. However, some amount of specificity is found in many of these broad-spectrum fungicides also. Pentachloro nitrobenzene fungicides are not effective against *Fusarium* wilts and seedling diseases caused by *Pythium* and *Phytophthora*. Captan is not effective against rusts, powdery mildews and downy mildews. Zineb and mancozeb are not effective against powdery mildews. Dinocap is very effective against powdery mildews. Nickel chloride is specific for the control of blister blight of tea.

High degree of specificity is seen in the case of systemic fungicides. While bezimidazoles and triazoles are broad-spectrum fungicides and can control many foliar diseases of several crops, some others act only against particular target pathogens. Carbendazim is highly effective against sheath blight of rice caused by *Rhizoctonia solani*. Carboxin is very effective against groundnut rust, blister blight of tea, powdery mildews and seedling blights. Oxycarboxin is very effective against rusts, powdery mildews and *Rhizoctonia* root rots of various crops. Fosetyl-Al is effective against oomycetous fungi. Edifenphos and Kitazin are specific against blast disease. Metalaxyl is quite specific against oomycetous fungi, such as *Pythium*, *Phytophthora* and the downy mildew pathogens. Pyrimidines are highly effective against powdery mildews of various crops. Tricyclazole and pyroquilon are very effective against blast disease.

Antibiotics are also quite specific in their action against pathogens. While many antibiotics are antagonistic towards bacterial pathogens, a few act against fungal pathogens. The antibiotics, which control bacterial diseases, are not effective against fungal diseases.

Fungicides act on the pathogens responsive to them through direct effect on the fungal cells or spores after entering them. Solubilization of fungicides on the surface of the host is ensured by free water, carbon dioxide and ammonia in rainwater or dew, exudates from plant parts etc. The fungal spores are capable of accumulating the fungicides from such very dilute fungicidal solutions. Some fungicides may exert their injurious effects on the fungal cell walls and interfere in cell division. Some may affect the permeability of fungal cell membrane, while a few others cause chelation of mineral nutrients inside the cells and aid their excretion or precipitate the chemicals inside the cells and make them unavailable. However, most of the fungicides interfere with the metabolic activities and physiological functioning of the pathogens. Thus, they may prevent sporulation of the pathogens or inhibit their growth or in some cases cause their death. The dithiocarbamates contain isothiocyanate, which inactivates the amino acids and enzymes within the pathogen cells and inhibits their functioning thus, arrests the growth of the mycelium. Cupric hydroxide, the active principle in Bordeaux mixture is toxic to the spores and the germ tubes. The copper ions in copper oxychloride are toxic to the spores and mycelium. Benzene compounds, captan etc. inhibit the production of amino acids and enzymes. Systemic fungicides inhibit the metabolic activities of the pathogens and suppress their growth or kill them.

Simple Diagnostic Techniques for Identification of Diseases

Visual symptoms. Almost all the diseases caused by fungal, bacterial and virus pathogens produce specific symptoms on various parts of the plants. Examining the symptoms closely helps to identify many of the pathogens causing the disease upto the level of Genera. Many of the leaf spots produced by various fungi, such as *Helminthosporium (Bipolaris), Cercospora, Phoma, Pestalotia, Macrophomina, Pyricularia, Septoria, Diplocarpon, Taphrina, Phyllachora* etc. are quite distinct and by examining the symptoms visually the causal organisms can be identified. Similarly the leaf blights caused by *Alternaria, Phytophthora* and *Ascochyta;* anthracnose caused by *Colletotrichum* and *Gloeosporium;* rusts caused by *Puccinia, Uromyces, Hemileia, Melampsora* and *Phragmidium;* blister blight of tea caused by *Exobasidium;* white rust caused by *Albugo;* powdery mildews caused by *Erysiphe, Leveillula, Phyllactinia, Uncinula* and *Oidium;* downy mildews caused by *Peronospora, Plasmopara, Sclerospora* and *Pseudoperonospora;* smuts caused by *Sphacelotheca, Ustilago, Tolyposporium* and *Urocystis;* false smut of rice caused by *Ustilaginoidea;* leaf smut caused by *Entyloma;* bunt caused by *Tilletia;* ergot caused by *Claviceps;* green ear caused by *Sclerospora;* wilt caused by *Fusarium;* seedling blight caused by *Pythium* and *Phytophthora;* root and collar rot caused by *Rhizoctonia, Sclerotium* and *Sclerotinia* and several other diseases can be identified by examining the symptoms produced by them. Many bacterial diseases, such as bacterial leaf blight, leaf streak of rice, black arm of cotton etc. caused by *Xanthomonas* sp.; bacterial soft rots caused by *Erwinia*; virus diseases, such as rice 'tungro', rice ragged stunt, mosaic, gingelly phyllody, banana bunchy top, papaya leaf curl, vein clearing of lady's finger, marble mosaic of cardamom etc. can also be identified from the visual symptoms. In the case of an already documented plant disease, if the Genus of the pathogen is identified then the species name can be included as recorded originally.

Examination of inner tissues. Many plant pathogens may cause distinct discoloration or other symptoms in the inner tissues of stems, collar region, roots etc., which can be seen visually. Thus, some of the vascular wilt diseases can be identified by splitting open the stem region longitudinally. In the case of *Fusarium* wilt, the vascular region is discolored brown to black extending from the root to the top of the stem, while in the case of *Verticillium* wilt the discoloration is discontinuous and may not extend up to the top of the stem. Further, in *Fusarium* affected plants, loose, whitish mycelial growth and pinkish, slimy spore masses are seen near the basal part of the stem of infected plants. In the case of *Fusarium* wilt of banana, the cut end of the stem of affected plants show brown discoloration, which is seen as brown dots and the discoloration is seen in the vascular bundles of the corm also as pinkish dots. In the case of bacterial wilt of banana, the vascular bundles show yellowish-brown discoloration and bacterial exudate drips down in drops from the cut end of the stem. Red rot of sugarcane caused by *Colletotrichum* shows reddening of the inner tissues, interspersed with whitish, transverse patches and in advanced cases, grayish mycelial growth is also seen in the pith region. In the case of sugarcane wilt caused by *Cephalosporium* the inner tissues show red coloration, but no transverse white patches are seen. In the later stages, the pith region becomes hollow and boat-shaped. Sett rot caused by *Ceratostomella* produces dark red discoloration of the inner tissues and small, black masses of microsclerotial bodies in the inner tissues, besides emitting a sweet smell of pineapple. Potato brown rot caused by *Pseudomonas* shows discoloration of vascular bundles, which is seen as a brown ring when the infected tuber is cut open. The sclerotia produced by the dry root rot pathogen of groundnut, *Rhizoctonia* is seen as minute, black, dot-like bodies on the collar and root regions of the infected plants, which can be seen if the bark is removed. In rice plants affected by sheath blight *(Rhizoctonia solani)*, the sclerotia can be seen adhering to the basal part of the stem, leaf sheaths and sometimes on the leaves of infected plants. Black spore masses are seen on the collar and root regions of groundnut plants affected by collar rot caused by *Aspergillus*.

Microscopic examination. Many pathogens produce spores on the surface of the affected parts. The spores are either dark-colored or light-colored or hyaline. Scrapings are taken from the surface of the spots or other affected areas by means of a blade or teasing needle and the scrapings examined under a microscope in plain water mounts. Colored spores of pathogens, such as *Helminthosporium, Alternaria, Curvularia, Pestalotia, Cercospora* etc.can easily be identified by this method. In the case of hyaline spores, such as *Pyricularia, Phoma, Diplodia, Colletotrichum, Gloeosporium* etc. simple stains like 'lactophenol-cotton blue' can be used, which will impart a bluish color to the spores and thus the diseases can be identified. Pathogens, such as downy mildews, powdery mildews etc. also produce their asexual spores on the surface of affected parts. They can also be identified from surface scrapings. *Rhizopus* soft rot of sweet potato, fruit rot of jack, soft rots and moulds of fruits caused by *Aspergillus, Penicillium* etc. can also be identified from surface scrapings. Different Genera of powdery mildews can be identified by examining the conidia produced by them. By examining the sporangiophores produced by the downy mildew pathogens, the organisms can be identified. Similarly, different Genera of rust pathogens can be identified by examining the uredospores or teliospores produced by them.

Sometimes, spores may not be present on the infected areas or spots. Under such conditions, the infected portions along with some healthy portion is placed on a moist filter paper and kept inside a moist chamber for a few days. The pathogen inside the tissues may grow within a few days and produce spores, which can then be examined under a microscope and the organism identified.

Many bacterial pathogens, such as bacterial leaf blight and bacterial leaf streak of rice, bacterial blight of cotton etc. produce large number of bacteria by rapid multiplication under favorable conditions and these bacteria come out of the tissues to the leaf surface as exudates. By taking scrapings, they can be examined under microscope after staining with methylene blue. A suspension of the bacterial culture is smeared on a clean, grease-free slide and fixed by passing the slide over a thin flame. Then a few drops of crystal violet is poured over the smear and allowed to stand for about 30 seconds, gently washed and then examined under a microscope. Cultures of bacteria may also be stained with other stains, such as carbol-fuchsin, gram stain, spore stain, capsule stain, negative stain etc. and identified

Taking sections. Many pathogens produce spores inside fruiting bodies, which may be embedded in the plant tissue with their mouth protruding out or completely inside the plant tissue without any opening. The phycomycetous fungi produce oospores inside the plant tissues as a result of sexual reproduction. The ascomycetous fungi produce perithecia, cleistothecia or apothecia as a result of sexual reproduction. The deuteromycetous fungi produce pycnidia and acervuli asexually. To identify these pathogens properly, it may be necessary to take either hand sections or microtome sections. Taking sections helps to identify some of the rust pathogens also. The sections may be stained with suitable stains, such as 'lactophenol-cotton blue', 'hematoxylin-orange G', 'light green-saffranin' etc.for proper identification. Spore characters, such as size, shape, septation etc. are also helpful in the identification of pathogens.

Culturing in artificial media. Facultative pathogenic microorganisms, which can be grown in artificial culture media, can be identified by this method. About 0.75 cm^2 pieces of the diseased plant parts, especially the leaves, with both diseased and healthy portions are cut and surface sterilized by immersing them in 1:1,000 mercuric chloride solution for about 10 seconds or 1.0 % sodium hypochlorite solution for about 60 seconds. The pieces are then removed and washed in 3 or 4 changes of sterile distilled water, till they are completely free from any trace of the sterilant. Then, the pieces are placed on sterilized filter paper to remove the water adhering to them.

The pieces are then slightly teased with sterilized teasing needle to expose the inner tissues and placed on solid agar medium in petridishes aseptically by means of an inoculating needle. 'Potato-dextrose agar' or 'oatmeal agar' medium may be used and one or two bits may be placed in each petridish. Within a few days the pathogen may start growing in the agar medium and after some more days start sporulating. The spores may then be identified by examining them under a microscope. However, it may take several days to identify a pathogenic organism by this method. In the case of suspected bacterial diseases 'nutrient agar' medium is used for culturing.

Test for identifying bacterial leaf blight of rice. A few leaves from the affected plant are cut and the cut ends are placed inside a test tube containing some water and left as such for some time. The bacterial exudate comes out through the vascular system and the water in the test tube becomes turbid. This test can be done using a microscope also. A small bit of leaf from the affected plant is taken and mounted on a slide in plain water and covered with a cover slip. The bacterial ooze coming out of the vascular system of the cut-end is clearly visible.

Test for identifying rice 'tungro' virus disease. 'Iodine test' using a mixture of potassium iodide and iodine crystals or commercial tincture of iodine solution is used to identify the disease.

Tetrazolium chloride test. This test can be done to detect banana bunchy top virus and cassava mosaic virus diseases. Thin sections of the leaf mid rib or petiole are taken from the suspected plants and they are kept immersed in a 1.0 % solution of '2,3,5 - triphenyle tetrazolium chloride' for 30 - 60 minutes and then examined for any color change. If the normal color of the tissues of the sections turns brick-red in color, that indicates the presence of the virus in the suspected plants. This method may be used to test the presence of viruses in the planting materials.

Seed Health Testing Methods

Seeds of several crop plants are affected by seed-borne pathogens. These pathogens may cause seed abortion, shrunken seeds, reduced seed size, seed rot, seed necrosis, seed discoloration, sclerotization of seed, loss of viability and some physiological changes. All these, ultimately result in reduced yield, poor grain quality and other products and germinability of grains used as seeds. The extent of seed-borne infection needed to cause appreciable damage to the crop and yield loss varies with the kind of crop and the pathogen involved. In the case of foot rot and blight of pea *(Ascochyta pisi)*, unless there is more than 20 % of seed infection, there is little loss in the stand of the crop. However, in the case of many seed-borne diseases, such as brown leaf spot of rice *(Helminthosporium oryzae)*, blast of rice *(Pyricularia oryzae)* etc., even a small percentage of seed infection may cause severe disease incidence in the crop. It is reported that in the case of chilli bacterial leaf spot *(Xanthomonas vesicatoria)*, even 1.0 % of seed infection is sufficient to cause severe incidence of disease under optimum conditions for the pathogen. This may also be true in the case of bacterial leaf blight of rice *(Xanthomonas campestris* pv. *oryzae)*, black arm of cotton *(Xanthomonas campestris* pv. *malvacearum)* and many other bacterial diseases.

Testing methods for identifying seed-borne fungal and bacterial pathogens is done by symptomatology, microscopically and by culturing the pathogens in general or selective nutrient media. The presence of pathogenic organisms on the surface of seeds, as well as inside the seeds may be detected by the following methods.

Visual examination of seeds. Fungal spores or other structures produced by some of the pathogens are quite visible to the naked eye. These are mostly found as seed contaminants or seed infestants. The sclerotia of ergot of pearl millet *(Claviceps fusiformis)* and rye *(Claviceps purpurea)*, bits of green ear containing oospores of downy mildew of pearl millet *(Sclerospora graminicola)*, wheat grains containing spore balls of bunt *(Tilletia caries)*, pearl millet grains with spore balls of smut *(Tolyposporium penicillariae)*, grains affected by false smut of rice *(Ustilaginoidea virens)* etc., are found admixed with seeds as contaminants and can be easily seen and identified. Several fungal pathogens are responsible for causing discoloration and loss of weight of seeds. *Helminthosporium oryzae* produces dark-brown to black spots on the surface of rice grains. Such discoloration in rice is also caused by other pathogens, such as *Curvularia, Pyricularia, Alternaria, Fusarium, Phoma* etc. Wheat seeds infected with *Alternaria triticina* are shrivelled and the seed surface becomes brown or black. *Macrophomina phaseolina* causes deformed and wrinkled seeds. Fruit rot of chillies *(Colletotrichum capsici)* develops loose, grayish mycelium on the seeds. Angular leaf spot of cotton *(Xanthomonas campestris* pv. *malvacearum)* forms brownish bacterial encrustation on the seeds.

The presence of sclerotial bodies admixed with pearl millet or rye seeds can be detected by a simple method. Sample of the seeds is put in 20 % common salt solution, stirred well and allowed to settle. The ergot sclerotia, if present float on the surface of the salt solution.

Microscopic examination. Fungal spores and bacteria adhering to the surface of seeds may be identified by direct examination of surface scrapings under a microscope and also by adopting different techniques given below: -

Seed washing technique. Seed sample is put in a small quantity of water in a flask and shaken well by hand or by means of a mechanical stirrer for 5 - 10 minutes. The supermatent liquid is centrifuged at 2,500 - 3,000 revolutions per minute for 10 - 15 minutes. Because of the centrifugal force the spores are carried to the bottom of the centrifuge tubes. The liquid at the top is gently poured out and the liquid at the bottom is examined under microscope. Presence of conidia of pathogens, such as *Helminthosporium, Alternaria, Curvularia, Pyricularia, Fusarium, Septoria, Cercospora, Tilletia,* oospores of downy mildew pathogens, teliospores of smut, bunt pathogens etc. can be detected.

Rolled paper towel test. Seeds are placed between layers of filter paper or blotter paper or paper towels, rolled and placed in germination trays and incubated at relative humidity between 90 - 95 %. Pathogens, which may be present on the surface of seeds or inside the seeds start growing on the surface of the emerging seedlings and produce spores, which can be identified microscopically. Seed-borne pathogens, such as *Alternaria, Ascochyta, Helminthosporium, Septoria* etc., can be detected by this method.

Blotter method. Two to three layers of sterilized blotting paper or filter paper are soaked in sterile distilled water and placed inside sterilized petridishes. Excess of water in the petridishes is drained off. On the surface of the filter paper, about 25 seeds are placed at equal distances from each other and then incubated at about 22°C. The growth of fungal colonies are observed regularly. The organisms start growing and develop colonies of mycelium and produce conidiophores and conidia directly or produce fruiting bodies, such as pycnidia, acervuli or sporodochia. These structures can be examined under microscope and the pathogens identified. In this method *Helminthosporium, Curvularia, Drechslera, Alternaria, Fusarium, Aspergillus, Penicillium, Ascochyta, Septoria* etc., can be identified.

Dilution plate method. A sample of the seeds is put into a small quantity of water and shaken well. A quantity of 10 ml. of the seed wash is taken and added on to 90 ml. of sterilized water and stirred well. From this, 10 ml. of the liquid is taken and added on to another 90 ml. of sterilized water. This dilution process is continued 4 or 5 times. Finally 10 ml. of the liquid is taken and poured over agar medium, such as oats agar or potato dextrose agar in petridishes and spread uniformly. Then the petridishes are incubated at 22°C. As a result of such serial dilutions, only a few spores will be present in the final dilution. Once the spores start growing and produce colonies, a small portion is removed and inoculated on agar slants and allowed to grow and produce fructifications. The pathogens can then be identified microscopically. However, this method will take many days for completion of the test.

Detection of seed-borne viruses. Seeds are sown in sand, soil or vermiculite and allowed to grow in insect-proof cages. The seedlings are examined periodically for the development of typical symptoms.

Biological Control of Crop Diseases

The object of biological control is to eliminate or suppress a particular plant pathogen by using another microorganism, which is antagonistic to the pathogen, but harmless to the host. According to Garrett (1965), Biological control of a plant disease is defined as 'any condition or practice whereby survival or activity of a pathogen is reduced through the agency of any other living organism, with the result there is a reduction in the incidence of the disease caused by the pathogen'. Biological control mechanisms include, competition, hyperparasitism, predation and production of extracellular metabolites, such as antibiotics, lytic enzymes, hydrogen cyanide and siderophores, which aid in combating low iron stress.

In some soils, even under normal conditions certain pathogens cannot develop and such soils, which are called **'suppressive soils'** contain naturally occurring microorganisms antagonistic to the pathogens. Sometimes the antagonistic organisms may consist of avirulent strains of the same pathogen that may destroy or inhibit the development of the pathogens. In some cases, the exudates released into the soil by some higher plants may be toxic to certain pathogens and may suppress their growth and development. The development of several soil-borne pathogens, such as *Fusarium oxysporum* causing vascular wilts, *Pythium* species causing damping off, *Phytophthora cinnamomi* causing root rot of many fruit and forest trees are much less in suppressive soils than in conducive soils. The presence of one or more microorganisms antagonistic to the pathogens may be the reason for the suppression of the pathogens. The toxic metabolites produced by the antagonistic microorganisms, such as antibiotics, lytic enzymes etc. may suppress the development of the pathogens.

Competition for food between the pathogen and the antagonistic microorganisms, as well as direct parasitization of the pathogen may also result in such suppression. Several antagonistic fungi, such as species of *Trichoderma, Gliocladium* and *Verticillium* and bacteria such as, species of

Streptomyces, Bacillus and *Pseudomonas* occur naturally in soils and may serve as biocontrol agents. *Trichoderma viride,* a common saprophytic fungus is able to parasitize the mycelia of many other fungi. The lethal action of *Trichoderma viride* is due to secretion of the antibiotic substances **'Gliotoxin'** and **'Viridin'**, which are lethal to the root rot pathogen of *Citrus, Armillaria mellea* and many other soil fungi.

Another important aspect in biological control is the production of **'biosurfactants'** or **'surface active agents'** by some microorganisms, such as *Pseudomonas, Bacillus, Arthrobacter, Rhodococcus, Actinobacter* and *Corynebacterium.* Fungal organisms, such as *Candida* and *Torulopsis* are also known to produce biosurfactants. The biosurfactant **'rhamnolipid'**, which is from a class of glycolipids secreted by *Pseudomonas aeruginosa* causes cessation of motility and lysis of zoospores of *Pythium aphanidermatum, Phytophthora capsici* and *Plasmopara lactucae-radicis.* So, there is a possibility of utilizing some of the non-pathogenic bacteria and fungi capable of producing biosurfactants for biological control of plant pathogens affecting the aerial parts.

Application of natural organic manure, such as farmyard manure, compost, green manure, oil cakes etc. to the soil helps to increase the multiplication and activity of several microorganisms present in the soil to a large extent. As a result of the respiratory activities of such microorganisms, large quantity of carbon dioxide is liberated. The pathogens sensitive to carbon dioxide are inhibited. Further, due to rapid multiplication of the microorganisms in the soil, the available nitrogen in the soil is utilized quickly by these microorganisms. This acute shortage in the available nitrogen also affects the development of many pathogens adversely. *Fusarium solani* cannot grow without nitrogen, which is required for spore germination and infection by the pathogen. Similarly, the germination of chlamydospores and further development of many fungal pathogens are adversely affected due to lack of nitrogen. This nitrogen shortage may be due to locking up of soil nitrogen inside the cells of the microorganisms, which are responsible for decomposing the organic matter present in the soil. *Rhizoctonia solani* can be effectively controlled by the addition of organic matter to the soil.

Biological control of plant diseases has received much attention in the recent past and several fungal and bacterial mycoparasites have been identified and some of them are being used in field scale with some amount of success for the control of specific pathogens. *Trichoderma viride, T. harzianum, Gliocladium virens, Agrobacterium radiobacter, Pseudomonas fluorescens, Bacillus subtilis* etc. are used for the control of several soil pathogens.

Biocontrol agents are mostly applied to the soil as soil drench or used as seed treatment and rarely as foliar sprays. *Trichoderma harzianum* controls several soil pathogens including wood decaying fungi. *Gliocladium virens* controls many seedling diseases. *Agrobacterium radiobacter* controls the crown gall pathogen *(Agrobacterium tumefaciens)*. *Pseudomonas fluorescens* controls many *Pythium* and *Rhizoctonia* species and *Bacillus subtilis* controls some of the root rot and vascular wilt pathogens. Treating the seeds or seed pieces with *Pseudomonas fluorescens* controls damping off and soft rot. Dry root rot of groundnut *(Rhizoctonia bataticola)* and *Fusarium* wilt of red gram *(Fusarium udum)* can be controlled by seed treatment with *Trichoderma viride* at 4.0 gm./ kg. of seeds. Cotton seedling blight *Rhizoctonia solani* and *R. bataticola* can also be controlled by treating the seeds with *Trichoderma viride* at 4.0gm./ kg. of seeds. Seed rot and damping off of seedlings of eggplant *(Pythium aphanidermatum, P. de baryanum and P. ultimum)* can be controlled by seed treatment with *Trichoderma harzianum, Penicillium oxalicum* and *Pseudomonas fluorescens*. Dipping betelvine cuttings in a spore suspension of *Trichoderma viride* controls foot rot of betelvine caused by *Phytophthora parasitica* var. *piperina*. Similarly, dipping transplants of pepper in a spore suspension of *Trichoderma viride* before planting controls foot rot of pepper caused by *Phytophthora capsici*. *Trichoderma lignorum* controls the root rot pathogen *(Rhizoctonia solani)* and the damping off pathogen *(Pythium de baryanum)*. Soil application of *Trichoderma viride* and *T. lignorum* is effective against black scurf of potato and sheath blight of rice *(Rhizoctonia solani)*. Betelvine *Sclerotium* wilt *(Sclerotium rolfsii)* can also be controlled by soil application of the antagonistic fungi, *Trichoderma viride* and *T. harzianum*. Soil application of *Trichoderma viride* controls the coconut root wilt pathogen *(Ganoderma lucidum)*. *Verticillium lacanii* can parasitize *Cercospora personata* and *C. arachidicola* causing 'tikka' disease of groundnut. *Trichoderma koningii, Chaetomium globosum, Aspergillus japonicus* and *Fusarium oxysporum* are mycoparasites of pearl millet rust pathogen *(Puccinia penniseti)*. 'Ergot' produced by the sugary disease pathogen is destroyed by *Fusarium roseum* and *F. sambucinum*. Spraying with a culture of *Pseudomonas fluorescens* at 200 gm. in 200 liters of water can control rice blast *(Pyricularia oryzae)*.

Many antagonistic fungi are known to affect the pathogens on the aerial plant parts. *Athelia bombacina* suppresses ascospore and conidia production of *Venturia inaequalis* on growing leaves of apple; *Verticillium lacanii* parasitizes several rusts; *Tilletiopsis* sp. parasitizes the cucumber powdery mildew fungus, *Sphaerotheca fuligena* and *Nectria inventa*. The mechanism of antagonistic microorganisms is attributed to direct penetration and lysis of the pathogen, competition for food with the pathogen, production of certain antibiotic substances and production of some toxic volatile substances, such as **'ethylene'**.

Vesicular-arbuscular mycorrhizae. 'Mycorrhiza' is a symbiotic association between roots of plants and a fungus, which may form a net work of mycelium outside the feeder roots (ectotrophic) or within the outer tissues (endotrophic). The infected feeder roots are transformed morphologically into unique structures called **'mycorrhizae'** or **'fungus roots'**. Mycorrhiza, which is common in forest trees are seen in the feeder roots of many plants including cereals, vegetables, ornamentals and trees especially in infertile soils.

'Ectomycorrhizae' roots are usually swollen and sometimes much branched than normal roots. Ectomycorrhizae are formed primarily on the roots of forest trees, mostly by mushroom and puffball-producing Basidiomycetes and by some Ascomycetes. The mycelium of the ectomycorrhizal fungi usually produces a closely interwoven fungus mantle around the feeder roots. Sometimes these fungi enter the roots and grow around the cortical cells. The color of the mycorrhizae may be white, yellow, brown or black depending on the color of the fungus growing on the roots.

In the case of **'endomycorrhizae'**, which is more common and more important, the fungus grows into the cortical cells of the feeder root by specialized feeding hyphae or haustoria called **'arbuscules'** or by forming large, swollen, food-storing hyphal swellings called **'vesicles'**. **'Vesicular-arbuscular mycorrhizae'** produce both vesicles and arbuscules. The surface of the feeder roots is surrounded only by a loose mycelial growth. Endomycorrhizae are produced in most of the cultivated plants and in some forest trees mostly by Zygomycetes, such as *Glomus* and *Acaulospora* and by some Basidiomycetes. Mycorrhizae make the feeder roots more resistant to invasion by soil fungi, such as *Pythium*, *Phytophthora* and *Fusarium* and by soil-inhabiting nematodes. The mechanisms of mycorrhizal control from soil-borne pathogens involve:

(i) creation of a mechanical barrier preventing the penetration of the pathogen into the roots
(ii) inducing thickening of cell walls through lignification and production of other polysaccharides
(iii) stimulating host roots to accumulate metabolites
(iv) increasing concentrations of phenolic substances in the roots
(v) producing antibacterial and antifungal antibiotics and
(vi) stimulating microbial activities in the root zone of the host plant.

Constraints in the implementation of biological control measures. There are several constraints in the implementation of biological control measures. Because of the high degree of specificity of the biocontrol agent towards the target pathogen, a thorough knowledge of the ecology and biology of both the biological agent and the pathogen, as well as the mechanism of microbial antagonistic action is of vital importance. Only the biocontrol agent, which has antagonistic effect against a particular pathogen can control that pathogen.

Biological control measures involve the use of microorganisms. So, technical expertise is required at each and every stage of operations, such as maintenance of cultures, subculturing, mass culturing etc. Though seed treatment with biocontrol agents is somewhat easy, soil application in field scale may be very difficult. When the biocontrol agent is applied to the soil, there should be sufficient moisture in the soil for its establishment in the soil. Further, biocontrol agents should have good competitive ability to compete against several other native microbes present in the soil. When soil drenching is done with fungicides or when soil fumigants are used for the control of soil-borne pathogens, the biological agents applied to the soil directly or as seed treatment, as well as those present in the soil naturally are affected adversely.

Biotechnological Approaches in Plant Disease Management

The methods normally used to breed crops for obtaining desirable agronomic qualities, such as high yield, better quality of grains or other produce, early maturity etc. are used for breeding resistant varieties also. However, in recent times biotechnological methods are being used with great success, especially in the case of some vegetatively propagated crops and against some particular diseases, mostly virus diseases. Several biotechnological methods for obtaining disease resistant varieties, such as tissue culture, plant cell culture, protoplast fusion, genetic transformation of plant cells etc. are being used all over the world. According to Agrios (1988), **'Biotechnology'** is defined as 'the manipulation, genetic modification and multiplication of any living organism through novel techniques, such as tissue culture and genetic engineering, resulting in the production of improved or new organisms and products that can be used in a variety of ways'. The term **'Genetic engineering'** is defined as 'the techniques designed to manipulate the genetic constitution of organisms'.

Tissue Culture Techniques

Tissue culture of disease resistant plants. In the case of many clonally propagated plants, such as sugarcane, potato, banana, cassava, apple, strawberry etc. tissue culture of disease resistant plants is very useful. Large number of plantlets produced from meristem and other tissue cultures facilitate the rapid propagation of plants with high degree of resistance.

Isolation of disease resistant mutants from plant cell cultures. Tissue culture of callus, single cells or protoplast, obtained from a genetically stable parent plant may be differentiated into new plants that are genetically different. The clones produced from these cultured plants are called **'somaclones'** and the variation exhibited is referred to as **'somaclonal'**. In some instances, such variation artificially induced has produced virus or other disease resistant somaclones, while such resistance was not present in the parent plant. In sugarcane, some clones produced from single cells derived from callus cultures have shown resistance to *mosaic virus* and *Fiji disease virus*. Similarly, plants exhibiting increased resistance to diseases caused by *Cochliobolus* and *Ustilago* have been obtained from tissue culture of sugarcane. Clones regenerated from single leaf cell protoplast of genetically stable potato variety 'Russet Burbank' have shown higher degree of resistance to *Alternaria* leaf blight *(Alternaria solani)* and late blight *(Phytophthora infestans)*.

Apical or meristem tissue culture. The differentiation of shoot or root tip apical meristem into plantlets by tissue culture technique gives rise to disease resistant plants. This is possible mostly in the case of virus diseases, because most of the viruses are unable to attack or reach the apical meristem of systemically infected plants.

Protoplast fusion. 'Protoplast fusion' involves the isolation of individual cells, removal of the cell walls, fusion of the protoplasts and regeneration of the somatic hybrid cell into a plant by tissue culture. Protoplast fusion and production of hybrid cells known as **'cybrids'** is possible, either from closely related species or from more distantly related species. The cybrid may contain the nuclei (chromosomes) and the cytoplasm of both protoplasts or the nucleus of one cell and the cytoplasm of the other cell. In such fusion, the genes responsible for imparting resistance against a particular disease may be incorporated in the genome. When the products of protoplast fusion are regenerated into plants, they may have resistance against that particular disease. Cultivars of *Brassica* resistant to club root *(Plasmodiophora brassicae)* have been evolved by protoplast fusion of *Brassica oleracea* and *Raphanus sativus*.

Gene transformation or genetic engineering. Insertion of foreign DNA into the genome of another cell **(DNA transformation)** has resulted in inducing resistance against certain diseases. Since 1980, emphasis was placed on determining the specific molecule and genetic connection involved in disease development. Many genes of several virus diseases have been transformed to the host plants for imparting resistance. It was shown that RNA of *Tobacco mosaic virus* was responsible for infection of plant cells, as well as for the reproduction of the virus. A major achievement was made, when transfer of the coat protein genes of the *Tobacco mosaic virus (TMV)* with *Alfalfa mosaic virus (AMV)* in tobacco, resulted in the protection against disease development. The purpose of introducing coat protein genes to impart resistance against the virus is that the multiplication of infecting viral RNA is somewhat checked by the coat protein synthesized in the plant cells. Similarly with the help of DNA technology, the chitinase genes of *Serratia marcescens* has been introduced into tobacco, potato, lettuce and sugarbeet. Such genetically engineered plants synthesize chitinase, which breaks down the fungal cell wall and thus kills the soil-borne pathogen *Rhizoctonia solani*.

Intensive studies on the soil-borne bacterium *Agrobacterium tumefaciens*, the causal agent of crown gall disease revealed that the bacterium caused tumors (galls) in plants by inserting a portion of the DNA (T-DNA) of its tumor inducing plasmid (Ti-plasmid) into the chromosomes of plant cells. Two genes in the T-DNA were found to code for growth regulators and these were responsible for the production of tumors in the infected plants. It was found that these two genes could be removed and replaced with one or more genes from other organisms, such as plants, other bacteria, viruses etc. thereby resistance could be induced. Some bacterial and fungal genes coding for enzymes that break down the cell wall of the pathogen have also been engineered into plants and have provided the plants with resistance against these pathogens. Foreign DNA can be introduced into plant cells by using viruses as vectors, bombarding plant cells with foreign DNA or by growing plant cells in the presence of foreign DNA. The molecular phase of Plant Pathology is still in the early stages of its development and has vast scope in the future in developing resistant varieties.

Use of Plant Products in Plant Disease Management

Several plant products behave as insecticides, nematicides, insect repellants and attractants, and some of them are being widely used for the control of some insect pests and nematodes. Plant products are mostly used for the control of vectors, which spread different virus diseases in various crops. However, a few of these products possess fungicidal and bactericidal properties and can control a few specific fungal or bacterial plant pathogens. Such plant products may act as a physical barrier and prevent entry of the pathogen into the host or the alkaloids and other toxic metabolites present in these products may enter the fungal cells and prevent their growth or trigger some type of resistant mechanism in the host, thereby ward off infection by the pathogen or encourage the growth of antagonistic microorganisms that may produce antibiotics, lytic enzymes, toxins etc. when applied to the soil, which may destroy soil-borne fungal, bacterial and nematode pathogens.

Spraying the crop at the time of panicle initiation with neem oil emulsion - 3.0 % (neem oil - 6.0 liters + liquid soap - 200 ml. in 200 liters of water / acre) is effective in controlling sheath rot of rice caused by *Acrocylindrium oryzae*. Rice blast *(Pyricularia oryzae)*, rust of groundnut *(Puccinia arachidis)* and powdery mildew of black gram *(Erysiphe polygoni)* can also be controlled by spraying with neem oil emulsion - 3.0 % at 15 days intervals. Spraying banana plants with neem oil emulsion - 3.0 % controls 'Sigatoka' disease caused by *Cercospora musae*. The oil coating on the surface of leaves prevents adherence of the fungal spores on to the leaf surface and prevents the germ tube from reaching the infection site.

Spraying rice crop at the time of panicle initiation with neem seed kernel extract - 5.0 % (neem seed kernel extract - 10 lit. + liquid soap 200 ml. in 200 lit. of water / acre) also controls sheath rot of rice. Neem seed kernel extract - 5.0 % is prepared by soaking 10 kg. of powdered neem seed kernel in 50 lit. of water for 12 hours. Then the soaked powder is crushed and squeezed and the extract filtered through a gunny cloth. To this extract 200 ml. of liquid soap is added and stirred vigorously. To this mixture, water is added to make the volume to 200 liters and then it is used for spraying.

Spraying *Citrus* crop with neem cake extract - 10 % is effective in controlling *Citrus* canker. Neem cake extract is prepared by soaking 1.0 kg. of powdered neem cake in 5.0 lit. of water for 7 days with frequent stirring, then filtering the extract through a cloth and making the volume of the filtered extract to 10 lit. by adding water. To this extract 10 ml. of liquid soap is added and then sprayed.

Incorporating powdered neem cake in the soil at the rate of 60 - 75 kg. / acre prior to sowing is effective in controlling sheath blight of rice *(Rhizoctonia solani)*, root rot of black gram *(Macrophomina phaseolina)*, wilt of chick pea *(Fusarium oxysporum* var. *ciceri)* and wilt of crossandra *(Fusarium solani)*. Basal stem rot or Thanjavur wilt of coconut *(Ganoderma lucidum)* can be controlled by soil application of neem cake at 5.0 kg. / tree in the initial stages of the disease. Neem cake application to the soil at 5.0 kg, / plant controls leaf and foot rot of betelvine caused by *Phytophthora parasitica* var. *piperina.*

Spraying with neem oil emulsion - 3.0 % or neem seed kernel extract - 5.0 % controls the leafhopper vectors *(Nephotettix* species*)* transmitting rice 'tungro' virus disease, and the white fly vector *(Bemisia tabaci)* transmitting yellow mosaic virus disease of green gram and black gram

Application of oilseed cakes of castor, neem, groundnut, mustard, cotton, pongam etc. to the soil encourages the rapid multiplication and increased activity of several antagonistic microorganisms, such as species of *Trichoderma, Gliocladium, Verticillium* etc. and bacteria, such as species of *Streptomyces, Bacillus, Pseudomonas* etc., which are present naturally in the soil. The toxic metabolites, such as antibiotics, toxins, lytic enzymes etc. produced by these antagonistic microorganisms may suppress the development of many plant pathogens. *Trichoderma viride* parasitizes the mycelium of several soil-inhabiting fungi. The toxins viz., **'Gliotoxin'** and **'Viridin'** produced by this fungus are lethal to the *Citrus* root rot pathogen *Armillaria mellea* and many other pathogens. Addition of green manure or oilseed cakes to the soil is very effective for the control of the root rot pathogen, *Rhizoctonia solani.*

Incorporating oilseed cakes of castor, mustard, groundnut, Indian beech *(Pongamia pinnata)* etc. into the soil is found to be effective in controlling pathogenic nematodes. Oilseed cakes remain effective for more than 6 months in the soil. Oilseed cakes are dry, devoid of water and oil content, light and produce heat in the immediate surroundings. As such, they cause some changes in the internal environment of the soil, which adversely affects the nematodes living in a somewhat moist environment in the soil.

Some bacterial species belonging to *Pseudomonas, Bacillus, Arthrobacter, Actinobacter Corynebacterium* etc., as well as some fungal species belonging to *Candida, Torulopsis* etc. produce biosurfactants. The biosurfactant **'rhamnolipid'**, a glycolipid secreted by *Pseudomonas aeruginosa* affects the motility and causes lysis of zoospores of *Pythium aphanidermatum, Phytophthora capsici* and *Plasmopara lactucae-radicis.*

Rapid multiplication of antagonistic microorganisms in the soil and consequent increase in their activity is detrimental to the survival of many pathogenic soil microorganisms. As a result of the respirative activities of these antagonistic microorganisms, large quantities of carbon dioxide is liberated. The pathogens sensitive to carbon dioxide are inhibited. Further, these antagonistic microorganisms utilize the available nitrogen in the soil quickly, as a result there is considerable depletion of available nitrogen, which affects the pathogens adversely. *Fusarium solani* cannot grow without nitrogen, which is required for spore germination and infection by the pathogen. Similarly the germination of chlamydospores and further development of many fungal pathogens are adversely affected due to lack of nitrogen. Competition for food between the antagonistic microorganisms and pathogens and direct parasitization of the pathogens by the antagonistic microorganisms may also result in the suppression of plant pathogens.

Soaking sprouted rice seeds in an extract of sweet flag *(Acorus calamus)* for 30 minutes prior to sowing increases the germination and seedling vigor, besides inducing resistance to a number of diseases. The extract is prepared by soaking 250 gm. of powdered sweet flag in 3.0 lit. of water for one full night. The extract is filtered the next morning. One liter of the extract is diluted in 5.0 liters of water. Rice seeds required for sowing in an acre is allowed to sprout and the sprouted seeds are soaked in the diluted extract for 30 minutes, shade dried for 10 minutes and then sown.

Leaf extracts of some plant species, such as *Agave americana, Bougainvillea spectabilis, Clerodendron fragrans, Azadirachta indica, Thevetia neriifolia, Carica papaya* etc. inhibit *Tobacco mosaic virus.* Prophylactic spraying with 1.0 % leaf extract reduces the incidence of the disease.

Application of 2 to 3 drops of neem oil or castor oil or linseed oil on young broomrape shoots *(Orobanche* sp.*)* before flowering kills the phanerogamic parasite.

Some plant products possess healing properties. A thin paste of Asafoetida *(Ferula asafoetida)* applied on the rotten portions of cucurbit fruits caused by *Pythium aphanidermatum* and *P. butleri* and then covered by a fine bandage cloth is effective in healing the wound.

Plant products obtained from a few other plants also have fungicidal properties and are used in plant disease management. Bulb extract of garlic *(Allium sativum)* controls early blight of tomato *(Alternaria solani)*, seedling blight and leaf blight of finger millet *(Helminthosporium nodulosum)* and rice blast *(Pyricularia oryzae)*. Banana rhizome extract controls basal stem rot of coconut *(Ganoderma lucidum)*. Leaf extract of mint *(Mentha piperita)* controls grain discoloration of rice caused by *Helminthosporium oryzae.* Leaf extracts of *Eucalyptus globulus, Punica granatum* and *Datura stramonium* help to prevent fruit rot of lemon. Leaf extract of tulsi checks infection by various rot pathogens. Groundnut, mustard and castor oils have also been found to be effective against *Rhizopus* rot of mango and *Aspergillus* rot of papaya.

Use of Antiviral Principles in Plant Disease Management

'Antiviral principle' (AVP), **'antiviral agent'** or **'antiviral factor'** (AVF) denotes the synthesis of certain complex organic substances by some plants or presence of such complex substances already in the virus inoculum, which either prevent infection or inhibit multiplication of pathogenic viruses in the host. They may be present in the host itself as a separate constituent or they may be activated in the host by the application of another extraneous substance. These complex organic substances may act directly by inactivating the virus pathogen or indirectly by inducing some type of resistance mechanism against the virus pathogen, thereby preventing infection.

Several non-host plants contain antiviral principles, which may exert their action on specific viruses in their respective host plants. Extracts from *Beta vulgaris, Bougainvillea spectabilis, Chinopodium ambrosioidea* etc. prevent infection by *Tobacco mosaic virus (TMV)* in the host plants. Leaf and root extracts of *Phyllanthus niruri* inhibit *TMV*, *Groundnut green mosaic virus* and *Tobacco ring spot virus.*

Leaf extract of *Sorghum* and coconut, which contains 'AVP' is effective against *Tomato spotted wilt virus (TSWV),* which causes bud necrosis, bud blight or ring mosaic of groundnut. To prepare sorghum or coconut leaf extract, freshly cut leaves are cut into small pieces, air-dried and powdered. To 1.0 kg. of the powder 2.0 lit. of water is added and heated for 1 hour at 60°C. Care should be taken to maintain the temperature at 60°C all the time, as the AVP may lose its effectiveness at higher temperatures. The extract is then filtered through a muslin cloth and the volume of the filtrate made up to 10 liters. This solution is sprayed on the leaves of groundnut crop on the 10th and 20th day after sowing. A quantity of 200 liters of the solution is required to cover one acre of the crop.

For testing the effectiveness of the 'AVP', the extract is added to the virus inoculum and then inoculated on the leaves of host plants and the antiviral activity of the AVP is determined or the extract may be sprayed on the leaves of the test host plants and then the test plants are inoculated with the virus inoculum at different periods and the antiviral activity ascertained.

The inhibitory substances present in the 'AVP' are mostly complex proteins containing many amino acids, while some other AVPs may contain glycoproteins, polysaccharides, flavones or glycoalkaloids. *Dianthus caryophyllus, Chenopodium album, Spinacia oleracea* etc. contain complex proteins, which inhibit *Tobacco mosaic virus. Cocos nucifera* contains a complex protein effective against *Tomato spotted wilt virus. Punica granatum* contains glycoprotein effective against TMV. *Beta vulgaris* and *Abutilon striatum* have polysaccharides that inhibit TMV. *Capsicum annuum* contains flavones, which are effective against *Potato virus Y. Solanum* have glycoalkaloids effective against *TMV*. Heating the extracts for 30 minutes at 70°C destroys the antiviral activity of AVPs.

The AVPs have certain common biological properties. They are active against viruses affecting other hosts and not against the viruses affecting the plant species from which the AVP is obtained. AVPs are more effective as pre-inoculation treatments, but not so effective in the case of post-inoculation treatments.

Besides plant extracts, antiviral agents are found in culture filtrates of some fungi and bacteria. Culture filtrates of *Sclerotinia fructigena, Venturia inaequalis* and *Trichoderma* species and the bacterium *Aerobacter aerogenes* inhibit *TMV* infection. The AVP from yeast culture also reduces *TMV* infection. The extract obtained from potato leaves infected by the late blight pathogen *(Phytophthora infestans)* has an AVP capable of inhibiting *Potato virus X*. The degree of resistance depends on the concentration of the extract. Liquid culture containing heat-killed bacteria - *Pseudomonas syringae*, when injected into plants inhibits *TMV* infection.

Some chemical substances also act as antiviral agents. Injecting polyacrylic acid into plants 2 to 3 days before inoculation induces systemic resistance to *TMV*. Similarly, salicylic acid is also known to induce resistance against some viruses.

References

Anaja, K.R. 2003. Experiments in Microbiology, Plant Pathology and Biotechnology. *New Age International (P) Limited., Publishers, New Delhi.* pp. 607.

Asoka Kumar Sinha. 2007. Fundamentals of Plant Pathology. *Kalyani Publishers, Ludhiana.* pp. 462.

Alexopoulos, C.J. and Mims, C.W. 1979. Introductory Mycology. *Wiley Eastern Limited, New Delhi*-110 002. pp. 631.

Alexopoulos, C.J., Mims, C.W. and Blackwell, M. 1996. Introductory Mycology. *John Wiley and Sons, Inc., New York.* pp. 869.

Chaube, H.S. and Pundir, V.S. 2005. Crop diseases and their management. *Prentice Hall of India Pvt., Ltd., New Delhi.* pp. 703.

Chopra, G.L. and Verma, V. 1985. A Textbook of Fungi. *Pradeep Publications, Jalandhar.* pp. 586.

Dubey, R.C. and Maheshwari, D.K. 2002. A Textbook of Microbiology. *S. Chand and Company Limited, New Delhi*-110 055. pp. 682.

Edwin J. Butler and Jones, S.G. 1955. Plant Pathology. *Macmillan and Company Limited, New York.* pp. 979.

Ganapathy, T, Rabindran, R. and Sabitha Doraiswamy. 2002. An Illustrated Glossary of Plant Pathology. *AE Publications, Coimbatore* - 41. pp 251.

Gaumann, E.A. and Wynd, F.L. 1952. The Fungi. *Hafner Publishing Company, New York.* pp. 420.

George N. Agrios. 1997. Plant Pathology. *Academic Press, California, USA.* pp. 635.

Haarer, A.E. 1956. Modern coffee production. *Leonard Hill (Books) Ltd., 9-Eden Street, N.W. 1, London.* pp. 467.

Mani, A., Selvaraj, A.M., Narayanan, L.M. and Arumugam, N. 1998. Microbiology (General and Applied). *Saras Publications, Nagercoil, Tamil Nadu.* pp. 166.

Mehrotra, B.S. 1967. The Fungi. *International Publishing House, Allahabad.* pp. 331.

Mehrotra, R.S and Ashok Aggarwal. 2003. Plant Pathology. *Tata McGraw-Hill Publishing Company Limited, New Delhi*-110 008. pp. 846.

Nita Bhal. 1988. Hand book on Mushrooms. *Oxford and IBH Publishing Co., Pvt., Ltd., New Delhi.* pp. 129.

Padoley, S.K. and Mistry, P.B. 1982. A manual of Plant Pathology. S. *Chand and Co., Ltd., New Delhi.* pp. 131.

Pandey, B.P. 2001. Plant Pathology. S. *Chand and Company limited, New Deli* - 110 055. pp. 492.

Powar, C.B. and Daginawala, H.F. 2001. General Microbiology (Vol. II). *Himalaya Publishing House, Mumbai*-400 004. pp. 680.

Rangaswami, G. 1988. Diseases of crop plants in India. *Prentice Hall of India Pvt., Ltd., New Delhi.* pp. 498.

Rangaswami, G. and Mahadevan, A. 1999. Diseases of crop plants in India. *Prentice Hall of India Pvt., Ltd., New Delhi.* pp. 536.

Salle, A.J. 1961. Fundamental Principles of Bacteriology. *McGraw-Hill Book Company, Inc., New York.* pp. 812.

Sarmah, K.C. 1960. Diseases of tea and associated crops in North-East India. *The Ganges Printing Company Ltd., Sirpur, Howrah.* pp. 68.

Sendilvel, V, Kavitha, K, Nakkeeran, S, Raguchander, T and Marimuthu, T. 2004. Glimpses of Plant Pathology. *AE Publications, Coimbatore*-41. pp. 276.

Seshagiri Rao, D. 1972. A hand-book of Plant Protection. *S.V.Rangaswamy & Co., Pvt., Ltd., Bangalore*-2. pp. 841.

Singh, R.S. 1983. Plant Diseases. *Oxford and IBH Publishing Co., New Delhi*-110 028. pp. 608.

Singh, R.S. 1984. Introduction to Principles of Plant Pathology. *Oxford and IBH Publication Co., New Delhi.* pp. 532.

Singh, R.S. 1989. Plant Pathogens - The Prokaryotes. *Oxford and IBH Publishing Co., Ltd., New Delhi*. pp. 215.

Singh, S.S. 1998. Crop Management. *Kalyani Publishers, Ludhiana*. pp. 507.

Verma, H.N. 2003. Basics of Plant Pathology. *Oxford and IBH Publishing Co. Ltd., New Delhi*. pp. 218.

Vidhyasekaran, P. 2006. Principles of Plant Pathology. *CBS Publishers and Distributors, New Delhi*-110 302. pp. 166.

Walker, J.c. 1969. Plant Pathology. *Tata Mc.graw Hill Publishing Co., Ltd., New Delhi*. pp. 819.

Glossary of Scientific Terms

Acervulus (pl. acervuli). A subepidermal, inverted saucer-shaped, asexual fruiting body, consisting of a cushion-like mass of hyphae and palisade-like conidiophores that cut off conidia from their tips; characteristic of Melanconiales of Fungi Imperfecti.

Achlorophyllous. Organisms, which do not posses chlorophyll.

Acropetal. Upward from the base to the apex of a shoot of a plant; in fungi, production of spores in succession in the direction of the apex, so that the apical spore is the youngest.

Acquired resistance (induced resistance or acquired immunity). A resistance response induced in a normally susceptible host following inoculation of the plant with certain microorganisms or treatment with certain chemical compounds. Genes do not govern this type of resistance.

Acuminate. Pointed; tapering at the end.

Aecium (pl. aecia). A cup-shaped fruiting body of the rust fungi consisting of binucleate hyphal cells, with or without a peridium that produces spore chains consisting of aeciospores, alternating with disjunctor cells, following successive conjugate division of the nuclei (Stage I of the rust fungus).

Aerobic. Organisms, which can live only in the presence of oxygen or a process that occurs in the presence of molecular oxygen.

Aflatoxin. Mycotoxins produced by the fungi belonging to Genus - *Aspergillus*, which is carcinogenic (substance or agent that causes cancer) and sometimes may even be lethal. They are the derivatives of coumarine.

Alate. Winged form in the life cycle of certain insects (e-g. aphids).

Amphitrichous. Single flagellum at both the polar ends.

Anaerobic. Organisms, which can live in the absence of oxygen or a process that occurs in the absence of molecular oxygen.

Anamorph. The asexual or imperfect stage of a fungus.

Annelides. A type of conidiogenous cells.

Annulus. The ring found on the stalk of certain species of mushroom; the remnant of the inner veil.

Antheridium. The male gametangium of heterogamous fungi.

Antibiotic. A chemical compound produced by a microorganism, which inhibits or kills some other specific microorganisms.

Aplanospores. Non-motile spores produced from sporangia.

Aplerotic. Oospores not filling the oogonium completely, as in Pythiaceae.

Apophysis. A slipper-like structure that produces secondary sporangium and secondary zoospores.

Apothecium (pl. apothecia). An open cup- or saucer-shaped ascocarp

Appressorium (pl. appressoria). A swollen, flattened, pressing organ arising from the tip of a hypha or germ tube that facilitates attachment to the host. A minute infection peg usually grows from it and enters the epidermal cell of the host.

Apterous. Wingless stage in the life cycle of certain insects (e-g. aphids).

Arbuscule. A branched, tuft-like haustorium produced by certain endomycorrhizal fungi inside the host root cells.

Arthrospores. The fungal hypha breaks up into its component cells and each cell serves as a conidium.

Ascocarp or **Ascoma.** Sexual fruiting body of an ascomycetous fungus that produces asci and ascospores.

Ascogonium (pl. ascogonia). The female gametangium of the Ascomycetes.

Ascostroma (pl. ascostromata). A stromatic ascocarp bearing asci directly in locules within the stroma.

Ascus. A sac-like cell containing a definite number of ascospores, usually eight, formed by free cell formation, mostly after karyogamy and meiosis; characteristic of Class - Ascomycetes.

Ascus mother cell. The binucleate crook cell in the Ascomycetes, in which karyogamy occurs and then develops into the ascus.

Aseptate. Non-septate, coenocytic; lacking cross walls.

Attenuated. Reduced virulence; reduced capacity of a pathogen to cause disease.

Atrophy. Reduction in size of an organ or the entire organism.

Autoecious (syn. monoecious). A parasitic fungus, which can complete its entire life cycle on a single host species as in the case of some rusts.

Avirulent. Lacking virulence; unable to cause disease.

Azoosporangium. A sporangium that contains non-motile spores.

Azygospore. A zygospore that develops parthenogenetically.

Bacteriophages. Bacterial viruses; Viruses, which are capable of lysing bacteria through multiplication within the bacterial cell and kill them; referred to as 'phage' also.

Basidiocarp. A sexual fruiting body of the basidiomycetous fungi in or on which basidia and basidiospores are produced.

Basidium (pl. basidia). A club-shaped, zygote cell bearing a definite number of basidiospores, usually four, at the end of minute sterigmata formed at the apex of the basidium, following karyogamy and meiosis.

Basipetal. Downward from the apex toward the base of a shoot; development in the direction of the base, so that the apical part is oldest; formation of spores in succession in which the apical spore is the oldest and the youngest spore is at the base.

Binary fission. A type of asexual reproduction in whish two cells, usually of similar size and shape are formed by the growth and division of one cell by a cross septum.

Biological control. Manipulation of natural enemies, such as parasitoids, predators or other microorganisms in an attempt at reducing the pest numbers and keep them at much reduced levels

Bioluminescence. Emission of light by living organisms.

Biotechnology. The use of genetically modified organisms and/or modern techniques and processes with biological systems for industrial production.

Biotroph. An organism that can live and multiply only on another living organism.

Biotype. A subgroup within a species or race usually characterized by the common possession of a single or a few new characters.

Bitunicate. An ascus with two walls, in which the inner wall is elastic and expands greatly beyond the outer wall at the time of spore liberation.

Blight. A disease characterized by general and rapid killing of tissues of leaves, flowers and stems.

Blotch. A disease characterized by large and irregularly shaped, necrotic spots or blots on leaves, shoots and stems.

Brachypterous (Alate). Winged; having wings.

Broad-spectrum pesticide. Pesticide effective against a variety of pests and diseases.

Budding. A form of asexual reproduction typical of yeast, in which a new cell is formed as a small outgrowth (bud) from the parent cell.

Bud wood. Wood consisting of strong, young shoots bearing buds suitable for use in budding.

Caeoma. An aecium that is surrounded by fungal filaments, but that has no distinct peridium

Callus. Superficial, unspecialized tissue produced by plants in response to wounding; parenchymatous tissue of cambial origin that are formed in response to wounding; a mass of thin-walled, undifferentiated cells usually developed as the result of wounding or infection; culture on nutrient media.

Campanulate. Bell-shaped.

Canker. A necrotic, often sunken or cracked lesion surrounded by callus on a stem, branch or twig of a plant.

Capitulum. A stalked, globose, apical apothecium.

Capsid. The protective protein shell, covering the virus nucleic acid.

Capsomere. The protein subunits of the capsid, enclosing the virus nucleic acid.

Capsule. A relatively thick, slimy layer of mucopolysaccharides that surrounds some kinds of bacteria.

Catenulate. Conidia produced in chains.

Cellulolytic. Enzymes, such as cellulase capable of decomposing cellulose.

Centriole. Short, cylindrical or bell-shaped cell

Centrum (pl. centra). An ascocarp, in which all the structures are enclosed by the ascocarp wall.

Chemotherapy. Control of plant diseases with chemotherapeutants, such as systemic fungicides and antibiotics that are absorbed and are translocated internally.

Chlamydospores. Thick-walled or double-walled, asexual resting spores formed from hyphal cells, either terminal or intercalary or by the transformation of one or more conidial cells, that can serve as an overwintering reproductive structure.

Chlorosis. Yellowing of normally green tissue due to chlorophyll destruction or failure of chlorophyll formation.

Chronic. Slow developing, persistent or recurring symptoms.

Circulative viruses. Viruses that are acquired by their vectors through their mouthparts, which multiply and accumulate in the intestine, then pass through the gut wall of the vectors into the haemolymph and salivary gland, and eventually are transmitted through the mouthparts into the host along with saliva.

Cirrhus. A curled, tendril-like mass of exuded spores, held together by a slimy matrix.

Clamp connection. A bridge-like hyphal connection formed at cell division and connecting the newly divided cells; characteristic of the secondary mycelium of many Basidiomycetes.

Clavate. Club-like; narrowing in the direction of the base.

Cleistothecium (pl. cleistothecia). A completely closed ascocarp containing asci and ascospores.

Cloning. Asexually producing multiple copies of genetically identical cells or organisms descended from a common ancestor.

Coalesce. Grow or join together into one body or spot.

Coat protein. The protective layer of protein molecules surrounding the nucleic acid core of a virus.

Coelomycetes. A group of fungi in the Deuteromycetes that produce their spores within covered or closed structures, such as pycnidia or acervuli.

Coenocytic. Non-septate or aseptate; the nuclei are distributed in the cytoplasm without being separated by cross walls.

Columella (pl. columellae). A sterile structure within a sporangium or other fructification; often an extension of the stalk bearing the sporangium or fructification.

Conidiophores. Simple or branched specialized hyphae arising from somatic hyphae, bearing at the tip conidiogenous cell that produces conidia singly or in chains.

Conidium (pl. conidia). Asexual, non-motile spore formed by abstriction and detachment of part of a hyphal cell at the end of a conidiophore that germinates by a germ tube. Sporangia germinating by issuing germ tubes, instead of producing zoospores may also be called as conidia.

Conjugate nuclear division. The simultaneous division of the two nuclei in a dikaryon, giving rise to four daughter nuclei; these generally become separated by a septum into two cells, with the sister nuclei migrating into different daughter cells.

Coremium (pl. coremia; syn. Synnema). Compact or fused, generally upright conidiophores with branches and terminal spores forming a head-like cluster.

Cotyledon. Seed leaf, one in monocots and two in dicots; primary embryonic leaf within the seed, in which nutrients for the emerging seedling are stored.

Cross protection. The process whereby a normally susceptible host is infected with a less virulent pathogen (usually a virus) and thereby the host becomes resistant to infection by a more virulent strain of the same virus.

Cystidium (pl. cystidia). A sterile element occurring in the hymenium of certain Basidiomycetes; cystidia are larger than the other hymenial elements and protrude beyond them.

Cystogenous plasmodia. Plasmodia that cleave into thick-walled cysts, each of which ultimately giving rise to a single zoospore.

Cystosori. The resting spores (cysts) adhering together forming more or less compact groups.

Deciduous. Detach and falling down easily.

Demicyclic. A rust fungus that lacks the uredinial (repeating) stage, but typically has stages 0, I, III and IV.

Determinate. The stage at which vegetative growth ceases and the reproductive structures are formed.

Detoxification. A process or processes of metabolism, which renders a toxic molecule, less toxic by removal, alteration or masking of the effect of the active functional groups.

Diagnosis. Identification of the nature and cause of the illness, ailment or disease on the basis of its signs, symptoms, etiology, pathogenesis etc.

Dichotomous branching. Pair-wise forking; often repeatedly.

Dictyospore. A spore with both vertical and horizontal septa.

Differential host. Indicator host; a test plant, which by positive or negative response to inoculation of a strain of a pathogen differentiates it from other strains of this pathogen.

Digitate. Having lobes radiating from a common center.

Dikaryon. Having two sexually compatible haploid nuclei per cell, either in the spores or in the hyphal cells, which is common in the Basidiomycetes.

Dikaryophase. The division of the dikaryon simultaneously, leading to the presence of such pair of sexually compatible nuclei in each of the cell.

Dimorphic. Having two distinct shapes or forms.

Dioecious. Species, which produce separate male and female thalli. The male thalli produce only male sex organs, while the female thalli produce only female sex organs. Single dioecious fungi cannot reproduce sexually by itself.

Diploid. Having two complete sets of chromosomes (2N chromosomes), as a result of fusion of two compatible male and female, haploid nuclei, each having N chromosomes.

Discomycetes. A group of Ascomycetes, in which the fruiting body is an apothecium or discocarp.

Disinfectant. A physical or chemical agent that frees a seed, plant, organ or tissue from infection.

Disinfestant. An agent that kills or inactivates pathogens in the environment or on the surface of a seed, plant or plant organ before infection takes place.

Dissemination (syn. Dispersal). The transportation of inoculum from diseased plants to healthy plants and from one location to another location

Echinulate. Having small, pointed processes or spines projecting from cell walls.

Ectal exipulum. The outermost layer of the apothecium.

Ectomycorrhiza (pl. ectomycorrhizae). A symbiotic association between a non-pathogenic or weak pathogenic fungus and the roots of plants, with the fungal hyphae only in-between and outside the root cells.

Ectoparasite. A parasite that lives externally and feeds from the host by sending in specialized organs to absorb nourishment from the host.

Effused. Growing freely and extensively.

Elicitor. A physical, chemical or biological stimulus that triggers defense responses in plants.

Ellipsoid. Spores, which are rounded-oblong; having long sides parallel and ends almost hemispherically.

Embryo. The rudimentary plant or germ as formed in a seed.

Enation. Tissue malformation or an abnormal outgrowth, usually on the leaf and sometimes on the stem also, often induced by virus infection.

Encapsidated. Virus particles covered by a protective protein coat.

Encapsulated. Covered by a relatively thick, slimy layer that forms a capsule in the case of some bacteria.

Endemic disease. A disease permanently established in moderate or severe form in a defined area, commonly a country or part of a country.

Endobiotic. Living entirely within the host tissues.

Endoconidium. A conidium produced inside a hypha or conidiophore.

Endoparasite. A parasite, which enters a host, lives within the host and feeds from the host.

Ephemeral. Short-lived.

Epibiotic. Producing only the reproductive organs on the outside of the host, while the thallus remains inside the host tissues.

Epidemic (epiphytotic). A sudden, rapid spread of a disease; a widespread and severe, temporary increase in the incidence of an infectious disease, particularly within a season.

Epidemiology. The study of factors influencing the initiation, development and spread of infectious diseases.

Epidermis. Outer layer of cells in plants below the cuticle and forming the external integument in plants.

Epiphyllous. Found on the upper surface.

Epiphytic. Living on the surface of plants, but not as a parasite.

Epithecium. A layer of tissue on the upper surface of the hymenium of an apothecium, formed by the union of the tips of the paraphyses over the asci.

Eradicants. Systemic fungicides that can enter into the host tissues to a little extent and eradicate the dormant structures of the pathogen, as well as the active pathogen from the host.

Erumpent. Bursting or erupting through the substrate surface.

Etiology. The determination and study of the cause of a disease.

Extramatrical. Outside the host tissues.

Eucarpic. The reproductive organs arise from only a portion of the thallus, while the remainder continues its normal somatic life.

Facultative parasites. Organism, which are normally saprophytes, but when environmental conditions are favorable to them and when suitable hosts are present they become parasites. They can be grown on artificial culture media.

Facultative saprophytes (facultative saprobes). Organisms, though normally parasitic on living hosts, start leading a saprophytic life in the absence of suitable hosts. They can be grown on artificial culture media.

Falcate. Curved; sickle-shaped.

Fascicle. A small group or bundle.

Fastidious. Prokaryotic organisms, which exhibit special growth and nutritional requirements.

Flaccid. Wilted; lacking turgor or rigidity.

Flagellum (pl. flagella). A whip-, hair- or tinsel-like structure that serves to propel a motile cell.

***Forma specialis* (f. sp.).** A group of races and biotypes of a pathogen species, differing in some physiologic properties, that can infect only plants within a certain host Genus or species.

Free cell formation. The process by which the eight nuclei, each with a portion of adjacent cytoplasm are cut off by walls in the immature ascus to become ascospores.

Free water. Unbound water; a film of water on the plant surface.

Fructification. Any complex fungal fruiting body that contains or bears spores.

Fungicidal. Fungicides capable of killing fungal spores or mycelium. Applicable also to physical agents, such as heat, ultraviolet light, X-rays, gamma radiation etc., as well as to chemicals that are lethal at low concentrations.

Fungicide resistance. A decrease in sensitivity to a fungicide due to selection or mutation following repeated and frequent exposure to the chemical compound.

Fungicide resurgence. A fungicide, which may lead to increase in the occurrence of other diseases or non-target pathogens.

Fungistatic. Fungicides, which are able to inhibit germination of fungal spores or development of mycelium, without killing the fungus.

Fusiform. Spindle-shaped, tapering at each end as in the case of macroconidia of *Fusarium*.

Gall (syn. Tumor). Abnormal swelling or localized outgrowth, often roughly spherical, but unlike any organ of the normal plant, produced by a plant as a result of attack by a fungus, bacterium, nematode, insect or other organisms.

Gametangia. Hyphal cells of Zygomycetes containing gametes or nuclei that behave as gametes, which fuse to form zygospores in sexual recombination.

Gamete. Sex cell; a reproductive cell usually haploid that unites with another gamete of the opposite sex to form a zygote.

Gemma (pl, gemmae). A thick-walled cell similar to a chlamydospore.

Genetic engineering. The deliberate alteration of the genetic composition of a cell by various techniques, such as transformation of genes, protoplasmic fusion, use of enzymes etc. by man in tissue culture.

Geniculate. Bent like a knee.

Genome. The complete genetic constitution of an organism; the nucleic acid component of a virus.

Genotype. Genetic constitution of an individual, a class or group of individuals sharing a specific genetic makeup; the genetic factor, which influences the phenotype.

Germ tube. The initial hyphal growth from a germinating fungal spore, which develops into a mycelium.

Girdle. To encircle and cut through a stem or the bark and outer few rings of wood, disrupting the phloem and xylem.

Glabrous. Smooth; without hairs.

Halo. A spot surrounded by a chlorotic zone, which is characteristic of spots produced by some microorganisms on the leaves and sometimes on the other plant parts.

Haploid. A cell or organism whose nuclei have a single complete set of chromosomes.

Haplophase. The part of the life cycle of the organism in which the cells are haploid.

Haustoria (sing. Haustorium). Specialized absorbing organs produced by the mycelium, which penetrate the host cells and absorb nourishment from the host; mostly produced by obligate parasites, but also produced by some facultative parasites.

Helicospore. A coiled or cork screw-like spore; helical spore.

Hermaphroditic. Species, which produce distinguishable male and female sex organs in the same thallus i.e. each individual produces both recognizable male and female sex organs. This is also called as **'monoecious'**.

Heteroecious. Organisms, which require two botanically different hosts for completion of its life cycle, as in the case of some rust fungi.

Heterogametangia and **heterogametes.** The gametangia and gametes produced by the sex organs, which are morphologically quite different.

Heterokont. A biflagellate zoospore with two flagella of unequal size or unequal length.

Heterothallic fungi. Fungi producing compatible male and female sex organs or gametes on morphologically similar but physiologically distinct mycelia.

Holocarpic. In the case of either asexual or sexual reproduction, the entire thallus is converted into reproductive structures and segmented into spores.

Homothallic fungi. Fungi producing compatible male and female sex organs or gametes on the same mycelium.

Honeydew. A sugary exudate secreted by the ergot pathogen and sucking insects, such as aphids, white flies and scale insects.

Horizontal resistance. Partial resistance equally effective against most or all races or strains of a particular pathogen.

Host plant. Living plant attacked by or harboring a parasite or pathogen and from which the invader obtains part or entire requirement of its nourishment.

Hyaline. Colorless, transparent.

Hyalosporae. Organims belonging to Fungi Imperfecti producing hyaline spores.

Hyalospore. A hyaline spore; categories of Fungi Imperfecti, having hyaline spores.

Hybridization (cross breeding). The crossing of two individuals of different genotypes to produce 'F1' progeny with genes for the characteristics of both the parents.

Hydathodes. Structures with one or more openings at the leaf edge that discharge water from the interior of the leaf to the surface.

Hymenium. A spore-producing layer of a fruiting body of Ascomycetes or Basidiomycetes.

Hyperplasia. Abnormal multiplication of cells

Hypertrophy. Abnormal enlargement of cells.

Hypersensitive. Extreme or excessive sensitivity of plant tissues to infection by certain pathogens, leading to death of a portion or the entire affected part, thus blocking the advance of the pathogen to other parts of the plant.

Hypha (pl. hyphae). One of the simplest branched filaments of the mycelium of a fungus that is composed of one or more cylindrical cells and that increases in length by growth at its tip. New hyphae arise as lateral branches.

Hypocotyl. Portion of the stem below the cotyledons and above the root.

Hypophyllous. Found on the lower surface.

Hypothecium. A thin layer of interwoven hyphae immediately below the hymenium of an apothecium.

Icosahedral. Viruses having a geometrical shape, with 20 triangular faces and 12 corners.

Immune. Cannot be infected by a specific pathogen; having absolute resistance to a particular disease.

Immunity. The state of being immune; total exclusion of the potential pathogen; may be natural due to innate genetic characters or may be acquired by some predisposing treatment that modifies the chemistry of the plant.

Immunization. The process of inducing complete resistance to a particular disease.

Imperfect stage (anamorph stage). The part of the life cycle of a fungus in which no sexual spores are produced.

Inclusion bodies. Virus induced crystalline or amorphous structures that may occur in the cytoplasm or nucleus of cells of virus-infected plants consisting largely of viruses and are visible under the compound microscope.

Incubation period. The period of time between penetration of a host by a pathogen and the first appearance of symptoms, signs of disease or both in the host.

Indefinite. Not sharply delimited.

Indeterminate. Continuing to grow vegetatively while producing reproductive structures or flowers.

Indigenous pathogen. Native to an area; not introduced.

Infection. The entry of a pathogenic organism or virus into a host and establishment of a permanent or temporary parasitic relationship.

Infection peg. A very fine hypha with a deposition of substances, such as lignin, callose, cellulose, suberin etc. around it, that is thrust through the cuticle or epidermis of a host cell to cause infection.

Inhibit. To hold in check; to arrest the growth.

Inner veil. The hyphal membrane covering the gills of young mushrooms.

Inoculate. To introduce the inoculum into the host for the purpose of producing infection for testing susceptibility to infection.

Inoculum. Infective material of the pathogen.

Inoculum potential. The number of infective agents or particles present in the environment of the uninfected host.

Inoperculate. Not opening by a lid or operculum.

***In situ*.** Study of microorganisms in their natural environment, such as soil, water and the host body.

Integrated control. An approach that attempts to use of all available methods of control of a disease or of all the diseases and pests of a crop plant, such as physical, cultural, biological and chemical methods.

Intercalary. Formed along and within the mycelium and not at the hyphal tips.

Intercellular. Hyphae growing in-between the host cells.

Interstitial cells. Small, flat, dikaryotic cells formed in-between pairs of true aeciospores that aid in the release of aeciospores.

Intracellular. Hyphae penetrating into the host cells.

Invasion. The penetration, establishment and colonization of a host by a pathogenic organism.

.***In vitro*.** Study of microorganisms under laboratory conditions, in test tubes, petridishes etc.

***In vivo*.** Experiment on living organisms living in their natural environment within a living organism, in contrast to *in vitro* studies.

Isogametangia and **isogametes.** The gametangia and the sex cells (gametes) produced by the sex organs of fungi, which are morphologically indistinguishable

Isoplanogametes. Motile gametes, presumably of opposite sex, that are morphologically indistinguishable

Isthmus. Pads of gelatinous material formed in-between every pair of sporangia, which function as disjunctor cells that aid in the discharge of the sporangia.

Karyogamy. The fusion of two nuclei brought together by plasmogamy into one diploid or zygote nucleus.

Kresek. The vascular wilt phase of bacterial leaf blight of rice caused by *Xanthomonas oryzae* pv. *oryzae*.

Lamella (pl. lamellae). A gill; the vertical radial plates or gills on the lower surface of the 'mushroom' of Agaricaceae, which bear basidia and basidiospores on both the surfaces.

Lamina (leaf blade). The expanded, flat part of a leaf

Latent infection. The state in which a host is infected by a pathogen, but does not show any visible symptoms.

Lenticel. A natural opening in the stem of woody plants, bark, tuber, some fruits or root for the exchange of gases between the plant and the atmosphere.

Lesion. Localized, usually well-defined, abnormal change in the structure of an organ or diseased tissue due to disease or injury.

Liberation. Spore discharge; the process of aerial dispersal, in which the fungal spores are freed from its sporophore, pustules or lesions; also called 'take-off'.

Life cycle. The stage or successive stages in the growth and development of an organism that occur between the appearance and reappearance of the same stage, such as spore of the organism.

Lignification. Hardening of tissue through the deposition of lignin in the cell wall.

Lipids. Substances, whose molecules consist of glycerin and fatty acids and sometimes certain additional types of compounds.

Local invasion. Infection involving only localized part of a plant.

Local lesions. Small, restricted lesion, often the characteristic reaction of different cultivars to specific pathogens, especially in response to mechanical inoculation with a virus.

Locule. A cavity within a stroma

Lophotrichous. Bacteria having tuft of flagella on both the polar ends.

Lumen (pl. lumina). Central cavity of a cell or other structure.

Lysis. The enzymatic dissolution of all or part of an uni- or multicellular structure; dissolution of a phage-infected bacterium.

Macerate. To cause disintegration of tissues by separation of cells; to soften by soaking.

Macroclimate. The climate operating over wider regions and larger heights.

Macroconidium (pl. macroconidia). The larger of the two kinds of conidium formed by certain fungi.

Macrocyst. A thick-walled structure representing the sexual stage of the Acrasiomycetes.

Macrocyclic. A rust fungus that typically exhibits all five stages of the rust life cycle, viz., stages 0, I, II, III and IV.

Macroscopic. Visible to the naked eye without the aid of a magnifying lens or a microscope.

Malignant. A cell or tissue that divides and enlarges autonomously and that its growth can no longer be controlled by the organism on which it is growing.

Mantle (syn. Hyphal sheath). A dense hyphal mass of ectomycorrhizal fungus enclosing the short feeder roots of plants (ectomycorrhiza)

Marsh spots. Disease of pea seed, due to deficiency of manganese, exhibiting an extended internal brown, necrotic area at the central part of the cotyledons.

Masked symptoms. Disease symptoms that are absent in virus-infected plant under certain environmental conditions, but appear when the host is exposed to certain other conditions of light and temperature.

Matrix. The material in which an organism or organ is embedded.

Mechanical transmission. Transmission of a virus from an infected host to a healthy one without the intervention of a vector, either by physical contact and friction of one plant against the other or experimentally by rubbing sap from an infected plant on the leaves of a healthy one or by other machanical means of inoculation.

Meiosis. Process of nuclear division, in which the number of chromosomes per nucleus is halved, i.e., converting the diploid state to the haploid state; a series of two nuclear divisions, usually in quick succession in which the chromosome number is reduced by one-half.

mm (millimeter). A unit of length 1/10 of a centimeter (cm).

ìm (micrometer) or **ì (micron).** A unit of length 1/1,000 of a mm or 1/10,000 of a cm.

nm (nanometer). A unit of length 1/1,000 of a micrometer or 1/10,00,00,0 of a mm.

Meristem. The undifferentiated, mitotically active plant tissues, characterized by frequent cell division, that produce cells that become differentiated into specialized tissues. The meristems at the tip of roots and shoots are referred to as apical meristems.

Mesobiotic factors. Organisms possessing characters of both living organisms and non-living entities, such as viruses.

Messenger RNA. A chain of ribonucleotides that codes for a specific protein.

Metabasidium. The portion of the basidium in which meiosis takes places; also called the promycelium in certain Basidiomycetes.

Metabolism. The process by which cells or organisms utilize nutritive substances to build up living matter and structural components or to break down complex cellular material into simple substances to perform special functions.

Metabolite. Any chemical substance participating in metabolism; a nutrient.

Microclimate. The climate of a usually small site or habitat, usually the weather conditions prevailing between and immediately above the plants of an individual crop.

Microconidium (pl. microconidia). A type of small conidium as compared to a macroconidium; some type of microconidia act as spermatia.

Microcyclic. A rust in which the teliospore is the only binucleate spore in the life cycle.

Microcyst. A small, encysted protoplast; usually an encysted myxamoeba of the Myxomycetes or Acrasiomycetes.

Microsclerotia. A type of small sclerotia produced by some fungi, such as *Verticillium alboatrum*, which can withstand adverse environmental conditions.

Middle lamella. The cementing layer between adjacent cell walls, generally consisting of pectinaceous materials, except in wood tissues, where pectin is replaced by lignin.

Mitochondrion (pl. mitochondria). Cellular organelle outside the nucleus that functions in respiration.

Mitosis. Nuclear division, in which the chromosome numbers remain the same in the daughter nuclei.

Mode of action. The way or method by which a pesticide may alter or adversely affect the physiological or biochemical events in an organism resulting in a toxic effect, usually ending in death.

Monocentric. A thallus radiating from a single point, at which a reproductive organ, either sporangium or resting spore is formed.

Monocyclic. Having one cycle per season or year.

Monoecious (syn. autoecious). Rust fungi that have all stages of their life cycle on a single species of plant; also refers to organisms that produce male and female reproductive organs on a single individual.

Monogenic resistance. Resistance determined by a single gene; gene to gene resistance.

Monokaryotic. A cell, which contains one nucleus, as in the monokaryotic phase.

Monotrichous. Bacteria having single polar flagellum.

Monopartite. Viruses consisting of a single molecule of nucleic acid

Morphology. Study of the form, structure, architecture and development of an organism.

Mosaic. Symptom of many virus diseases, appearing as an abnormal pattern of dark green, light green and chlorotic or yellow areas on leaves.

Motile. Ability of self-propulsion by means of flagella, cilia or amoeboid movement.

Mottling. A leaf symptom showing small but numerous spots or blotches of more distinct discoloration comprising of light and dark green areas in an irregular pattern.

Mould. Any profuse or wooly, conspicuous, superficial fungus growth, consisting of mycelium and/or spore masses on damp or decaying matter or on surfaces of plant tissue.

Multipartite. Segmented genomes with more than one molecule of nucleic acid, encapsidated within a single protein shell (capsid) or within separate protein shells.

Mummy. A dried, shrivelled fruit.

Mutant. An individual acquiring a new, heritable characteristic, as a result of mutation, caused accidentally or due to exposure to certain mutagenic substances or irradiation, which causes a change in the gene of chromosome that leads to one or more discrete heritable differences from the parent.

Mycelium (pl. mycelia). Mass of hyphae constituting the body (thallus) of a fungus.

Mycorrhiza (pl. mycorrhizae). A symbiotic association between the hyphae of certain fungi and the absorptive organ, typically the roots of some plants.

Mycotoxins. Toxic substances produced by several fungi in infected seeds, feeds or foods, capable of causing illness of varying severity and even death to animals and humans that consume such substances.

Myxamoeba (pl. myxamoebae). An amoeboid cell particularly produced by the Myxomycetes.

Naked virion. A non-enveloped virus.

Natural host (syn. typical host). A host in which the pathogenic microorganism or parasite is commonly found and in which the pathogen can complete its development.

Natural immunity. Immunity due to innate genetical character.

Natural opening. Stomata, lenticels, hydathodes and nectarthodes.

Necrosis. Disintegration and death of cells or tissue, usually a clearly delimited part of a plant, accompanied by black or brown darkening due to local toxic or microbiological action.

Nectarthode. An opening at the base of a flower from which nectar exudes.

Non-persistent virus. A virus that persists in its vector only for a short period of time, usually less than 4 hours after acquisition.

Nucleic acid. Genetic material of all living organisms, including RNA (ribonucleic acid) and DNA (deoxyribonucleic acid), which may be composed of one or two strands.

Nucleocapsid. The structure composed of the capsid with the enclosed viral nucleic acid.

Nucleotide. A sub-unit of a nucleic acid, either DNA or RNA; the building blocks of DNA and RNA.

Obligate parasite. An organism that can live and obtain food only from living protoplasm of the host; it cannot be grown in artificial culture media.

Obligate saprophyte. An organism that can obtain its food only from dead organic matter, and is incapable of infecting another living organism.

Oblong. About twice as long as broad, with the sides nearly parallel through the middle portion.

Obtuse. With rounded or blunt tips.

Oedema. Blisters produced on leaves and other plant parts under conditions of high moisture stress and restricted transpiration.

Oidiophore. Specialized hypha that bears oidia.

Oidium (pl. oidia). Spores produced by simultaneous segmentation of oidiophore.

Oligogenic. Inheritance of characters determined by a few genes.

Oogamous. A type of fertilization, in which two heterogametangia viz., an antheridium and an oogonium come in contact and the contents of one flow into the other through a pore or tube.

Oogonium. The female gametangium of heterogamous fungi belonging to Oomycetes, containing one or more gametes.

Oosphere. An undifferentiated egg; a female gamete.

Oospore. A thick-walled, resting spore produced by sexual reproduction in the Oomycetes.

Operculum. Well-defined circular hinged cap-like structure at the tip of the discharge papilla, through which the spores escape outside.

Organelle. A membrane-bound, organized microstructures within cell cytoplasm, having specialized biochemical functions (e-g) chloroplasts, mitochondria.

Ostiole. A pore-like opening in the papilla or neck of a perithecium, pseudothecium, pycnium or pycnidium, through which spores are released.

Ovary. The female reproductive structure of organisms; in plants, enlarged basal portion of a pistil containing the ovule that develops into a fruit.

Ovule. Enclosed structure or egg contained within the ovary that becomes a seed after fertilization.

Pandemic. A widespread and destructive outbreak of a disease simultaneously in several countries.

Panicle. An indeterminate inflorescence, the main axis of which is branched, with pedicillate flowers on the secondary branches.

Papilla. A nipple-like elevation.

Paraphysis (pl. paraphyses). A sterile, upward growing, basally attached hyphal element present in the hymenium of the fruiting bodies, such as perithecium apothecium etc. produced by some fungi.

Parasexuality. A process in which plasmogamy, karyogamy and haploidisation take place in sequence, but not at specified points in the life cycle of an individual.

Parenchyma. Soft tissue of living plant cells with undifferentiated, thin, cellulose walls.

Parthenogenesis. Reproduction by the development of an unfertilized female gamete alone.

Pathogen. A disease-producing organism, agent or factor.

Pathogenic. Capable of causing disease in a host or range of hosts.

Pathogenicity. The ability of the pathogen to cause disease.

Pathovar (pathotype). A subdivision of a plant pathogenic bacterial species characterized by its pattern of virulence or avirulence to a series of differential host varieties; pathovar (pv) for bacteria is equivalent to *forma specialis* for fungi.

Pedicel. Small slender stalk bearing an individual flower, inflorescence or spore.

Peduncle. Stalk or main stem of an inflorescence, part of an inflorescence or a fructification.

Penetration peg. The specialized, narrow, hyphal strand produced from the underside of an appressorium that penetrates the host cells.

Perennial mycelium. Mycelium, which persists in a host plant from one season to the next.

Perfect stage. The sexual stage in the life cycle of a fungus.

Peridium (pl. peridia). The outside covering or wall of a fructification.

Periphyses (sing. periphysis). Short hair-like growth, in the form of a fringe lining the inside of an ostiole or of a pore in the stroma.

Periphysoids. Lateral periphyses.

Periplasm. A layer of protoplasm surrounding the oosphere of certain Oomycetes.

Perisporium. A thin, hyaline, membranous sheath surrounding the spore.

Perithecium (pl. perithecia). A flask-shaped or subglobose, thin-walled, fungus fruiting body (ascocarp), containing asci and ascospores, which are expelled or released through a pore (ostiole) at the apex.

Peritrichous. Having a number of flagella distributed over the whole surface.

Persistent transmission. A type of insect transmission in which the viruses are acquired by the vector after a long acquisition feeding period, and in which there may be a latent period following acquisition before the vector can transmit the viruses. The vector remains viruliferous for a long period, often throughout its life span. The virus sometimes multiplies within the vector.

Phanerogamic parasites. Parasites belonging to the higher plants.

Phaeosporae. Organisms belonging to Fungi Imperfecti producing colored spores.

Phaeospore. A dark colored or brown spore; categories of Fungi Imperfecti, having dark colored spores.

Phagotrophic. Feeding by engulfing the food.

Phialide. A type of bottle-shaped conidiogenous cell, that produces blastic conidia in a basipetal fashion without detectably increasing in length.

Phialoconidium. A conidium formed from a phialide.

Phragmobasidium (pl. phragmobasidia). A basidium typically divided into four cells by transverse septa.

Phloem. Food-conducting tissues, consisting of sieve tubes, companion cells, phloem parenchyma and fibers.

Phyllody. The replacement of floral parts by leaf-like structures.

Physiologic race. Biotype of a species of a pathogen that are alike in morphology, but differ in certain cultural, physiological, biochemical, pathological or other characteristics and hence differ in their ability to infect particular varieties of the susceptible host species; they are also referred to as ‘physiological form’ or ‘biological form’.

Phytoalexin. A substance in plants that inhibits the development of pathogenic microorganisms, that is produced in response to physical or chemical injury or when they are infected by a pathogen.

Phytosanitation. Measures requiring removal or destruction of infected or infested plant material likely to form source of reinfection or reinfestation.

Phytotoxic. Harmful or poisonous to plants or plant growth (usually used to describe toxicity developed due to application of herbicides, insecticides, fungicides etc.)

Pileus. The fleshy, umbrella-like or cap-like upper portion of the basidiocarp of Agaricales.

Pistil. The ovule bearing organ of the plant consisting of the ovary and its appendages, such as style, stigma etc.

Pith. Parenchymatous tissue occupying the center of the stem.

Planogamete. A motile gamete.

Plasmids. Small, spherical, genetic material containing several non-essential genes as are found in bacteria

Plasmodium. A naked, multinucleate mass of protoplasm moving and feeding in amoeboid fashion; the somatic phase of the Myxomycetes.

Plasmodesma (pl. plasmodesmata). A fine strand of cytoplasm passing through a pore in the cell wall, thus usually connecting the protoplasm of adjacent cells.

Plasmogamy. Union of protoplasts of two cells that brings the nuclei close together within a single cell.

Plectenchyma. The general term employed to designate all types of fungal tissues; the two most common types of tissues are prosenchyma and pseudoparenchyma.

Pleomorphic. Having more than one independent form or spore stage in the life cycle.

Polycentric. A thallus resulting from many centers at which reproductive organs, either sporangia or resting spores are formed.

Polycyclic. Completing many life cycles or disease cycles in one year

Polyploid. Having three or more complete sets of chromosomes.

Poroconidium or **porospore.** A conidium produced through a pore in the conidiogenous cell.

Predisposing factors. Factors, such as genetic-, cultural- or environmental factors, which by their actions render an organism susceptible to a particular disease.

Prevalence of a disease. The total number of cases of a particular disease at a given moment of time, in a given population.

Primary inoculum. Propagules of the pathogen, usually from an overwintering source, that cause primary infection or that initiate the disease in the field, as opposed to inoculum that spreads the disease during the season.

Primordium (pl. primordia). The rudiment of beginning stage of any structure.

Probasidium. The primary basidial cell; in Uredinales, the teliospore; in Ustilaginales, the chlamydospore.

Prokaryotes. Organisms that lack a clearly defined nuclear membrane.

Proliferation. A rapid multiplication of microorganisms; formation of new cells, tissues or organs, specifically a hyperplastic development of plant diseases in which organs continue to develop after they have reached the point beyond which they normally do not grow.

Promycelium (pl. promycelia). A germ tube or a short hyphal filament or basidium produced meiotically by the germinating rust spores (teliospores) or smut spores (chlamydospores), from which basidiospores are formed.

Propagative virus. A virus, which multiplies in its vector.

Propagule. Any unit or part of an organism, such as spore, sclerotium, chlamydospore, mycelial fragment etc., capable of independent growth to produce a new individual.

Protein subunit. A small protein molecule that is the structural and chemical unit of the protein coat of a virus; a capsomere.

Protogyny stage. Fertilization stage, but before actual fertilization has taken place.

Protoplast fusion. Any induced or spontaneous union between two or more protoplasts to produce a single binucleate or multinucleate cell; in particular, a tissue culture procedure for somatic hybridization that is used in cell manipulation studies.

Pseudomorph. An indefinite or irregular structure.

Pseudomycelium. A series of cells adhering end to end to form a chain; produced by some yeast.

Pseudoparaphyses. Sterile threads attached to both to the roof and to the base of an ascocarp.

Pseudopodia. False plasmodia or plasmodia-like structure.

Pseudoseptae (sing. pseudoseptum). Plug-like partition of cellulin or other substances in the hyphae or spores of some fungi, such as in the case of conidia of *Drechslera*

Pseudothecium (pl. pseudoperithecia). Contraction of pseudoperithecium; an ascocarp similar to a perithecium, but unlike typical perithecium, produces asci in unwalled locules.

Pulverulent. Break through the host surface.

Pustule. A blister-like fungal spore mass, breaking through a plant epidermis; a pimple-like, eruptive fruiting structure, such as a uredinium of a rust fungus.

Pycnidiospore. An asexual spore produced in a pycnidium.

Pycnidium (pl. pycnidia). An asexual, thick-walled, spherical or flask-shaped fruiting body, lined inside with conidiophores that produce pycnidiospores (conidia) from their tips; characteristic of Sphaeropsidales of Fungi Imperfecti.

Pycnium or **spermagonium.** Globose or flask-shaped, haploid fruiting body of rust fungi bearing pycniospores or spermatia and receptive hyphae, representing stage '0' of the rust fungi.

Pycnosclerotium (pl. pycnosclerotia). A more or less hard-walled structure resembling a pycnidium but containing no spores.

Quarantine. Legislative control of the transport of plants or plant parts from one particular place to another, from one state to another or from one country to another to prevent the spread of pests or pathogens.

Quiescent. Dormant or inactive.

Race. A subgroup or biotype within a species or variety, morphologically identical, but distinguished from other races by virulence, symptom expression or host range.

Race specific resistance. Having resistance only to some races of a pathogen, but not to other races.

Rachis. Elongated main axis of an inflorescence.

Radicle. Part of the plant embryo that develops into the primary root.

Reactive oxygen radicals. Oxygen radicals, much more reactive than molecular oxygen (O_2) that on contact with a pathogen in the resistant plant cell reacts quickly and oxidize various cellular components into compounds toxic to the pathogen.

Recognition factor. Specific receptor molecules or structures on the host (or pathogen) that can be recognized by the pathogen (or host).

Receptive hyphae. A simple or branched hypha-like structure of a rust fungus produced by a pycnium (spermagonium) that protrudes out through the ostiole and receives the nucleus of a pycniospore (spermatium) of the opposite mating type so as to initiate the dikaryotic phase of the fungus.

Relative humidity (RH). The ratio of the amount of water (water vapor) contained in a sample of air to that which could be contained in the same volume, if it were saturated.

Resistance. The ability of an organism to exclude or overcome completely or to some extent the harmful effect of the pathogen or other damaging factor.

Resistance - horizontal. Resistance, which is evenly spread against all races of a particular pathogen.

Resistance - vertical. Resistance only to some races or strains of a pathogen and not to others; resistance that completely protects the host, but only against specific races or strains of a pathogen.

Respiration. A metabolic process occurring in all living organisms, in which a series of chemical reactions take place and by the oxidation of carbohydrates and fat, energy is made available to the organisms.

Resting spore. A sexual or asexual, thick-walled spore of a pathogen that is resistant to extremes of temperature and moisture, and which often germinates only after a period of dormancy from its formation.

Reticulate. Covered with net-like ridges; reticulate spore.

Rhizoid. A short, thin hypha growing in a root-like fashion toward the substratum and obtain nourishment for the fungus.

Rhizome. A mostly horizontal, jointed, fleshy, often elongated, usually underground stem with nodes, buds and scales.

Rhizomorph. An aggregation of hyphae resembling a root and having a well-defined apical meristem capable of transporting nutrients and spreading infection over long distances; often differentiated into a rind of small, dark cells surrounding a core of elongated, colorless cells.

Rhizomycelium. Fungal, rhizoidal strands produced by some lower fungi, which resemble mycelium superficially, but grossly dissimilar to that of true mycelium and are of varying diameter and are anucleate.

Rhizosphere. The soil closely surrounding a living plant roots, in which the microbial population is enhanced by root exudates.

Rhynchosporous. Having beaked spores.

Ribosome. A subcellular protoplasmic particle, made up of one or more RNA molecules and several proteins, involved in protein synthesis.

Rickettsia (pl. rickettsiae). Parasitic or symbiotic microorganisms, similar to bacteria in most respects, but generally capable of multiplying only inside living host cells.

Ring spot. A disease symptom characterized by yellowish or necrotic, more or less concentric rings, enclosing green tissue, as in some virus diseases.

Root exudates. The various compounds that are exuded from growing or expanding sections of roots.

Rosette. Short, bunchy habit of plant growth with small, circular, cluster of leaves.

Rouging. Removal of undesirable plants, usually off-types or infected plants in a crop.

Rugose. Wrinkled and roughened appearance of leaves as in some virus diseases.

Russetting. Formation of brownish, roughened areas on the skin of fruits or tubers because of hyperplastic reaction caused by some fungi.

Rust. A disease caused by one of the Uredinales that often produces spores of a rusty color.

Saccate. Sac-like; bag-like.

Saltation. A mutation within an isolate known to be of pure genotype.

Sanitation. The removal and burning of infected plant parts, decontamination of tools, equipment, hands etc.

Saprobe (syn. saprophyte). Organism that obtains nourishment from non-living organic matter.

Scab. A hyperplastic symptom characterized by rough, crusty lesions formed by excessive cork production.

Scald. A lesion that appears as an apparent consequence of scalding with hot water, usually bleached and sometimes translucent.

Scar. A clear mark in the conidiophore at the place from where a conidium had dislodged.

Schizogamy. A multinucleate plasmodium breaking up into daughter plasmodia by a budding-like process. This type of vegetative multiplication is called 'schizogamy'; the mother plasmodium is called a 'schizont' and the daughter plasmodia as 'meronts'.

Scion. A detached shoot or bud used in vegetative propagation for grafting or budding.

Sclerenchyma. A protective or supporting plant tissue made up of cells with thickened, lignified walls.

Sclerotium (pl..Sclerotia). A hard resting body, resistant to adverse environmental conditions and may remain dormant for long period of time and germinates when favorable conditions return; it consists of closely packed, more or less isodiametric or oval cells, known as pseudoparenchyma tissue.

Scolecospore. An elongated, slender, needle- or worm-like or filiform spore.

Scorch. Any symptom that suggests the action of fire or flame on the affected part, often seen at the leaf margins.

Secondary inoculum. Inoculum produced by infection that took place during the same growing season or inoculum produced as a result of infection by the primary inoculum, which causes secondary infection

Secondary mycelium. The dikaryotic mycelium of Basidiomycetes, resulting from plasmogamy.

Seed. Ripened ovule consisting of an embryo and stored food enclosed by a seed coat.

Seed-borne. Carried on or in the seed.

Seed dressing or **seed treatment.** Application of a biological agent, chemical substance or physical treatment to seed, seed material or plant, so as to protect the seed, seed material or plant from pathogens, or to stimulate germination or plant growth.

Seed transmission. Perpetuation of a pathogen from one generation to another through seed.

Senescence. Process or state of growing old.

Septate. Hyphae divided into compartments or cells by cross walls or septae.

Serology. A laboratory method using the specificity of the antigen-antibody reaction for the detection and identification of antigenic substances and the organisms that carry them.(e-g. microorganisms and viruses).

Sessile. A leaf, leaflet, flower, floret, fruit, ascocarp, basidiocarp etc. without a stalk, petiole, pedicel, stipe or stem.

Seta (pl. setae). Stiff, bristle-like or hair-like structure occurring in the fruiting body (acervuli) of some fungi belonging to Melanconiales.

Sett. A piece of stem with nodal buds or tuber used as planting material.

Sexual reproduction. Reproduction involving fusion of two haploid nuclei of opposite mating types (karyogamy) to form a diploid nucleus followed by meiosis back to the haploid nuclei state at some point in the life cycle, resulting in genetic recombination.

Sexual spore. A spore produced as a result of sexual reproduction.

Sieve tubes. Plant cells specialized for the transport of sugars, connected to one another by perforated **'sieve plates'** to form a **'cellular tube'** or **'sieve element'** in the phloem vessels of leaves, stems and roots.

Shock symptoms. The severe, often necrotic symptoms produced on the first new growth following infection with some viruses; also called acute symptoms.

Shoestring. A symptom of extreme narrowing of the leaf often caused by a virus.

Shot holes. A symptom of certain leaf spotting pathogens, in which necrotic areas of limited size fall out from the lesions on the leaf lamina, leaving small, almost circular holes, which appear as holes made by bullet shots.

Siderophores. Specific agents produced, which aid in combating low iron stress.

Signal molecule. Host molecules that react to infection by a pathogen and transmit the signal to other plant parts and activate proteins and genes in the cells of other parts and of the plant so they will trigger the defense reaction.

Slime layer. A diffused layer of polysaccharides exterior to the cell wall in some bacteria.

Slime mould. Saprophytic organisms that form vegetative amoeboidal plasmodia and spores as found in the Myxomycetes.

Smut. A group of fungi in the Basidiomycetes that typically releases masses of black, thick-walled, dusty, resting spores or teliospores or smut spores at maturity.

Soft rot. A rot of fleshy fruit, vegetable or ornamental, in which the tissue becomes macerated by the enzymes of the pathogen.

Soil inhabitant. Pathogenic microorganism that maintains its population in soil over a long period of time.

Soil transients. Pathogenic microorganisms that can live in the soil only for short periods.

Soma. Filamentous body or thallus of a fungus.

Somaclones. Plants produced by a genetic engineering technique, by which single cells or protoplasts are cultured to produce new individuals, which are genetically variable from their genetically stable parent. The variations thus induced are called **'somaclonal'**.

Somatic cell. Cells of the thallus that are not involved in sexual reproduction.

Somatic hybridization. The formation of hybrids by fusion of somatic cells, as opposed to the fusion of gametes. The term is generally applied to fusion of plant protoplasts.

Sooty mould. A black, sooty coating on the foliage and fruits formed by the dark hyphae of the fungi that live in the honeydew secreted by insects, such as aphids, planthoppers, mealy bugs, scales etc.

Sorus. Compact fruiting structure, especially spore masses produced in or on the host plant, such as the rust and smut fungi.

Source of inoculum. The place, which harbors inoculum and from which it may be disseminated and cause infection.

Spawn. Fungal culture consisting of mycelium that is used for mushroom cultivation.

Species (pl. species). The unit of classification of a group of closely related individuals resembling one another in certain inherited characteristics of structure and behavior, and relatively stable in nature; the individuals belonging to a species are subordinate to a Genus, but above a race; it is designated by a binomial consisting of the generic name and the specific epithet.

Spermagonium (pl. spermagonia, syn. pycnium). A globose or flask-shaped, haploid, fruiting body composed of receptive hyphae and spermatia (pycniospores) produced by the rust fungi (stage '0' of the rust fungi.

Spermatium (pl. spermatia). A non-motile, haploid, male sex gamete produced in the spermagonium.

Spermatization. Plasmogamy by the union of a spermatium with a receptive hypha of the opposite mating type to form a dikaryoyic mycelium.

Spike. An elongated inflorescence with sessile flowers on the main axis.

Spikelet. Spike-like appendage comprising of one or more reduced flowers and associated bracts, an unit of inflorescence as in grasses.

Spindle-shaped. Fusoid, narrowing towards the tip.

Sporadic. A disease that breaks out occasionally without being constantly destructive.

Sporangiolum (pl. sporangiola). A small sporangium containing few spores.

Sporangiophore. A specialized branch of fungal hyphae bearing sporangia.

Sporangiospore. Asexual spores borne inside a sporangium.

Sporangium (pl. sporania). A sac-like fungal structure, in which the entire contents are converted into an indefinite number of asexual spores.

Spores. Reproductive structures produced by fungi and some other organisms constituting of one or more cells, which are synonym to seeds of higher plants; a bacterial cell modified to survive any adverse environmental conditions.

Sporidium (pl. sporidia). The basidiospores of the rust or smut fungi.

Sporocarp. Spore-bearing fruiting body.

Sporodochium (pl. sporodochia). A specialized, cushion-shaped stroma consisting of masses of closely woven hyphae without any lateral union and produce conidia from the hyphal tips.

Sporophore. A spore-producing or spore-bearing structure, such as a conidiophore, ascocarp or basidiocarp.

Sporogenous plasmodia. Plasmodia giving rise to thin-walled sporangia containing many zoospores.

Sporulation. The process of producing spores.

Spot. A disease symptom characterized by a limited necrotic or chlorotic area on leaves, flowers and stems of plants.

Spreader. A chemical substance added to a spray fluid to assist even and uniform distribution of the active material on the sprayed surface.

Stamen. Male reproductive structure of a flower, composed of a pollen-bearing anther and a stalk.

Staphylococcus. Spherical bacteria occurring in irregular or grape-like clusters.

Starch. A polysaccharide consisting of glucose units, the principal food substance stored in plants.

Stem pitting. A symptom of some virus diseases characterized by longitudinal depressions or pits on the stem of plants.

Sterigma (pl. sterigmata). A small, slender, pointed projection that supports a spore.

Sterile fungi (Fungi Imperfecti). A group of fungi that are not known to produce any kind of spores.

Sterilization. Total destruction of all living organisms by various means, including heat, chemicals or irradiation.

Sticker. A chemical substance added to a plant protectant spray fluid to increase its tenacity.

Stigma. The portion of a flower that receives pollen and on which the pollen germinates and cause fertilization.

Stipe. Stalk.

Stipule. Small, leaf-like appendage at the base of a leaf petiole, usually occurring in pairs.

Stolon (syn. runner). Fungal hyphae that grow horizontally along the surface of the substratum (e-g. *Rhizopus* species).

Stoma (pl. stomata). Structure composed of two guard cells and the opening in-between them in the epidermis of a leaf or stem.

Stone fruit. Fruit with a stony endocarp (e-g. cherry, peach, plum etc.)

Strain. A distinct form of an organism or virus within a species, differing from other forms of the species biologically, physically or chemically, but identical morphologically.

Streak. A disease characterized by elongated lesions or areas of discoloration, usually of limited length, on leaves with parallel venation or on stems.

Streptobacilli. Bacilli in chains.

Streptococci. Cocci in chains

Streptomycete. A filamentous eubacterium, which forms extensive mycelium, aerial hyphae and conidia similar to moulds; many produce antibiotics.

Striate. Marked with delicate lines, grooves or ridges.

Stripe. A disease characterized by elongated areas of discoloration of indefinite length on stems or on leaves with parallel venation.

Stroma (pl. stromata). A compact somatic structure, much like a mattress or cushion made up of loosely woven hyphae, more or less parallel to one another known as prosenchyma tissue; usually fructifications are formed on or in the stroma.

Stunting. Reduction in height of a vertical stem axis, resulting from a progressive reduction in the length of successive internodes or a decrease in their number.

Style. Slender part of many pistils located between the stigma and the ovary, through which the pollen tube grows

Stylet. Stiff, slender, hollow feeding organ of plant parasitic nematodes or sap-sucking insect pests, such as aphids, leafhoppers etc

Stylet-borne transmission (syn. non-persistent transmission). A type of virus transmission, in which the virus is acquired and transmitted by the vector after a short feeding period and is transmitted by the vector for only a short period of time.

Suberization. The formation of cork layer by the action of cork cambium, or the deposition of suberin in or on cell walls as in cork cells.

Subglobose. Almost global or spherical.

Subhyaline. Somewhat or imperfectly clear or nearly colorless.

Substrate. The material or substance on which a microorganism feeds and develops.

Sunscald. Injury of plant tissues, burnt or scorched by direct sun.

Suppressive soil. Soil in which certain diseases are suppressed because of the presence of microorganisms antagonistic to the pathogen in that soil.

Surfactant (surface active agent). A material affecting surface tension, used as a wetting agent in spray fluids.

Susceptible. Capable of being easily infected or infested; lacking the inherent ability to resist disease or attack by a pathogen; not immune.

Swarm spores. Zoospores or swarm cells or swarmer.

Symbiosis. Mutually beneficial association of two different kinds of organisms.

Sympodial. Proliferation of axes, in which each successive spore or branch develops behind and to one side of the previous apex where growth has ceased.

Symptom. A visible abnormal change in a host and its behavior as a result of infection by a pathogen.

Symptomless carrier. A plant although infected with a pathogen (usually a virus) produces no obvious symptoms.

Syndrome. The totality of effects produced in a host by one disease, whether simultaneously or successively, including effects not directly detectable to the unaided eye.

Synergism. The association of two or more pathogenic organisms acting at one time and effecting a change, which one only is not able to make; the increased efficacy of some of the plant protection chemicals, when combined with certain non-toxic chemicals

Synnema (pl. synnameta). A group of conidiophores unite more or less closely and cemented together into columns and produce conidia from the tips of the conidiophores. It is also called as **'coremium'** (pl. coremia)

Systemic. Term applied to pesticide when applied to soil or foliage are absorbed by the plant parts and translocated to other parts through the vascular system, and imparting protection from disease or pest attack; in case of plant disease, a single infection at one site resulting in the spread of the pathogen throughout the plant.

Systemic acquired resistance. A broad, physiological immunity in plants that results by the application of certain natural and systemic chemical compounds, which trigger some kind of responses in the host and induce resistance.

Target spot. A lesion consisting of a dark brown circular area containing a series of brown, concentric rings appearing like a target board; typical of infection caused by *Alternaria* sp.

Teliomorph. The sexual form in the life cycle of a fungus.

Teliospore or **teleutospore.** A thick-walled resting spore of the rusts and smuts in which karyogamy occurs.

Telium (pl. telia). The fruiting body (sorus) of a rust fungus that produces teliospores (stage III of the rust fungus).

Thallus. Vegetative body of fungi devoid of stem, roots and leaves or soma of a fungi.

Therapeutants. Systemic fungicides that can enter into the host, translocated inside the host tissues, affect the deep-seated infection and can cure diseases even after their establishment in the host.

Thermophiles. Organisms, which require an optimum temperature of 55°-65°C. for their growth.

Thigmotropism. Response caused as a result of contact.

Tiger-stripe. A leaf symptom in which clear yellow areas separate marginal and intervenal areas of dark necrotic tissue from the normal green areas along and adjacent to the main veins, giving a yellow and black appearance as in a tiger skin.

Tinsel. Ciliated.

Tissue. Group of cells, usually of similar structure, that performs the same or related functions.

Tissue culture. *In vitro* method of propagating healthy cells from plant tissues.

Tolerance. Ability of a plant to endure an infectious or non-infectious disease, adverse conditions or chemical injury without serious damage or yield loss.

Tomentose. Densely covered with short, fine hairs.

Toxin. An organic poisonous substance produced by a microorganism, which even at low concentrations deleteriously and irreversibly affects the normal processes of living organisms.

Tracheid. Elongated conducting cell of the xylem, with tapering or oblique end walls and pitted walls.

Translocation. Movement of water, nutrients, chemicals or food materials within a plant.

Translucent. Semi-transparent; allowing some light to pass through, but not transparent

Transmission. The transfer of an infectious agent (usually a virus) from one plant to another; the dissemination of pathogens and the inoculation of suscepts.

Transovarial transmission. Transmission of the virus through the eggs of the insect vector to its progeny.

Transpiration. The evaporation of moisture from a living plant, mainly through the stomata of the leaf. This moisture represents a surplus from that taken in by the roots and which is not required for photosynthesis.

Transstadial. Viruses retained in the vector even after moulting.

Trichogyne. The receptive neck of the ascogonium, which is often long and hair-like.

Truncate. Ending abruptly as though with the end cut off horizontally.

Tuberculate. Having small swellings or warts on the surface.

Tumor. An uncontrolled overgrowth of tissue or tissues.

Turbinate cells. The diploid, intramatrical mycelium produced from the zygote as a result of copulation of two compatible zoospores, enlarge in portions of the hypha, which later become large cells (turbinate cells); a group of such turbinate cells are linked together by hyphae called rhizomycelium; these cells may become resting spores; meiosis occurs during germination of the resting spores; this is characteristic of the Genus - *Physoderma*.

Turgidity. State of being rigid or swollen as a result of internal water pressure.

Tylosis (pl. tyloses). A balloon-like overgrowth of the protoplast of a parenchyma cells into adjacent xylem vessels that partially or completely block them.

Umblicate. With a central depression.

Undulate. With wavy surface at the margin.

Unilocular. Single, simple, undivided cavity.

Uniseriate. Arrangement of ascospores in an ascus in one series.

Unitunicate. An ascus in which both the inner and outer walls are more or less rigid and do not separate during spore ejection.

Universal veil. A thin, veil-like membrane that covers certain types of young mushrooms; upon expansion of the mushroom, the universal veil tears and its remnants may be seen in the form of scales on the pileus and in the form of a volva at the base of the stipe.

Uredinium or **uredium.** The sorus of rust fungi producing urediniospores or uredospores (stage II of the rust fungus)

Uredospore. A binucleate, repeating spore.

Ustilospore. Smut spore of Ustilaginales, from which the basidiospores are formed.

Utriform. Bag-like.

Vacuole. Mostly spherical organelle within a cell bound by a membrane and containing dissolved materials, such as metabolic precursors, storage materials or waste products.

Variegation. Pattern of two or more colors in a plant part, as in a green and white leaf.

Vascular bundle. A strand of conductive tissues, usually composed of xylem and phloem vessels (in leaves, small bundles are called veins).

Vascular wilt disease. A disease in which the pathogen is almost entirely confined to the vascular system of the host during pathogenesis, and in which wilting is a characteristic symptom.

Vector. A living organism (e-g. insects, mites, birds, higher animals, nematodes, parasitic plants, humans etc.) capable of carrying and transmitting a pathogen and disseminating diseases.

Vegetative. Asexual, somatic.

Vegetative phase. Refers to a non-reproductive phase in fungi and plants.

Vein. A vascular bundle forming the framework of fibrous tissue in a leaf; a blood vessel conducting blood towards the heart in any animal.

Vein banding. Symptom of a virus disease, in which the areas adjacent to the veins of the leaf remain green, in contrast to the remaining areas, which may be chlorotic.

Vein clearing. The cells adjacent to the veins become translucent or chlorotic, while the interveinal areas remain green.

Vermiform. Worm-shaped.

Vertical resistance. Resistance only to some races of a pathogen and not to others; resistance that completely protects the host, but only against specific races or strains of a pathogen.

Verrucose. Slightly roughened surface.

Vesicle. In fungi, a small, thin-walled, bladder-like structure produced by a zoosporangium and in which the zoospores are released or are differentiated.

Vesicular-arbuscular mycorriyza (syn. endomycorrhiza). Symbiotic association between a non-pathogenic or weakly pathogenic fungus and the roots of plants, in which the fungal hyphae invade cortical cells of the root and produce vesicles and arbuscules.

Vessel. A xylem element or series of such elements whose function is to conduct water and mineral nutrients to the aboveground parts of the plant through pit openings in end walls.

Viable. Capable of living; able to germinate.

Vigor (of seeds). Physiological potential for rapid and uniform germination and fast growth of seedlings under normal field conditions.

Viricide. A substance that completely and permanently inactivates a virus.

Virion. A complete, matured virus particle consisting of the nucleic acid and protein shell.

Viroid. A pathogenic agent that has virus-like properties composed of a circular, single-stranded molecule, containing a small amount of RNA and lacking a protective protein coat.

Virulence. Degree or qualitative measure of pathogenicity; relative capacity of a pathogen to cause disease.

Virulent. Capable of causing a severe disease; strongly pathogenic.

Viruliferous. An insect vector containing virus and is capable of transmitting it into a susceptible host.

Virus. An ultramicroscopic, filterable, intracellular, infective agent of obligate nature, consisting of a core of infectious nucleic acid, either RNA or DNA, usually surrounded by a protein coat and capable of causing various diseases in living organisms.

Virusoids. Similar to viroids, but smaller than viroids, consisting of a circular RNA and are often associated with another virus RNA and behaves as a satellite of the virus, and can replicate only in the presence of the associated virus.

Volatile. A chemical compound, which evaporates or vaporizes from a liquid to a gaseous state at ordinary temperatures on exposure to the air.

Volunteer plant. A plant from a previous season's crop that regenerates in a subsequent crop or plant developing from a self-sown or lost seed.

Vulnerability. Inability of a plant or organism to resist attack by a phytopathogen or parasite and to counteract the effect of the attack.

Volva (pl. volvae). A cup-like structure at the base of the stipe of certain mushrooms, which is a remnant of the universal veil.

Warty. Having a rough surface.

Water-soaked. A disease symptom, in which an area of plant cells becomes darker in color owing to the filling of intercellular air spaces with cell sap, which comes out to the surface as exudate.

Wet rot. Any kind of rot, in which the tissue is rapidly and completely disintegrated, with release of water from the lysed cells.

Wetting agent. A substance added to the plant protection chemical formulations or to the pesticidal spray fluid to reduce the surface tension of the applied droplets and to facilitate uniform spreading of the spray fluid over the sprayed surface.

Whiplash. Long, whip-like flagellum.

White rot. Rotting of wood in trees invaded by lignin-destroying fungi that leaves a white cellulose residue.

Wilt. Loss of turgidity and drooping of plant parts generally caused by insufficient water or excessive transpiration in the plant; a disease symptom caused by the wilt fungi.

Witches' broom. Broom-like growth or massed proliferation caused by the dense clustering of branches of woody plants.

Wound parasite. A parasitic organism, which can invade a host only if it can first become established in damaged tissues.

Xylem. A plant tissue consisting of tracheids, vessels, parenchyma cells and fibers; wood.

Xylem-limited bacteria (XLB). Endophytic bacterial parasites that live in plants exclusively in the xylem cells or tracheary elements (e-g. *Clavibacter xyli* sub sp. *xyli)*

Yellows. A plant disease characterized by pronounced chlorosis, yellowing and stunting of the host plant.

Zoogametes. Motile gametes; planogametes.

Zoospores. Non-sexual, motile sporangiospores or swarm spores produced from either sporangia or zygosporangia, having one or two flagella.

Zygosporangium. A sporangium that arises from a zygospore at the end of a stalk, which contains non-motile spores.

Zygospore. The zygote formed as a result of fusion of two morphologically identical gametangia.

Zygote. A diploid cell formed by the union of two, compatible gametes, as well as the individual produced from such a cell.

Subject Index

A

Abiotic 2
Abnormal cell proliferation 264, 265
Abscesses 6
Abscission layers 294, 302
Acaricides 362
Acaulospora 387
Aceria tulipae 244
Acetic acid 373
Achlorophyllous 10, 38
Acholeplasma 14
Acorus calamus 392
Acquisition 15, 16, 241, 242, 243, 244
Acquisition feeding time 241, 243, 244
Acrasiogymnomycotina 34, 38
Acrasiomycetes 34, 38
Acrocylindrium (sarocladium) 159
Acrocylindrium (sarocladium) oryzae 159
Acropetal 331
Acropetally 161, 292
Acrosporium 96, 159
Actidione PM 339
Actidione RZ 339
Actinobacter 384, 392
Actinomyces 171, 210, 213, 219
Actinomyces bovis 219
Actinomycetaceae 217
Actinomycetes 338, 373
Actispray 339
Activator virus 235
Active or Autonomous Dispersal 174
Active oxygen radicals 304
Active stage 365
Acylalanines 334
Acylon 335
Aecial base 130, 156
Aecial stage 128
Aeciospore initial 129
Aeciospores 127, 128, 129, 133, 167

Aerated steam treatment 289
Aerial spraying 354, 356
Aerial spraying and dusting 356
Aerobic 214, 215, 216, 219
Aeroscope studies 182, 205
Afugan 334
Agallol 326, 327, 365
Agar agar 6
Agar media 210
Agaricaceae 37, 125
Agaricales 37, 121, 122
Agaricus 115, 121, 125
Agaricus brunnescens (bisporus) 121
Agaricus campestris 125
Agave americana 392
Agents of inoculation 178
Agitator 346, 349
Agonomycetales 38, 145, 156, 164
Agrimycin 338
Agrobacterium 215, 225, 284, 290, 385, 389
Agrobacterium radiobacter 290, 385
Agrobacterium tumefaciens 215, 225, 284, 385, 389
Agrochemicals 358
Agrosan GN 326
Agrozim 332
Air-borne 369
Air-chamber 348
Albuginaceae 35, 63
Albugo 65, 66, 168, 370, 377
Alcohol 67, 70, 80, 292
Alfacryptovirus 249
Alfalfa latent virus 252
Alfalfa mosaic virus 232, 246, 247, 257, 389
Alfamovirus 247, 256
Aliette 334
Alkaloids 99, 101, 390
Allium sativum 393
Almond leaf scorch 16
Alternaria 161, 168, 176, 177, 288, 289, 297, 322, 323, 329, 330, 332, 339, 360, 365, 368, 370, 378, 379, 382, 383, 389, 393
Alternaria blight of cotton 161
Alternaria blight of potato and eggplant 288
Alternaria brassicae 289
Alternaria brassisicola 369
Alternaria macrospora 364, 369
Alternaria melangenae 168, 369
Alternaria solani 161, 339, 369, 388, 393
Alternaria triticina 369, 382
Alternate hosts 128, 167, 183, 283, 286, 287
Amanita 121, 122
Amastigomycota 35, 38, 67
Amino acids 171, 214, 377
Ammonia 376
Ammonium bicarbonate 341
Amoeboid 20, 49
Amorphous 39, 237
Anaerobic 290, 373
Anaerobic bacteria 290
Anaerobic growth 5
Anguina tritici 179
Angular cells 108
Angular leaf spot 166, 178, 299
Angular leaf spot of cotton 180, 221, 382
Angular leaf spot of cucumber 178
Aniline dyes 6
Animal kingdom 207
Animals 1, 6, 10, 11, 14, 67, 72, 86, 101, 145, 161, 179, 210, 215, 216, 217, 219, 221, 231, 235, 239, 268, 274
Annelides 154
Annuals 13, 165
Antagonistic 287, 371, 373, 376, 383, 384, 385, 386, 388, 391, 392, 393
Antagonistic microorganisms 370, 382, 385, 389, 390, 391
Anthocyanins 266
Anthracnose 31, 98, 153, 154, 166
Anthracnose of banana 361
Anthracnose of cabbage and cauliflower 154
Anthracnose of cashew 153
Anthracnose of citrus 153, 322
Anthracnose of coffee 153
Anthracnose of eggplant 153
Anthracnose of grapevine 154
Anthracnose of guava 154
Anthracnose of mango fruits 153
Anthracnose of papaya 153

Anthracnose of pea 153
Anthracnose or pollu disease of pepper 153
Anthrax (bacillus anthracis) 6
Antibacterial antibiotics 217, 219
Antibiotic substances 384, 386
Antibiotics 15, 17, 86, 87, 145, 217, 219, 233, 289, 291, 292, 295, 309, 311, 314, 315, 337, 338, 340, 341, 357, 359, 360, 366, 367, 375, 376, 383, 384, 387, 390, 391
Antibodies 15
Antifungal antibiotics 387
Antimicrobial 303, 304
Antimicrobial substances 303, 304
Antirabies vaccine 5
Antiserum 15
Anti-transpirant polymers 342
Ants 178, 291
Aphanomyces 52, 55, 371
Aphanomyces euteiches 52, 55
Aphelenchoides 179
Aphids 111, 113, 171, 178, 188, 240, 241, 242, 252, 254, 256
Aphids 178, 241, 253, 257, 285, 291
Aphyllophorales 119
Apical bulbs 47
Apical meristem 238, 389
Apogamously 61
Apophysis 45
Apothecial fundamentals 107, 108
Apparent resistance 296
Appendages 70, 94, 96, 97, 154, 182, 210
Apple blotch 150
Apple chlorotic leaf spot virus 246
Apple mosaic virus 256
Apple rough skin virus 266
Apple scab 33, 115, 148, 159, 166, 185, 190, 205, 327, 328, 336, 360, 372
Apple scar skin 13
Apple stem grooving 246, 252, 265
Apple stem grooving virus 252
Apron 335
Arachis mosaic virus 255
Arasan 321
Arathane 328
Arbitrary system of classification 146, 148
Arbuscules 386
Aretan 326, 327, 365
Armillaria 123, 125, 172, 175, 290, 318, 384, 391
Armillaria mellea 123, 125, 290, 384, 391
Armillaria tubescens 125
Arthrobacter 384, 392
Arthrospores 24, 79, 83, 118
Artificial culture 6, 15, 380
Artificial culture media 15, 378
Artificial inoculation 305
Ascal apex 89
Asci of the hook-type 94
Ascigerous stages 144, 158
Ascocarp centrum 85, 89
Ascochyta 111, 152, 172, 289, 365, 366, 369, 377, 381, 382, 383
Ascochyta pinodella 111, 172
Ascochyta pinodes 289, 369
Ascochyta pisi 381
Ascochyta rabiei 152, 365, 369
Ascochyta tenuis 365
Ascogenous cell 81, 83, 92
Ascogenous hyphae 77, 83, 87, 92, 94, 97, 111
Ascogonial coil 98
Ascogonial nuclei 87
Ascomycetes 26, 36, 73, 75, 76, 77, 79, 81, 99, 104, 116, 119, 144, 146, 148, 151, 152, 153, 156, 157, 168, 336, 370, 386
Ascomycetous fungi 181, 379
Ascomycetous yeasts 80
Ascomycotina 36, 67, 73
Ascospore stages 184
Ascospores 67, 73, 75, 77, 79, 80, 81, 83, 85, 86, 87, 89, 90, 94, 97, 98, 99, 101, 103, 104, 105, 107, 108, 109, 110, 111, 113, 114, 115, 168
Ascosporogenous species 80
Ascosporogenous yeasts 80
Ascostroma 77, 110, 114
Ascus mother cell 75, 83
Aseptate hyphae 65
Asexual fruiting bodies 73
Asexual process 145

Asexual reproduction 23, 39, 65, 67, 70, 73, 79, 80, 81, 107, 111, 118, 121, 137
Asexual spores 72, 73, 377
Asexually 11, 25, 26, 80, 144, 168, 174, 212, 379
Ashbya 81
Aspergillaceae 87
Aspergilli 85, 86, 87
Aspergillic acid 341
Aspergillus 85, 86, 88, 156, 157, 282, 333, 341, 362, 368, 378, 379, 383, 386, 393
Aspergillus colonies 86, 157
Aspergillus flavus 85, 86, 341
Aspergillus fumigatus 86, 157
Aspergillus japonicus 386
Aspergillus niger 85, 86, 157, 282, 368
Aspergillus rot of papaya 393
Aspergillus terreus 86
Assessment of disease intensity 194, 197, 198, 200
Aster yellows 268, 269
Athelia bombacina 386
Atrophy 265
Attached gills 123
Aureofungin 311, 338, 339, 340, 341, 366, 367
Aureofungin Sol. 339
Aureomycin (chlorotetracycline) 340, 341
Autoecious 131
Avena 184
Avirulent 303, 304, 383
Azadirachta indica 392
Azotobacter 7, 213, 214
Azotobacter agilis 214
Azotobacter chroococcum 214
Azotobacteriaceae 213

B

Bacillaceae 216
Bacilliform 109, 232, 249, 254, 257, 259, 267
Bacillus 6, 210, 212, 216, 217, 232, 290, 309, 340, 341, 374, 384, 385, 391, 392
Bacillus brevis 216, 341
Bacillus cereus 217
Bacillus subtilis 216, 290, 340, 341, 374, 385
Bacterial blight of cabbage 371
Bacterial blight of pea 367, 371
Bacterial blight of rice 177, 180, 185, 186, 190, 191
Bacterial blight of rice 284, 288, 371, 375
Bacterial blight of tomato 180
Bacterial brown rot of potato 177
Bacterial canker of tomatoes 178
Bacterial cell 12, 211, 212, 232
Bacterial diseases 177, 178, 180, 188, 189, 216, 220, 221, 222, 223, 278, 288, 315, 338, 367, 373, 375, 376, 377, 380, 381
Bacterial exudate 182, 223, 227, 378, 380
Bacterial fermentation 210
Bacterial forms 210
Bacterial leaf blight of rice 206, 223, 281, 380, 381
Bacterial leaf spot 339, 381
Bacterial leaf spot of mango 216
Bacterial leaf streak of rice 223, 367, 379
Bacterial ooze 189, 380
Bacterial soft rot of carrot, onion 371
Bacterial species 211, 392
Bactericidal 330, 390
Bactericidal effect 330
Bactericidal properties 390
Bactericides 310, 315, 316
Badnavirus 249, 259
Banana bunchy top 251, 259, 284, 285, 365, 377, 380
Banana streak virus 259
Banana wilt 30, 175
Banol 336
Banrot 336
Barberries 133, 184
Bardew 334
Bark boring insects 178
Barley covered smut pathogen 363
Barley loose smut pathogen 363
Barley rust 171, 183
Barley stripe mosaic 240, 246, 252, 262
Barley stripe mosaic virus 240, 246, 252
Barley yellow dwarf 262, 264
Barley yellow dwarf disease 262

Barley yellow dwarf virus 247, 254
Barley yellow mosaic 262
Barley yellow mosaic virus 247, 253
Basal rot and fruit rot of pineapple 369
Basal stalk cell 83
Basal wilt of pepper 164
Basidial fungi 126
Basidioles 118, 123
Basidiomycetes 20, 26, 34, 37, 115, 116, 118, 119, 127, 145, 164, 176, 332, 336, 370, 386, 387
Basidiomycetous yeast 126
Basidiomycotina 37, 67, 115
Basidiophora 63
Basidiophora entospora 63
Basidium bearing hyphae 120
Basipetal 331
Basipetally 128, 292
Bavistin 332
Baycor 335
Bayleton 335
Bayton 335
Beacon 334
Beam 337
Bean common mosaic virus 240
Bean golden mosaic virus 251, 259
Bean leaf roll virus 254, 265
Bean leaf roll virus in Vicia faba 265
Bean mosaic virus 247, 254, 255, 266, 365
Bean rugose mosaic virus 255
Bean rust 30
Bean summer death virus 259
Bean yellow mosaic 240, 262, 286
Bean yellow mosaic virus in soybean 240
Beet curly top virus 239, 251, 259
Beet leaf curl virus 257
Beet mild yellowing virus 254
Beet mosaic 262
Beet mosaic virus 253
Beet yellow stunt virus 253
Beet yellow virus 247
Beet yellows 241, 253, 264
Beetles 178, 240, 241, 243, 254, 255, 291
Bemisia tabaci 243, 391
Beneficial fungi 310
Beneficial insects 194
Bengard 332
Benlate 331
Benomyl 292, 311, 315, 317, 319, 331, 332, 360, 361, 362, 366
Benzimidazoles 331
Benzoic acid 319
Beta vulgaris 393
Betacryptovirus 249
Betelvine foot rot 365
Betelvine sclerotium wilt 385
Bicellular conidia 109
Biennials 165
Bim 337
Binary fission 208, 210, 212, 215
Binomial nomenclature 38
Binucleate (dikaryotic) cells 129
Binucleate (dikaryotic) mycelium 138
Binucleate cells 83, 116, 130
Binucleate conidia 138
Binucleate hypha 116
Binucleate urediniospores 133
Binucleate, homokaryotic basidiospores 115
Biochemical agents 295
Biochemical defenses 300, 303
Biochemical reactions 298, 303, 304
Biochemical substances 287, 309
Biocontrol agents 308, 365, 384, 385, 387
Biological control mechanisms 383
Biological control of crop diseases 383
Biological specialization 135
Biological subspecies 135
Bioluminescent 125
Biotechnological approaches 387
Biotechnological methods 388
Biotic 2
Biotypes 8
Bipartite genome 251, 259
Biphenyl 319, 328
Bipolaris (helminthosporium) 114
Bird's eye spot 162
Bird's eye spot of rubber 162
Bird's nest fungi 115
Birds 10, 26, 86, 179, 274, 276

Bitetanol 335
Bitunicate asci 77, 110, 111
Black leg 150, 178
Black leg of crucifers 150
Black leg of potato 178, 338
Black mould 158
Black pod of cacao 58
Black root 161
Black root of tobacco 161
Black rot of apple 151, 172
Black rot of Chrysanthemum 152
Black rot of grapevine 173
Black rust or stem rust of wheat 132
Black shank of tobacco 58, 369
Black spot of apple 108
Black spot of rose 108, 154
Blackberry rust 136
Blast 101, 156, 167, 172, 180, 185, 186, 188, 190, 199, 200, 201, 204, 205, 206, 207, 279, 281, 283, 284, 286, 289, 306, 307, 322, 330, 332, 337, 361, 369, 372, 374, 375, 376, 381, 386, 390, 393
Blast disease of rice 156, 330, 332, 334, 340, 341, 366
Blasticidin 338, 340, 341
Blastocladiales 35, 40, 47
Blastomycetidae 145, 148
Blastospores 81, 83, 125
Blicin 322
Blight 3, 5, 6, 7, 29, 32, 33, 58, 107, 114, 126, 150, 154, 158, 161, 177, 178, 180, 181, 182, 185, 186, 188, 190, 191, 198, 203, 204, 205, 206, 207, 215, 216, 223, 227, 269, 279, 281, 282, 284, 288, 289, 296, 297, 299, 307, 308, 322, 323, 329, 330, 332, 333, 334, 335, 338, 339, 340, 360, 361, 364, 365, 367, 368, 369, 371, 372, 373, 375, 376, 377, 378, 379, 380, 391, 393
Blight and fruit rot of eggplant 151
Blight of castor 161, 168, 369
Blight of cotton 161, 206, 281, 333, 338, 369, 375, 379
Blight of gram 152
Blight of grapevine 152
Blight of mango 150, 154
Blight of pea and lentil 152
Blight of peonies 158
Blight of rapeseed 369
Blight of redgram 150
Blight of tulip 158
Blight of yam 369
Blister blight 32, 33, 126
Blister blight of tea 33, 126, 181, 322, 330, 333, 334, 376, 377
Blisters 32, 33, 81
Blitane 322
Blitox 325
Blossom blight 153
Blotch 150, 152
Blotter method 382
Blue copper 325
Blue mould rot of Citrus and grape fruits 158
Blue moulds 85
Blue-green algae 11, 207
Boletes 121, 123
Boll rot of cotton 152
Borax 120, 319
Bordeaux mixture 6, 271, 290, 318, 323, 324, 325, 329, 336, 357, 359, 374, 375, 377
Bordeaux paste 313, 318, 325
Botran 328
Botryotinia 158
Botrytis 158, 166, 297, 299, 328, 329, 332, 339, 341, 360, 361, 362
Botrytis cinerea 166, 341, 361
Botrytis tulipae 158
Bougainvillea spectabilis 392, 393
Brachiaria mutica 286
Bracket fungi 67, 116
Branched filaments 219
Branched hyphae 73, 148
Branching bacteria 217, 338
Brassica oleracea 389
Brassicol 327, 366, 374
Bravo 328
Bread moulds 67
Breaking down of resistance 184
Breeding for resistance 135

Bremia 63
Bremia lactucae 63
Brestan 330, 376
Brestanol 330
Broad bean mosaic virus 255
Broad bean mottle virus 256
Broad bean wilt virus 247, 255
Broad-spectrum activity 365
Broad-spectrum fungicides 321, 376
Brome mosaic virus 247, 256
Bromoviridae 247
Bromovirus 245, 247, 256
Brown leaf spot of grapevine 162
Brown leaf spot of rice 27, 161, 162, 185, 206, 361
Brown root disease of rubber 371
Brown rot of peach and other fruits 159
Brown rot of potato 177, 281, 371
Brown spot of broad bean 152
Brown spot of maize 370
Bucket sprayers 345, 346
Bud 23, 32, 58, 72, 81, 83, 85, 130, 137, 166, 178, 213, 268, 269, 393
Bud necrosis, bud blight or ring mosaic of groundn 393
Bud rot of coconut 58
Budding knives 267
Buffer solution 239
Build-up of inoculum 279, 287, 373
Built-in fluid tank 346
Bulbiformin 338, 340, 341
Bunchy top of banana 179, 191, 266
Bunt 4, 30, 142, 176, 282, 287, 339, 364, 370, 382
Bunt of wheat 30, 142, 166, 173, 339, 364, 370
Bunt or covered smut 142
Bunt or kernel smut or black smut 142
Bunt pathogens 381
Bunyaviridae 249
Bupirimate 335
Burgundy mixture 325
Burning symptoms on fruits 360
Butrizol 335
By using tubers, corms, bulbs, rhizomes 238
Bymovirus 247, 253

C

Cabbage leaf necrosis 264
Cabbage necrotic ring spot 264
Cacao necrosis 264
Cacao necrosis virus 255
Cacao swollen shoot 243, 259, 265, 283
Cacao swollen shoot virus 243, 259
Cacao yellow mosaic virus 255
Calcium 324, 359
Calixin 334
Calocybe 115, 125, 126
Calocybe indica 126
Camembert cheese 87, 158
Candida 65, 66, 80, 168, 370, 384, 392
Candida utilis 80
Canker 32, 58, 103, 150, 151, 152, 154, 163, 164, 165, 171, 173, 178, 216, 217, 225, 338, 339, 371, 391
Canker of apple 32, 154
Canker of lady's finger 150
Canker of peach 32
Capillovirus 246, 252
Capitate paraphyses 108
Capnodiaceae 37, 110, 113
Capnodium 113
Capnodium ramosum 113
Capryl 328
Capsid 232, 233, 235, 237, 245, 251
Capsomeres 245
Capsule 211, 273, 379
Capsule stain 379
Capsules 271, 272
Captafol 329
Captan 50, 289, 290, 311, 317, 318, 319, 328, 329, 330, 335, 357, 360, 362, 366, 367, 374, 376, 377
Carbaryl 356
Carbendazim 289, 292, 311, 331, 332, 366, 376
Carbofuran 290
Carbohydrates 38, 80, 122, 211, 214, 233, 274, 374
Carbol-fuchsin 379
Carbon dioxide 10, 80, 290, 372, 373, 376, 384, 392

Carbon disulfide 289
Carbonate compounds 341
Carboxin 289, 292, 311, 315, 317, 330, 333, 366, 374, 376
Cardamom mosaic virus 267
Carica papaya 392
Carlavirus 245, 246, 252
Carmovirus 247
Carnation Italian ring spot 264
Carnation Italian ring spot virus 254
Carnation latent virus 246, 252
Carnation mottle virus 247
Carnation necrotic fleck 253, 264
Carnation necrotic fleck virus 253
Carnation vein mottle virus 253
Carpenteles 158
Carrot yellow leaf virus 253
Cassava latent mosaic virus 259
Cassava mosaic 262, 380
Cassava vein mosaic virus 259
Cattle 3, 85, 87, 101, 175, 180, 219, 227, 274, 326, 337
Cauliflower mosaic 249, 259, 262
Cauliflower mosaic virus 249, 259
Caulimovirus 246, 249, 259
Caused by other pathogens 225, 380
Cela 336
Celery mosaic virus 253
Cell division 25, 117, 376
Cell membranes 163, 303, 304
Cell sap 212
Cell wall defense structures 301
Cell wall of the pathogen 390
Cellular metabolism 232
Cellulosic walls 302
Cell-wall degrading enzymes 368
Cenchrus ciliaris 286
Central stalk 125
Centrioles 39
Cephaleuros 18, 271
Cephaleuros parasiticus 18
Cephalosporium 103, 104, 156, 365, 378
Cephalosporium sacchari 365
Cephalosporium ulmi 156
Ceratiomyxomycetidae 34
Ceratobasidiaceae 37, 126
Ceratobasidium 126, 127
Ceratobasidium anceps 127
Ceratocystis (ceratostomella) paradoxa 156
Ceratocystis (ophiostoma / ceratostomella) 89
Ceratocystis fagacearum 89
Ceratostomella 89, 156, 161, 173, 178, 378
Ceratostomella (ophiostoma) ulmi 156
Cercobin 333
Cercobin M 333
Cercospora 111, 162, 168, 173, 330, 332, 360, 369, 377, 378, 382, 386, 390
Cercospora arachidicola 111
Cercospora beticola 173
Cercospora musae 111, 390
Cercospora personata 162, 168, 369
Cereal rust 135
Cereal tillering disease virus 258
Ceresan dry 326, 365
Ceresan wet 326
Chaetomium 386
Chaetomium globosum 386
Chanterelles 119
Charcoal rot 151, 164, 368, 369
Cheese 87, 158, 239
Chelation of mineral nutrients 376
Chemical action 172
Chemical defenses 299
Chemical formulations 291, 310, 311, 314, 316, 344, 365, 373
Chemical heterogeneity of the host 183
Chemical methods 187
Chemical molecules 231, 233
Chemical reaction 324, 357
Chemical sterilants 268
Chemical treatment 3, 282, 286, 288, 289, 291, 292
Chemical treatment of seeds 282, 286
Chemical treatment of soil 289
Cherry leaf roll virus 266
Cherry rugose mosaic virus 256
Cheshunt compound 325
Chicken cholera 5
Chinopodium ambrosioidea 393

Chipco 26019 329
Chitinase genes 389
Chloranil 327, 366
Chlorinated water 319
Chloromycetin 341
Chloroneb 317, 328, 366
Chlorophyceae 18
Chlorophyll 10, 17, 18, 19, 27, 262
Chloropicrin 290, 313, 318
Chlorosis 262, 266, 268
Chlorothalonil 328, 375
Chlorotic 9, 13, 243, 246, 247, 254, 256, 257, 260, 262, 264, 267, 268, 269
Choanephora 69, 70
Choanephora cucurbitarum 70
Choanephoraceae 36, 70
Chromista 34
Chytridiaceae 35, 47
Chytridiales 35, 40, 47, 245
Chytridiomycetes 35, 39
Chytrids 40
Cilia 245
Circular leaf spot of bean 221
Circulifer tenellus 270
Citric acid 68
Citrus exocortis 267
Citrus greasy spot 111
Citrus leaf rugose mosaic virus 256
Citrus root rot pathogen 390
Citrus scab 33
Citrus stubborn 9, 269, 270
Citrus tristeza virus 253, 265, 310
Citrus variegation virus 256
Cladochytriaceae 35, 43
Cladosporium 171, 333
Cladosporium fulvum 171
Clamp connections 126
Classification of Bacteria 213
Classification of Fungicides 314
Classification of Plant Viruses 245
Clavibacter 9, 17, 177, 178, 179, 217, 223, 270, 289, 365
Clavibacter (corynebacterium) 217, 223
Clavibacter diphtherae 217
Clavibacter michiganense sub sp. Sepidonicum 177
Clavibacter tritici 179
Clavicepitaceae 36, 98
Clavicepitales 36, 97, 98
Claviceps 21, 98, 99, 100, 164, 168, 178, 184, 278, 286, 364, 377, 381
Claviceps ciliaris 286
Claviceps fusiformis 364, 381
Claviceps microcephala 168, 178, 278
Claviceps purpurea 99, 100, 164, 178, 184, 364, 381
Claviceps sorghi 164
Cleistothecial fungi 85
Clerodendron fragrans 392
Clones 388
Closed ascocarp 90
Closed fructifications 87
Closterovirus 247, 253
Clostridium 6
Clostridium titani 6
Clover necrotic mosaic 256, 264
Clover yellow mosaic 252, 262
Club root 33, 49, 297, 370, 372, 389
Club-shaped basidia 119
Coalesce 27, 29, 33, 108, 199, 223
Coat protein genes 389
Cobb scale 198
Cochliobolus 114, 161, 162, 165, 304, 388
Cochliobolus carbonum 304
Cochliobolus miyabeanus 114, 165
Cochliobolus setariae (=helminthosporium setariae) 114
Cochliobolus spicifer 162
Coconut stem bleeding pathogen 165
Coelomycetidae 38, 148
Coenocytic hyphae 56, 67
Coffee ring spot 264
Coffee ringspot virus 257
Coffee rust 30, 136, 180, 185, 205, 284
Coleoptile 173, 174
Collar region 31, 173, 378
Collar rot 164, 167, 168, 175, 179, 279, 282, 361, 368, 377, 378
Collar rot or crown rot of groundnut 158

Collar rot pathogen 175
Collateral hosts 165, 174, 186, 187, 192, 286, 308, 372
Colletotrichum 98, 153, 154, 165, 166, 173, 180, 185, 279, 299, 319, 332, 360, 361, 364, 365, 369, 377, 378, 382
Colletotrichum capsici 364, 369, 382
Colletotrichum circinans 299
Colletotrichum coffeanum 98
Colletotrichum falcatum 98, 153, 165, 173, 185, 279, 365
Colletotrichum gloeosporioides 98
Colletotrichum graminicola 98
Colletotrichum lindemuthianum 98, 166
Colletotrichum zingiberis 369
Commercial yeast 80
Common scab of potato 33, 219, 371
Comoviridae 247
Comovirus 245, 247, 255
Compatibility 131, 138, 142, 300, 357, 358, 359
Compatibility of fungicides with other agrochemica 357, 358
Compatibility of plant protection chemicals 357
Compatible 25, 26, 73, 75, 116, 122, 129, 131, 137, 140, 142, 144, 322, 323, 329, 336, 357, 359
Compatible basidiospores 142, 144
Compatible homokaryotic mycelium 115
Competition 383, 384, 386, 392
Complementary MRNA strand 235
Concentric leaf spot of sweet potato 152
Conidial chains 87, 92, 158
Conidial characters 146
Conidial stages 79, 85, 89, 103, 110, 144, 146, 148
Conidial succession 145
Conidiogenous cells 73, 86, 154, 157
Conidiophore and conidial development in eysiphac 91
Coniothyrium 107, 151
Coniothyrium fuckelii 151
Coniothyrium minitans 107
Conjugate division 83, 129
Conjugate divisions of the nuclei 116
Conjugately 129, 142, 144
Contact fungicides 366, 373, 375
Contaminants 176, 364, 381
Contaminated field 279
Control measures 192, 194, 205, 278, 283, 285, 291, 292, 372, 387
Control of post-harvest diseases 319, 328
Copper 290, 311, 314, 318, 325
Copper carbonate 323
Copper ions 377
Copper oxide 323
Copper oxychloride 271, 290, 318, 322, 323, 325, 338, 375, 377
Copper sulfate 4, 323, 324, 325, 340, 367
Coprinus lagopus 122
Coral fungi 116, 119
Coremia of conidiophores 154
Cork cells 302
Cork layers 294, 302
Corky tissues 271
Corn leaf blight 114, 296
Corn yellow leaf blight 296
Coromet 322
Cortical cells 386
Corticiaceae 120, 121
Corticium 121, 172, 329, 334, 371
Corticium koleroga 172
Cortizone 70
Corynebacterium (clavibacter) 17, 270
Corynebacterium fascians 179
Corynebacterium michiganensis 180
Coryneform group of bacteria 217
Coryneum 154
Coryneum beijerinckii 154
Cosan 320
Cotton seedling blight 368, 385
Cotton wilt 10, 30, 318
Covered smut of barley 140, 286, 288
Covered smut or bunt of wheat 30
Cowpea chlorotic mottle 256, 264
Cowpea chlorotic mottle virus 256
Cowpea mild mottle virus 252
Cowpea mosaic virus 247, 255
Cowpea severe mosaic virus 255

Crop inspection 282
Crop rotation 187, 286, 287, 298, 308, 372, 373
Cross protection 187, 310
Crotothane 328
Crown rust of oats 132
Crown wart of alfalfa 45
Crown wart of lucerne 33
Croziers 98
Crystal violet 379
Crystalline inclusion bodies 237
Cucumber beetles 178
Cucumber green mottle mosaic 262
Cucumber mosaic 239, 240, 247, 254, 255, 256, 262, 265, 266
Cucumber mosaic virus 239, 240, 247, 254, 255, 256, 262, 265, 266
Cucumber necrosis 264
Cucumber necrosis virus 245
Cucumber wilt pathogen 178
Cucumovirus 245, 247, 256
Cultural methods 187
Cultural practices 3, 191, 192, 278, 281, 298, 372
Culture media 6, 15, 210, 378
Culturing in artificial media 380
Culturing the pathogens 381
Cuman L 313, 322
Cup fungi 67, 104, 109
Cup-like appressorium 276
Cupramar 325
Cupric hydroxide 324, 377
Cuprocide 325
Cuprous oxide 325
Cup-shaped aecium 133
Curamil 334
Curative fungicides 333
Curative measures 277
Curling 83, 272
Curvularia 114, 162, 332, 333, 378, 382, 383
Cuscuta 18, 170, 180, 245, 272, 273
Cuscuta gronovii 18, 170
Cuscutaceae 272
Cushion 21, 25, 103, 114, 146, 331
Cushion-shaped stroma 25, 103, 113, 145
Cuticularised epidermal wall 172
Cutting knives 270
Cycloheximide 338, 339, 341
Cylindrical asci 89, 97, 99, 104
Cylindrical sporangia 52
Cylindrosporium 154
Cylindrosporium hiemale 154
Cyprex 336
Cystidia 118, 123
Cytoplasm 14, 42, 43, 49, 50, 54, 65, 83, 211, 212, 213, 237, 296, 301, 303, 389
Cytoplasmic membrane 211, 212
Cytorhabdovirus 249, 257
Cytospora 101

D

Daconil 328
Dahlia mosaic virus 259
Damping off 31
Damping off of seedlings 31, 56, 166, 168, 281, 290, 324, 335, 335, 369, 374, 385
Datura stramonium 393
Daughter cells 23, 86, 116, 212
Decay 30, 56, 85, 87, 119, 127, 158, 171, 293, 316
Decay of apples in storage 87, 158
Deciduous 59
Deep-seated infection 291, 292
Defective nucleic acid 235
Defense mechanism 6, 298, 301, 303, 304, 337
Defense mechanism of white blood cells 6
Deficiency diseases 280, 287
Deficient nucleic acid 235
Definition of viruses 229
Dehiscence 85, 104, 110
Delicate filaments 219
Delivery hose 347, 349, 350, 352, 353, 354
Delivery tube 346, 347, 349, 350, 352, 353, 354
Dematiaceae 38, 156, 161
Demicyclic rusts 128
Demosan 328
Dendrophoma 151

Dendrophoma obscurans 151
Dendrophthoe 19, 274, 276
Dendrophthoe falcata 19, 276
Dermataceae 37, 108
Derosal 332
Detection of seed-borne viruses 383
Determinate 59
Detoxification of phenolic toxins 305
Deuteromycetes (Fungi Imperfecti) 26
Deuteromycetous fungi 336, 379
Deuteromycotina 38, 67, 85, 89, 98, 108, 110, 111, 144, 145
Dexon 328, 374
Diamethirimol 335
Dianthovirus 247, 255
Diaporthaceae 36, 101
Diaporthales 36, 101, 156
Diaporthe 90, 101, 151
Diaporthe citri 151
Diaporthe type centrum 90, 101
Diaporthe vexans 101, 151
Diazinon 356
Dichlone 327, 366
Dichloran 319, 328, 362, 375
Dichloropropane-dichloropropene (DD mixture) 290
Didymella 150, 152
Didymella (mycosphaerella) arachidicola 150
Didymella lycopersici 152
Die-back and fruit rot of chillies 369
Die-back of maple 163
Die-back of rose 152
Difenoconazole 335
Differential hosts 183
Difolatan 329, 357
Difosan 329
Digitaria marginata 286
Dikaryon 92, 94, 116
Dikaryotic 25, 73, 75, 79, 80, 83, 94, 117, 121, 122, 126, 129, 130, 133, 135, 137, 138, 140
Dikaryotic cell division by clamp connection 117
Dikaryotic cells 75, 129, 137
Dikaryotic hyphae 122, 125
Dikaryotic phase 80, 133
Dikaryotic stage 73, 79
Dikaryotization 129, 138
Dikaryotization process 129
Dilution plate method 383
Dinocap 328, 361, 376
Dioecious 26, 274
Diphtheria 6, 217
Diplobacilli 210
Diplobiontic (diploid) 39
Diplocarpon 108, 154, 341, 377
Diplocarpon rosae 108, 154, 341
Diplococci 210
Diplodia 152, 328, 378
Diplodia natalensis 152
Diploid cells 81
Diploid stage 132
Diplomastigomycotina 35, 52
Diplophase 81
Dipotassium phosphate 341
Dipping transplants 385
Direct Control 310
Direct parasitization 384, 392
Direct penetration 109, 170, 225, 299, 386
Discharge papilla 40
Discomycetes 104, 109
Disc-shaped apothecia 104
Disease control 294, 310, 338, 341, 360, 365
Disease escape 296, 297, 298
Disease forecasting 204, 205
Disease infection 175, 194, 199, 204, 298, 310
Disease occurrence 174, 185, 186, 192, 194, 195, 200, 207
Disease occurrence 308
Disease resistant mutants 388
Disease symptoms 2, 170, 198, 199, 204
Disease tolerance 298
Disease-free nurseries 281
Disease-stress conditions 305, 306
Disjunctor cells 130, 133
Disjunctors 65
Dispersal by soil 174, 175
Dispersal of inoculum 189

Dispersal through birds and animals 179
Dispersal through humans 177
Dispersal through insects 178
Dispersal through nematodes 179
Dispersal through plants and vegetative plant prop 177
Dispersal through seed 176
Dispersal through soil 174
Dispersal through water 180
Dissemination of spores 181
Dithane D 14 323
Dithane M 22 322
Dithane Z 78 322
DNA transformation 389
DNA viruses 249
Dodine 315, 336, 360, 361
Dormancy 68, 276
Dormant stage 365, 368
Dormant state 170, 176, 372
Dormant structures 288, 291, 314
Dormant structures of the pathogen 291
Dothidiales 110
Double annulus 123
Double stranded 233, 235, 237
Downy mildew (blue mould) of soybean 61
Downy mildew and green ear disease of pearl millet 370
Downy mildew of corn 63
Downy mildew of lettuce 63
Downy mildew of millets and grasses 63
Downy mildew of onion 29, 61
Downy mildew of peas 168
Downy mildew of rapeseed 370
Downy mildew of sorghum 63, 317, 370
Downy mildew of sugarcane 370
Downy mildew of sunflower 61
Downy mildew of tobacco 61
Downy mildews 3, 52, 59, 180, 181, 182, 188, 190, 206, 297, 318, 322, 324, 328, 333, 334, 335, 336, 362, 366, 376, 377, 379
Drechslera 114, 162, 332, 334, 383
Drechslera (bipolaris) rostrata 162
Drechslera (bipolaris) spicifera 162
Drechslera (helminthosporium) maydis 162
Drechslera (helminthosporium) sacchari 162
Drenching 290, 317, 318, 321, 323, 325, 327, 329, 331, 332, 335, 342, 374, 387
Drenching the soil 317, 323, 327, 374
Dry root rot 151, 364, 378, 385
Dry root rot and stem blight of cowpea 151
Dry root rot of groundnut 151, 385
Dry seed treatment 327, 366
Dusters 312, 345, 354, 356
Dutch elm disease 156
Du-ter 330, 376
Dwarfing 260, 268

E

Early blight 161, 339, 393
Early blight of tomato and potato 161
Earth tongues 104
Earthstars 119
EBP 334
Ectomycorrhizal fungi 386
Ectoparasite 179
Ectotrophic 386
Ectotunica 110
Edible 73, 120, 121, 123, 124, 125, 337, 363
Edifenphos 315, 334, 344, 361, 376
Eggplant mild mottle virus 252
Eggplant mosaic virus 255
Electron microscope 8, 13, 229, 230
Elicitors 303
Elongate 13, 20, 83, 87, 148, 154, 232
Elosan 320
Elsinoe 154
Elsinoe ampelina 154
Embryo 20, 176, 240, 268, 364, 365
Embryo of the nuts 268
Emericella 157
Emulsifiable concentrates 312, 346
Emulsifiers 312, 313
Enamovirus 247, 256
Endemic 185, 191, 206
Endoascus 77
Endobiotic 40, 43, 45

Endogenously 105
Endomyces 80, 161
Endomyces geotrichum 80
Endomycetaceae 36, 80
Endomycetales 36, 79, 161
Endoparasitic 39, 47
Endoparasitic slime moulds 39, 47
Endosperm 176, 364
Endospore formation in bacteria 208
Endospore forming rods 216
Endosulfan 356
Endotrophic 386
Endotunica 77, 110
Enovit 333
Enovit methyl 333
Enterobacteriaceae 215
Entities 12, 231
Entomophthora 72, 74
Entomophthora muscae 72
Entomophthoraceae 36, 72
Entomophthorales 36, 67, 72
Entomosporium 108
Entyloma 144, 377
Entyloma oryzae 144
Envelope 233, 245, 257
Environment 1, 184, 187, 189, 278, 308, 309, 392
Environmental conditions 20, 26, 59, 73, 174, 175, 176, 181, 185, 188, 189, 191, 205, 210, 212, 221, 227, 233, 278, 280, 282, 291, 295, 296, 297, 298, 368, 370
Environmental factors 1, 2, 185, 189, 191, 194, 297
Environmental pollution 194
Enzymatic activities 86
Ephemeral 43
Epibiotic 40, 43
Epicotyl 172
Epidemic proportion 189
Epidemiology 185
Epidemiology of Crop Diseases 185
Epigean 104
Epiphytic 113
Epiphytotics 54
Eradicant fungicides 291, 292
Eradicate 291, 292, 314, 318, 337, 373
Eradication 187, 278, 283, 285, 286, 289, 291, 294, 308
Eradication of alternate and collateral hosts 286, 308
Eradication of vectors 291
Erect conidiophores 87, 92, 158
Eremothecium 81
Eremothecium ashbyi 81
Ergot 31, 73, 99, 101, 168, 176, 178, 278, 279, 280, 286, 287, 364, 377, 381, 382
Ergot or sugary disease 31
Ergot pathogen of pearl millet 176, 363
Erumpent 75, 108, 110, 136, 146, 153
Erwinia 7, 178, 180, 215, 223, 225, 300, 360, 361, 371, 373, 377
Erwinia amylovora 7, 361
Erwinia carotovora 178, 225, 300, 371
Erwinia carotovora pv. carotovora 225, 371
Erwinia soft rots 373
Erwinia tracheiphila 178, 215, 223
Erysiphaceae 36, 90, 91, 92, 95
Erysiphales 36, 90
Erysiphe 90, 92, 94, 96, 168, 172, 184, 284, 334, 361, 370, 377, 390
Erysiphe cichoracearum 96, 284, 361, 370
Erysiphe graminis 94, 172, 184, 334, 361, 370
Erysiphe graminis f.sp. tritici 370
Erysiphe lamprocarpa 92
Erysiphe martii 92
Erysiphe polygoni 90, 370, 390
Erythromycin 341
Establishment 10, 127, 189, 190, 205, 298, 387
Establishment in the host 190
Etaconazole 335
Ethirimol 335, 366
Ethyl alcohol 292
Ethyl methane 307
Ethylene 322
Eubacteriales 213
Eucalyptus globulus 393
Eukaryonta 34
Evanescent 85, 86, 89, 101

Examination of inner tissues 378
Exclusion 187, 278, 282, 285, 291, 294, 308
Exipulaceae (discellaceae) 150
Exoascus 77
Exobasidiaceae 37, 126
Exobasidiales 37, 126
Exobasidium 126, 181, 334, 377
Exobasidium vexans 126, 181, 334
Exogenously 105
Externally seed-borne pathogens 289, 316, 365
Extracellular metabolites 383
Extramatrical 45, 96
Exudates 113, 171, 228, 276, 287, 376, 379, 384
Exudation 223, 281
Eye spot of sugarcane 162

F

Fabavirus 247, 255
Factors responsible for disease epidemics 187
False smut of rice 140, 377, 381
Farmatin 330
Fastidious vascular bacteria 16, 17, 270, 340
Favorable environmental conditions 174, 176, 189, 297, 368
Feeding hyphae 172, 175, 385
Female nucleus 56, 65, 111
Fenarimol 292, 315, 335
Ferbam 321, 322, 375
Fermate 322
Fermentation 5, 67, 210
Fermenting sugars 215
Fermocide 322
Fertilization tube 56, 58, 59, 65
Fertilizing tubes 54
Ferula asafoetida 393
Field sanitation 187, 288, 308, 371
Fijivirus 249, 258
Filamentous 14, 20, 38, 39, 52, 207, 210, 212, 213, 216, 217, 219, 246, 247, 257
Filamentous forms 14, 219
Filamentous hyphae 212
Filamentous Particles 246
Filaments 16, 20, 217, 219, 247
Filiform (scolecospore) 152
Film Forming Compounds 342
Filters 7, 229, 346
Fire blight 178, 180, 182, 215, 227, 338, 339, 340, 360, 361
Fire blight bacteria 180
Firmibacteria 216
Firmicutes 216
Fission 14, 16, 24, 73, 80, 81, 208, 210, 212, 213, 215
Fixed plot survey 191, 192, 193
Flag smut 284
Flagella laterally inserted 52
Flagellate 16, 38, 39, 54, 73
Flagellate cells 39, 73
Flagellate fungi 39, 73
Fleshy basidiocarp 121
Flexuous threads 232
Flies 72, 178, 188, 240, 242, 258, 285, 291
Florets 31
Flutolanil 329
Foci of infection 286, 302
Folded 247, 257
Foliar sprays 331, 336, 385
Folpet 330
Foltaf 329
Fomes 120, 175, 371
Fomes annosus (Heterobasidium annosum) 120
Fomes noxius 371
Food yeast 80
Foot rot 58, 281, 288, 289, 339, 365, 367, 369, 381, 385
Foot rot of betelvine 369, 385, 391
Foot rot of pea 111, 152
Foot rot of pepper 385
Foot rot of rice 163
Fosetyl-Al 331, 334, 374
Fosetyl-Al 334, 376
Fragmentation 23, 24, 67, 73, 118, 156, 164
Fragmentation of mycelium 118, 156
Fragmentation of somatic hyphae 23
Frankliniella fusca 244

Free ascus 80
Free cell formation 73, 87
Free conidiophores 110
Free gills 125
Fresh infection 107, 174, 239, 292, 316, 367, 369
Frog's eye leaf spot of tobacco 162
Frog's eye spot 150, 151
Frog's eye spot of red gram 150
Frost 359, 360
Fructifications 21, 27, 31, 34, 87, 89, 94, 126, 165, 271
Fructifications 271, 383
Fruit canker of guava 154
Fruit decay 127
Fruit dipping 332
Fruit diseases 322
Fruit rot 101, 127, 150, 151, 153, 154, 157, 161, 163, 322, 340, 364, 369, 379, 382, 393
Fruit rot of apple 161
Fruit rot of banana 154, 163
Fruit rot of chillies 161, 322, 369, 382
Fruit rot of citrus 157, 161, 369
Fruit rot of eggplant and tomato 153
Fruit rot of grapes 154
Fruit rot of lemon 393
Fruit rot of pomegranate 163
Fruit spot 154
Fumaric acid 67, 70
Fumigants 290, 313
Functional gametangia 86, 87
Fungal cell membrane 376
Fungal cell walls 376
Fungal cells 334, 376, 390
Fungal colonies 382
Fungal diseases 27, 219, 285, 310, 318, 333, 339, 376
Fungal etiology 6
Fungal genes 390
Fungal organisms 167, 310, 384
Fungal parasites 309
Fungal plant pathogens 219
Fungal species 10, 21, 145, 392
Fungal vectors 245
Fungicidal 290, 292, 311, 313, 317, 318, 321, 323, 324, 325, 326, 327, 328, 329, 332, 334, 337, 341, 342, 343, 344, 345, 354, 356, 363, 364, 374, 376, 393
Fungicidal pastes 318
Fungicidal properties 323, 337
Fungicidal solutions 317, 325, 363, 376
Fungicidal toxicity 312
Fungicides 289, 291, 292, 293, 295, 305, 310, 311, 313, 314, 315, 316, 318, 319, 320, 321, 322, 323, 324, 325, 326, 327, 329, 330, 331, 333, 334, 335, 336, 337, 344, 345, 357, 358, 359, 360, 361, 362, 363, 366, 373, 374, 375, 376, 377, 387
Fungicidin 340
Funginex 336
Fungitoxic substances 299
Fungo 333
Furovirus 246, 252
Fusaria 163
Fusarium 23, 103, 157, 163, 167, 175, 177, 179, 180, 188, 190, 195, 279, 281, 282, 286, 287, 290, 297, 306, 319, 323, 327, 332, 336, 339, 340, 360, 364, 367, 368, 371, 372, 373, 392
Fusarium oxysporum 163, 167, 175, 364, 368, 371, 384, 386, 391
Fusarium oxysporum f.sp. ciceri 368
Fusarium oxysporum f.sp. cubense 368
Fusarium oxysporum f.sp. lycopersici 368
Fusarium oxysporum f.sp. vasinfectum 364, 368
Fusarium roseum 386
Fusarium solani 175, 368, 385, 391, 392
Fusarium solani f.sp. Phaseoli 368
Fusarium udum 340, 368, 385
Fusarium wilt of banana 163, 177, 282, 373, 378
Fusarium wilt of tomato 373
Fusarium wilts 336, 372, 373, 376
Fytolon 325

G

Galls 33, 45, 137, 140, 170, 171, 215, 225, 389
Ganoderma 120, 175, 340, 371, 386, 391, 393
Ganoderma lucidum (Polyporus lucidus) 120
Gasteromycetes 119
Gelatin 6
Geminiviridae 249
Geminivirus 249, 258
Gene for avirulence 296
Gene recombination 138
Gene transformation 389
Genes 138, 211, 233, 294, 295, 296, 298, 304, 305, 306, 360, 389, 390
Genes located in the chromosomes 296
Genetic engineering 305, 307, 388, 389
Genetic engineering techniques 307
Genetic material 211, 213, 229, 231, 232, 233, 296, 307
Genetic resistance 280
Genetical agents 278
Genetically evolved 306
Genetically stable parent plant 388
Genome 9, 231, 232, 251, 252, 253, 254, 255, 256, 257, 258, 259, 389
Geotrichum 80, 161
Geotrichum candidum 80
Germ theory of diseases 5
Germinating seedlings 316, 317, 319, 359, 360, 365, 373
Germination of fungal spores 190
Germplasm 270
Gibberella 72, 103, 163
Gilbertella persicaria 72
Gills 121, 123, 125
Gingelly phyllody 377
Ginger rhizome rot 365
Gliocladium 107, 290, 384, 385, 391
Gliocladium roseum 107
Gliocladium virens 290
Gloeosporium 98, 153, 154, 180, 377, 378
Gloeosporium album 154
Glomerella 97, 98, 152, 153, 154, 165, 166
Glomerella (physalospora) 98, 153
Glomerella gossypii (thanatephorus cucumeris) 152
Glomerella manihotis 153
Glomus 387
Gluconic acids 86
Glycolipids 384
Glycophene 329
Glycoprotein molecules 304
Glycoproteins 233
Gnomonia fragariae (fructicola) 153
Gracilicutes 213
Grade values 198, 200, 203
Grades 204
Grain discoloration 334, 393
Grain discoloration in rice 162
Grain smut 140, 142, 197, 200, 367
Gram stain 379
Graminella nigrifrons 243
Granular chemicals 290
Granules 120, 211, 212, 303, 312, 313
Grape fan leaf virus 244
Grapevine chrome mosaic virus 255
Grapevine fan leaf virus 179, 255
Grass hosts 133, 167, 183
Gray blight 154
Gray blight of tea 154
Gray blight or leaf spot of mango 154
Gray leaf spot or blight of coconut and oil palm 154
Gray mould rot fungus 166
Greasy spot 111
Green ear 31, 32, 180, 286, 370, 377
Green mould of citrus fruits 171
Griphosphaeria 173
Griphosphaeria corticolax 173
Griseofulvin 87, 338, 339, 341
Groundnut green mosaic virus 393
Groundnut rosette 267
Groundnut rosette virus 267
Groundnut rust 30, 376
Growth promoting substance 103
Guanidol 336
Guard cells of the stoma 171
Gum secretion 303

Gummosis 152, 339, 369
Gummy substances 302
Gymnomycota 34, 38

H

Haemolymph 15
Hair-like setae 153
Hairs on the scale leaves 173
Halo 27, 221, 225, 288
Halo blight 288
Halo blight of runner beans 288
Hand rotary duster 355
Haplobiontic (haploid) 39
Haploid basidiospores 119, 122, 127
Haploid condition 25
Haplomastigomycotina 35, 39
Haplophase 81
Harpochytridiales 35, 47
Haustorial sinkers 274
Head smut of sorghum 140, 367, 370
Heat energy 289, 291, 293
Heat inactivation 268
Heavy metals 359
Helical 9, 13, 15, 16, 232, 245, 249, 251, 252, 256, 257
Helical RNA molecule 251
Helminthosporium 114, 161, 162, 166, 174, 176, 185, 288, 296, 330, 332, 333, 361, 364, 369, 377, 378, 381, 382, 383, 393
Helminthosporium avenae 174
Helminthosporium gramineum 174
Helminthosporium maydis 296
Helminthosporium nodulosum 393
Helminthosporium oryzae 166, 185, 361, 364, 369, 381, 393
Helminthosporium solani 161
Helminthosporium turcicum 369
Helotiales 37, 104, 105
Helper virus 235
Hemiascomycetidae 36, 79
Hemileia 136, 180, 185, 284, 377
Hemileia vastatrix 284
Heterodera 170, 182, 284
Heterodera major 182
Heterodera rostochiensis 170, 284
Heterodera schachtii 170
Heteroecious 131, 167, 183, 186
Heteroecious fungi 167
Heteroecious rusts 131, 186
Heteroecism 131, 167
Heterokaryosis 8
Heterothallic 81, 92, 98, 119, 144
Heterothallic species 119
Heterothallism 122
Heterozygous 184
Hexacap 329
Hexaferb 322
Hexathane 322
Hexathir 321
Hexazir 322
Hibiscus latent ring spot 264
Hibiscus tetraphyllus 286
Higginsia 154
Higginsia hiemalis 154
High reproductive capacity 188
Higher fungi 73
Hinosan 334
Histological defense structures 302
Histological defense structures 302
Holobasidiomycetidae I 119
II 121
III 126
Holocarpic 40, 42
Homokaryotic basidiospores 115
Homothallic 92, 119
Honeybees 16, 291
Honeydew 30, 113, 163, 178
Hoof-shaped fruiting bodies 120
Hordeivirus 246, 252
Hordeum 184
Hormonal imbalance 264, 265
Hormones 122
Hose 346, 347, 349, 350, 352, 353, 354
Host cells 21, 33, 40, 42, 47, 49, 90, 107, 126, 127, 171, 229, 231, 232, 233, 237, 242, 244, 271, 299, 300, 301, 302, 303
Host genera 132, 183
Host penetration 190
Host plants 5, 11, 16, 18, 19, 31, 32, 40, 123,

127, 170, 171, 174, 176, 179, 180, 182, 186, 214, 223, 225, 228, 232, 272, 276, 278, 301, 302, 305, 308, 309, 314, 315, 319, 337, 338, 359, 370, 375, 389, 393
Host root 245, 276
Host root cell 245
Host xylem 276
Host-pathogen combination 302
Host-specific toxin 300
Hot air treatment 289
Hot water treatment 15, 288, 289, 367
Human activities 185, 190, 191
Hyaline apical appendages 154
Hyaline conidiophores 156
Hyaline hyphae 90, 154
Hyaline spores 148, 377
Hyaline wall 50
Hybrid morel 110
Hybridization 135, 187, 188, 294, 306
Hybridize 183
Hydathodes 170, 226
Hydrogen cyanide 383
Hydrolytic enzymes 300
Hymenial holobasidiomycetes 119, 121
Hymenial layer 97, 109, 119
Hymenium of a basidiomycete 117
Hymenoascomycetidae I 90
II 97
III 101
IV 104
Hymenomycetes I 119
II 121
III 126
Hyperparasitism 383
Hyperplasia 42, 43, 47, 214, 265
Hypersensitive response (HR) 303
Hyphal bodies 72
Hyphomycetidae 38, 157
Hyphomycetous fungi 154
Hypochytridiales 35
Hypochytridiomycetes 35
Hypocotyl 52
Hypocrea 103, 156
Hypocrea citrina 103
Hypocrea rufa 103, 156
Hypocreaceae 36, 101, 103
Hypocreales 36, 101
Hypoplasia 265

I

Icosahedral 232, 233, 245, 257, 258
Icosahedral virus 232
Identical gametangia 73
Identical gametes 67
Identifying bacterial leaf blight of rice 379
Ilarvirus 247, 256
Imazalil 319, 336
Immunization 6, 187, 278, 293, 294
Immunization technique against 6
Imperfect stage 72, 79, 94, 96, 97, 101, 103, 107, 146, 156
Incubation 15, 170, 226, 241, 251, 276, 370
Indar 335
Indeterminate growth 58
Induced biochemical defenses 303
Induced resistance 303
Inert carriers 312
Infected 280, 283, 284, 290, 295, 298, 302, 303, 305, 308, 316, 376, 377, 378, 380, 385, 388, 389, 393
Infected pollen 235, 240
Infected seeds 280
Infection 7, 13, 27, 29, 31, 32, 40, 42, 43, 47, 50, 58, 59, 73, 90, 97, 98, 104, 105, 107, 108, 125, 128, 132, 133, 164, 165, 167, 168, 170, 171, 172, 173, 174, 175, 178, 179, 180, 185, 186, 188, 189, 190, 194, 196, 199, 205, 206, 207, 225, 226, 228, 231, 232, 235, 238, 239, 240, 241, 242, 245, 248, 249, 253, 254, 257, 258, 260, 262, 264, 265, 267, 269, 274, 276, 277, 286, 289, 291, 292, 293, 294, 296, 297, 298, 299, 300, 301, 302, 303, 304, 308, 310, 312, 314, 315, 316, 331, 336, 338, 344, 369, 377, 379, 380, 383, 390, 393
Infection by plus and minus strands of ssrna virus 248
Infection by the pathogen 298, 299, 301, 303, 337, 383, 389, 391

Infection hyphae 108
Infection of the host 128, 189, 243
Infectious communicable disease 228
Infectious diseases 6, 12, 288
Infectious living fluid 7, 228
Infestation 7, 292, 363
Infested 42, 272, 283
Inflorescence 31, 136, 153, 163, 172, 267, 269, 367
Inflorescence rot 369
Inhibit the production of amino acids and enzymes 375
Inhibitory substances 299, 300, 393
Initial inoculum 186, 190, 369
Inoculation 236, 305, 310, 373, 393
Inoculum 73, 123, 164, 165, 166, 186, 188, 189, 190, 191, 204, 205, 278, 279, 282, 285, 286, 287, 288, 289, 315, 368, 369, 371, 372, 392, 393
Inoculum potential 282, 287, 372
Inoculum production 286
Inoperculate 40, 104
Inoperculate chytridiales 40
Inoperculate discomycetes 104
Inorganic chemical compounds 313
Inorganic chemicals 373
Inorganic sulfur fungicides 320, 375
Insect repellants 389
Insecticides 292, 356, 357, 360, 389
Insect-proof cages 382
Intensity of the disease 195
Intercalary 23, 92, 129
Intercellular fashion 127
Intercellular hyphae 21, 81
Intercellular mycelium 22, 59
Intercellularly 15, 18, 140, 226
Internal infection 47, 165
Internally seed-borne pathogens 282, 317
Intramatrical 45, 97
Intramatrical growth 45
Introduction 6, 7, 135, 177, 190, 191, 272, 276, 283, 284, 285, 306, 323, 330, 331, 359
Introduction of new diseases 190
Ion movement 303
Iprobenphos 334
Iprodione 329, 361, 362, 376
Iron 293, 324, 359, 383
Irregular spots of tomato fruits 221
Irrigation 179, 180, 189, 190, 227, 270, 271, 281, 308, 309
Irrigation water 179, 189, 227, 309
Isoprothiolane 337, 375

J

Jelly fungi 115
Juvenile moults 244

K

Karathane 328
Karbation 323
Karyogamy 33, 40, 72, 73, 79, 81, 83, 87, 114, 116, 118, 125, 126, 127, 130, 132
Kasugamycin 341
Kasumin 341
Kavach 328
Kernel bunt of wheat 142
Kernel smut of rice 30
Kingdom 11, 17, 20, 34, 38, 207, 212, 245
Kingdom - Prokaryotae 207, 212
Kindom - Viruses 245
Koban 336
Koch's postulates or Koch's rules 19

L

Lack of sexual reproduction 207
Lactic acid 67, 70
Lagenidiales 35, 52
Late blight 5, 6, 29, 58, 166, 168, 177, 180, 185, 186, 188, 190, 191, 198, 205, 206, 207, 279, 284, 297, 307, 322, 330, 335, 339, 361, 365, 369, 388
Latent infection 255
Lateral stalk 123
Leaf blight 29, 107, 114, 150, 151, 153, 206, 223, 281, 296, 299, 307, 322, 338, 339, 364, 369, 379, 380, 381, 388, 393
Leaf blight of arecanut 151

Leaf blight of corn 150, 162
Leaf blight of jasmine 153, 161
Leaf blight of sorghum 369
Leaf blight of wheat 369
Leaf blotch of wheat and other cereals 152
Leaf curl 75, 81, 83, 166, 179, 257, 259, 265, 327, 377
Leaf curl fungi 75
Leaf distortion 243, 265
Leaf fall 58, 108, 374
Leaf fall, canker and pod rot of rubber 58
Leaf miner 178
Leaf mould of tomato 171
Leaf necrosis of gingelly 216
Leaf roll of potato 179
Leaf rot 281, 288
Leaf rust or brown rust of wheat and rye 132
Leaf rusts 340
Leaf scald 16
Leaf scorch 16, 17
Leaf smut 30
Leaf smut of dahlia 144
Leaf smut of rice 30, 144
Leaf spot 27, 185, 206, 216, 221, 246, 361
Leaf spot of apple 150, 152
Leaf spot of beet 173
Leaf spot of betelvine 150
Leaf spot of blackberry 152
Leaf spot of brinjal (eggplant) 162
Leaf spot of castor 150
Leaf spot of cherries 154
Leaf spot of chillies 162
Leaf spot of chrysanthemum 152
Leaf spot of coconut 27, 162
Leaf spot of corn 304
Leaf spot of cotton 162, 180, 221
Leaf spot of cotton 382
Leaf spot of cowpea 150, 152
Leaf spot of finger millet 114
Leaf spot of ginger 153, 369
Leaf spot of grapevine, pomegranate, tapioca etc. 162
Leaf spot of jasmine 152
Leaf spot of oil palm 162
Leaf spot of onion 153
Leaf spot of papaya 150
Leaf spot of pomegranate 153
Leaf spot of rose 152, 153
Leaf spot of safflower 152
Leaf spot of sapota 154
Leaf spot of sunflower 152
Leaf spot of tapioca 151, 163
Leaf spot of tobacco 150, 152, 162
Leaf spot of tomato 152, 339, 371
Leaf spot of turmeric 83, 153, 369
Leaf spots of crucifers 161
Leaf stomata 223
Leaf streak 223, 338, 339, 364, 367, 377, 379
Lenticels 226, 298
Lenticels 58, 61, 170, 171, 225, 226
Leprosy in humans 219
Leptomitales 35, 52
Leptosphaerella 146
Leptosphaerella avenaria 146
Leptosphaeria 150, 151, 164
Leptosphaeria coniothyrium 151
Leptosphaeria maculans 150
Leptosphaeria salvinii 164
Leptostromataceae 148
Lettuce mosaic virus 252, 266
Lettuce necrotic yellow virus 256, 257
Lettuce vein clearing 264
Leucostoma (Valsa) 101, 102
Leveillula 90, 94, 96, 376
Leveillula taurica 90, 96
LHT system of classification 245
Lichens 26
Life cycle 39, 40, 41, 42, 43, 44, 45, 46, 48, 49, 50, 52, 53, 55, 56, 57, 58, 60, 61, 62, 65, 66, 71, 74, 79, 81, 83, 84, 99, 100, 105, 106, 107, 108, 111, 112, 115, 127, 128, 129, 131, 132, 133, 134, 140, 141, 142, 143, 144, 164, 167, 174, 190, 285, 286, 297, 372
Lignification 385
Lignin 119
Lime-sulfur 320, 321, 329
Lindane 356

Linseed oil 391
Linseed rust 136
Lipid bilayer 232
Living cells 227, 231, 238
Local infections 136
Local lesions 219, 259
Local systemic action 333
Localized lesions 27, 33
Locally systemic 330
Locules 109, 110
Loculoascomycetidae 37, 109
Logos 1
Long appendages 181
Long smut of sorghum 142, 370
Longidorus 244
Longitudinal fission 210
Lonocol 322
Loose smut of barley 140, 173, 288, 365
Loose smut of corn 140
Loose smut of oats 140
Loose smut of sorghum 140, 370
Loose smut of wheat 140, 173, 177, 286, 288, 317, 365, 366, 367
Loose smuts of cereals 166
Low volume spraying 343, 344, 354
Lysergic acid 101
Lysis of the pathogen 385
Lysis of zoospores 383, 390
Lytic enzymes 383, 384, 390, 391

M

Machlomovirus 246
Macroconidia 163
Macrocysts 38
Macrophomina 150, 164, 368, 376, 380, 390
Macrophomina phaseolina 150, 164, 368
Magnaporthe 101, 156
Magnaporthe grisea 156
Mahonia 128
Maize chlorotic dwarf virus 242, 253
Maize chlorotic mottle virus 246
Maize rayado fino virus 247
Maize rough dwarf 258, 264
Maize rough dwarf virus 257
Maize smut 173
Maize streak 251, 258, 262
Maize streak virus 249, 258
Maize stripe virus 257
Malathion 356
Malformation of organs 268
Management of foliar diseases 373
Management of seed-borne diseases 363, 366
Mancozeb 317, 322, 323, 334, 335, 361, 362, 375, 376
Maneb 317, 322, 323, 360, 366, 375
Manipulation of the disease triangle 187
Manually operated dusters 312, 356
Manually osperated hydraulic sprayers 347
Manually operated mist blowers 350
Manually operated pneumatic sprayers 349
Manually operated sprayers 346, 347
Manzate 200 , 322, 323
Marafivirus 247
Marasmius 123
Marasmius oreades 123
Marssonina 108, 153
Marssonina rosae 108, 153
Marssonina-type conidia 108
Mastigomycota 34, 38, 39
Mating type (-) 128
MBC 332
Mealy bugs 178, 240, 243
Mechanical pressure 171, 172
Mechanical transmission 238
Mechanism of dikaryotic cell division by clamp con 117
Mechanism of disease resistance 298
Mechanisms for persistence 164
Meiosis 25, 40, 49, 50, 65, 68, 73, 80, 87, 114, 116, 118, 121, 125, 126, 127, 130, 132, 137, 138, 140, 142
Meiosis 45, 73
Melampsora 136, 376
Melampsoraceae 37, 136
Melanconiaceae 38, 152
Melanconiales 38, 148, 149, 152

Melanconium 153
Melanconium fuligineum 153
Melprex 336
Membrane lipids 232
Mentha piperita 392
Mercuric chloride 292, 313, 316, 365, 366, 373, 374, 378, 380
Mercuric chloride 357, 359
Mercury 314, 326, 359
Meristem 237, 270, 306, 387
Meristem tissue culture 387
Merpan 329
Messenger RNA (MRNA) 233
Metabasidium 126, 127
Metabolic activities of the pathogens 375
Metabolites 290, 300, 302, 304, 382, 385, 389, 390
Metalaxyl 292, 305, 311, 315, 317, 318, 334, 335, 360, 361, 362, 366, 373, 374, 376
Metal-based fungicides 314
Methods of application of plant protection chemica 315
Methyl bromide 290, 313, 319
Methyl demetan 356
Methylene blue 378
Micop 325
Microascales 83, 87
Microbial activities 386
Microbial antagonistic action 387
Microbial population 372
Microclimatic factors 170, 189
Microconidia 105, 106, 108, 111, 163
Micronutrients 293
Microsclerotial bodies 377
Microscope 4, 5, 8, 11, 13, 229, 230, 231, 377, 378, 379, 381
Microscopic 11, 20, 83, 114, 115, 116, 207, 373, 377, 381
Microscopic examination 377
Microscopically 380, 381, 382
Microsphaera 92, 94, 96
Microsphaera alni 96
Microsphaera alphitoides 92
Milbam 322
Milcurb 335
Mild mosaic of potato 252
Mildews 3, 5, 52, 59, 65, 72, 89, 90, 180, 182, 183, 184, 186, 188, 189, 197, 205, 297, 314, 317, 320, 322, 323, 327, 328, 331, 332, 333, 334, 335, 336, 339, 360, 365, 373, 374, 375, 376, 377
Mildex 328
Milstem 335
Miltox 322
Mineral elements 304
Mineral oils 341, 342
Minus (negative) polarity 233
Miscellaneous systemic fungicides 336
Mites 178, 188, 238, 243, 285, 291
Mitotically 83, 142
Mixia 79
Mode of action of fungicides 313
Mode of entry of bacterial pathogens 224, 225
Mode of entry of fungal pathogens 170
Mode of infection 277
Mode of spread 2, 188
Moist chamber 378
Moist hot air treatment 289
Moisture 116, 170, 171, 189, 190
Mollicutes (mycoplasma-like organisms) 207
Moncut 329
Monilia 107, 158
Monilia sitophila 158
Moniliaceae 38, 79, 101, 103, 154, 155, 159
Moniliales 38, 101, 146, 154
Monilinia (sclerotinia) fructicola 105
Monobacilli or microbacilli 210
Monococci 210
Monocrotophos 356
Monoculture 190
Monocyclic 188
Monopartite genome 251, 258
Monopotassium phosphate 341
Morchella 72, 109
Morchellaceae 37, 109
Morphological defense structures 298, 299
Morphologically identical 26, 67, 131, 182

Morphology of bacteria 211
Mosaic 7, 8, 11, 12, 177, 178, 179, 228, 229, 232, 238, 239, 240, 243, 244, 245, 246, 247, 248, 249, 251, 252, 253, 254, 255, 256, 257, 259, 262, 264, 265, 266, 267, 286, 289, 307, 308, 310, 365, 377, 380, 388, 389, 391, 392, 393
Mosaic and grassy shoot of sugarcane 179
Mosaic streak of wheat 179
Mother cells 83, 86, 87
Mother plant 238, 239, 240, 265
Motile bacteria 210, 213
Motile cells 42, 47, 67
Motile nature 207, 227
Motile zoospores 40
Motility 383, 390
Mottle 9, 13, 246, 251, 252, 253, 255, 260, 262, 264
Moulds 30, 34, 38, 39, 40, 47, 52, 67, 85, 181, 379
Mucor 68, 69, 70, 332
Mucoraceae 35, 68
Mucorales 35, 67
Mulberry ring spot 264
Multicomponent viruses 231
Multinucleate ascospores 109
Multinucleate cells 83
Multinucleate gametangia 68, 75
Multinucleate vesicle 85
Multiple genes 295
Multispored sporangia 68
Mung yellow mosaic 179
Mungbean yellow mosaic virus 258
Mushroom proper 121
Mutagenic chemicals 306
Mutation 135, 186, 233, 294, 306, 331, 360, 306
Mycelial fragments of fungi 179
Mycelial hyphae 170, 172
Mycelial infection 172
Myceteae (fungi) 34, 38
Myclobutanil 335
Mycobacteriaceae 217
Mycobacterium tuberculosis 6, 219
Mycoparasites 105, 290, 383, 384
Mycoplasma 8, 13, 14, 165, 207, 285, 364
Mycoplasmal bodies 8
Mycoplasmataceae 8, 14
Mycorrhizal control 385
Mycorrhizal forms 121
Mycorrhizal fungi 67, 122
Mycosphaerella 110, 112, 146, 150, 151, 152, 159, 162, 172, 341, 360, 369
Mycosphaerella areola 159
Mycosphaerella rabiei 151
Mycostatin 340
Mycotoxins 157
Mylone 313
Myxamoeba 38, 39
Myxogastromycetidae 34
Myxomycetes 34, 39, 369
Myxomycota 20

N

Nabam 323
Naked zoospores 42
Narrow brown leaf spot of rice 162
Narrow-spectrum activity 374
Natural openings 170, 173, 223, 225, 226
Necrosis 14, 17, 31, 216, 246, 247, 254, 255, 264, 266, 374, 381, 393
Necrotic 31, 221, 225, 240, 247, 253, 255, 256, 257, 259, 262, 264, 302, 303, 357, 359
Necrotic lesions 302
Nectria 89, 99, 101, 102, 103, 162, 171, 384
Nectria type centrum 90, 103
Nectriaceae 101
Nectrioidaceae 37, 148, 152
Nemagon 290
Nematicidal 327
Nematicides 310, 360, 389
Nematode pathogens 174, 175, 189
Nematode pathogens 389
Nematodes 7, 8, 144, 154, 168, 179, 181, 188, 223, 228, 238, 244, 251, 255, 285, 290, 291, 298, 302, 303, 309, 313, 318, 323, 370, 371, 372, 385, 389, 390
Nematospora 81

Neoaliturus haemoceps 269
Neomycin 341
Nephotettix 242, 390
Nephotettix cincticeps 242
Nepovirus 245, 247, 254
Neurospora 158
New binucleate cells 116
Nickel chloride 313, 330, 376
Nickel fungicides 330
Nicotiana rustica 266
Nimrod 335
Nitrogen fixation 213
Nodules 214
Non-ascocarpic sac fungi 79
Non-flagellate 16, 38
Non-host plants 392
Non-pathogenic bacteria 384
Non-persistent manner 178, 252, 253, 256, 258
Non-stromatic 148, 311, 313, 330, 334, 360
Non-systemic 316
Non-systemic fungicides 311, 331
Notification of diseases 283
Nozzle 345, 347, 348, 349, 350, 352, 362
Nuarimol 335
Nuclear division 39, 67
Nucleic acid 8, 9, 15, 229, 231, 232, 233, 235, 237, 245
Nucleic acid techniques 15
Nucleorhabdovirus 247, 257
Number of rainy days 192, 205
Nursery beds 289, 290, 318
Nutrient media 14, 16, 17, 19, 27, 380
Nutrients 18, 19, 121, 170, 187, 223, 226, 272, 274, 276, 281, 287, 293, 295, 302, 309, 370, 371, 372, 375
Nystatin 338, 340, 341

O

Oat mosaic virus 253
Oat necrotic mottle 253, 264
Oat necrotic mottle virus 253
Oats agar 381
Obclavate 156, 161
Oidiophores 118
Oidium (acrosporium) 158
Oils 336, 341, 346, 362, 392
Okra mosaic virus 254
Olpidiaceae 35, 40
Onion downy mildew pathogen 165
Oogamous 52
Oomycetes 35, 50, 52, 54, 175, 331, 334, 365, 369, 373
Oomycetous pathogens 333, 359
Oosphere 56, 65
Oospora 332
Open ascocarp 104
Operculate asci 109
Operculate chytridiales 47
Operculate discomycetes 104, 109
Operculum 40, 47, 75, 77, 104, 108
Ophiobolus 172
Ophiostomataceae 36, 87
Opposite mating type 97, 121, 128, 130, 131, 142
Organic acids 288
Organic sulfur fungicides (dithiocarbamates) 321
Organic systemic fungicides 330
Organisms 2, 3, 8, 9, 11, 12, 13, 14, 16, 18, 23, 26, 33, 34, 38, 39, 40, 54, 80, 85, 87, 111, 113, 144, 157, 165, 167, 173, 177, 180, 186, 189, 207, 211, 212, 217, 229, 231, 277, 302, 309, 364, 367, 368, 369, 371, 373, 374, 376, 377, 380, 381, 382, 383, 387, 388
Organo-phosphorus fungicides 334
Ornalin 329
Orobanchaceae 274
Orobanche 19, 168, 180, 274, 275, 276, 370, 391
Orthocide 329
Orthostreptomycin 338
Oryzavirus 249, 257
Ostiolate fructifications 89
Ostiolate or astomous ascocarps 97
Ostiole 77, 87, 96, 99, 103, 113, 128, 132, 145, 146, 148
Overhead irrigation 180, 190, 227

Ovulariopsis 96
Oxalic acid 68
Oxidative reactions 303
Oxycarboxin 289, 330, 333, 366, 333, 376
Oyster shell-shaped caps 123

P

Panicum antedotale 286
Panicum repens 286
Panoram 321
Papaya leaf curl 377
Papaya mosaic 252, 262
Papaya ring spot 264, 310
Papovovirus 233
Paralongidorus 244
Paramyxoviruses 233
Paraphyses 77, 96, 97, 98, 101, 104, 108, 109
Parasitic 10, 11, 18, 21, 26, 30, 34, 39, 40, 47, 54, 68, 70, 72, 81, 85, 97, 114, 120, 121, 125, 126, 144, 146, 148, 159, 161, 162, 171, 172, 188, 207, 229, 274, 293, 309
Parasitic higher plants 188
Parasitic invasion 171
Parasitic plants 18
Parasitic relationship 293
Parasitically 18, 56, 104, 219
Parasitism 18, 26, 90, 131, 135, 184
Parasitism 26
Paratrichodorus 244
Parenchyma cells 238, 301
Parenchymatous cells 302
Parenchymatous tissues 367
Parrycop 325
Parsnip yellow fleck virus 246
Partitiviridae 247
Parts of a typical sprayer 345
Parzate 322, 323
Passive or indirect dispersal 177
Pastes 3, 313, 318
Pasteurella 6
Pasteurization 5
Pathogen gene 303
Pathogen-derived genes 305
Pathogenecity 6
Pathogenic bacteria 182, 210, 297, 367, 384
Pathogenic fungi 11, 174, 182, 189, 291, 309, 310, 339, 372
Pathogenic prokaryotes 207
Pathogenic viruses 188, 392
Pathos 1
PCNB 289, 327, 336
Pea enation mosaic 246, 247, 256, 264
Pea enation mosaic virus 245, 247, 255
Pea leaf roll 254, 265
Pea stunt disease 262
Peach leaf curl 166, 327
Peanut stunt disease 262
Peanut stunt virus 256
Pear leaf spot 152
Pear scab 115
Pebrine disease of silk worm 5
Pellicularia 120, 164
Pencozeb 323
Penetration sites 170
Penicillia 86, 338, 341
Penicillin 9, 16, 17, 87
Penicillium 83, 85, 86, 87, 88, 144, 154, 157, 171, 309, 328, 338, 340, 359, 360, 361, 373, 377, 381, 384
Penicillium strains 359
Percentage disease index (PDI) 85, 203
Perennial fruiting bodies 120
Perenox 325
Peridium 83, 87, 89, 92, 94, 97, 129
Periphysoids 77
Peritrichous flagella 213, 215
Peronosclerospora 63
Peronospora 61, 165, 168, 369, 376
Peronosporales 35, 52, 54, 59
Persistence in the soil 175
Persistence of the races 183
Persistent asci 101
Persistent manner 178, 243, 252, 253, 254, 255, 256, 257, 258, 291
Petunia asteroid mosaic virus 253
Phagotrophic 38
Phaltan 330
Phenol-oxidizing enzymes 304
Phenyl mercury acetate 326

Phialides 86, 148, 150, 152, 156, 157, 162
Phloem 7, 8, 9, 13, 14, 16, 17, 19, 237, 239, 241, 242, 243, 245, 260, 264, 268, 269, 276, 292
Phloem cells 241, 242, 243, 269, 276
Phloem of the dodder plant 245
Phloem tissues 13, 16, 242, 264
Phloem vessels 260, 292
Phloem-inhabiting bacteria 16, 17, 19
Phoma 110, 150, 332, 376, 377, 380
Phomopsis 101, 150, 328
Phosphamidon 356
Phospholipid bilayer 304
Phosphorus compounds 341
Photosynthesis 10, 18, 19, 274, 373
Photosynthetic 11, 30, 113, 207, 211
Photosynthetic bacteria 211
Phycomycetous fungi 328, 379
Phygon 327
Phyllactinia 89, 90, 92, 94, 96, 376
Phyllactinia type centrum 90
Phyllanthus niruri 392
Phyllosticta 27, 110, 148, 150, 151, 152, 180, 296, 368
Phyllosticta leaf spot of banana 27
Phyllosticta leaf spot of ginger 369
Phylogenetically related group 33
Physalospora 98, 151, 152, 153
Physical agents 364
Physical barrier 389
Physical methods 274
Physiological characters 182
Physiological functioning of the host 295, 336
Physiological specialization of pathogenic fungi 182
Physoderma 43, 46, 369
Phytoalexins 304
Phytomonas staheli 9
Phytomycin 338
Phytophthora 5, 56, 58, 60, 61, 156, 165, 167, 168, 173, 174, 178, 179, 180, 181, 185, 284, 297, 327, 329, 332, 333, 334, 335, 360, 361, 364, 365, 367, 368, 370, 371, 373, 374, 375, 376, 382, 383, 384, 385, 387, 390, 393
Phytophthora citrophthora 368
Phytoplasmas 13, 14, 15, 16, 19, 269
Phytoreovirus 249, 257
Phytosanitary certificate 284, 285
Phytotoxicity 322, 357
Phytotoxicity of fungicides 357
Picornaviruses 233
Pink disease of rubber 121, 334
Pink rot of potatoes 173
Planococcoides citri 243
Planococcoides njalensis 243
Plant borers 178
Plant cell cultures 387
Plant cells 12, 58, 226, 232, 237, 238, 244, 300, 307, 388, 389, 390
Plant disease surveillance 191, 194
Plant diseases 1, 2, 3, 5, 7, 8, 10, 12, 13, 16, 34, 87, 176, 185, 190, 214, 231, 238, 267, 268, 278, 291, 294, 296, 305, 308, 310, 313, 330, 337, 341, 344, 363, 374, 383
Plant growth hormones 122
Plant hoppers 14, 15, 240, 257
Plant introduction 306
Plant Kingdom 11, 17, 207
Plant parasitic 126, 148, 161
Plant pathogens 2, 10, 110, 114, 136, 137, 156, 159, 161, 164, 165, 173, 215, 219, 283, 286, 288, 289, 290, 310, 359, 376, 383, 389, 390, 391
Plant protection 193, 194, 204, 284, 291, 310, 315, 344, 354, 356, 357, 359, 361, 362, 363
Plant virus vectors 240
Plant Viruses 7, 40, 178, 228, 230, 231, 232, 235, 238, 239, 240, 241, 242, 243, 244, 245, 249, 250, 256, 258
Plantvax 333
Plasmodesmata 237, 260
Plasmodiogymnomycotina 34, 38
Plasmodiophoraceae 35, 47
Plasmodiophorales 35, 47, 244
Plasmodiophoromycetes 20, 35, 47
Plasmopara 59, 185, 284, 361, 376, 383, 390

Pleospora bjorlingii 150
Pleospora type centrum 114
Pleosporaceae 37, 113
Pleosporales 37, 110, 113
Pleurotus 115, 123
Plum leaf scald 16
Plum pockets 83
Plus (positive) polarity 233
Pod spot 111, 152
Podosphaera leucotricha 90, 341
Poisonous chemicals 360
Pollen grains 40, 176
Pollen transmission 240, 255
Polyhedral 233
Polyhedral-shaped viruses 244
Polymycin 341
Polymyxa 244
Polyporales 119
Polyram 322
Polysaccharides 385, 393
Pore fungi 119
Post-emergence damping off 31, 374
Post-harvest dip of fruits 329
Post-harvest diseases of Citrus 328
Post-harvest diseases of fruits 319
Post-harvest rots 339
Post-harvest spray 328
Potassium bicarbonate 341
Potato black scurf 176, 326, 364
Potato leaf roll 229, 254, 265
Potato ring disease bacterium 165, 168
Potato scab 326
Potato virus A 252
Potato virus T 252
Potato virus X 238, 246, 252, 393
Potato virus Y 246, 252, 393
Potato wart disease 176, 181
Potato yellow dwarf 249, 257, 264
Potato yellow dwarf virus 247, 257
Powdery mildew of apple 96, 166, 342
Powdery mildew of barley 361
Powdery mildew of cucumbers 339, 361
Powdery mildew of eggplant 361
Powdery mildew of grapes 29
Powdery mildew of jasmine 159
Powdery mildew of lady's finger 29
Powdery mildew of mango 173
Powdery mildew of mulberry 97
Powdery mildew of papaya 159
Powdery mildew of pea, cabbage etc. 370
Powdery mildew of pomegranate 159
Powdery mildew of rose 96, 339, 370
Powdery mildew of rubber 159, 191
Powdery mildew of shade and forest trees 97
Powdery mildew of tobacco and cucurbits 370
Powdery mildew of wheat 288, 370
Powdery mildews 3, 73, 89, 90, 181, 182, 188, 190, 198, 206, 297, 314, 320, 321, 322, 323, 328, 329, 331, 333, 334, 335, 336, 339, 366, 374, 376, 377, 379
Power dusters 312, 354
Power source 346
Power sprayers 344, 346, 347
Predators 194
Pre-existing chemical defenses 299
Preparation of fungicidal spray fluid 342
Pressure gauge 346, 349
Previcur 336
Primary host 165, 166, 167, 240, 286
Primary infection 59, 73, 108, 165, 167, 168, 185, 259, 260, 369
Primary inoculum 73, 164, 165, 166, 286, 368
Primary symptoms 259
Probenazole 305
Prochloraz 336
Promycelium 126, 130, 131, 132, 137, 138, 140, 142
Propamocarb 318, 336
Properties of plant viruses 232
Prophylactic spraying 391
Propiconazole 335
Prostar 329
Protectant fungicides 291, 311, 362, 375
Protein capsid 237, 251
Protein coat 12, 232, 235, 237
Protein synthesis 233, 328, 338

Proteins 211, 213, 215, 231, 232, 237, 300, 303, 304, 393
Proteobacteria 210, 213
Protomyces 79
Protomycetales 36, 79
Protoplast 40, 43, 49, 50, 56, 137, 140, 211, 301, 302, 386, 387
Protoplast fusion 386, 387, 388
Prototunicate asci 77
Protozoa 4, 9, 19, 20, 34, 39, 47, 168, 219, 228
Pruning cuts 172
Pruning shears 267
Pseudomonas 165, 167, 168, 176, 177, 215, 221, 223, 225, 288, 290, 364, 366, 370, 373, 377, 383, 384
Pseudomonas fluorescens 225, 290, 366, 373, 383, 384
Pseudomonas syringae 215, 221, 370, 393
Pseudoparenchymatous 83, 89, 98, 99, 109, 145
Pseudoparenchymatous sclerotium 98
Pseudoperidium 131, 136
Pseudoperonospora 61, 361, 369, 376
Pseudopodia 38, 49
Pseudosepta 161
Puccinia graminis hordei 135
Puccinia graminis tritici 134, 135, 167, 181, 183, 185, 280, 289
Pucciniaceae 37, 131
Puckering 81, 83
Puffballs 65, 116, 118
Punica granatum 392, 393
Pure cultures 6
Pycnidial stage 101
Pycniospores 128
Pyrazophos 334
Pyrenophora 161
Pyrethroids 356
Pyricularia 101, 156, 166, 172, 185, 361
Pyricularia setariae 368
Pyrimidines 331, 335, 376
Pyroquilon 289, 337, 366, 376
Pythiaceae 35, 54, 58
Pythium 56, 57, 61, 156, 166, 167, 174, 175, 179, 281, 297, 318, 322, 327, 329, 332, 333, 334, 335, 360, 364, 365, 367, 368, 370, 371, 372, 373, 374, 375, 376, 382, 383, 384, 385, 390, 391

Q

Qualities of an ideal systemic fungicide 336
Quarantine laws 283, 284, 285
Quarantine measures 282
Quarantine regulations 283, 285
Quinone fungicides 327
Quintozene 327

R

Rabbits 179
Race numbers 182
Races 8, 90, 135, 137, 182, 183, 184, 294, 295, 296
Radish mosaic virus 254
Rainfall 180, 189, 192, 205, 279, 280, 297
Rally 335
Ratoon crop 269
Ratoon stunting disease of sugarcane 289
Rats 179
Receptive hyphae 118, 127, 128, 129, 130, 132, 133
Recognition of the pathogen by the host 296, 300, 301
Red blood cells 4
Red clover vein mosaic virus 260
Red pumpkin beetle 178
Red rot 98, 120, 153, 165, 173, 177, 180, 185, 186, 195, 279, 281, 282, 286, 287, 288, 289, 307, 365, 366, 373, 378
Red rust 18, 271
Reduced seed size 381
Regular hosts 165
Reinforcement of host cell walls 304
Relative humidity 58, 189, 192, 205, 297, 381
Reoviridae 249
Replicase enzyme 12
Replicative form 233
Reproductive capacity of a pathogen 186

Reproductive cycles 188
Reproductive stages 127
Residual fumes 318, 359, 373
Residual toxicity 290, 313, 317, 327, 329, 330, 331, 337, 338, 362, 364
Resistance 7, 8, 135, 171, 172, 175, 181, 184, 186, 191, 207, 269, 280, 288, 292, 293, 294, 295, 296, 297, 298, 299, 303, 304, 305, 306, 308, 310, 331, 332, 334, 336, 359, 365, 387, 388, 391, 392, 393
Resistance of plant pathogens to plant protection 359
Resistant genes (R genes) 296
Resistant genotypes 305, 306, 307
Resistant hosts 260
Resistant strains 331, 359, 360
Resistant varieties 7, 135, 184, 186, 287, 294, 299, 300, 304, 305, 306, 307, 310, 386, 388
Rhabdoviridae 247
Rhinoceros beetle 178
Rhizobiaceae 213
Rhizoctonia 21, 120, 126, 150, 156, 163, 166, 167, 168, 174, 175, 176, 179, 280, 282, 297, 318, 322, 323, 327, 329, 331, 332, 333, 335, 359, 361, 363, 365, 366, 367, 371, 372, 373, 375, 376, 377, 383, 384, 388, 390
Rhizoctonia root rots 373, 376
Rhizoctonia seedling blight 367
Rhizoidal branches 68
Rhizome rot 281, 365, 369
Rhizome rot of ginger 281, 369
Rhizomorphs 20, 123, 172, 173
Rhizopus rot of mango 393
Rhodococcus 383
Rice bacterial leaf blight pathogen 165
Rice dwarf disease 262
Rice fiji disease virus 249
Rice ragged stunt 249, 258, 377
Rice ragged stunt virus 257
Rice tungro virus disease 259
Rice yellow dwarf 229
Ridomil 335
Ridomil MZ 335
Rigid rods 232
Ring disease of potato 170
Ring rot of potato 177, 217
Ring spot of cabbage and cauliflower 150
Ripe fruit rot and die-back of chillies 153
Rodenticides 360
Ronilan 329
Root and stem rot of gingelly 151
Root dips 331
Root exudates 171, 276, 287
Root nodules 214
Root parasite 172, 274, 275, 276
Root rot of citrus 290
Root rot of tobacco 164, 371
Root rot pathogen of Citrus 383
Root rots 56, 114, 119, 127, 164, 180, 317, 323, 328, 332, 373, 376
Roquefort cheese 158
Rose mosaic virus 256
Rot of mango 158, 161, 340, 393
Rot of onion 151, 164
Rots 56, 73, 114, 119, 120, 127, 164, 180, 195, 215, 225, 317, 319, 323, 325, 327, 328, 329, 332, 333, 335, 336, 339, 340, 366, 373, 376, 377, 379
Rouging 285, 286
Rubigan 335
Runner hyphae 172
Running water 179
Rust fungi 5, 29, 119, 127, 128, 132, 138, 170, 183
Rust of barley 132
Rust of bean 136
Rust of broad bean and lentil 136
Rust of carnation 136
Rust of castor 136
Rust of corn 132
Rust of cowpea 136
Rust of gram 136
Rust of groundnut 132, 390
Rust of pea 136
Rust of rose 136
Rust of soyabean 136
Rusts 3, 30, 52, 63, 67, 115, 116, 118, 127, 128, 130, 131, 136, 167, 168, 181, 183,

184, 186, 188, 207, 281, 297, 318, 320, 321, 322, 325, 328, 329, 330, 333, 335, 336, 339, 340, 374, 376, 377, 386
Ryegrass mosaic virus 246, 253

S

Sac fungi 73, 79
Saccharomyces 74, 80
Sampling 199
Sanitation 187, 190, 270
Sanitation 285, 288, 298, 308, 371
Sanspor 329
Sap transmissible viruses 177
Saponins 300
Saprol 336
Saprolegniaceae 35, 52
Saprolegniales 35, 52
Sartorya 157
Scab 33, 50, 115, 148, 159, 166, 171, 185, 190, 205, 219, 297, 326, 327, 328, 331, 336, 360, 361, 371, 372
Scab of apple 331, 361
Scientific monitoring 193
Sclerenchyma cells of the veins 299
Sclerospora graminicola 63, 168, 179, 185, 363, 369, 380
Sclerotial bodies 380
Sclerotiniaceae 36, 104
Sclerotium root rot pathogen 371
Sclerotium rot of potato 164
Sclerotium stem rots 373
Scorch 16, 17
Score 7, 335
Scrutiny of the survey reports 193
Scurf 121, 127, 161, 164, 176, 326, 364, 368, 385
Secondary host 167
Secondary infection 73, 221, 239
Secondary inoculum 164, 165
Secondary spores 72, 369
Sections 212, 378, 379
Seed abortion 381
Seed certification 282
Seed discoloration 381
Seed health testing methods 379
Seed infection 239, 379, 380
Seed necrosis 381
Seed pelleting 366
Seed potatoes 364
Seed rhizomes 364
Seed rot 282, 327, 368, 381, 385
Seed treating chemicals 317
Seed washing technique 381
Seed-borne infection 379
Seed-borne inoculum 372
Seedling blight and leaf blight of finger millet 393
Seedling blight of apple 164
Seedling blight of castor 168, 369
Seedling disease of rice 103
Seedling diseases 328, 336, 376, 385
Seedling rot 58, 167, 327, 369
Seed-treating fungicides 363
Selective nutrient media 380
Semi stem parasite 272
Semi-persistent manner 253
Semi-persistent viruses 242
Septate hyphae 68, 79, 81, 105, 113, 144, 163
Septoria 146, 152, 180, 329, 332, 376, 381
Septoria pyricola 152
Sequiviridae 246
Sessile 65, 75, 107, 119, 123, 130, 136
Sett rot 89, 152, 163, 366, 369, 378
Sett rot of tapioca 152
Sett rot or stem rot of sugarcane 163
Sett-borne diseases 326, 366
Sexual compatibility 130, 138, 140
Sheath blight of rice 29, 164, 279, 288, 332, 368, 375, 376, 385, 391
Sheath rot of rice 159, 334, 390, 391
Shelf fungi 115
Shoot and inflorescence blight of mango 154
Shoot eyes 173
Shoot proliferation 17
Shot holes 27, 29, 374
Shrunken seeds 381
Silver scurf disease of potato 161

Simple microscope 4
Single isometric particles 249
Single mutation 360
Single stranded 9, 237
Single-site chemicals 360
Single-site fungicide 360
Slime moulds 34, 38, 39, 47
Slurry 313, 317, 322
Smallpox virus 231
Smudge 299
Smudge disease of onion 153
Smut 4, 5, 10, 30, 137, 127, 136, 137, 138, 140, 142, 144, 173, 177, 191, 195, 196, 197, 200, 278, 279, 280, 282, 284, 286, 289, 307, 317, 327, 333, 335, 339, 361, 364, 365, 366, 367, 370, 371, 373, 377, 381, 382
Smut balls 138
Smut disease of barley and cholam 327
Smut fungi 30, 127, 136
Smut of barley 140, 173, 286, 288, 365, 361
Smut of onion 144, 370
Smut of pearl millet 142, 279, 286, 287, 370
Smut of sugarcane 140, 373
Smuts of cereals 136, 166, 184
Sobemovirus 247, 254
Sodium bicarbonate 341
Soft rot of citrus 361
Soft rot of potato, onion and fleshy fruits 225
Soft rot of sweet potato 369, 379
Soft rots 215, 225, 373, 377, 379
Soil drench 318, 324, 326, 333, 334, 336, 340, 385
Soil drenching 290, 321, 325, 327, 329, 332, 335, 342, 374, 387
Soil fungi 188, 384, 387
Soil management 187, 308, 309
Soil sickness 372
Soil treatment 286, 289, 317, 323, 336, 359, 372, 374
Soil-borne 52, 164, 174, 179, 190, 246, 251, 279, 287, 288, 289, 290, 297, 309, 316, 317, 318, 324, 326, 327, 328, 331, 332, 334, 366, 367, 370, 371, 372, 373, 382, 385, 386, 388, 389
Soil-borne bacterium 389
Soil-borne fungal pathogens 318, 367
Soil-borne fungi 319
Soil-borne pathogens 174, 179, 190, 279, 287, 289, 290, 297, 316, 318, 324, 334, 372, 373, 382, 385, 386
Soil-borne phycomycetous fungi 328
Soil-fumigants 318
Soil-inhabiting insects 178
Soil-inhabiting nematodes 179, 255, 290, 323, 385
Solanum virus 2 252
Solar heat energy 291
Solubilization of fungicides 375
Solutions 292, 313, 317, 324, 325, 341, 363, 376
Solvents 312, 313
Somaclones 388
Somatic cells 23, 24, 81
Somatic hyphae 21, 23, 56, 58, 73, 85, 87, 90, 94, 96, 122, 144, 145, 157, 161
Somatic hyphal cells 80
Sooty mould 30, 113
Sooty mould of coffee 114
Sorghum rust 132
Source of primary infection 185
Southern bean mosaic virus 247, 254
Southern corn leaf blight 114
Soybean mosaic 253, 262
Soybean mosaic virus 252
Specialization of parasitism 184
Specialized parasites 182
Specialized pathogen 184
Specific symptoms 12, 266, 373, 376
Speldrum 328
Spergon 327
Spermagonial stage 127
Spermatia 106, 107, 108, 110, 118, 121, 127, 128, 130, 131, 132, 133, 183
Spermatial nuclei 75
Spermophthoraceae 36, 80
Sphacelia 163, 178
Sphacelotheca reiliana 369

Sphacelotheca sorghi 140
Sphaeropsidaceae 37, 148
Sphaeropsis 151
Sphaerotheca 92, 94, 341, 360, 361, 369, 384
Sphaerotheca fuliginea 92, 361
Sphaerulina 152, 162
Spherical 12, 14, 15, 21, 33, 42, 45, 52, 68, 75, 81, 94, 108, 110, 140, 145, 148, 150, 151, 158, 162, 164, 207, 210, 211, 217, 232, 246, 253, 259, 266, 271
Spilocea 146, 148
Spinacia oleracea 393
Spindle tuber 9, 12, 267
Spirillum 211, 213
Spiroplasmas 9, 14, 15, 16
Spiroplasmataceae 9, 14
Splashing of raindrops 180
Spoilage 293
Spongospora subterranea 50
Spontaneous generation 5
Sporangiophore characteristics 64
Sporangiophores 25, 54, 58, 59, 61, 63, 65, 67, 68, 70, 271, 377
Spore balls 144, 176, 363, 380
Spore discharge 104, 180
Spore dispersal 21
Spore formation 212
Spore masses 136, 376, 377
Spore release 47, 75, 188
Sporidial infection 170
Sporodesmium sclerotiorum 105
Sporogenous cells 72, 129, 130
Sporophores 25, 38, 116, 119, 120, 121, 181
Sporulation 21, 137, 190, 277, 281, 313, 315, 375
Spots 18, 27, 29, 30, 32, 43, 73, 108, 114, 150, 153, 154, 161, 173, 188, 189, 196, 198, 199, 200, 216, 221, 223, 227, 268, 271, 281, 301, 302, 314, 322, 324, 325, 327, 328, 329, 331, 332, 333, 335, 336, 339, 360, 374, 377, 378, 379, 382
Spray fluid 311, 319, 320, 322, 325, 329, 330, 331, 332, 333, 334, 335, 341, 342, 344, 345, 346, 347, 348, 349, 350, 352, 353, 354, 356
Spray fluid tank 345, 346, 352
Sprayers 344, 345, 346, 347, 348, 349, 350, 352, 353, 354
Spreaders 311
Squirrels 179, 276
Stages in the development of a basidium 117
Stalk cell 83, 92
Stalked 75, 99, 107, 109, 119, 130, 131, 135
Staphylococcus 6
Stellaria media 239
Stem and pod blight of soybean 151
Stem and pod rot of groundnut 164
Stem and root rot of tobacco 164
Stem bleeding disease of coconut 89, 161
Stem blight of bean 164
Stem blight of Citrus plants 151
Stem blight of cowpea 151
Stem canker 32, 150
Stem canker of apple 154
Stem canker of potato 32, 164, 371
Stem canker of redgram 150, 152
Stem canker of rose and raspberry 151
Stem cuttings 176, 364
Stem pitting 253, 265
Stem rot of rice 30, 164, 337, 368
Stem rots 120, 164, 336, 373
Stemonitomycetidae 34
Stereum 172
Sterigmata 59, 61, 63, 118, 126, 132, 136, 140, 142, 157
Sterile hyphae 120, 138, 140
Sterile structures 116
Sterility 14, 266, 268
Stickers 311
Stink horns 114
Stinkhorns 118
Stomata 58, 61, 90, 158, 162, 170, 171, 173, 223, 225, 294, 298, 299
Storage rots 319, 332
Storite 332
Strains 68, 138, 140, 186, 210, 233, 361
Strap-like 264, 265
Strawberry latent ring spot virus 265
Strawberry ring spot 264

Strawberry vein banding virus 259
Streptobacilli 210
Streptococci 210
Streptomyces aureofaciens 340
Streptomyces erythreus 219, 340
Streptomyces griseochromogenes 340
Streptomyces griseus 219, 338, 340
Streptomyces rimosus 340
Streptomyces venezuelae 340
Streptomycetaceae 219
Streptomycin 311, 315, 337, 338, 339, 340, 341, 360, 366, 367
Streptoverticillium cinnamomeum var 339
Striga 19, 180, 275, 276, 370
Stripe 132, 174, 240, 246, 249, 252, 258, 262, 288, 340
Stripe disease of barley 340
Stromatic hyphal bed 145
Structure of a typical bacterium 208
Sub-epidermal 131, 135, 152
Suckers 280, 288, 316, 364
Suction hose 346, 347, 350, 352
Sudden death 103
Sudden death of clove 103
Sugarcane fiji disease 258
Sugarcane mosaic 177, 253, 262
Sugarcane mosaic virus 252
Sugarcane red rot pathogen 165, 279
Sugarcane smut 173, 366
Sugary disease of sorghum 164
Sulfex 320
Sulfur dioxide gas 319
Sulfur fungicides 314, 320, 321, 375
Summer sporangia 50
Super tin 330
Suppressive soils 382
Surface water 180
Surfactants 341, 342
Surgery 3, 286, 292, 313
Survey 191, 192, 193
Survey reports 193
Survival 2, 164, 165, 167, 168, 173, 174, 175, 211, 239, 285, 313, 371, 382, 391
Susceptible host 19, 107, 127, 171, 172, 174, 186, 189, 226, 228, 260, 272, 276, 279
Sweet clover necrotic mosaic virus 255
Syllit 336
Symbiotic association 385
Symbiotically 214
Symbiotically with bean 214
Sympodially branched 58
Symptoms 2, 3, 7, 8, 12, 13, 14, 15, 16, 17, 18, 27, 30, 31, 32, 33, 85, 108, 137, 138, 170, 194, 197, 198, 199, 203, 207, 219, 220, 222, 225, 226, 237, 239, 242, 252, 254, 255, 259, 260, 261, 262, 263, 264, 265, 266, 267, 268, 269, 270, 277, 296, 300, 359, 363, 373, 376, 382
Symptoms of viroid diseases 266
Synchytriaceae 35, 40
Synchytrium 44, 284, 369, 370
Synchytrium endobioticum 42, 44, 176, 181, 284, 369
Syphilis bacteria 6
Systemic 136, 236, 237, 260, 289, 291, 292, 295, 305, 311, 312, 313, 314, 315, 316, 317, 327, 328, 330, 331, 332, 333, 334, 335, 336, 337, 359, 360, 365, 373, 374, 375, 393
Systemic action 315, 333
Systemic antibiotic 360
Systemic chemicals 311, 317
Systemic fungicides 289, 292, 295, 311, 315, 318, 330, 331, 333, 335, 336, 337, 360, 362, 366, 374, 375, 376, 377
Systemically infected 239, 264, 387
Systemically translocated 292

T

Tagstin 332
Talaromyces 157
Tapping knives 268
Tar spot 32
Tar spot of grasses 32
Taxonomy of Fungi 33
Tecto 332
Teliomycetidae 37, 118, 126
Teliospore stage 130
Temperature 11, 15, 21, 58, 170, 189, 190, 192, 205, 207, 211, 239, 280, 297, 320, 357, 370, 392

Terminal flagellum 40, 42
Terrazole 336
Terrestrial orchids 126
Tersan 321, 322, 331
Tetanus 6
Tetracyclines 340
Tetrahydro dimethyl thiazene thione 318
Tetrazolium chloride test 379
Thalli of the bacteria 217
Thallobacteria 217
Therapeutant fungicides 291
Therapeutants 311, 314, 315, 374
Thermophilic 21
Thiabendazole 292, 311, 315, 332, 366
Thiophanate 331, 333, 336, 361, 366, 374
Thiophanate-methyl 333
Thiovit 320
Thiram 289, 290, 317, 318, 321, 333, 335, 360, 366
Thiride 321
Thread-like ascospores 99
Thrips 178, 188, 240, 243, 256, 257, 285, 291
Thryothrycin 341
Tilletia indica 369
Tilletia strains 359
Tilletiopsis 384
Tilt 335
Tinmate 330
Tinsel-type flagellum 47
Tissue culture 306, 386, 387
Tissue culture techniques 306, 387
Tissues 7, 8, 13, 16, 20, 21, 27, 29, 30, 32, 40, 42, 43, 50, 52, 59, 65, 75, 81, 83, 98, 103, 104, 105, 107, 116, 122, 125, 132, 136, 137, 138, 142, 162, 165, 167, 168, 170, 172, 174, 179, 181, 214, 221, 223, 225, 226, 238, 241, 242, 264, 265, 268, 270, 271, 272, 291, 292, 293, 295, 297, 298, 299, 302, 305, 306, 310, 311, 312, 314, 315, 337, 363, 364, 367, 374, 376, 377, 378, 379, 385
Tobacco chlorotic ring spot 264
Tobacco leaf curl 179, 259, 265
Tobacco leaf curl virus 258
Tobacco mosaic disease 229
Tobacco necrosis 245, 246, 247, 254, 264
Tobacco necrosis virus 244, 245, 246, 254
Tobacco rattle virus 244, 246, 251
Tobacco streak virus 247, 256
Tobamovirus 245, 246, 249
Togaviruses 233
Tolyposporium ehrenbergii 369
Tomato golden yellow mosaic virus 258
Tomato mosaic 228, 251, 262
Tomato ring spot 264
Tomato shoe string 265
Tomato spotted wilt virus 243, 245, 247, 257, 392
Tombusviridae 246
Tombusvirus 245, 246, 253
Tooth fungi 119
Topsin M 333
Toxic 287, 290, 291, 292, 294, 298, 299, 300, 302, 304, 318, 319, 323, 324, 325, 337, 338, 357, 359, 360, 362, 365, 367, 371, 372, 375, 382, 385
Toxic 85, 216, 223
Toxic effects 357
Toxic metabolites 290, 300, 302, 382, 389, 390
Toxic residues 325, 360
Toxic symptoms on fruits 360
Toxicity 281, 289, 290, 312, 313, 316, 317, 319, 321, 324, 327, 329, 330, 331, 337, 338, 357, 359, 362, 364
Toxicity diseases 281
Toxins 85, 157, 163, 225, 226, 231, 300, 305, 330, 367, 389, 390
Tractor-mounted power-operated 345
Tractor-mounted sprayers 354
Transcriptase 237
Translocated 292, 310, 311, 312, 314, 315, 330, 337, 339, 374
Transovarial passage 243
Transplants 176, 364, 384
Transverse fission 14, 80
Treating diseased plants 292
Treating tubers 339
Treatments 3, 291, 305, 330, 393
Tree hoppers 240

Tree surgery 3, 292, 313
Triadimefon 315, 317, 318, 335, 361
Triazoles 305, 335
Trichloromethyl thiocyclohexene 328
Trichoderma 103, 105, 156, 290, 309, 366, 373, 382, 383, 384, 390, 393
Trichodorus 244
Trichogyne 73, 111, 128
Tricholomataceae 37, 123
Trichometasphaeria 161
Trichovirus 246
Tricyclazole 289, 337, 366, 376
Tridemorph 292, 333, 334, 366
Trimidal 335
Triticum 183
Trolley-mounted sprayers 354
Tropical rust of cotton 136
Truban 336
True bugs 240
True conidia 105
True fungi 34
True mycelium 20, 81
Truffles 103
Trunk injections 331
Tuber eyes 267
Tuberculariaceae 38, 154, 160, 162
Tubers 32, 33, 42, 58, 163, 165, 173, 176, 177, 178, 186, 194, 223, 226, 237, 238, 267, 280, 285, 316, 326, 338, 364
Tubotin 330
Tubular hyphae 111
Tucumanensis (=colletotrichum falcatum) 98
Tulasnellales 37, 126
Tulip mosaic virus 265
Tumor-like outgrowths 264
Tumors 43, 45, 215, 225, 258, 264, 265, 389, 390
Turmeric rhizome rot 365
Turnip mosaic virus 264
Turnip yellow mosaic 247, 255, 262
Twig blight 107
Twig cankers 107
Two-celled ascospores 103
Two-celled or more than two-celled conidia 152
Tymovirus 245, 247, 254
Types of asci 77, 82
Types of haustoria 22, 90
Types of sprayers 344, 347, 354
Typhula 329
Typical cells 233
Typical symptoms 16, 137, 382

U

Ubiquitous 11, 156, 207
Ubiquitous soil fungus 156
Ultra low volume spraying 344
Uncinula necator 96, 165, 361
Undifferentiated cells 276
Unequal, anterior, whiplash flagella 49
Unicellular 11, 38, 39, 73, 79, 80, 83, 86, 118, 130, 157, 158, 207, 211
Uninucleate cells 94, 116, 131
Uninucleate microconidia 111
Unitunicate asci 77, 89, 97
Universal veil 122
Uredinales 37, 127, 130, 131, 139
Uredinial cells 130
Uredinial initial 129
Uredinial stage 127
Urocystis 144, 284, 369, 376
Uromyces 131, 135, 376
Uromyces appendiculatus 135
Uromyces fabae 131
Ustilagic acid 341
Ustilaginaceae 37, 137, 140
Ustilago 138, 141, 173, 361
Ustilago hordei 288, 363
Ustilago nuda 173, 288, 363
Ustilago segetum 363
Ustilago zeae 173, 340

V

Valves 346, 348, 350
Vancide 329
Vangard 335
Vapam 290, 313, 318, 323, 359, 374

Vascular bundles 223, 270, 377
Vascular cells 227
Vascular fungi 189
Vascular system 20, 260, 276, 379
Vascular system of the parasite 276
Vascular tissues 8, 162, 272
Vascular wilt diseases 376
Vector transmitted virus diseases 186
Vegetable oils 342
Vegetative propagants 173, 227
Vegetatively propagated plant parts 177
Veil 122
Vein clearing 307, 377
Venturia 114, 146, 159, 185, 359, 361, 384, 393
Venturiaceae 37, 113, 114
Vertical elongation of the sporophores 181
Verticillate stalks 156
Verticillium 101, 156, 167, 179, 188, 287, 288, 318, 331, 332, 360, 363, 366, 367, 370, 371, 373, 376, 382, 384, 390
Verticillium wilt disease 156, 157
Verticillium wilt of potato, chrysanthemum 157
Viable state 107, 167, 168, 175, 367, 368, 369
Vibrio 211, 213
Vinclozolin 329
Viral nucleic acid 233, 237
Viral RNA replication 234
Viricides 310
Virulent strain 186, 309, 310
Viruliferous 241, 242, 243, 244
Virus diseases 176, 178, 186, 189, 195, 205, 228, 239, 241, 242, 259, 260, 261, 262, 263, 281, 286, 289, 373, 376, 379, 386, 387, 388, 389
Virus induced hormonal imbalance 264
Viscaceae 272
Vitafume 323
Vitamin 81
Vitavax 333
Volatile chemicals 318
Volvariaceae 37, 123
Volvariella volvaceae 121, 125
Vorlan 329
Vorlex 313

W

Waikavirus 246, 253
Wart disease of potato 33, 370
Wasps 291
Water 4, 11, 15, 17, 19, 26, 31, 32, 39, 40, 42, 43, 47, 52, 54, 58, 59, 63, 65, 77, 90, 107, 132, 138, 144, 163, 170, 174, 179, 180, 189, 207, 211, 215, 216, 221, 223, 225, 226, 227, 244, 270, 271, 272, 274, 276, 343
Water management 187, 308, 309
Water moulds 40, 52
Watermelon mosaic virus 252
Watery soft rot 105
Wavy stalk 162
Weak parasites 67, 144
Weather parameters 192, 204, 205, 206
Web blotch or leaf spot of groundnut 150
Web-like fructifications 126
Wedge-shaped haustoria 274
Weed hosts 286
Weedicide spraying 357
Well-defined nucleus 11, 207, 211, 213
Wet ceresan 290, 318, 326, 327, 365
Wet seed treatment 366
Wettable powders 311, 312, 313, 316, 346, 362
Wettable sulfur 320, 359, 357, 359, 375
Wheat chlorotic streak virus 257
Wheat mosaic virus 244, 246, 251
Wheat spindle streak 253, 262
Wheat spindle streak mosaic virus 253
Wheat stinking smut pathogen 363
Wheat streak mosaic virus 243, 253
Wheat yellow leaf 253, 264
Wheat yellow leaf virus 253
Whip smut 177, 365
Whip smut of sugarcane 373
Whiplash flagellum 39
White clover cryptic virus 1, 249
White flies 178, 188, 240, 242, 258, 285, 291
White rot of onion and garlic 164
White rust 65, 168, 374, 377
White rust of eggplants 65

White rust of rapeseed, cabbage etc. 370
Wide-spectrum protectant fungicides 362
Wild cucumber mosaic virus 254
Wild rice hosts 286
Wilt disease of sugarcane 156
Wilt of betelvine 164
Wilt of brinjal, potato, chillies etc. 163
Wilt of cabbage 371
Wilt of cotton 163, 279, 286, 288, 323, 339, 368, 371, 372
Wilt of crossandra 391
Wilt of cucurbits 163, 223
Wilt of gingelly 163
Wilt of ginger 371
Wilt of pea 163
Wilt of tomato 163, 223, 373
Wilts 195, 215, 223, 227, 317, 336, 372, 373, 376, 384
Wind dissemination 182
Wind velocity 192, 205
Wind-borne 98, 127
Witchweed 276
Wither tip of mango 153
Wood decaying fungi 385
Wood rotting fungi 175
Wound infection 172
Wound tumor virus 249, 258, 264
Wounded oil vesicles 171

X

Xanthomonas 165, 176, 178, 179, 180, 185, 215, 216, 221, 225, 226, 284, 288, 296, 363, 364, 370, 376, 380
Xanthomonas campestris 165, 176, 178, 179, 180
Xanthomonas campestris pv. campestris 216, 370
Xanthomonas campestris pv. citri 165, 178, 216, 225
Xanthomonas malvacearum 176, 215
Xanthomonas translucens f.sp. oryzae 363
Xiphinema 244
Xylaria type centrum 90, 97
Xylariales (sphaeriales) 97
Xylella 16, 17
Xylella fastidiosa 17
Xylem bundle sheath 299
Xylem vessels 16, 17, 162, 223, 226, 260, 276, 292, 302
Xylem-feeding insects 17
Xylem-inhabiting bacteria 16, 17
Xylem-inhabiting coryneform bacterium 270

Y

Yeast cake 80
Yeast culture 393
Yeast-like cells 145
Yeasts 23, 24, 65, 72, 73, 75, 77, 79, 80, 81
Yellow ear rot of wheat 179
Yellow leaf blight of corn 150
Yellow mosaic of lab lab, double bean 179
Yellowing 13, 14, 242, 254, 268, 269
Yellows 241, 249, 253, 264, 268, 269

Z

Zerlate 322
Zineb 322, 335, 366, 367, 375, 376
Ziram 322, 375
Ziride 322
Zone of inhibition 316, 372
Zoom 332
Zoospore infection canal 244
Zygomycetes 35, 65, 67, 70, 175, 385
Zygote cell 26, 80, 81
Zygote nucleus 25, 73, 118
Zygote thallus 43